NANOTECHNOLOGY-BASED APPROACHES FOR TARGETING AND DELIVERY OF DRUGS AND GENES

T0311901

NANOTECHNOLOGY-BASED APPROACHES FOR TARGETING AND DELIVERY OF DRUGS AND GENES

Edited by

VIJAY MISHRA
Lovely Professional University, Phagwara, Punjab, India

PRASHANT KESHARWANI
Pharmaceutics Division, CSIR-Central Drug Research Institute (CDRI), Uttar Pradesh, India

MOHD CAIRUL IQBAL MOHD AMIN
Universiti Kebangsaan Malaysia, Kuala Lumpur, Malaysia

ARUN IYER
Wayne State University, Detroit, MI, United States

ACADEMIC PRESS
An imprint of Elsevier

Academic Press is an imprint of Elsevier
125 London Wall, London EC2Y 5AS, United Kingdom
525 B Street, Suite 1800, San Diego, CA 92101-4495, United States
50 Hampshire Street, 5th Floor, Cambridge, MA 02139, United States
The Boulevard, Langford Lane, Kidlington, Oxford OX5 1GB, United Kingdom

Notices
Knowledge and best practice in this field are constantly changing. As new research and experience
broaden our understanding, changes in research methods, professional practices, or medical treatment may
become necessary.

Practitioners and researchers must always rely on their own experience and knowledge in evaluating and
using any information, methods, compounds, or experiments described herein. In using such information
or methods they should be mindful of their own safety and the safety of others, including parties for
whom they have a professional responsibility.

To the fullest extent of the law, neither the Publisher nor the authors, contributors, or editors, assume any
liability for any injury and/or damage to persons or property as a matter of products liability, negligence
or otherwise, or from any use or operation of any methods, products, instructions, or ideas contained in
the material herein.

British Library Cataloguing-in-Publication Data
A catalogue record for this book is available from the British Library

Library of Congress Cataloging-in-Publication Data
A catalog record for this book is available from the Library of Congress

ISBN: 978-0-12-809717-5

For Information on all Academic Press publications
visit our website at https://www.elsevier.com/books-and-journals

Working together
to grow libraries in
developing countries

www.elsevier.com • www.bookaid.org

Publisher: Mica Haley
Acquisition Editor: Kristine Jones
Editorial Project Manager: Molly McLaughlin
Production Project Manager: Chris Wortley
Designer: Christian Bilbow

Typeset by MPS Limited, Chennai, India

DEDICATION

I dedicate this work to the sacrifices and love of my wife Dr. Yachana Mishra and my adorable children Vidhi and Jay; to the love of my parents; and to the encouragement and confidence bestowed on me by my mentor Professor N.K. Jain, who has always helped me in my quest for learning.

Vijay Mishra

I would like to dedicate this book to my parents, my sister Dr. Poonam, and my brother Er. Pankaj who always encouraged me throughout the journey. I also dedicate this book to the love and sacrifices of my wife Garima, my sweet daughter Yashsavi, and finally my mentor Professor N.K. Jain for believing in me and always being there for me.

Prashant Kesharwani

This book is purely dedicated to my wife Maheran Mossadeq and my children Qothrunnadaa, Muhammad Fakhrurrazi, Saffiyyah Khadijah, and Muhammad Fakhrullah. May this book inspire our family for a successful life in the future and in the hereafter. I thank to Allah for His mercy and providing me with continuous well-being, energy, and dynamism in completing this book. May Allah accept this as one of my contributions to the scientific community.

Mohd C.I. Mohd Amin

This book is dedicated to my parents Mr. and Mrs. R.K. Iyer, my wife Thivya, and my adorable son, Athindra; their selflessness and love will always be remembered.

Arun Iyer

CONTENTS

Section I   Introduction and Issues in Nanomedicine

Section II   Current Technologies in Nanomedicine

Section III   Future Developments and Challenges in Nanomedicine

LIST OF CONTRIBUTORS

Sara Abdelghany
Princess Al-Jawhara Centre for Molecular Medicine, Manama, Kingdom of Bahrain

Muhammad W. Amjad
Universiti Kebangsaan Malaysia, Kuala Lumpur, Malaysia

Firoz Anwar
King Abdulaziz University, Jeddah, Kingdom of Saudi Arabia

R. Athawale
Prin. K. M. Kundnani College of Pharmacy, Mumbai, Maharashtra, India

Rimesh Augustine
Pusan National University, Busan, Republic of Korea

Sarwar Beg
Panjab University, Chandigarh, India

Adeel M. Butt
Universiti Kebangsaan Malaysia, Kuala Lumpur, Malaysia

Sonam Chaudhary
Central University of Rajasthan, Ajmer, Rajasthan, India

Mahavir B. Chougule
University of Hawaii at Hilo, Hilo; University of Hawaii at Manoa, Honolulu, HI, United States

Mohammad A. Derakhshan
Tehran University of Medical Sciences, Tehran, Iran

Virendra Gajbhiye
Agharkar Research Institute, Pune, Maharashtra, India

Namita Giri
University of Missouri at Kansas City, Kansas City, MO, United States

Avinash Gothwal
Central University of Rajasthan, Ajmer, Rajasthan, India

Khaled Greish
Princess Al-Jawhara Centre for Molecular Medicine, Manama, Kingdom of Bahrain

Vivek Gupta
Keck Graduate Institute, Claremont, CA, United States

Umesh Gupta
Central University of Rajasthan, Ajmer, Rajasthan, India

Ekta Gurnany
Ravishankar College of Pharmacy, Bhopal, Madhya Pradesh, India

Atul Jain
University of Central Lancashire, Preston, United Kingdom

Ashay Jain
Dr. Hari Singh Gour University, Sagar, Madhya Pradesh, India

Anfal Jasim
Princess Al-Jawhara Centre for Molecular Medicine, Manama, Kingdom of Bahrain

Johnson V. John
Pusan National University, Busan, Republic of Korea

Ripandeep Kaur
University of Central Lancashire, Preston, United Kingdom

Prashant Kesharwani
Pharmaceutics Division, CSIR–Central Drug Research Institute (CDRI), Uttar Pradesh, India

Iliyas Khan
Central University of Rajasthan, Ajmer, Rajasthan, India

Rajneet K. Khurana
Panjab University, Chandigarh, India

Vali Kiani
Islamic Azad University (IAUPS), Tehran, Iran

I.L Kim
Pusan National University, Busan, Republic of Korea

Lalit Kumar
Shivalik College of Pharmacy, Nangal; I.K. Gujral Punjab Technical University, Jalandhar, Punjab, India

Rajendra Kumar
University of Central Lancashire, Preston, United Kingdom; Panjab University, Chandigarh, India

Vikas Kumar
Sam Higginbottom Institute of Agriculture, Technology & Sciences (SHIATS), Allahabad, Uttar Pradesh, India

N. Kurup
Prin. K. M. Kundnani College of Pharmacy, Mumbai, Maharashtra, India

Shikha Lohan
Panjab University, Chandigarh, India

Rahul Maheshwari
BM College of Pharmaceutical Education & Research, Indore, Madhya Pradesh, India

Mohammad H. Mansoori
Adina Institute of Pharmaceutical Sciences, Sagar, Madhya Pradesh, India

Vijay Mishra
Lovely Professional University, Phagwara, Punjab, India

Mohd C.I. Mohd Amin
Universiti Kebangsaan Malaysia, Kuala Lumpur, Malaysia

Mahfoozur Rahman
Sam Higginbottom Institute of Agriculture, Technology & Sciences (SHIATS), Allahabad, Uttar Pradesh, India

Sarita Rani
Central University of Rajasthan, Ajmer, Rajasthan, India

Seyed M. Rezayat
Tehran University of Medical Sciences; Islamic Azad University (IAUPS), Tehran, Iran

Sumant Saini
Panjab University, Chandigarh, India

Abdus Samad
Fortis Clinical Research Ltd., Faridabad, Haryana, India

Ashok K. Sharma
Central University of Rajasthan, Ajmer, Rajasthan, India

R. Shegokar
Freie Universität Berlin, Berlin, Germany

Bhupinder Singh
Panjab University, Chandigarh, India; University of Central Lancashire, Preston, United Kingdom

Namrata Soni
Sam Higginbottom Institute of Agriculture, Technology and Sciences (Deemed University), Allahabad, Uttar Pradesh, India

Behnaz Tavakol
Kashan University of Medical Sciences, Isfahan, Iran

Shima Tavakol
Iran University of Medical Sciences; Tehran University of Medical Sciences, Tehran, Iran

Muktika Tekade
Technocrats Institute of Technology, Bhopal, Madhya Pradesh, India

Rakesh K. Tekade
International Medical University, Kuala Lumpur, Malaysia; National Institute of Pharmaceutical Education and Research (NIPER), Ahmedabad, Gujarat, India

Bhuvaneshwar Vaidya
Keck Graduate Institute, Claremont, CA, United States

Amita Verma
Sam Higginbottom Institute of Agriculture, Technology & Sciences (SHIATS), Allahabad, Uttar Pradesh, India

Shivani Verma
I.K. Gujral Punjab Technical University, Jalandhar; Rayat Bahra College of Pharmacy, Hoshiarpur, Punjab, India

R. Yang
University of Hawaii at Hilo, Hilo, HI, United States

BIOGRAPHIES

Vijay Mishra Dr. Vijay Mishra is an Assistant Professor of Pharmaceutical Sciences at the Department of Pharmaceutics, Lovely Institute of Technology (Pharmacy), Lovely Professional University in Phagwara, Punjab, India. Dr. Mishra earned his doctoral degree in pharmaceutical sciences from the Department of Pharmaceutical Sciences, Dr. H.S. Gour Central University, Madhya Pradesh, India. He was the recipient of an AICTE Graduate Research Fellowship and UGC–BSR Senior Research Fellowship. He has coauthored more than 26 international publications in reputed journals, 2 international book chapters, and 1 book on pharmaceutical dosage forms. His current research interests encompass surface-engineered dendrimers, carbon nanotubes, quantum dots, siRNA delivery, as well as controlled and novel drug delivery systems.

Affiliation

Lovely Professional University, Phagwara, Punjab, India

Prashant Kesharwani Dr. Prashant Kesharwani is a Lecturer in the Department of Pharmaceutical Technology, School of Pharmacy, International Medical University, Malaysia. He received his doctoral degree in Pharmaceutical Sciences from the Dr. H.S. Gour Central University (Sagar, India) with Professor N.K. Jain's group. He is a recipient of several internationally acclaimed fellowships and awards, namely, Excellence Research Award 2014 (USA), Young Innovator Award (Gold medal) 2012 (India), International Travel Award/Grant from DST, New Delhi 2012 (India), and INSA, CCSTDS, Chennai 2012 (India). He received an ICMR Senior Research Fellowship (for his PhD) and an AICTE Junior Research Fellowship (for his MSc). After his doctorate, he worked as a postdoctoral fellow in The Wayne State University in Detroit, Michigan, United States. An overarching goal of his current research is the development of nanoengineered drug delivery systems for cancer with a prime focus on dendrimer-mediated drug delivery systems. Dr. Kesharwani has coauthored 2 books, 3 book chapters in International Reference Books, and authored more than 65 publications in peer-reviewed international journals. He is an Editorial Board Member for the Journal *Pharmaceutical Medicine and Outcomes Research*.

Affiliation

Pharmaceutics Division, CSIR-Central Drug Research Institute (CDRI), Uttar Pradesh, India

Mohd C.I.M. Amin Dr. Mohd C.M. Amin is a Professor of Pharmaceutics in the Faculty of Pharmacy, Universiti Kebangsaan Malaysia (UKM) in Kuala Lumpur, Malaysia. He is the youngest professor appointed in the faculty and the only expert in the area of pharmaceutics in UKM. He earned his BSc (Hons) in Pharmacy in 1996 and PhD in 2001 both from the University of Manchester, United Kingdom. He is currently the Head of Drug Delivery and Novel Targeting Research Group in UKM. He also serves as a panel expert in nanocellulose for drug delivery for the National Nanotechnology Centre, Ministry of Science, Technology and Innovation, Malaysia. His current research interests encompass responsive-hydrogel biomaterials for oral protein and peptide delivery, cell delivery in tissue engineering, and nanodrug delivery systems, in particular polymeric micellar conjugates for cancer research. Dr. Amin has authored more than 60 publications in high-impact factor journals, a number of book chapters, 5 patents issued/pending, and routinely consults in the area of drug delivery and pharmaceutical technology. He has wide expertise in biomaterials, nanomedicine, polymer chemistry, micro- and nanotargeting delivery systems, as well as oral and transdermal formulations and delivery systems.

Affiliation

Universiti Kebangsaan Malaysia, Kuala Lumpur, Malaysia

Arun Iyer Dr. Iyer is an Assistant Professor of Pharmaceutical Sciences and Director of Use-inspired Biomaterials and Integrated Nano Delivery (U-BiND) Systems Laboratory at Wayne State University in Detroit, Michigan, United States. He also serves as a Scientific Member of the Molecular Therapeutics (MT) Program at the Barbara Ann Karmanos Cancer Institute in Detroit, Michigan, United States. Dr. Iyer received his PhD in 2008, in Polymer Engineering, under the mentorship of world-renowned scientist, Professor Hiroshi Maeda at Sojo University in Kumamoto, Japan. He was the recipient of the Controlled Release Society's (CRS) T. Nagai Research Achievement Award in 2012. Dr. Iyer completed his postdoctoral training in Cancer Radiology at the University of California, San Francisco (UCSF) in California, United States, and trained as an Associate Research Scientist and Research Assistant Professor in Pharmaceutical Sciences at Northeastern University in Boston, Massachusetts, United States. Dr. Iyer has authored more than 70 publications in peer-reviewed international journals and books of high repute. He has more than 100 scientific presentations and invited talks at International Conferences and Workshops. He has five patents issued/pending. His areas of research are broadly focused on designing use-inspired bio- and nanomedical technologies aimed toward clinical translation using biocompatible delivery systems that have enhanced disease targeting with reduced toxicity burden to patients. He has wide expertise in biomaterials and nanomedicine, polymer chemistry and formulation development, drug and gene delivery systems, molecular and functional imaging, and micro- and nanoparticles for treating diseases such as infection, inflammation, and cancer. His laboratory is funded by agencies such as the National Institutes of Health (NIH), Wayne State University, and other private and nonprofit organizations.

Affiliation

Wayne State University, Detroit, MI, United States

PREFACE

Nanotechnology literally means any technology on a nanoscale that has applications in the real world. Extrapolating from known physical laws, Nobel Laureate Richard P. Feynman envisioned a technology using the ultimate toolbox of nature, building nanoobjects atom by atom or molecule by molecule. The use of nanoscale technologies to design novel drug delivery systems and devices is a rapidly developing area of biomedical research that promises breakthrough advances in therapeutics and diagnostics. Over the last few years, numerous breakthroughs in nanotechnology have made great impacts on different fields of scientific research. Out of these many breakthroughs, some have proved to be very promising for the diagnosis and treatment of diseases. It is widely felt that nanotechnology will be the next Industrial Revolution. However, there is a clear need for innovative technologies to improve the targeting and delivery of therapeutics as well as diagnostic agents in the body. Recent advancements in nanomedicine are now making it possible to deliver drugs, genes, and therapeutic agents to local areas of disease to maximize clinical benefit while limiting unwanted side effects.

There is an increasing need for a multidisciplinary, system-oriented approach to the manufacturing of nanodevices, which function reliably. This can only be achieved through the cross-fertilization of ideas from different disciplines and the systematic flow of information among different research groups. This book provides an overview of different aspects of nanomedicine, which help the readers to design and develop novel drug delivery systems and devices that take advantage of recent advances in nanomedical technologies. The organization of the book is straightforward. The book is divided into three major parts: SECTION I: Introduction and Issues in Nanomedicine; SECTION II: Current Technologies in Nanomedicine; and SECTION III: Future Developments and Challenges in Nanomedicine.

Focusing on the design, synthesis, and application of different nanocarriers in drug and gene delivery, this book will be a valuable resource for graduates, pharmaceutical scientists, clinical researchers, and anyone working to tackle the challenges of delivering drugs and genes in a more targeted and efficient manner. In totality, this book will prove to be one of the most comprehensive books available that combines both the fundamental pharmaceutical principles of nanocarriers along with the most important applications of nanotechnology in targeting and drug delivery. Featuring contributions from field experts and researchers in industry and academia *Nanotechnology-Based Approaches for Targeting and Delivery of Drugs and Genes* provides state-of-the-art information on nanocarriers and their use in targeting, as well as drug and gene delivery.

We hope this book will stimulate further interest in the drug delivery field, and that the readers of this book will find it useful.

Vijay Mishra, Prashant Kesharwani, Mohd C.I. Mohd Amin, and Arun K. Iyer

ACKNOWLEDGMENTS

We wish to sincerely thank the authors for offering to write comprehensive chapters on a tight schedule. This is generally an added responsibility in the hectic work schedules of researchers. We express our earnest gratitude to the reviewers, who provided their critical views for the improvement of the book chapters. We would like to thank reviewers of our book proposal for their suggestions in the framing of the chapters. We also thank Molly M. McLaughlin (Editorial Project Manager, Elsevier), whose efforts during the preparation of this book were very useful.

Editors

Introduction and Issues in Nanomedicine

CHAPTER 1

Nanotechnology for the Development of Nanomedicine

Rakesh K. Tekade, Rahul Maheshwari, Namrata Soni, Muktika Tekade and Mahavir B. Chougule

Contents

Nanotechnology-Based Approaches for Targeting and Delivery of Drugs and Genes.
DOI: http://dx.doi.org/10.1016/B978-0-12-809717-5.00001-4

Disclosures: There are no conflict of interest and disclosures associated with the manuscript.

1.1 NANOTECHNOLOGY AND NANOMEDICINE: APPROACHING THE IDEAL SCALE

Recent years have witnessed exceptional research attention in the area of nanoscience and nanotechnology (Tekade et al., 2015a,b,c; Malinoski, 2014). There is growing optimism that nanotechnology will bring key advances in the diagnosis, treatment, and prevention of diseases (Maheshwari et al., 2015a,b). All features of nanomedicine rely on progress in nanomaterials research and the nanoengineering essential to create devices to recognize their aspirations (Tekade et al., 2016). The term "nanotechnology" has been derived from the Latin and Greek words "nanus" and "nanos," respectively, meaning "dwarf." In general, nanotechnology is concerned with dimensions with tolerance limits of 1–100 nm as well as with the manipulation of materials at atomic and molecular levels (Tekade and Chougule, 2013; Soni et al., 2016). Hence, nanotechnology is often defined as the intentional design, characterization, production, and application of materials, structures, devices, and systems by controlling their size and shape in the nanoscale range (Fig. 1.1). Since nanomaterials are similar in scale to biological molecules, they can be engineered to have various useful medical functions (Tekade, 2014; Tekade et al., 2014a,b; Binns, 2010).

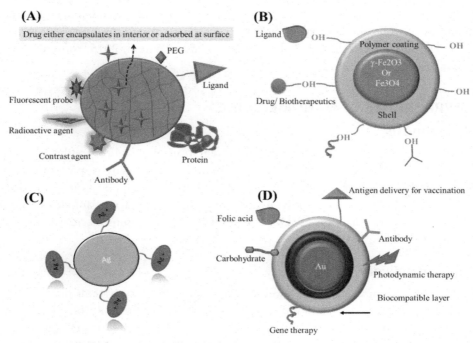

Figure 1.1 Schematic representation of (A) surface-modified polymeric nanoparticles, (B) superparamagnetic iron oxide nanoparticles (SPIONs), (C) silver nanoparticles, (D) surface-modified gold nanoparticles.

The field of nanomedicine aims to use the properties and physical characteristics of nanomaterials for the diagnosis and treatment of diseases at the molecular levels (Tekade et al., 2015c). The application of nanotechnology concepts in medicine joins multidisciplinary fields with unparallel social and economic potential for achievements in these fields. The common basis evolves from the molecular-scale properties relevant to the two fields. For example, regiospecific probes along with molecular imaging techniques allow efficient imaging of surface and interface properties at a predefined location (Widom et al., 2014; Chopdey et al., 2015; Chougule et al., 2014). On the other hand, chemical techniques offer the opportunity to investigate the surfaces for targeted drug delivery, enhanced biocompatibility, and neuroprosthetic purposes (Tekade et al., 2008a,b, 2009).

Nanomedicine permits the cure of disease within the body and at the cellular or molecular level; and is one of the most promising fields within the potential new technological advances in medicine. This technology is also revolutionizing medical areas such as monitoring, tissue repair, disease evolution control, protection and improvement of human biological systems, diagnosis, treatment and prevention, pain relief, health prevention, delivery of drugs to cells, positioning it as a revolution in the medical scientific and healthcare fields (Weissig and Villanueva, 2015; Youngren et al., 2013; Maheshwari et al., 2012).

1.2 APPROACHES OF NANOMEDICINES

1.2.1 Diagnosis

Nanodiagnostics improve the sensitivity and integration of analytical methods to yield a more coherent result. They have applications in genomic analysis, proteomics, and molecular diagnostics. Some nanotechnologies with potential application in molecular diagnostics are nanochips, nanoarrays, nanoparticles (gold NP), nanobarcodes, quantum dot (QD), nanowires, nanopores, DNA nanomachines, nanosensors, resonance light scattering, etc. (Binns, 2010).

1.2.2 Drug delivery

A complementary approach is the development of drug delivery systems that can act as a vehicle to carry and guide the bioactive agents to their desired site of action. Today, nanotechnology presents a great hope to improve available treatment strategies by acting at least at two main levels: (1) conferring new properties to a pharmaceutical agent, i.e., increased stability, modified pharmacokinetics, decrease toxicity; and (2) targeting the agent directly to the site of action (Mody et al., 2014; Mansuri et al., 2016; Moeendarbari et al., 2016; Prajapati et al., 2009).

1.2.3 Tissue engineering and implants

Engagement of the nanotechnology approach in this field is a vibrant area of research currently. Approach to the design of biomimetic nanomaterials capable of directing direct new tissue formation can be achieved by incorporating cell-binding peptides via chemical or physical bonding. This technology also offers the ability to produce tissue engineering vascular grafts that can be incorporated into host blood vessels. This strategy potentially resolves several issues associated with conventional vascular grafts (Neacsu et al., 2016; Sharma et al., 2015).

1.2.4 Bioavailability improvement

Nanoparticulate drug delivery has become an area of extensive research as these systems enable bioavailability improvement of poorly water-soluble compounds as well as targeted delivery of active pharmaceutical ingredients to various tissues and organs (Hock et al., 2011).

1.2.5 Medical devices

Microdevices are those materials that are fabricated with the aid of technologies such as microfabrication, surface patterning, and microfluidics. These techniques are often integrated with cell and tissue cultures for advancing the microbiological assays. Advancements in methods of microdevice fabrication and application could address the challenges faced in the development and assessment of nanomedicines (Boisseau and Loubaton, 2011).

1.3 HISTORY OF NANOMEDICINES

The production and use of nanosized particles was initiated hundreds of years ago in primordial times, however, the past few decades in particular have witnessed notable progress in the interdisciplinary application of this technology (Gandhi et al., 2014). In the early 2000s, it was the optimism of scientists in nanotechnology that prompted governmental and funding organizations to undertake strategic reviews of the present significance of nanomedicine (crucial objectives, potential opportunities for improved health care as well as the risk–benefit analysis of these new technologies) to determine priorities for future funding. The European Science Foundation launched its Forward Look on Nanomedicine (2003), which was the first foresight study on medical applications of nanosciences and nanotechnology. In the same year, the UK government carried out a study to investigate whether nanotechnology could raise new ethical, health, and safety issues. The final report of this investigation was published in 2004 with generally regarded as safe recommendations for a sure, safe, and responsible development of nanotechnology (Binns, 2010; Jain and Tekade 2013; Kurmi et al., 2010; Kayat et al., 2011).

The European Technology Platform Nanomedicine was launched in 2004 from the ingenuity of the European Commission. This group of experts from industry, research centers, and academia convened to prepare the vision regarding future research priorities in nanomedicine. In 2005, its vision paper and strategic research agenda for nanomedicine were released, as a first step toward setting up an international platform on nanomedicine, aiming at improving the quality of life and health care of patients. The European Foundation for Clinical Nanomedicine was established in Basel (Switzerland; in 2007) as a nonprofit organization which aimed at advancing medicine for the benefit of individuals and society through the application of nanoscience. The Foundation reached its goals through support of clinicians and this led to rapid advancement in the medical applications of nanosciences. This Foundation created the environment for focused research, interdisciplinary interaction, and information flow between clinicians, researchers, the public, and allied stakeholders (Lehner and Hunziker, 2012; Ghanghoria et al., 2016; Huang et al., 2015).

1.4 RATIONALE FOR THE DEVELOPMENT OF NANOMEDICINE

The goal of nanomedicine is to monitor, control, construct, repair, defend, and improve human biological systems at the molecular level using engineered devices and nanostructures, to achieve medical benefit. For this, the nanocarriers should be engaged with active components such as therapeutic agents, radionuclides, and gene/genome siRNA in the size range from one nanometer to hundreds of nanometers (Dhakad et al., 2013; Dwivedi et al., 2013). These may be included in a microdevice (that have a macrointerface) or in a biological environment. The focus, however, is always on nanocarrier interactions within the framework of a larger device or directly within a cellular or biological system of the body.

Nanotechnology is applied extensively to provide targeted drug therapy, diagnostics, tissue regeneration, cell culture, biosensors, and other tools in the field of molecular biology. Various nanotechnology platforms like fullerenes, nanotubes, QDs, nanopores, dendrimers (Kesharwani et al., 2015a,b), liposomes (Maheshwari et al., 2012, 2015a,b), magnetic nanoprobes (Liao et al., 2013), and radio-controlled nanoparticles (Stanley et al., 2012) readily interact with biomolecules located on both the cell surface and inside. Thus, nanocarriers acting as a vehicle can not only transport encapsulated or grafted small chemotherapeutic drugs, but also convey them inside cells once they have penetrated those (Tekade et al., 2016; Gajbhiye et al., 2007, 2009). Such systems can also be anchored with ligand, peptide, and fragments of antibodies on their surface to target-specific tissues, thus improving the specificity of the delivered drug molecule.

1.5 ADVANTAGES OF NANOMEDICINE OVER CONVENTIONAL DRUG DELIVERY SYSTEMS

Novel drug delivery systems offer unique advantages for cancer therapy as compared to free drug administration. Nanomedicines augment drug concentration in the site of action through passive and active targeting, which reduces the drug concentration in normal tissues. The availability of drug specifically at the site of action reduces the toxic side effects, improving the pharmacokinetics and pharmacodynamics profiles of the drugs (Soni et al., 2016). Nanomedicine approaches improve the solubility and drug stability (by reducing its peripheral drug degradation in the systemic circulation) and improve cellular internalization of the loaded drug by adoption of well-established surface functionalization strategies. These unique advantages opened up the opportunities to reevaluate potent drugs with poor pharmacokinetics that had been previously discarded. Over the last two decades, a large number of nanoparticulate drug delivery systems consisting of organic or inorganic materials have been evaluated for cancer therapy. Some have translated to the clinic or are being investigated in advanced preclinical stages of development (Nguyen, 2012).

Nanoparticulate drug delivery systems for anticancer applications include liposomes, polymeric nanoparticles, micelles, nanoshells, dendrimers, inorganic/metallic nanoparticles, and magnetic and bacterial nanoparticles (Tekade and Chougule, 2013; Tekade et al., 2013). Owing to the abnormalities of tumor vasculature, most nanocarriers with a size range of 10–500 nm can eventually accumulate at the tumor site. However, the nanoparticles have to unload their payloads at the site of disease for effective drug action, more specifically, the encapsulated drug has to successfully reach its subcellular target. Therefore, the drug-loaded nanocarriers have to overcome systemic, extracellular, and intracellular barriers to deliver the drug to the specific organelles for effective therapeutic benefits (Biswas and Torchilin, 2014).

More recently developed nanoparticles now being investigated for cancer therapy have been designed with multifunctional capabilities such as long systemic circulation, tumor targeting, cytosolic translocation, and organelle-specific targeting for effective cytotoxic effect. Functionalization of nanocarriers for organelle-specific targeting of bioactive molecules to specific intracellular compartments can sharply increase the efficiency of various treatment protocols. Generally, nanocarriers are internalized by receptor-mediated endocytosis (RME) and reach lysosomal compartments where the nanocarriers and any released drug molecules encounter numerous degradative enzymes (Yameen et al., 2014). Lysosomal degradation allows only a fraction of intracellular drugs or nanoparticles to become available to a site of action or specific organelles. Bypass of the endocytic pathway or disruption of the endosomes could be a powerful strategy for improved cytosolic delivery that enhances the therapeutic efficacy of the loaded drug (Toy and Roy, 2016). Therapeutic efficacy of drugs could further be improved if the nanocarriers were actively targeted to the specific organelle, and this is of particular importance in regards to development of anticancer therapies for cancer.

1.6 IMPACT OF NANOMEDICINE APPLICATIONS ON HEALTHCARE COSTS

Nanomedicine will be important to improve health care in all phases of the care process. New in vitro diagnostic tests will shift diagnosis to an earlier stage, hopefully before symptoms really develop and allow preemptive therapeutic measures. In vivo diagnosis will become more sensitive and precise thanks to new imaging techniques and nanosized targeted agents. Therapy as well could be greatly improved in efficacy by new systems that allow targeted delivery of therapeutic agents to the diseased site, ideally avoiding conventional parenteral delivery (Garg et al., 2016). Regenerative medicine may provide a therapeutic solution to revitalize tissue or organs, which may make life-long medication unnecessary. While the diseases vary in their pathways, and often demand very different levels of maturity from the proposed technologies, they also share some common clinical needs. Those activities which could be applied broadly should have top priority. For example, in all diseases new in vitro diagnostic tests are generally required that allow rapid, sensitive, and reliable detection of a broad set of disease-indicative biomarkers (Lam et al., 2016).

The discovery of disease-specific biomarkers itself is beyond the scope of nanomedicine and should be the focus of medical research. Following the same line of thought, research on multitasking agents for in vivo use and aspects of regenerative medicine that could offer broad applications in different diseases should be supported. Additionally, research is needed on clinical needs, which are specific to one disease. For example, the clinical need for noninvasive measurement of blood glucose levels or the need for agents that cross the blood–brain barrier are unique aspects to diabetes and neurodegenerative diseases, respectively (Pautler and Brenner, 2010).

1.7 NANOMATERIALS USED FOR THE DEVELOPMENT OF NANOMEDICINES

Use of novel bionanomaterials like nanoparticles, liposomes, metal nanoparticles, dendrimers, and carbon nanotubes (CNTs), nanoshells, nanopores, nanorobots, and nanosuspension for drug delivery purposes constitutes a burgeoning new field called "nanomedicines" that seeks to address the issue in order to maximize the therapeutic response with improved patient compliance (Ranganathan et al., 2012).

1.7.1 Nanosuspensions

Nanosuspensions are submicron colloidal biphasic dispersions of pure drug particles (smaller than 1 μm) that are stabilized by a small percentage of excipients, such as surfactants and polymers, without any matrix material, which could dramatically enhance the saturated solubility, dissolution rate, and adhesion of drug particles to cell membranes. Nanosuspensions are most suitable for drugs that require high dosing or have limited administrative volume. Nanosuspensions have been demonstrated to be a superior substitute over other approaches currently available for improving bioavailability of a number of poorly soluble drugs due to higher dissolution rate, amplified rate and extent of absorption, area under plasma versus time curve, onset time, peak drug level, reduced variability, and reduced fed/fasted effects. The best option is for topical preparations, i.e., the higher penetration capability. For oral preparations nanoparticles can adhere to the gastrointestinal mucosa prolonging the contact time of the drug thus enhancing its absorption (Attari et al., 2016).

Stabilizers play an important role in the formulation of nanosuspensions. In the absence of an appropriate stabilizer, the high surface energy of nanosized particles can induce agglomeration or aggregation of the drug crystals. The main function of a stabilizer is to wet the drug particles thoroughly, and to prevent Ostwal's ripening and agglomeration of nanosuspensions in order to yield a physically stable formulation by providing steric or ionic barriers. The type and amount of stabilizer has a pronounced effect on the physical stability and in vivo behavior of nanosuspensions. In some cases, a mixture of stabilizers is required to obtain a stable nanosuspension (Karakucuk et al., 2016).

Organic solvents may be required in the formulation of nanosuspensions if they are to be prepared using an emulsion or microemulsion as a template. As these techniques are still in their infancy, elaborate information on formulation considerations is not available. The acceptability of organic solvents in the pharmaceutical area, their toxicity potential, and the ease of their removal from the formulation need to be considered when formulating nanosuspensions using emulsions or microemulsions as templates (Mishra et al., 2016).

The choice of cosurfactant is critical when using microemulsions to formulate nanosuspensions. Since cosurfactants can greatly influence phase behavior, the effect of cosurfactant on uptake of the internal phase for selected microemulsion composition and on drug loading should be investigated. Although the literature describes the use of bile salts and dipotassium glycerrhizinate as cosurfactants, various solubilizers, such as transcutol, glycofurol, ethanol, and isopropanol, can be safely used as cosurfactants in the formulation of microemulsions (Singare et al., 2010).

Nanosuspensions are prepared by two reverse methods "bottom–up" and "top–down" technologies. Conventional methods of precipitation are called "bottom–up technology." The "top–down technologies" are the disintegration methods and are preferred over the precipitation methods. These include media milling (nanocrystals), high-pressure homogenization in water (dissocubes), high-pressure homogenization in nonaqueous media (nanopure), and a combination of precipitation and high-pressure homogenization (nanoedge) (Mahesh et al., 2014).

Various antibiotics, such as atovaquone and buparvaquone, reflect poor oral absorption; nanosuspension of these drugs can lead to an incredible increase in their oral absorption and subsequent bioavailability. Poor oral bioavailability of the drug may be due to poor solubility, poor permeability, or poor stability in the gastrointestinal tract (GIT), with the nanosuspension leading to improved bioavailability of a number of poorly soluble drugs due to higher dissolution rate, amplified rate and extent of absorption, area under plasma versus time curve, onset time, peak drug level, and reduced variability. One of the important applications of nanosuspension technology is the formulation of intravenously administered products. Intravenous administration results in several advantages, such as administration of poorly soluble drugs without using a higher concentration of toxic solvents, improving the therapeutic effect of the drug available as conventional oral formulations, and targeting the drug to macrophages. Nanosuspensions of poorly soluble drug tarazepide have been prepared to overcome the limited success achieved using conventional solubilization techniques, such as use of surfactants, cyclodextrins to improve bioavailability (Mou et al., 2011).

Nanosuspensions may prove to be an ideal approach for delivering drugs that exhibit poor solubility in pulmonary secretions. Aqueous nanosuspensions can be nebulized using mechanical or ultrasonic nebulizers for lung delivery. Because of their small size, it is likely that in each aerosol droplet at least one drug particle is contained, leading to a more uniform distribution of the drug in the lungs. The nanoparticulate nature of the drug allows rapid diffusion and dissolution of the drug at the site of action (Zhang and Zhang, 2016).

Nanosuspensions could prove to be a boon for drugs that exhibit poor solubility in lachrymal fluids. Nanosuspensions, by their inherent ability to improve the saturation solubility of the drug, represent an ideal approach for ocular delivery of hydrophobic drugs and the nanoparticulate nature of the drug allows its prolonged residence in the cul-de-sac, giving sustained release of the drug. Nanosuspensions can be used for targeted delivery as their surface properties and in vivo behavior can easily be altered by changing either the stabilizer or the milieu. The engineering of stealth nanosuspensions (analogous to stealth liposomes) by using various surface coatings for active or passive targeting of the desired site is the future of targeted drug delivery systems (Hong et al., 2016).

Drug nanoparticles can be incorporated into creams and water-free ointments. The nanocrystalline form leads to increased saturation solubility of the drug in the topical dosage form and thus enhances the diffusion of the drug into the skin. Nanosuspensions afford a means of administering increased concentrations of poorly water-soluble drugs to the brain with decreased systemic effects. Significant efficacy has been shown with microparticulate busulfan in mice administered intrathecally. Several concepts of targeting of nanosuspension dosage forms for treatment of bioweapon-mediated diseases have been developed at the Baxter Healthcare Corporation. Alterations of pharmacokinetic profiles of existing antibiotics can lead to enhanced efficacy with reduced side effects. This has been shown for a nanosuspension formulation of the antifungal agent itraconazole (Sutradhar et al., 2013).

1.7.2 Polymeric nanoparticles

Polymeric nanoparticles are colloidal solid particles prepared from biodegradable polymers such as chitosan and collagen or nonbiodegradable polymers such as poly(lactic acid) (PLA) and poly(lactic co-glycolic acid) (PLGA). The schematic representations of polymeric nanoparticles are shown in Fig. 1.1.

Polymeric nanoparticle-based materials have gained much deliberation due to advancements in polymer science and technology. Polymeric nanoparticles are playing an eccentric role in a broad spectrum of applications. The importance of these materials lies in idiosyncratic features associated with polymeric nanoparticles. Biodegradable polymeric nanoparticles are highly preferred because they show promise in drug delivery systems. Such nanoparticles provide controlled/sustained release properties, subcellular size, and biocompatibility with tissue and cells. Apart from this, these nanomedicines are stable in blood, nontoxic, nonthrombogenic, nonimmunogenic, noninflammatory, do not activate neutrophils, are biodegradable, avoid the reticuloendothelial system, and are applicable to various molecules such as drugs, proteins, peptides, or nucleic acids. The drug molecules either bind to the surface as nanospheres or are encapsulated inside as nanocapsules (Guerrero-Cazares et al., 2014).

The unique sizes of nanoparticles are amenable to surface functionalization or modification to achieve desired characteristics. This was achieved by various methods to form the corona to increase drug retention time in blood, reduce nonspecific distribution, and target tissues or specific cell surface antigens with targeting ligands (peptide, aptamer, antibody, and small molecule). Different materials/particles are used for the preparation of nanoparticles leading to a distinction in surface properties (Kumari et al., 2010). The most commonly and extensively used biodegradable polymeric nanoparticles, gelatin, albumin, and chitosan and their therapeutic advantages, general synthesis, and encapsulation of various disease-related drugs are described in this part of the chapter.

1.7.2.1 Gelatin nanoparticles

Gelatin is extensively used in food and medical products and is attractive for use in controlled release due to its nontoxic, biodegradable, bioactive, and inexpensive properties. It is a polyampholyte having both cationic and anionic groups along with a hydrophilic group. It is known that its mechanical properties, swelling behavior, and thermal properties depend significantly on the crosslinking degree of gelatin. Gelatin nanoparticles can be prepared by desolvation/coacervation or an emulsion method. Desolvation/coacervation is a process during which a homogeneous solution of charged macromolecules undergoes liquid–liquid phase separation, giving rise to a polymer-rich dense phase at the bottom and a transparent solution above. The addition of natural salt or alcohol normally promotes coacervation and the control of turbidity/ crosslinking that resulted in desired nanoparticles. Various drugs such as the anticancer drug paclitaxel, anti–HIV drug (didanosine), and antimalarial drug (chloroquine phosphate) have been encapsulated in gelatine nanoparticles (Azimi et al., 2014).

1.7.2.2 Chitosan nanoparticles

Chitosan is a modified natural carbohydrate polymer prepared by the partial N-deacetylation of crustacean-derived natural biopolymer chitin. Chitosan is one of the most commonly used natural polymers in the production of nanomedicines, because it displays very attractive characteristics for drug delivery and has proved very effective when formulated in a nanoparticulate form. Properties such as its cationic character and its solubility in aqueous medium, have been reported as determinants of the success of this polysaccharide. However, its most attractive property relies on the ability to adhere to mucosal surfaces leading to prolonged residence time at drug absorption sites and enabling higher drug permeation. Chitosan has further demonstrated capacity to enhance macromolecules' epithelial permeation through transient opening of epithelial tight junctions. In addition, the polymer is known to be biocompatible and to exhibit very low toxicity (Rampino et al., 2013).

There are at least four methods reported for the preparation of chitosan nanoparticles: ionotropic gelation, microemulsion, emulsification solvent diffusion, and polyelectrolyte complex (PEC). Ionotropic gelation is based on an electrostatic interaction between the amine group of chitosan and negatively charged groups of polyanion such as tripolyphosphate. Chitosan is dissolved in acetic acid in the absence or presence of stabilizing agent. Polyanion was then added and nanoparticles were spontaneously formed under mechanical stirring. In the microemulsion method, a surfactant was dissolved in an organic solvent like n-hexane. Then chitosan in acetic acid solution and glutaraldehyde were added to the surfactant/hexane mixture under continuous stirring at room temperature. PEC or self-assembled polyelectrolyte is a term to describe complexes formed by self-assembly of the cationic charged polymer and plasmid DNA. The mechanism of PEC formation involves charge neutralization between cationic polymer and DNA, leading to a fall in hydrophilicity as the polyelectrolyte component self-assembly. In this method nanoparticles are spontaneously formed after addition of DNA solution into chitosan dissolved in acetic acid solution under mechanical stirring at or under room temperature (Jayakumar et al., 2010).

1.7.2.3 Albumin nanoparticles

Albumin, a versatile protein carrier for drug delivery, has been shown to be nontoxic, nonimmunogenic, biocompatible, and biodegradable. The albumin-based nanoparticle product, Abraxane, is in clinical use. Therefore, it is an ideal material to fabricate nanoparticles for drug delivery. Albumin nanoparticles have gained considerable attention owing to their high binding capacity of various drugs and being well tolerated without any serious side effects. Major outcomes of in vitro and in vivo investigations as well as site-specific drug targeting have been obtained using various ligands modifying the surface of albumin nanoparticles with special insights into the field of oncology. Specialized nanotechnological techniques like desolvation, emulsification, thermal gelation, and recently nanospray drying, nab-technology, and self-assembly have been investigated for fabrication of albumin nanoparticles (Kumari et al., 2010).

1.7.3 Superparamagnetic iron oxide nanoparticles

Superparamagnetic iron oxide nanoparticles (SPIONs) are novel drug-delivery vehicles. SPIONs are small synthetic γ-Fe_2O_3 (maghemite) or Fe_3O_4 (magnetite) particles with a core ranging between 10 and 100 nm in diameter. These magnetic particles are coated with certain biocompatible polymers, such as dextran or polyethylene glycol, which provide chemical handles for the conjugation of therapeutic agents and also improve their blood distribution profile. SPIONs (Fig. 1.1) with appropriate surface chemistry have been widely used experimentally for numerous in vivo applications such as magnetic resonance imaging (MRI) contrast enhancement, tissue repair, immunoassay, detoxification of biological fluids, hyperthermia, drug delivery, and in cell separation. Delivery of anticancer drugs by coupling with functionalized SPIONs to their targeted site is one of the most pursued areas of research in the development of cancer treatment strategies. SPIONs have also demonstrated their efficiency as nonviral gene vectors that facilitate the introduction of plasmids into the nucleus at rates multifold those of routinely available standard technologies. SPION-induced hyperthermia has also been utilized for localized killing of cancerous cells (Mahmoudi et al., 2011).

Despite their potential biomedical application, alteration in gene expression profiles, disturbance in iron homeostasis, oxidative stress, and altered cellular responses are some SPION-related toxicological aspects which require due consideration. SPION unique to nanoparticles is very important for their use as drug-delivery vehicles because these nanoparticles can literally drag drug molecules to their target site in the body under the influence of an applied magnet field. Moreover, once the applied magnetic field is removed, the magnetic particles retain no residual magnetism at room temperature and hence are unlikely to agglomerate (i.e., they are easily dispersed), thus evading uptake by phagocytes and increasing their half-life in the circulation. Moreover, due to a negligible tendency to agglomerate, SPIONs pose no danger of thrombosis or blockage of blood capillaries. SPION fabrication involves three steps, i.e., core fabrication, coating on the core, and finally drug loading. The core is fabricated by nucleation and crystal growth. The most commonly used methods for preparation of a uniform iron-based nanoparticle core in solution are coprecipitation and microemulsion. The next step after fabrication of SPION cores is their coating.

Coating with suitable polymers endows some important characteristics to these nanoparticles that are essential for their use as drug-delivery vehicles. Coating of SPIONs is essential because it: reduces the aggregation tendency of the uncoated particles, thus improving their dispersibility and colloidal stability; protects their surface from oxidation; provides a surface for conjugation of drug molecules and targeting ligands; increases the blood circulation time by avoiding clearance by the reticuloendothelial system; makes the particles biocompatible and minimizes nonspecific interactions, thus reducing toxicity; and increases their internalization efficiency by target cells. The final step of drug loading can be achieved either by conjugating the therapeutic molecules on the surface of SPIONs or by coencapsulating drug molecules along with magnetic particles within the coating material envelope. A number of approaches have been developed for conjugation of therapeutic agents or targeting ligands on the surface of these nanoparticles. They can be grouped under two categories, i.e., conjugation by means of cleavable covalent linkages and by means of physical interactions (Mahmoudi et al., 2011).

1.7.4 Metal nanoparticles

Metal nanoparticles are nanosized inorganic particles, especially nanoparticles of the alkali metals and the noble metals, copper, silver, and gold of either simple or composite nature; they display unique physical chemical and optical properties and represent an increasingly important material in the development of novel nanodevices which can be used in numerous physical, biological, biomedical, and pharmaceutical applications. Generally, metal nanoparticles can be prepared and stabilized by chemical, physical, and biological methods; for the chemical approach, such as chemical reduction, electrochemical techniques, photochemical reduction, and pyrolysis and physical methods, such as Arc–discharge and physical vapor condensation are used (Kulkarni and Muddapur, 2014).

1.7.4.1 Silver nanoparticles

Colloidal silver is of particular interest because of its distinctive properties, such as good conductivity, chemical stability, and catalytic and antibacterial activity. Silver nanoparticles have many important applications that include spectrally selective coating for solar energy absorption and intercalation material for electrical batteries, as optical receptors, polarizing filters, catalysts in chemical reaction, biolabeling, and as antimicrobial agents. The preparation of uniform nanosized drug particles with specific requirements in terms of size, shape, and physical and chemical properties is of great interest in the formulation of new pharmaceutical products. It is well known that silver ions and silver-based compounds are highly toxic to microorganisms, showing strong biocidal effects on as many as 16 species of bacteria including *Escherichia coli*. Thus, silver ions, as an antibacterial component, have been used in the formulation of dental resin composites and ion exchange fibers and in coatings of medical devices. Nanosilver is commercial pure deionized water with superfine silver in suspension. Nanoparticle sizes from 5 to 50 nm in the form of metallic silver nanoparticles provide greater effectiveness then silver solution in the body (Rauwel et al., 2015).

Nanocrystalline silver products (Acticoat) are used in wound management and utilize nanotechnology to release nanocrystalline silver crystals. They release 30 times less silver cations than silversulfadiazine cream or 0.5% silver nitrate solution but more of the silver is released (by Acticoat). Silver-impregnated slow-release dressings release minute concentrations of silver which are quickly bound up by the chloride in the wound exudate. In vitro and animal studies suggests Acticoat is effective against most common strains of wound pathogens, can be used as a protective covering over skin grafts, has a broader antibiotic spectrum activity, and is toxic to keratinocytes and fibroblasts. Animal studies suggest a role for nanocrystalline silver in altering wound inflammatory events and facilitation of the early phase of wound healing. Acticoat in wound management is cost-effective, reduces wound infection, decreases the frequency of dressing changes and pain levels, decreases matrix metalloproteinase activity, wound exudate, and bioburden levels, and promotes wound healing in chronic wounds. The schematic representation of silver nanoparticles is shown in Fig. 1.1.

1.7.4.2 Gold nanoparticles

Gold nanoparticles based on gold cores are another versatile platform that provides desirable qualities for biomedicinal applications such as theragostics, photothermal therapy, and tissue imaging. They are prepared with core size from 1.5 to 10 nm, providing a large area for efficient drug and ligand conjugations. Gold nanoparticles are commonly synthesized by chemical treatment of hydrogen tetrachloroaurate. Gold nanoparticles can be conjugated with drug and targeting ligand as advanced theranostics that specifically recognizes the target receptor for active targeting. The therapeutic loading is achieved by either noncovalent interaction (e.g., via electrostatic interaction) or covalent chemical conjugation (i.e., organic drug). The inherent features of gold nanoparticles include diagnostic property, tunable core size, monodispersity, low toxicity, large surface to volume ratio, surface plasmon absorption, ability to bind to biomolecules via Au–S bonds, light-scattering properties, and ease of fabrication.

Recently, gold nanoparticles showed the possibility of treating multidrug-resistant tumors by targeted photothermal treatment in combination with a chemotherapeutic agent. Gold nanoparticles surface-functionalized with PEG, biotin, paclitaxel, and rhodamine B linked beta-cyclodextrin (beta-CD) as a theranostic platform. Paclitaxel formed an inclusion complex with beta-CD which was then conjugated with gold nanoparticles. In vitro studies suggest that gold nanoparticles have higher affinity toward cancer cells such as HeLa, A549, and MG63 in comparison with NIH3T3 cells. Furthermore, gold nanoparticles displayed a significant cytotoxic effect against HeLa cancer cells. The preliminary work indicates that the gold nanoparticles are feasible for use as theranostic nanomedicine (Zhou et al., 2015). The structure variant of surface-modified Gold nanoparticles is depicted in Fig. 1.1.

1.7.5 Carbon nanotube and fullerenes

CNTs are unique sp^2 hybridized three-dimensional tubular structures, consisting exclusively of carbon atoms with C–C distance of ~1.4A° arranged in a series of graphene sheets rolled up into a seamless tubular cylinder with open ends and a diameter of around tens of nanometers. This carbon-based novel nanomaterial belongs to the fullerenes family, i.e., the third allotropic form of carbon along with diamond and graphite. CNTs' unique physicochemical properties make them an ideal candidate for drug delivery and targeting; however, poor dispersibility has been the greatest obstacle to their use in nanomedicines. In recent years, various functionalization approaches have been developed to solubilize CNTs and increase their biocompatibility. Several unique physicochemical properties of CNTs, such as ultralight weight, high surface area, high aspect ratio (length/diameter), excellent surface chemistry, nonimmunogenicity, biocompatibility, photoluminescence, nanoneedle, biliary excretion, rapid uptake by cells due to anisotropic "needle-like" morphologies, and extremely high drug entrapment ability, are expected to make them attractive nanovehicles for drug delivery.

Additionally, functionalized CNTs mimic a nanomatrix, wherein drug molecules get entrapped well and hence release can be controlled temporally or spatially. CNTs have been classified depending on the number of graphene layers into single-, double-, triple-, and multiwalled carbon nanotubes (SWCNTs, DWCNTs, TWCNTs, and MWCNTs). The CNTs–hybrid conjugates attract great attention in the current scenario for promising delivery of bioactives in the pharmaceutical, biotechnological, and biomedical arena. The CNTs–hybrid minimizes the toxicities and improves the safety and efficacy of the pharmaceutical products (Li et al., 2015a,b). Many CNTs and dendrimer hybrids are used for delivery of antifungal and anticancer bioactives, while QD and CNT hybrids are used for photothermal therapy and imaging of tissues.

Fullerenes, a carbon allotrope, also called "bucky balls," were discovered in 1985. The buckminster fullerene is the most common form of fullerene, measuring about 7Å in diameter with 60 carbon atoms arranged in a shape known as a truncated icosahedron.

Fullerenes are classified as alkali-doped fullerenes, endohedral fullerenes, endohedral metallo fullerenes, and exohedral fullerenes. Fullerenes are being investigated for drug transport of antiviral drugs, antibiotics, and anticancer agents. Fullerenes can also be used as free radical scavengers due to the presence of a high number of conjugated double bonds in the core structure. These are found to have a protective activity against mitochondrial injury induced by free radicals. However, fullerenes can also generate reactive oxygen species during photosensitization that kills the target cells. This technique is employed to evaluate antimicrobial potential in some bacteria and mycobacterium and also used in cancer therapy. Fullerenes have the potential to stimulate host immune response and produce fullerene-specific antibodies. Animal studies with C60 fullerene conjugated with thyroglobulin have produced a C60-specific immunological response which can be detected by ELISA with IgG-specific antibodies. This can be used to design methods of estimation of fullerene levels in the body when used for therapeutic or diagnostic purposes.

1.7.6 Quantum dot

Semiconductor QDs (Fig. 1.2A) are a generation of superior optical property fluorophores that have captivated researchers in the biomedical field over the last decade. QDs have unique optical properties such as tunable emission spectra, improved brightness, superior photostability, and simultaneous excitation of multiple fluorescence colors as compared to organic dyes and fluorescent proteins. QDs are colloidal semiconductor nanocrystals with excellent photoluminescent properties, high

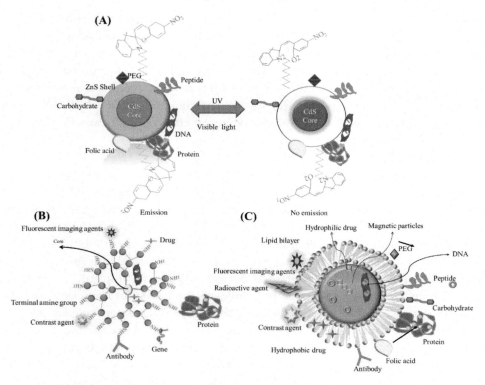

Figure 1.2 Schematic representation of (A) QD, (B) dendrimer, and (C) liposomes.

quantum yields, and high resistance to photo bleaching. The term "quantum dot" was coined in 1988 and generally produced from groups III–V and II–VI elements of the periodic table. QDs must be considerably expanded to include wider categories of nanoparticles more recently established by researchers and based on carbon, silicon, gold, molybdenum, sulfate, and analogous materials, all of which exhibit the quantum confinement event associated with a spectacular change of electron behavior kept at the boundaries of the Bohr radius in ultrasmall objects with size below 10 nm (Wang and Hu, 2014).

At present, QDs have been widely used in biological and medical studies because of their superior photoemission and photostability. Due to the smaller size they can penetrate deeper tissues' intracellular barrier, as well as having efficient excretion if introduced in vivo, be as bright and stable as possible for more sensitive and reliable detection, and last but not the least, show minimal signs of toxicity if applied in vivo or introduced into the human organism following unintentional exposure. Nanocrystal QDs are widespread as novel tools in various fields, not only in materials engineering, electronics, plastics, and the automobile and aerospace industries, but also in molecular biology and medicine. The design of QDs for efficient imaging of specific tissue, cellular, and subcellular targets is achieved via their functionalization with biologically active molecules, such as antibodies, drugs, peptides, and small molecular moieties, and micelles. In recent years, near-infrared (NIR) QDs have emerged as a promising tool in analytical applications, especially for in vivo imaging and therapy. NIR QDs allow photon penetration through tissue and minimize the effects of tissue autofluorescence.

1.7.7 Nanopores

Nanopores, designed in 1997, consist of wafers with a high density of pores (20 nm in diameter). The pores allow entry of oxygen, glucose, and other products like insulin to pass through. Nanopores have potential applications in biosensing, diagnostics, and separation, and are based on a remarkably simple concept of threading single molecules through a nanometer-sized hole in a solid-state, biological, or biomimetic membrane (Liu et al., 2014). However, they do not allow immunoglobulin and cells to pass through them. Nanopores can be used as devices to protect transplanted tissues from the host immune system, at the same time utilizing the benefit of transplantation. Beta cells of the pancreas can be enclosed within the nanopore device and implanted in the recipient's body. This tissue sample receives the nutrients from the surrounding tissues and at the same time remains undetected by the immune system and hence they are not rejected. This could serve as a newer modality of treatment for insulin-dependent diabetes mellitus.

Nanopores can also be employed in DNA sequencing. Modified nanopores have the ability to differentiate DNA strands based on differences in base pair sequences. Nanopores are also being developed with the ability to differentiate purines from pyrimidines. Further, incorporation of electricity-conducting electrodes is being designed to improve longitudinal resolution for basepair identification. Such a method could possibly read a thousand bases per second per pore. These can be used for low-cost high-throughput genome sequencing, which would be of great benefit for application of pharmacogenomics in drug development process (Liu et al., 2014).

1.7.8 Nanoshells and magnetic nanoprobes

Nanoshell particles constitute a special class of nanocomposite which are thin coatings deposited around the core particles of different materials. Nanoshell particles are highly functional materials with tailored properties, which are quite different to either of the core or of the shell material. Core particles of different morphologies, such as rods, wires, tubes, rings, cubes can be coated with a thin shell to get the desired morphology in core shell structures. Nanoshell materials can be synthesized practically using any material, like semiconductors, metals, and insulators. Usually dielectric materials such as silica and polystyrene are commonly used as core because they are highly stable, chemically inert, and water-soluble. Metal nanoparticles show optical absorption in the visible range of the electromagnetic spectrum, luminescence enhancement, surface chemical and catalytic properties, and magnetic property. Numerous techniques have been developed to synthesize nanoshell particles. Preparation of nanoshell particles involves a multistep synthesis procedure. It requires highly controlled and sensitive synthesis protocols to ensure complete coverage of core particles with the shell material. There are various methods to fabricate core shell structures, e.g., precipitation, grafted polymerization, microemulsion, reverse micelle, sol–gel condensation, and layer-by-layer adsorption technique (Coughlin et al., 2014).

Nanoshells have gained considerable attention in clinical and therapeutic applications such as photothermal therapy, colorimetry and biosensing, chemical library, gene screening, barcoding and biological imaging, and coloidal stability. Nanoshells are strong absorbers that can be used in photothermal therapy, while efficient scatters can be used in imaging applications (Coughlin et al., 2014). Magnetic nanoprobes are used for cancer therapy. Iron nanoparticles coated with monoclonal antibodies directed to tumor cells can generate high levels of heat and after that accumulate in target sites by means of an alternating magnetic field applied externally. This heat kills the cancer cells selectively. This method, designed by Triton Biosystems, is under clinical trials for solid tumors (Zabow et al., 2015).

1.7.9 Dendrimers

Among the available polymeric nanocarriers, dendrimer is one of the most widely explored polymeric nanocarriers (Tekade et al., 2015d). Dendrimers are novel three-dimensional highly branched polymeric nanocarriers that are synthesized in a reiterative trend. The globular shape with well-defined multivalent structure, monodispersity, and highly controlled architecture of dendrimers render them an inimitable carrier system which could be successfully explored for targeted drug delivery. These structures have diameters ranging from 1 to 10 nm. The presence of a large hydrophobic cavity can be used for the entrapment of bioactives, providing opportunities for controlled and sustained drug release.

The architecture of dendrimers comprises of three distinct domains, namely the central core, branches, and many terminal functional groups (Fig. 1.2B). The core of a dendrimer consists of a single atom or an atomic group having at least two identical chemical functions, whereas branches, deriving from the core, comprise of repeat units having at least one branch junction, whose repetition is organized in a geometrical progression that results in a series of radially concentric layers called "generations." This repetition leads to the construction of subsequent higher generations. With each subsequent generation, the number of end groups increases exponentially. Dendritic macromolecules tend to linearly expand in diameter and assume a more globular shape with increasing dendrimer generation. Addition of successive layers gradually increases molecular size and amplifies the number of surface groups present (Tekade et al., 2015d).

Amine terminal functionalities of dendrimers can enter into the cell due to the positive charge itself, while lipid membranes of the cells have a negative charge and dendrimers can cross several delivery barriers by active as well as passive targeting. These properties make dendrimers incomparable and most advantageous carriers in nanomedicine because the objective of nanomedicine is to control and manipulate biomacromolecular constructs and supramolecular assemblies that are critical to living cells in order to improve the quality of human health.

1.7.10 Liposomes

Liposomes are microparticulate or colloidal carriers, usually 0.05–5 µm in diameter which form spontaneously when certain lipids are hydrated in aqueous media (Fig. 1.2C). They have recently undergone vibrant technical advances such as remote drug loading, extrusion for homogeneous size long-circulating liposomes, triggered release liposomes, liposomes containing nucleic acid polymers, ligand-targeted liposomes, and liposomes containing drug combinations. These advances including the safety profile of lipids have led to numerous clinical trials in such diverse areas as the delivery of anti-cancer, antifungal, and antibiotic drugs, the delivery of gene medicines, and the delivery of anesthetics and antiinflammatory drugs (Bozzuto and Molinari, 2015).

The major advances in liposome technology resides in its ability to produce well defined spherical shape particles averaging less than 100 nm in diameter, composed of a wide variety of phospholipids with different physical and chemical properties with high drug-entrapment efficiency. These physical and chemical properties have been shown to significantly affect the stability and pharmacokinetics of liposomes.

A number of procedures have been established to produce well-defined liposomes. These include extrusion, where the liposomes are forced through filters with well-defined pore sizes under moderate pressures, reversed-phase evaporation, sonication, and detergent. It soon became clear that there were a number of problems associated with the in vivo use of the first-generation liposomes, such as drug release was shown to be affected by exposure to serum protein sometimes, rapid clearance, how to deliver molecules across cell membranes to intracellular sites of action, and passive targeting. These problems can be overcome by designing liposomes to achieve optimum properties. Changing the content of the liposome bilayer, in particular by incorporation of cholesterol was shown to "tighten" fluid bilayers and reduce the leakage of contents from liposomes, switching from a fluid-phase phospholipid bilayer to a solid-phase bilayer also reduced leakage. Addition of the mono-asialoglycoprotein GM1 to liposomes composed of egg phosphatidylcholine (egg PC), in combination with cholesterol for membrane rigidity, resulted in the first long-circulating liposomes that did not require MPS blockade to achieve the effect. Substitution of sphingomyelin for egg PC resulted in even longer circulation half-lives, and lower uptake of liposomes into liver drugs (Bozzuto and Molinari, 2015).

1.7.11 Self-microemulsifying drug delivery systems (SMEDDS)

In recent years, self-emulsifying drug delivery systems (SEDDS) and SMEDDS have revealed a rational accomplishment in convalescing oral bioavailability of poorly water soluble and lipophilic drugs. SEDDS are ideal isotropic mixtures of oils, surfactants, solvents, and sometimes cosolvents/surfactants, which emulsify under conditions of gentle agitation, can be used for the design of formulations in order to improve the oral absorption of highly lipophilic drug compounds. SEDDS are physically stable formulations that are easy to manufacture. SEDDS can be orally administered in soft or hard gelatin capsules and form fine relatively stable oil-in-water emulsions upon aqueous dilution owing to the gentle agitation of the gastrointestinal fluids (Wu et al., 2015).

Several unique physicochemical properties of SEDDS have been investigated, such as enhanced oral bioavailability, more consistent temporal profiles of drug absorption, selective targeting of drug(s) toward specific absorption window in the GIT, protection of drug(s) from the hostile environment in the gut, control of delivery profiles, protection of sensitive drug substances, and high drug payloads. SEDDS consist of an oil phase, surfactant, cosurfactant, and cosolvents. The oils can solubilize the lipophilic drug in a specific amount. It is the most important excipient because it can facilitate self-emulsification and increase the fraction of lipophilic drug transported via the intestinal lymphatic system, thereby increasing absorption from the GIT. Long-chain triglyceride and medium-chain triglyceride oils with different degrees of saturation have been used in the design of SEDDS. Modified or hydrolyzed vegetable oils have contributed widely to the success of SEDDS owing to their formulation and physiological advantages. Nonionic surfactants with high hydrophilic–lipophilic balance (HLB) values are used in the formulation of SEDDS (e.g., Tween, Labrasol, Labrafac CM 10, Cremophore).

Cosurfactant/cosolvents like Spans, capyrol 90, Capmul, lauroglycol, diethylene glycol monoethyl ether (transcutol), propylene glycol, polyethylene glycol, polyoxyethylene, propylene carbonate, tetrahydrofurfuryl alcohol, polyethylene glycol ether (Glycofurol), etc., may help to dissolve large amounts of hydrophilic surfactants or the hydrophobic drug in the lipid base. These solvents sometimes play the role of the cosurfactants in the microemulsion systems. SMDDS have been potentially applied for enhancement of solubility and bioavailability of class Π drug (low solubility/high permeability) protection against biodegradation in degrading environments, such as acidic pH in stomach, enzymatic degradation, or hydrolytic degradation (Wu et al., 2015).

1.8 NANOTECHNOLOGY AND MEDICINAL APPLICATIONS

Nanostructured biomaterials have inimitable physicochemical properties such as being ultrasmall with controllable size, large surface area to mass ratio, high reactivity, and functionalizable structure. They modify and widen the pharmacokinetic and pharmacodynamic properties of various types of drug molecules (Malinoski, 2014). Therefore, nanomaterials with such incredible properties have been comprehensively investigated in a broad array of biomedical applications, in particular targeted delivery of both imaging agents and anticancer drugs and early detection of cancer lesions, determination of molecular signatures of the tumor by noninvasive imaging and, most importantly, molecular targeted cancer therapy, regenerative medicine, and tissue engineering.

1.8.1 Nanomedicine in drug delivery and therapy

1.8.1.1 Targeted drug delivery

Targeted drug delivery is a strategy that selectively and preferentially delivers the therapeutic agents to the target site concurrently failing access to the nontarget site. For this purpose, one of the attractive strategies is use of "ligand" that will facilitate the homing of the therapeutic moieties to the target tissues. Ligand–targeted chemotherapeutics are based on the exploitation of antigens or receptors that are either inimitably expressed or overexpressed on the tumor cells comparative to normal tissues in order to protect normal tissues from the reach of anticancer drugs and to deliver the anticancer drug selectively to tumor tissue. These ligands can be small molecules (e.g., carbohydrates, folate), low–density lipoprotein (LDL), or peptides and proteins (e.g., transferrin or antibodies). Some important considerations govern the choice of the ligand that is used to target the drug molecule (Fig. 1.3A).

The anticancer potential of the SQV-loaded folic acid (FA) conjugated PEGylated and nonPEGylated poly(D,L–lactide-coglycolide) (PLGA) nanoparticles (NPs) (SQV–Fol–PEG–PLGA and SQV–Fol–PLGA) reported by Singh et al. on PC-3 (human prostate) and MCF-7 (human breast) cancer cell lines. SQV–Fol–PEG–PLGA showed enhanced cytotoxicity and cellular uptake and was most preferentially taken up by the cancerous cells via folate RME mechanism. At 260 mM concentration, SQV–PLGA NPs and SQV–Fol–PEG–PLGA NPs showed 20%, and 23% cell growth inhibition in PC-3 cells, respectively, whereas in MCF-7 cells it was 15%, and 14% cell growth inhibition, respectively (Singh et al., 2015).

1.8.1.2 Radiotherapy

Radiation sensitization is a process of enhancing the susceptibility of tumor tissues to injury by radiation exposure. Therapeutic beams, employed in the form of radiation therapy, do not distinguish between normal and cancerous cells and must rely on targeting the radiation beams to specific cells. However, the application of nanoscale

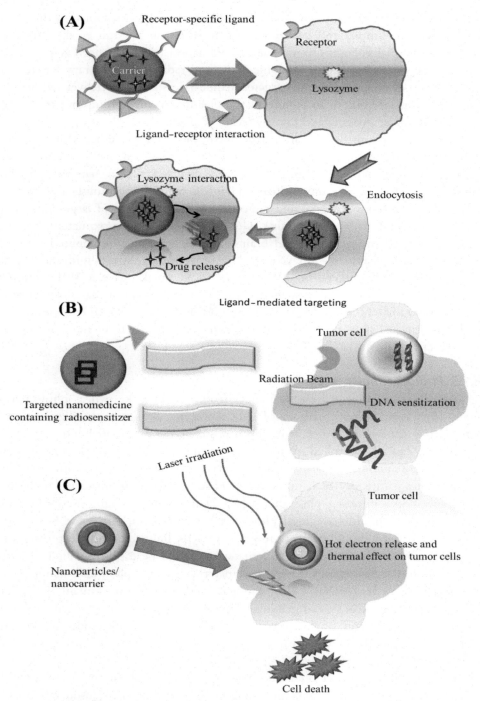

Figure 1.3 Schematic representation of (A) ligand receptor-mediated interaction, (B) radiotherapy, and (C) photothermal therapy.

particles in radiation therapy has aimed to develop outcomes in radiation therapy by rising toxicity in tumors and lowering it in normal tissues (Atun et al., 2015). The use of nanomaterial radiosensitizers is also called nanoparticle enhanced X-ray therapy (Fig. 1.3B). Current progress in nanomedicine and radionuclide therapy has revealed the possibility of designing tumor-targeted nanocarriers that can deliver radionuclide payloads to a specific site or in a molecularly selective manner to improve the efficacy and safety of cancer imaging and therapy (Atun et al., 2015).

Nanoparticles consisting of rare earth metals and high Z elements such as using hafnium oxide (HfO_2), gadolinium, QDs made from CaF, LaF, ZnS, or ZnO, superparamagnetic iron oxide, and amino silanized oxidized silicon nanoparticles have been suggested for use as radiosensitizers. Hainfeld tested intravenously injected gold nanoparticles for X-ray imaging and radiotherapy enhancement of large, imminently lethal, intracerebral malignant gliomas. Gold nanoparticles approximately 11 nm in size were injected intravenously and brains imaged using microcomputed tomography. A total of 15 h after an intravenous dose of 4 g Au/kg was administered; brains were irradiated with 30 Gy 100 kVp X-rays. Gold uptake gave a 19:1 tumor-to-normal brain ratio with 1.5% w/w gold in tumor, calculated to increase local radiation dose by approximately 300%. Mice receiving gold and radiation (30 Gy) demonstrated 50% long-term (>1 year) tumor-free survival, whereas all mice receiving radiation died (Hainfeld et al., 2013).

1.8.1.3 Photothermal therapy

The use of photo-induced heat for cancer management is known as photothermal therapy (Fig. 1.3C). These methods of light-absorbing dyes were reutilized to attain photothermal damage of tumor. It has been postulated as a potential anticancer treatment since it can be controlled spatiotemporally, thus avoiding damage to nontargeted regions. As most photothermal-inducing materials are nanoscaled, photothermal therapy constitutes a category of nanomedicine. First, template nanomaterials to be used for therapy should possess a photothermal effect that converts light energy into heat energy, i.e., they should have high photothermal conversion efficiency. Second, the photothermal effects should occur in response to NIR light to ensure deep tissue penetration. Finally, the surface of the nanomaterials should be easily modified to enable efficient photothermal therapy. Many nanomaterials such as plasmonic nanoparticles, i.e., gold nanoparticles (AuNPs), silver nanoparticles (AgNPs), sp^2 domain-rich carbon nanoparticles such as CNTs, and graphene are well known as photothermal converting nanoparticles. In addition, single-layered transition metal dichalcogenides and melanin structure have recently shown great potential for application as photothermal therapeutic agents (Kim et al., 2016).

Sun et al. (2016) reported salt-induced aggregation of gold nanoparticles in biological media to form extremely biocompatible NIR photothermal transducers for PTT and photothermal/photoacoustic (PT/PA) imaging of cancer. The GNP depots in situ formed by salt-induced aggregation of GNPs show strong NIR absorption induced by plasmonic coupling between adjacent GNPs and very high photothermal conversion efficiency (52%), enabling photothermal destruction of tumor cells. More interestingly, GNPs aggregate in situ in tumors to form GNP depots, enabling simultaneous PT/PA imaging and PTT of the tumors (Sun et al., 2016).

1.8.1.4 AIDS management vaccination

In 1984, HIV was identified as the cause of AIDS 1 year after the virus was isolated. Nanotechnology-based antiretroviral drug delivery is embraced in the search for a cure for HIV, because it could alter tissue distribution by targeting drugs to HIV reservoirs and by increasing the half-lives of drugs. Immunotherapeutic nanomedicines are new, complex, multimodular vaccines that are endowed with advanced therapeutic effects for HIV treatment. A particulate vaccine based on HIV VLPs (40 nm) has been recently introduced, made up of noninfective viruses, viral envelope proteins without the accompanying genetic material, having both humoral and cellular immune responses against HIV.

Various biodegradable and nonbiodegradable polymeric and liposomal delivery systems have been explored for transforming HIV antigens to synthetic NPs to increase their immunogenicity and to protect them against extra- and intracellular degradation. Targeting dendritic cells (DCs) that are essential for initiating immune responses can be achieved by Dermavir, which is the first nanomedicine developed for the treatment of HIV/AIDS that has demonstrated encouraging phase II clinical safety, immunogenicity, and efficacy results (Kumar et al., 2015).

Vela Ramirez et al. designed carbohydrate-functionalized nanovaccines (polymeric nanoparticles) against viral pathogens, such as HIV-1, due to the important role of DCs and macrophages in viral spread. The carbohydrate-functionalized nanoparticles preserved antigenic properties upon release and also enabled sustained antigen release kinetics. Particle internalization was observed to be chemistry-dependent with positively charged nanoparticles being taken up more efficiently by DCs. Upregulation of the activation makers CD-40 and CD206 was demonstrated with carboxymethyl-α-D-mannopyranosyl-(1,2)-D-mannopyranoside-functionalized nanoparticles. The secretion of the cytokines IL-6 and TNF-α was shown to be chemistry-dependent upon stimulation with carbohydrate-functionalized nanoparticles (Vela Ramirez et al., 2014).

1.8.1.5 Fungal infection

Fungal infections can attack epithelial tissues, deeper organs, as well as the immunological state of the patient. Topical therapy is desirable since, in addition to targeting the site of infection, it reduces the risk of systemic side effects and increases patient compliance. Nontoxic nanoobjects were also included because they improve the ocular delivery of antifungals. The nanoparticulate agents against cutaneous and ocular mycosis are metallic nanoparticles and nonmetallic nanoparticles (Perera et al., 2015).

Sanchez et al. investigated the efficacy of nanoparticle-encapsulated AmB (AmB-np) as a topical therapeutic against *Candida* species. Clinical strains demonstrated equal or enhanced killing efficacy with 72.4%–91.1% growth reduction by 4 h. AmB-nps resulted in statistically significant reduction of fungal biofilm metabolic activity ranging from 80% to 95% viability reduction ($p < 0.001$). Using a murine full-thickness burn model, AmB-np exhibited a quicker efficiency in fungal clearance versus AmB-sol by day three, although wound-healing rates were similar (Sanchez et al., 2014).

1.8.2 Nanotechnology in protein, peptide, and gene therapy

1.8.2.1 Gene angiogenesis inhibitors

Angiogenesis, the formation of new capillaries from preexisting vessels, is crucial for ensuring normal embryonic vascular development of all vertebrates, as well as regulating physiological processes such as menses and wound healing in adults. The development of therapies targeting tumor-associated angiogenesis is aimed as harnessing the potential of nanotechnology to advancement in the pharmacology of chemotherapeutics, including antiangiogenic agents. Nanoparticles present numerous advantages above those of free drugs, together with their capability to carry high payloads of therapeutic agents, confer increased half-life and reduced toxicity to the drugs, and provide means for selective targeting of tumor tissue and vasculature. The plethora of nanovectors available, in addition to the various methods available to combine them with antiangiogenic drugs, allows researchers to fine-tune the pharmacological profile of the drugs ad infinitum. The use of nanovectors has also opened up novel avenues for noninvasive imaging of tumor angiogenesis. As such, the first US Food and Drug Administration (FDA)-approved antiangiogenic therapy was the monoclonal antibody Bevacizumab (Avastin), that targets VEGF proteins overexpressed in colorectal cancer cells and their vasculature. The tumor vessels have increased permeability due to aberrant angiogenesis, thus allowing nanoparticles with diameters less than 200 nm to passively extravasate into the tumor sites through the enhanced permeability and retention effect (Massadeh et al., 2016).

Li et al. reported a novel tumor vascular-targeting multidrug delivery system using mesoporous silica nanoparticles as carrier to coload an antiangiogenic agent (combretastatin A4) and a chemotherapeutic drug (doxorubicin) and conjugate with targeting molecules (iRGD peptide) for combined antiangiogenesis and chemotherapy. Such a dual-loaded drug delivery system is capable of delivering the two agents at the tumor vasculature and then within tumors through a differentiated drug release strategy, which consequently results in greatly improved antitumor efficacy at a very low doxorubicin dose of 1.5 mg/kg. The fast release of the antiangiogenic agent at tumor vasculatures led to disruption of the vascular structure and had a synergetic effect with the chemotherapeutic drug slowly released in the following delivery of chemotherapeutic drug into tumors (Li et al., 2015a,b).

1.8.3 Nanomedicine in imaging and diagnostics

Nanomedicine is the application of nanotechnology in the medical field. As physiological processes at cellular and subcellular levels occur on a nanoscale, nanomedicine holds great promise for improving medical diagnostics and therapeutics (Durymanov et al., 2015). The technology is expected to create innovations and play a critical role in various biomedical applications, not only in drug delivery, but also in molecular imaging, biomarkers, and biosensors. Target-specific drug therapy and methods for early diagnosis of pathologies are the priority research areas where nanotechnology could play a vital role.

1.8.3.1 Multifunctional nanomedicines as theranostics

The advances in genomics, proteomics, and bioinformatics have directed the development of novel anticancer agents to diminish drug abuse and augment safe and specific drug treatment. Theranostics, combining therapy and diagnosis, is an interesting approach for chemotherapy in medicine which displays enhanced biodistribution, selective cancer targeting ability, reduced toxicity, masked drug efficacy, and minimum side effects. The role of diagnosis tools in theranostics is to collect the information of the diseased state before and after specific treatment.

Nanotheranostics is applying and further developing nanomedicine strategies for advanced theranostics. Advanced theranostic nanomedicine is multifunctional in nature, capable of diagnosis and delivery of therapy to the diseased cells with the help of targeting ligand and biomarkers, various nanocarriers such as polymer conjugations, dendrimers, micelles, liposomes, metal and inorganic nanoparticles, CNTs, and nanoparticles of biodegradable polymers, which are used as nanotheranostics for sustained, controlled, and targeted codelivery of diagnostic and therapeutic agents for better theranostic effects with fewer side effects. Theranostic nanomedicine can achieve systemic circulation, evade host defenses, and deliver the drug and diagnostic agents at the targeted site to diagnose and treat the disease at the cellular and molecular levels.

The therapeutic and diagnostic agents are formulated in nanomedicine as a single theranostic platform, which can then be further conjugated to biological ligand for targeting. Nanotheranostics can also promote stimuli-responsive release, synergetic and combinatory therapy, siRNA codelivery, multimodality therapies, oral delivery, delivery across the blood–brain barrier, as well as escape from intracellular autophagy. The fruition of nanotheranostics will be able to provide personalized therapy with bright prognoses, which will make even the fatal diseases curable or at least treatable at the earliest stage (Muthu et al., 2014).

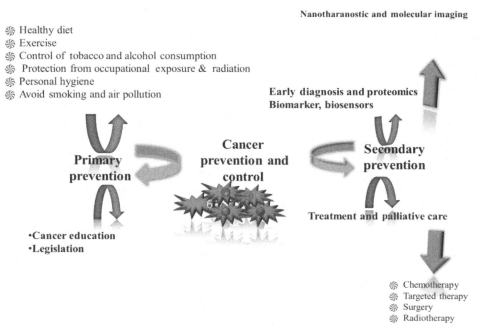

Figure 1.4 Prevention and control of cancer.

1.8.3.2 Nanotechnology-based biosensors

Detecting cancer at an early stage is one of the most important factors associated with the increase in the survival rate of patients. A complete schematic representation of cancer proteomics is presented in Fig. 1.4. During the last two decades there have been considerable advances in the field of nanotechnology-based biosensors and diagnostics (NBBD), including the production of nanomaterials, employing them for new biosensing and diagnostic applications, their widespread characterization for in vitro and in vivo applications, and toxicity analysis. All these developments have led to a tremendous technology push and flourishing demonstrations of several promising NBBD. During the last decade, nanomaterials have been widely used in the fields of in vitro diagnostics, imaging, and therapeutics. They have enabled the simultaneous multiplex detection of many disease biomarkers and the diagnosis of diseases at a very early stage. They have also opened the possibility to explore the detection of ultra-trace concentrations of target analytes and have led to ultrasensitive, rapid, and cost-effective assays requiring minimum sample volume (Arduini et al., 2016).

The nanomaterials are being seen as the most promising candidates for the development of high-throughput protein arrays. Nanomaterials such as QDs and NPs are good imaging agents due to their enhanced performance and functionality. They can

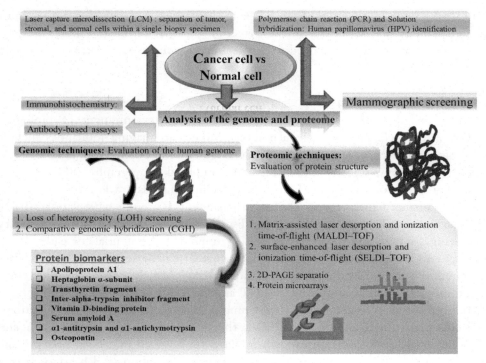

Figure 1.5 Early detection and proteomics.

be targeted to the specific disease sites in the body by conjugating them to biomarker-specific vectors. The nanomaterial-based imaging agents provide additional information pertaining to the physiology and function apart from the anatomical information, which enables more accurate and early disease diagnosis, such as the highly sensitive detection of early-stage cancer, thereby leading to better therapy. Similarly, the effectiveness of treatments can be monitored more rapidly and accurately (Huang et al., 2012). Early detection and proteomics regarding cancer is demonstrated in Fig. 1.5.

Ma et al. (2015) synthesized a highly sensitive electrochemical immunosensor, i.e., trimetallic AuPdPt nanoparticles for detection of typical bladder cancer by using oxide-tetraethylene pentamine (rGO-TEPA) for signal amplification and AuPdPt NP immobilization due to its excellent conductivity and large surface area. An effective platform was constructed for antibody anchoring by using AuPdPt NPs, which kept the antibodies' high stability and bioactivity. Moreover, AuPdPt NPs could accelerate the electron transfer and enhance the signal response, which was assisted by the synergistic effect of the three different metals (Au, Pd, and Pt). The proposed immunosensor showed satisfactory performance such as simple fabrication, low detection limits

(0.01 U/mL), wide linear range (from 0.040 to 20 U/mL), short analysis time (2 min), and high stability and selectivity in the detection of NMP22. Furthermore, the proposed immunosensor was employed to test real urine samples with satisfactory results (Ma et al., 2015).

1.8.3.3 Molecular imaging

Molecular imaging refers to the characterization and measurement of biological processes at the cellular and molecular levels. Molecular imaging modalities include optical bioluminescence, optical fluorescence, targeted ultrasound, MRI, magnetic resonance spectroscopy, single-photon-emission computed tomography (SPECT), and positron emission tomography (PET). Molecular imaging plays a key role in molecular or personalized medicine in patient management. Recently, nanoplatform-based molecular imaging has emerged as an interdisciplinary field, which involves chemistry, engineering, biology, and medicine. Possessing unpredicted potential for early detection, accurate diagnosis, and personalized treatment of diseases, nanoplatforms have been employed in every biomedical imaging modality, namely, optical imaging, computed tomography, ultrasound, MRI, SPECT, and PET (Kompella et al., 2013).

Multifunctionality is the genus development of nanoplatforms above traditional approaches. Targeting ligands, imaging labels, therapeutic drugs, and many other agents can be integrated into the nanoplatform to allow for targeted molecular imaging and molecular therapy by encompassing many biological and biophysical barriers. In this chapter, we summarized the current state of the art of nanoplatforms for targeted molecular imaging in living subjects.

1.8.3.4 Biomarker mapping

Early detection of cancer biomarkers with high accuracy is vital for cancer therapy. A variety of sensors based on different nanostructured materials have attracted intensive research interest due to their potential for highly sensitive and selective detection of cancer biomarkers. A variety of nanostructured materials including CNTs, silicon nanowires, gold nanoparticles, and QDs have been explored for the fabrication of cancer biomarkers. Cancer biomarkers can be substances other than just an antigen and the binding of these biomarkers may not be necessarily based on antigen–antibody interactions. For example, the folate receptor is a prototypic cancer biomarker that is overexpressed in many carcinoma cell lines. Nanotechnology pertains to nanostructured materials or materials with nanostructured components that usually possess physical and chemical properties different from their bulk counterparts. One intriguing benefit of nanotechnology involved in cancer diagnostics is the development of nanostructured materials-based detection systems that can be applied to detect cancer biomarkers with an extremely high sensitivity and an ultralow detection limit (Chinen et al., 2015).

1.8.4 Nanotechnology in cell repair and tissue engineering

1.8.4.1 Cell repair

Nanotechnologists have become involved in regenerative medicine via creation of biomaterials and nanostructures with potential clinical implications. Their aim is to develop systems that can mimic, reinforce, or even create in vivo tissue repair strategies. In fact, in the last decade, important advances in the field of tissue engineering, cell therapy, and cell delivery have already been achieved. Cell and/or tissue repair devices are nano-engineered scaffolds designed to replace or restore the following tissues: (1) skin; (2) cartilage; (3) bone; (4) nerve; and (5) cardiac. Nanoscale technologies are emerging as powerful enabling tools for tissue repair and drug discovery (Huang et al., 2016).

Nanotechnologies can be used in tissue repairing to fabricate biomimetic scaffolds, with increased complexity and vascularization. Over the last decade, nanomaterials have been highlighted as a promising aspirant for convalescing traditional tissue repairing materials. The nanomaterials display superior cytocompatible, mechanical, electrical, optical, catalytic, and magnetic properties compared to conventional (or micronstructured) materials. These unique properties of nanomaterials have helped to improve various tissue growth over what is achievable today. Biocomposite nanofibrous scaffolds made from synthetic and natural polymeric blends provide suitable substrate for tissue engineering and it can be used as nerve guides, eliminating the need for autologous nerve grafts. Nanotopography or orientation of the fibers within the scaffolds greatly influences the nerve cell morphology and outgrowth, and the alignment of the fibers ensures better contact guidance of the cells. The use of a degradable, nanofibrous scaffold made by electrostatic fiber spinning has been postulated to be a feasible method to produce cardiac grafts with clinically relevant dimensions (Shi et al., 2010).

1.8.5 Nanotechnology in tissue engineering

Tissue engineering is a very fast-growing scientific area in the current era which is used to create, repair, and/or replace cells, tissues, and organs by using cells and/or combinations of cells with biomaterials and/or biologically active molecules and it helps to produce materials which very much resemble the body's native tissue/tissues. Tissue engineering is the connecting discipline between engineering materials science, medicine, and biology. In typical tissue engineering cells are seeded on biomimicked scaffolding, providing adhesive surfaces, then cells deposit their own proteins to make them more biocompatible, but being unable to vascularize properly, due to lack of functional cells, low mechanical strength of engineered cells, not being immunologically compatible with the host, and nutrient limitations are a classical issue in the field of tissue and tissue engineering (Dvir et al., 2011).

Nanotechnology, or the use of nanomaterials, may have the answers since only these materials can mimic surface properties (including topography, energy, etc.) of natural tissues. For these reasons, over the last decade, nanomaterials have been high-lighted as promising candidates for improving traditional tissue engineering materials. Importantly, these efforts have highlighted that nanomaterials exhibit superior cyto-compatible, mechanical, electrical, optical, catalytic, and magnetic properties compared to conventional (or micron-structured) materials. These unique properties of nanoma-terials have helped to improve various tissue growths over what was previously achiev-able (Cunha et al., 2011).

1.8.6 Nanotechnology-based active implant

1.8.6.1 Sensory aids

Nano- and related microtechnologies are being used to develop a new generation of smaller and potentially more powerful devices to restore lost vision and hearing func-tions. Degenerative diseases of the retina, such as retinitis pigmentosa or age-related macular degeneration, decrease night vision and can progress to diminishing peripheral vision and blindness. The artificial retina uses a miniature video camera attached to a blind person's eyeglasses to capture visual signals. The signals are processed by a micro-computer worn on the belt and transmitted to an array of electrodes placed in the eye. The array stimulates optical nerves, which then carry a signal to the brain. Another approach by Optobionics makes use of a subretinal implant designed to replace photo-receptors in the retina. In severe hearing loss, cochlear implants include an electronic circuit that is surgically placed in the skull behind the ear on the mastoid process of the temporal bone. A new generation of smaller and more powerful cochlear implants are intended to be more precise and offer greater sound quality. An implanted trans-ducer is pressure-fitted onto the incus bone in the inner ear. The transducer causes the bones to vibrate and move the fluid in the inner ear, which stimulates the auditory nerve (Sedaghati et al., 2011).

1.8.6.2 Assessment and treatment devices

Nanotechnology offers sensing technologies that provide more accurate and timely medical information for diagnosing disease, and miniature devices that can administer treatment automatically if required. Health assessment can require medical profession-als, invasive procedures, and extensive laboratory testing to collect data and diagnose. Some medical information is extremely time-sensitive, such as finding out if there is sufficient blood flow to an organ or tissue after transplant or reconstructive surgery, before irreversible damage occurs. Nanotechnology can new offer new implantable and/or wearable sensing technologies that provide continuous and extremely accurate medical information.

Complementary microprocessors and miniature devices can be incorporated with sensors to diagnose disease, transmit information, and administer treatment auto-matically if required. Micro- and nanosized sensors can make use of a wide range of technologies that most effectively detect a targeted chemical or physical property. Implantable medical devices that are greater than 1 mm in diameter might unfavorably alter the functions of surrounding tissue. Smaller implantable devices with nonintrusive or minimally intrusive systems will likely contain nanoscale materials and smaller systems approaching the nanoscale. Implantable sensors can also work with a series of medical devices that administer treatment automatically if required (Lee et al., 2016).

1.8.6.3 Bone substitute materials

Cartilage injuries lead to joint pain and loss of function. Mature hyaline cartilage has a very low self-repair potential due to its intrinsic properties. For this reason, researchers have focused on the search for methods to reproduce the tissue characteristics of hyaline cartilage and induce complete cartilage repair. A new approach for the treatment of articular cartilage defects is the use of biocompatible scaffolds. There are plenty of polymers, but only some are suitable for cartilage tissue engineering. Natural materials used in the field of cartilage engineering include alginate, agarose, chitosan, chondroitin sulfate, hyaluronan, collagen, fibrin gelatine, and silk fibroin. It has been demonstrated that these natural materials could potentiate the production of collagen type II and sulfated glycosaminoglycans by both chondrocytes and stem cells (Gong et al., 2015).

Allograft bone, dematerialized bone matrix, and calcium-based synthetic materials have long been used as bone graft substitutes. First-generation bone graft substitutes as standalone graft substitutes have not been developed as hoped. It remains a great challenge to design an ideal bone graft that emulates nature's own structures or functions. To further improve the performance of such bone graft substitutes, scientists are investigating biomimetic processes to incorporate the desirable nanofeatures into the next generation of biomaterials. In this regard, nanostructured biomaterials less than 100 nm in at least one dimension, in particular nanocomposites, are perceived to be beneficial and potentially ideal for bone applications, owing to their nanoscale functional characteristics that facilitate bone cell growth and subsequent tissue formation. In fact, bone itself is a nanocomposite system with a complex hierarchical structure (Kolk et al., 2012).

1.8.7 Nanotechnology in cosmetics

The increasing uses of nanoparticulate engineered materials poses the question of the safety of those materials. A paradigmatic case is their use in cosmetics, in principle because those materials are in direct contact with the body and because probably cosmetic usage of nanotechnology will anticipate the use of nanoparticles in medicine. Now, nanometer-sized materials have characteristics that improve the existing cosmetics materials, or at least the product appeal (Morganti et al., 2014). Nowadays, there are almost 600 manufacturers of nanotechnology-based consumer products on the market.

In the year 2009, nanotechnology put $30 billion of manufactured products (a number predicted to grow to $2.6 trillion by 2014) on the market and, moreover, the National Science Foundation estimated that by 2015 the nanotechnology sector would employ more than a million workers. Now, nanometer-sized materials have characteristics that improve the existing cosmetics materials, or at least the product appeal.

Nanoemulsions are commonly used in certain cosmetic products, such as conditioners or lotions. Nanoemulsions combine traditional cosmetic ingredients, such as water, oils, and surfactants, in a two-phase system in which droplets sized 50–100 nm are dispersed in an external (aqueous) phase. The small droplet size renders nanoemulsions transparent and pleasant to touch; their texture and rheological properties have yet to be obtained by other formulation methods. Liposomes and niosomes are globular vesicles with diameters between 25 and 5000 nm and are composed of amphiphilic molecules which associate as a double layer (unilamellar vesicles) or multiple double layers (multilamellar vesicles) (Sonneville-Aubrun et al., 2004).

Vesicle formulations are important in cosmetic applications because they may improve the stability and skin tolerance of ingredients, such as unsaturated fatty acids, vitamins, or antioxidants and thereby contribute to the safety of cosmetics. Sunscreens contain insoluble, mineral-based materials whose performance depends on their particle size. Mineral particles, such as TiO_2, reflect and scatter UV light most efficiently at a size of 60–120 nm. The surface of these particles is frequently treated with inert coating materials, such as aluminum oxide or silicon oils, in order to improve their dispersion in sunscreen formulations. The transparency of nm-sized particles of titanium or zinc oxides results in better consumer acceptance/compliance and thus improves the protection of human skin against UV-induced damage (Morganti et al., 2014).

1.8.8 Nanotechnology in dentistry

Nanotechnology is thought to offer particular advances in dentistry and innovations in oral health-related diagnostic and therapeutic methods. Resembling nanomedicine, the improvement in nanodentistry will allow nearly perfect oral health by the use of nanomaterials and biotechnologies, including tissue engineering and nanorobots. A recent development was a drug-delivery system for the treatment of periodontal diseases by producing nanoparticles impregnated with triclosan. The toothpaste containing nanosized calcium carbonate enabled remineralization of early enamel lesions. The silver nanoparticles had an antimicrobial effect in lower concentrations and with lower toxicity. Artificial teeth made by nanocomposite containing nanodimensions inorganic fillers are diffused homogeneously without any accumulation in the matrix (Ozak and Ozkan, 2013).

Dental nanorobots are able to move through the teeth and surrounding tissues by using specific movement mechanisms. Titanium implants treated with a nanostructured calcium surface coat were inserted into rabbit tibias, and their effect on osteogenesis was investigated; the nanostructured calcium coat increased the responsiveness of the bone around the implant (Ozak and Ozkan, 2013). Advances in digital dental imaging techniques are also anticipated with nanotechnology. In digital techniques, such as nanophosphor scintillators, the radiation dose is diminished and is of high quality. Silver nanoparticles were synthesized using several biomaterials, since their small size provides great antimicrobial effect, at low filler level. Hence, these nanoparticles have been applied in dentistry, due to their antimicrobial potential, mechanical properties, cytotoxicity, and long-term effectiveness, preventing or reducing biofilm formation over dental material surfaces (Tansık et al., 2016).

1.9 NANOMEDICINE RESEARCH, DEVELOPMENT, AND CURRENT CLINICAL STATUS

1.9.1 Publications and patents on nanomedicine

More than 12,000 and more than 10,000 results were obtained when the term "nanomedicince" was entered into PubMed and Science Direct, respectively. A tenfold increase in publication data has been noticed over the span of 10 years from 2005 to 2015. Clearly there are possibilities and growth pertaining to the application and research area of nanomedicines. Even the big pharma majors are investing a large amount to carry out R&D activities in order to get nanomedicines into the commercial market (Weissig et al., 2014). Worldwide researchers are focusing on developing nanopharmaceuticals using nanotechnology-based vectors such as liposomes, dendrimers, nanoshells, CNTs, RNA-based carriers, metallic nanoparticles, and solid lipid NPs. Apart from nanovectors, researchers are also engaged in treating complex conditions like psychiatric illnesses, cancer, endothelial disorders, genetic disorders, photosensitivity, and cardiovascular diseases using nanomedicines (Finch et al., 2014).

Recently, researchers at the University of Southern California explored the therapeutic potential of elastin-like polypeptides (genetically engineered protein polymers obtained from human tropoelastin) as substrate for proteolytic degradation (a process required for cellular homeostasis and control of the cell cycle as well as genetic expression) as anticancer agents, ocular agents, and protein trafficking modulators (Despanie et al., 2015). Another collaborative research revealed the emergence of a fluorescent cancer stem cell (CSC) model as a nanodevice that permits tagging CSC employing an ALDH1A1/fd tomato reporter vector. The developed model is exclusively used to differentiate between fluorescent marking between CSC subpopulation and differentiated cells. This CSC nanodevice can be used to measure the potential of nanocarrier-loaded drugs, as in this case paclitaxel was delivered through polymeric micelles (Gener et al., 2015).

In comparison to data on publications rationally fewer patents on the nanomedicines were found. Only 884 patents granted by the USPTO were found when a search was made using the text "nanomedicne" on the USPTO website, and the data in European patent office (worldwide.espacenet.com) reflect the granting of around 50 patents only. To date, approximately 1200 patent applications were pending related to nanomedicine in the USPTO. Moreover, due to many reasons, patenting nanomedicines is not an easy task. The reasons includes post patent protection, some technical confusions (as the definition of nanotechnology cited by US national nanotechnology initiatives is not appropriate in regards to nanomedicines), and an inadequate patent classification system. Collectively these problems will result in poor patent quality, which ultimately affects the commercial viability of an invention. The PTO needs to urgently address all these problems to boost the morale of competing industries and to increase the number of commercially viable products. Nanomedicines will ultimately improve patient compliance due to the unique characteristics they possess.

A recently granted US patent (9211346) combines the two novel approaches of liposomes (as carrier) and magnetic NPs (attached with prodrug and incorporated into liposomes) and the delivery of this complex system within the subject is accelerated by an alternating magnetic field. This invention is leading to an era of combining nanomaterials and targeting moieties in order to achieve less toxic and targeted delivery of therapeutic agents. Another US patent (9226972) explores the delivery of siRNA (silencing RNA) to the desired site using targeted NPs made of fusion protein and heterologous nucleic acid. This complex is effective in modulating genetic expression in cells. A complete summary of the products approved by US FDA or European medicines agency (EMA) is presented in Table 1.1.

1.9.2 Strategic initiatives and government funding

Globally many local and international funding agencies are transferring a huge amount of capital in nanomedicine-based product projects through intra/intercollaboration and establishing a center of excellence (Hafner et al., 2014). The editorial report published by Nature materials in the year 2006 reveals that around 130 nanobased medicines including delivery systems were being developed worldwide (Nanomedicine: a matter of rhetoric, 2006). Various funding agencies worldwide are taking initiatives in carrying out high-level research to invent nanomedicines in order to support better patient care and to fight against life-threatening diseases. National Institute of Health (US Department of Health and Human Services) initiates a program especially to promote nanomedicines with the following objectives: (1) to understand basic biological machinery within living cells and their construction; (2) applying this knowledge to fabricate novel tools and medicines to combat diseases.

Table 1.1 List of some approved nanomedicines available commercially

Tradename	Nanodevice and active agent	Application	Approval/ clinical status	Company	Administration route	References
Abelcet	Liposomes/lipid NPs Amphotericin B	Systemic fungal infections when amphotericin B is not recommended	FDA (1995)	Sigma–Tau, Cephalon, Enzon, Elan/Alkermes	Intravenous	Tinkle et al. (2014), Sainz et al. (2015)
Abraxanes (ABI–007, nab–paclitaxel, NSC–736631)	Paclitaxel (Taxol) bound albumin nanoparticles	Metastatic breast cancer, nonsmall-cell lung cancer	FDA (2005) and EMA (2008)	Abraxis BioScience, AstraZeneca, Celgene	Intravenous	Tinkle et al. (2014), Sainz et al. (2015)
Adagens	Pegylated bovine adenosine deaminase (polymer–protein conjugate)	Severe combined immuno deficiency disease	FDA (1990)	SigmaTau, Enzon	Intravenous	Tinkle et al. (2014), Sainz et al. (2015)
AmBisome	liposomes/lipid NPs Amphotericin B	Fungal infections	FDA (1997) and EMA (1990)	Astellas/Gilead	Intravenous	Tinkle et al. (2014), Sainz et al. (2015)
Amphotec	Lipid-based nonliposomal nanoformulation (Amphotericin B)	Invasive aspergillosis when amphotericin B is not recommended	FDA (1996)	Alkopharma, Three Rivers/Alza	Intravenous	Tinkle et al. (2014), Sainz et al. (2015)
Avinza	Nanocrystal (Morphinesulfate)	Moderate/severe pain	FDA (2002)	Elan/Alkermes, Pfizer	oral	Tinkle et al. (2014), Sainz et al. (2015)
BIND–014ACURINS	Liposome Doxetaxel	Solid Tumor	Phase I	BIND Biosciences, Inc.	Intravenous	Sanna et al. (2014)

(Continued)

Table 1.1 List of some approved nanomedicines available commercially (Continued)

Tradename	Nanodevice and active agent	Application	Approval/clinical status	Company	Administration route	References
CALAA –01	Cyclodextrin containing polymeric Nanoparticles Si RNA	Solid Tumor	Phase I	Calando Pharmaceuticals	Intravenous	Sanna et al. (2014)
Cimzia	PEGylated antibody (Certolizumab pegol)	Crohn's disease, rheumatoid arthritis, psoriatic arthritis, ankylosing spondylitis	FDA (2008)	UCB	Intravenous	Sainz et al. (2015)
Copaxone (glatiramer acetate, copolymer 1)	Polymeric nanoformulation (Glatiramer acetate) Copolymeric mixture of L–glutamic acid, L-alanine, L-tyrosine and L-lysine (polypeptide colloidal suspension)	Multiple sclerosis	FDA (1996)	Teva Pharma	Subcutaneous	Tinkle et al. (2014), Sainz et al. (2015)
Curosurf	Liposome (Poractant alfa)	Respiratory distress syndrome (RDS) in premature infants	FDA (1999)	Chiesi	Intrathecal	Tinkle et al. (2014), Sainz et al. (2015)
DaunoXome	Liposome/NPs (Daunorubicin citrate)	HIV-related Kaposi's sarcoma	FDA (1996)	NeXstar, Gilead Sciences, Galen, Teva	Intravenous	Tinkle et al. (2014), Sanna et al. (2014), Sainz et al. (2015)
Definity	Liposome (Perflutren)	Contrast agent	FDA(2001)	Lantheus, BristolMyers Squibb	Intravenous	

Name	Description	Indication	Approval	Company	Route	References
DepoCyt	Sustained release cytarabine liposomes/lipid NPs	Lymphomatous malignant meningitis	FDA (1999) and EMA (2001)	Sigma-Tau, Skye/Enzon, Pacira	Intravenous	Tinkle et al. (2014), Sanna et al. (2014), Sainz et al. (2015)
Diprivan	Propofol liposomes/lipid NPs	Anesthetic	FDA (1989)	Astra Zeneca Pharma	Intravenous	Tinkle et al. (2014), Sainz et al. (2015)
DepoDur	Liposome (Morphine sulfate)	Chronic pain	FDA (2004)	Pacira	Intrathecal	Sainz et al. (2015)
Doxil/Caelyx in EU	Pegylated doxorubicin (Adriamycin) HCl Liposome/Lipid NPS	HIV-related Kaposi's sarcoma, ovarian cancer, myeloma, breast cancer	FDA (1995) and EMA (1996)	Centocor Ortho Biotech	Intravenous	Tinkle et al. (2014), Sanna et al. (2014), Sainz et al. (2015)
Elestrin	Estradiol gel (0.06%) incorporating calcium phosphate nanoparticles	Hot flashes during menopause	FDA (2006)	Meda, BioSante	Transdermal	Tinkle et al. (2014), Sainz et al. (2015)
Eligard	Polymeric nanoformulation (Leuprolide acetate)	Advanced prostate cancer	FDA (2002)	Atrix, Tolmar	Subcutaneous	Sainz et al. (2015)
Epaxal (HAVpur, VIROHEP-A)	Hepatitis A vaccine adjuvanted with immunopotentiating reconstituted influenza virosomes (IRIV)	Active immunization against hepatitis A for adult and children >12 months (age may vary and depend upon the country)	Approved in Canada and elsewhere	Berna Biotech Crucell	Intramuscular (in the deltoid muscle)	Tinkle et al. (2014), Sainz et al. (2015)

(Continued)

Table 1.1 List of some approved nanomedicines available commercially (Continued)

Tradename	Nanodevice and active agent	Application	Approval/clinical status	Company	Administration route	References
Emend	Nanocrystal (Aprepitant)	Emesis, antiemetic for chemotherapy patients	FDA (2003) and EMA (2003)	Merck, Elan Corp	Oral capsule Intravenous	Tinkle et al. (2014), Sainz et al. (2015)
Estrasorb	TM Surfactant-based nanoformulation (Estradiol hemihydrate)	Reduction of vasomotor symptoms during menopause	FDA (2003)	Medicis, Novavax/Espirit, Graceway	Topical	Sainz et al. (2015)
Feraheme	Metal nanoformulation (Ferumoxytol)	Treatment of iron deficiency anemia in adults with chronic kidney disease	FDA (2009)	AMAG	Intravenous	Sainz et al. (2015)
Ferrlecit	Metal nanoformulation (sodium ferric gluconate complex)	Iron deficiency anemia	FDA (1999)	Sanofi–Aventis	Intravenous	Sainz et al. (2015)
Fungizone	Surfactant nanoformulation (Amphotericin B)	Systemic fungal infections	FDA (1966)	Bristol–MyersSquibb, Apothecon	Intravenous	Sainz et al. (2015)
Genexol– PM	PLGA-PEG Micell Paclitaxel	Metastatic breast cancer	Phase II	(Samyang Bio–pharmaceuticals Corporation, Jongno-gu, Seoul, Korea	Intravenous	Sanna et al. (2014)
Invega	Nanocrystal (Paliperidone)	Schizophrenia	FDA (2006) and EMA (2007)	Janssen	Intramuscular	Sainz et al. (2015)
Kadcyla	Protein–drug conjugate (Ado–Trastuzumab Emtansine)	Metastatic breast cancer	FDA (2013)	Genentech	Intravenous	Sainz et al. (2015)

Name	Composition	Indication	Approval	Company	Route	References
Macugen	siRNA anti-VEGF inhibitor (PEG) aptanib sodium (polymer–aptamer conjugate)	Neovascular age-related macular degeneration	FDA (2004) and EMA (2006)	OSI/Pfizer, Valeant	Intravitreal	Tinkle et al. (2014), Sainz et al. (2015)
Marqibo	Liposome (Vincristine sulfate)	Philadelphia chromosome and acute lymphoblastic leukemia	FDA (2012)	Talon Therapeutics	Intrathecal	Sainz et al. (2015)
MBP–426	Liposome Oxaliplastin	Gastric, easophageal, gastric esophageal adenocarcinoma	Phase Ib/II	Mebiopharm Co. Ltd.	Intravenous	Sanna et al. (2014)
MCC–465	PLGA-PEG nanoparticles Doxorubicin	Metastatic stomach cancer	Phase I (not continued)	Osaka Japan	Intravenous	Sanna et al. (2014)
Megace ES	Nanocrystal (megestrol acetate)	Anorexia, cachexia, breast and endometrial cancer	FDA (2005)	Par	Oral suspension	Sainz et al. (2015)
Mepact TM	Liposome (Mifamurtide)	Osteo sarcoma	EMA(2009)	Takeda	Intravenous, intramuscular	Sainz et al. (2015)
Mircera	PEGylated epoetin beta (methoxy polyethylene glycol-epoetin beta)	Anemia associated with chronic renal failure	FDA (2007) and EMA (2007)	Hoffman-La Roche	Intravenous	Sainz et al. (2015)
Myocet	Liposome/lipid NPs (Doxorubicin)	Cardioprotective formulation of doxorubicin used in late metastatic breast cancer	EMA (2000)	Cephalon/Zeneus, Elan, Sopherion Therapeutics	Intravenous	Tinkle et al. (2014), Sainz et al. (2015)

(Continued)

Table 1.1 List of some approved nanomedicines available commercially (Continued)

Tradename	Nanodevice and active agent	Application	Approval/ clinical status	Company	Administration route	References
Naprelan	Nanocrystal (naproxenesodium)	Rheumatoid arthritis and osteoarthritis, gout	FDA (1996)	Almatica, Elan/ Alkermes, Wyeth	Oral tablet	Tinkle et al. (2014), Sainz et al. (2015)
Neulasta	PEGylated filgrastim/ PEG-G-CSF or pegfilgrastim (covalent conjugate of recombinant methionyl human G-CSF [filgrastim] and monomethoxypoly- ethylene glycol) (polymer–protein conjugate)	Febrile neutropenia	FDA (2002) and EMA (2002)	Amgen	Subcutaneous	Tinkle et al. (2014), Sainz et al. (2015)
Oncaspar (PEG- lasparginase)	PEGylated L-asparaginase/ Pegasparginase (polymer–protein conjugate)	Lymphoblastic leukemia	FDA (1994)	Enzon/Schering- Plough, Sigma- Tau	Subcutaneous	Tinkle et al. (2014), Sainz et al. (2015)
Ontaks	Protein–drug conjugate (Denileukin difitox)	Persistent or recurrent cutaneous T-cell lymphoma	FDA (1999)	Eisai		Sainz et al. (2015)
Pegasys	PEGylated interferon alfa-2b/PEG interferon alfa-2b (polymer–protein conjugate)	Hepatitis B and C infection	FDA (2002) and EMA (2002)	Hoffmann–La Roche/Nektar Genentech	Subcutaneous	Tinkle et al. (2014), Sainz et al. (2015)

Name	Description	Indication	Regulatory status	Company	Route of administration	References
Pegintron	PEGylated interferon alfa-2b/ PEG interferon alfa-2b (polymer–protein conjugate)	Hepatitis C in patients with compensated liver disease (Merck's Sylatron approved for melanoma with nodal involvement after surgical resection)	FDA (2001) and EMA (2000)	Schering-Plough Merck	Subcutaneous	Tinkle et al. (2014), Sainz et al. (2015)
Paclical	Micellar retinoid-derived Paclitaxel	Ovarian cancer	Phase III	Oasmia Pharmaceutical AB, Uppsala, Sweden	Intravenous	Sanna et al. (2014)
ProLindac	ProLindac	Ovarian cancer	Phase II	Access Pharmaceuticals Inc. (Dallas, TX, USA)	Intravenous	Sanna et al. (2014)
Rapamune	Nanocrystal (Sirolimus)	Immunosuppressant (kidney transplants)	FDA (2002) and EMA (2001)	Wyeth/Alkermes, Elan, Pfizer	Oral solution/ oral tablets	Tinkle et al. (2014), Sainz et al. (2015)
Renagel	Polymeric nanoformulation Crosslinked poly(allylamine) resin (Sevelamer hydrochloride)	Hyper phosphatemia in patients with chronic kidney disease on dialysis	FDA (2000) and EMA (2000)	Genzyme	Oral tablets	Tinkle et al. (2014), Sainz et al. (2015)
Ritalin LA	Nanocrystal (methylphenidate hydrochloride) Attentiondeficit	Hyperactivity disorder	FDA (2002)	Novartis	Oral	Sainz et al (2015)

(Continued)

Table 1.1 List of some approved nanomedicines available commercially (Continued)

Tradename	Nanodevice and active agent	Application	Approval/ clinical status	Company	Administration route	References
Somavert	PEGylated human growth hormone receptor agonist (Pegvisomant) (PEG–hGH) (polymer–protein conjugate)	Acromegaly	FDA (2003) and EMA (2002)	Pharmacia and Upjohn, Nektar, Pfizer	Subcutaneous	Tinkle et al. (2014), Sainz et al. (2015)
SGT53-01	Liposome (P^{53} gene)	Solid tumor	Phase I		Intravenous	Sanna et al. (2014)
Triglide	Nanocrystal Fenofibrate	Lipid disorders; reduces elevated plasma concentrations of triglycerides, LDL, and total cholesterol and raises abnormally low levels of HDL	FDA (2005)	Skye Pharma First Horizon Sciele Pharma	Oral tablets	Tinkle et al. (2014)
TriCor	Nanocrystal fenofibrate	Primary hypertrigly ceridemia hyper cholesteria, mixed lipidemia	FDA (2004)	Abbot Labs, Elan Corp.	Oral tablet	Tinkle et al. (2014)

In order to promote the research specific to nanomedicine the department has established several nanomedicine development centers amongst which five are functional. These are: (1) nanomedicine centers for nucleoprotein machines; (2) a center for protein-folding machinery; (3) a nanomedicine center for mechanobiology directing the immune response; (4) NDC for optic control of biological function; and (5) engineering cellular control synthetic signaling and systems. A huge amount ($31.311 billion) was requested by the NIH director for the upliftment of biomedical science in the United States (https://www.nih.gov/about-nih/who-we-are/nih-director/fiscal-year-2016-budget-request-0). The expenditure shows the commitment of the leading country in developing technology for health betterment.

Apart from United States, the European funding agency (European Commission) is tremendously supporting projects based on "nanomedicines" through its different programs, namely FP7 and current horizon-2020. FP7 is comprised of FP7-health (approximately 85 projects of about 400 million Euros) and FP7-MNP (around 31 projects of about 150 million Euros), whereas current Horizon-2020 funding is based on excellent science, industrial leadership, and societal challenges (Hafner et al., 2014). Other than this the Japanese Agencies Ministry of Health Labor & Welfare (MHLW) governs nanomedicine technology in Japan. Furthermore, worldwide initiatives are taken by the many other funding agencies like Institute Gustave Roussy (France), Karolinska Institute (Sweden), Agency for Science Technology and Research (Singapore), and Department of Science & Technology (India).

1.9.3 Current clinical status and regulatory considerations of nanotherapeutics

As per the report of the BCC research, the value estimated for the worldwide nanomedicine market was $214.2 billion in 2013. It was expected to rise by more than 13% in 2014 to over $248 billion and it is expected to grow at a compound annual growth rate of 16.3% from 2014 to 2019 to reach a total of over $528 billion by 2019 (Hafner et al., 2014). The data show that nanomedicine has tremendous pharmaceutical and scale-up potential, a revolutionary alarm for the management of several diseases. Moreover, a recent report investigated more than 245 nanobased products that have been approved by US FDA and are presently under clinical trials (Desai, 2012). The result of this report concluded that more than 785 significant clinical studies were found with regards to nanomedicine-based products.

Recently, Celgene Corporation developed paclitaxel nanoparticles to manage stent restenosis and completed phase II of clinical stage. Similarly, Barbara Ann Karmanos Cancer Institute in collaboration with the National Cancer Institute (NCI) developed multidrug-loaded albumin-stabilized NPs to treat metastatic pancreatic malignancy. They used paclitaxel along with gemcitabin HCL and selinexor as multidrug chemotherapy for cancer and completed phase II clinical trials.

Another clinical trial study completed phase I based on a combinational liposomal formulation for the treatment of AIDS-associated non-Hodgkins lymphoma, sponsored by the University of Alabama at Birmingham in collaboration with the NCI. However, the field of nanomedicine clinical trials is suffering from many obstacles as revealed by a recent survey (Goldberg et al., 2013). The researchers conducted face-to-face discussions with 46 potential stakeholders from Europe and North America employing qualitative research methodology. The long discussion concluded that the major diseases targeted were cancer, immunological, and infectious diseases, with a focus on nanoparticles like advanced liposomes, metallic nanoparticles, and siRNA polymeric micelles. The stakeholders admitted that clinical trials for nanobased medicines are more time-consuming, more expensive, and need high-level quality input and collaborative studies between academia and industries with proper coordination. For example, chemical vapor deposition (CVDs) demand 10,000 patient (volunteers) to carry out trials with a general cardiovascular drug, but the fact is that this is not possible when it comes to nanomedicines due to going over budget (too expensive). Also, the approval processes for nanobased medical products are much technical, which further increases complexity.

Overall, the commercialization and growth rate of nanomedicines are encouraging, showing that big pharma companies are meeting with the challenges in clinical trials, and need to focus more on translational nanomedicine research.

1.10 CURRENT CLINICAL STATUS AND REGULATORY CONSIDERATIONS

1.10.1 Regulation and approval of nanomedicines

The encourageable research funding and successful clinical translation resulted in the approval of a significant number of nanobased biomedical products in the last few years (Heath, 2015). However, differences reside in the approval process of conventional and nanomedicine-based products. The protocols for conventional dosage form are well defined and established in approving authorities like the FDA, but the lack of specific protocols due to complex manufacturing procedures as well as sophisticated and not fully validated methods for characterization of nonomedicines urgently need to be addressed. Although several attempts have already been made, regulatory aspects for nanobased products still need to be defined. While formulating nanomedicine some points related to manufacturing and characterizations are very important to be considered in order to ensure final safety, toxicity, compatibility, and storage life.

In general, the major obstacle resides in characterization due to the diverse properties of nanoformulations such as PEGylation, conjugation, surface modification, entrapment efficiency, hydration phenomenon, and morphology. These factors even in minute changes lead to significant difference in clinical outcomes (Kamaly et al.,

2016). Also, the toxicity of gold and silver NPs is much dependent on size. In addition, validated methodology and protocols are required for different aspects like functional group of excipients, modification of surface properties, target group attachment, combination of delivery system and encapsulation of drug(s), because apart from physicochemical characteristics (size, shape, morphology, electric charge) makers need an insight into the performance to check drug permeation, absorption, metabolism, protein binding, cellular uptake, and release profile. It is fundamental to address the control parameters for these critical points and other in-process variables.

Recently the concept of quality by design (QbD: online/at line quality assessment) was also applied to performance to check nanoformulations' in-process variables and final product characterization (Sangshetti et al., 2014). QbD involves the complete understanding of procedures and predefined objectives (knowledge-rich environment) which further assure the physical, chemical, and biological properties of an intended nanobased formulation in order to ensure final product characteristics. Apart from QbD, the implementation of ICH (Q_8, Q_9, Q_{10}) also nurtures innovative healthcare product development regulations and further advances in the field of novel pharmaceutical formulations. However, the evaluation of a very small quantity of nanoformululations, separation of entrapped and unentrapped cargo, and surface attachment miniaturizations still present limitations. However, the emergence of newer imaging/detection/microscopic techniques restricts these limitations to a greater extent.

Stockholders, pharma companies, and regulators, especially in Europe, the United States, and Japan, are taking serious initiatives to develop distinct, validated, and redefined regulatory approaches employing ICH and QbD advancement. Recently the EMA introduced orientation documents as a guideline to explore essential perspectives pertaining to nanoproducts and these should be followed by manufacturers. Apart from pharmacokinetics and pharmacodynamics, pharmacogenomics should also be considered while developing nanomedicines to make them affordable as a part of the patient budget.

1.10.2 US FDA-approved nanomedicine

After the approval of the first nanomedicine Doxil (known as Caelyx in the EU) by the US FDA in 1995, more than 250 nanobased pharmaceutical products are in the pipeline and in different stages of clinical studies. After the success of Doxil, which was basically based on physicochemical properties of active moiety, doxorubicin researchers are concentrating on a more profound strategy which combines the advantages of targeting and limiting the side effects of chemotherapy. Moreover doxorubicine-based nanoformulation lipodox was considered by US FDA (in 2013), which was generic liposomal injection introduced by Sun Pharma global FZE, a unit of Sun Pharma Industries, India. However, it may be noted that the same product was not approved by the EMA. Azaya and Sorrento Therapeutics are the other companies who are conducting bioequivalence studies for doxorubicin (Doxil generic version) and Paclitaxel (alternatively, IG-001), respectively.

Apart from liposomes, the FDA has approved many nanoproducts including polymeric micelles, nanocrystals, nanoparticles, and nanotubes utilized for diverse applications such as fungal infection, cancer, skin infection, iron deficiency, antiemetic, and immunosuppressant CVDs.

1.10.3 European medicines agency

In 2006, the EMA constituted a dialogue forum to enhance experience and knowledge in the development of nanotechnology-based medicinal products amongst regulatory agencies from the EU, the United States, Canada, Japan, and Australia. Moreover, since 2009 the EMA has chaired an international regulators expert group consisting of Japan, the United States, Canada, and Australia. In 2010, the EMA organized an international workshop on nanomedicine in order to discuss different regulatory issues with global stakeholders. There are many EMA-approved nanomedicines as listed in Table 1.1.

Apart from collaboration with EU universities and organizations, the EC is also developing various training centers to train regulatory officials in nanomedicines concerns. The EMA is also promoting worldwide conversations on nanomedicine regulatory approaches under the banner of the International Pharmaceutical Regulatory Forum.

1.10.4 National cancer institute nanotechnology initiative

Nanotechnology will have a great impact on how cancer is diagnosed and treated in the future. New technologies to detect and image cancerous changes and materials that enable new methods of cancer treatment will radically alter patient outcomes (Roco et al., 2010; Roco, 2011). The NCI of the National Institutes of Health explores innovative approaches to multidisciplinary research allowing for a convergence of molecular biology, oncology, physics, chemistry, and engineering and leading to the development of clinically worthy technological approaches. These initiatives include programmatic efforts to enable nanotechnology as a driver of advances in clinical oncology and cancer research, known collectively as the NCI Alliance for Nanotechnology in Cancer (NCI-ANC).

In recent time, ANC has demonstrated that a multidisciplinary approach catalyzes scientific developments and advances clinical translation in cancer nanotechnology. The research conducted by ANC members has improved diagnostic assays and imaging agents, leading to the development of point-of-care diagnostics, identification, and validation of numerous biomarkers for novel diagnostic assays, and the development of multifunctional agents for imaging and therapy. Numerous nanotechnology-based technologies developed by ANC researchers are entering clinical trials. NCI has reissued the ANC program after 2010, signaling that it continues to have high expectations for cancer nanotechnology's impact on clinical practice.

The goals of the next phase will be to broaden access to cancer nanotechnology research through greater clinical translation and outreach to the patient and clinical communities and to support the development of entirely new models of cancer care (Roco et al., 2010; Roco, 2011). The Fourth Annual NCI Alliance Principal Investigator Meeting focused on the research highlights in the areas of in vitro diagnostics, targeted delivery of anticancer and contrast enhancement agents, and nanotherapeutics and therapeutic monitoring. The phase I funding period (2005–10) involved funding a constellation of eight Centres for Cancer Nanotechnology Excellence (CCNEs) and 12 Cancer Nanotechnology Platform Partnerships (CNPPs), together with 11 Multidisciplinary Research Training and Team Development awards. CCNE teams were focused on developing integrated nanotechnology solutions with future potential for clinical applications. The CCNEs evolved into research organizations having distinct area(s) of technical excellence and core resources (e.g., fabrication and materials development, diagnostic assays, toxicology, drug delivery, in vivo technology validation, informatics).

1.11 CHALLENGES IN COMMERCIALIZATION OF NANOTHERAPEUTICS

1.11.1 Toxicity issues and health risks of nanomedicine

Despite the many proposed advantages of nanomaterials, increasing concerns have been expressed on their potential adverse human health effects. Because nanoparticles possess different physicochemical properties than their fine-sized analogues due to their extremely small size and large surface area, they need to be evaluated separately for toxicity and adverse health effects. These nanoparticulate carriers themselves may be responsible for toxicity and interaction with biological macromolecules within the human body. Second, insoluble nanoparticulate carriers may accumulate in human tissues or organs (Oberdorster et al., 2010).

A key issue for nanomaterial toxicity is their ability to bind and interact with biological matter and as a result change their own properties. The toxicity of nanomaterials to human health is how they biological compartment (in the body or within). One of the greatest potential dangers may be the propensity of some nanomaterials to cross biological barriers in a manner not predicted from studies of larger particles of the same chemical composition. The possible effects of nanomaterials on the developing fetus are of special concern, with gold nanoparticles and QDs having been shown to cross the maternal–fetal barrier and cause teratogenic and carcinogenic effects. Another potential exposure route in humans is through the skin. Access by fullerenes and QDs has been reported, dependent on size and surface coatings. Gastrointestinal assimilation and movement from there into the bloodstream has also been demonstrated.

Nanomaterials have been found to accumulate in low concentrations in the liver, spleen, heart, and brain.

1.11.2 Effects on the environment and other species

Globally, hundreds of thousands of tonnes of nanomaterials are being released into the environment, both intentionally and unintentionally, throughout the lifecycle of nanomaterial production, use, and disposal. Possible entry routes for nanomaterials into the environment include waste disposal (waste water treatment, storm water run-off, such as through abrasion and wear, as well as emissions from manufacturing and accidental spills). Unsatisfactory removal of emerging contaminants in general is a recognized worldwide problem and nanomedicines may exacerbate this issue.

When nanomedicine products enter the environment it becomes crucial to understand their fate and effects in terms of bioavailability, bioaccumulation, toxicity, environmental transformation, and interaction with other environmental contaminants. Nanodrugs will be able to travel for long periods in the external environment, will contaminate soils via sewage and spills, and ultimately migrate into groundwater and river systems and into the food chain. Exposure to carbon nanoparticles may harm earthworms by slowing population growth, increasing mortality, and damaging tissue (Lopez-Serrano et al., 2014).

1.11.3 The industrial and scale-up perspective

Nanomedicine industrialization: Biopraxis is specialized scaling up the manufacturing of different nanoformulations from milligram-scale laboratory synthesis up to multigram-scale production to generate sufficient material for clinical and regulatory assays. When standardizing the up-scale production of nanoparticles under GMP the main bottleneck aspects to be considered are: reproducibility, stability, and nonimmunogenicity (sterility and nonpyrogenicity). At the same time, we consider critical aspects of the GMP design such as continuous quality control, risk assessment for manufacturing process, specifications for excipients, intermediates, and finished products; room classification, equipment, supplies (water, heat, stirring, gases; Charitidis et al., 2014).

Although a number of methods are available for the production of nanomedicines, the two most familiar/major categories that require scale-up technologies are the emulsion-based method and the nanoprecipitation method. The technical steps (e.g., introducing organic solvent into aqueous system, solvent evaporation, stirring rate, and size reduction by forces) behind these two methods are similar to the steps of most of the other methods available for the fabrication of nanomedicines (e.g., solvent evaporation techniques, gelation technique, and emulsion polymerization technique).

The scale-up of nanomedicines includes the integration of methods as well as the transfer of technology for their large-scale industrial production. The process limitations in the small-scale preparation may lead to failure of translating any preparative method from the laboratory scale to an industrial scale. Also, a well-designed scale-up procedure will assure the quality of the nanomedicine, cost-effectiveness, and a timely product launch. In the literature, there are only a few reported laboratory-scale manufacturing processes of nanomedicines, especially using the emulsion-based and nanoprecipitation methods, which are supported by scale-up aspects for industrial manufacturing.

The effects of scale-up and process limitations of preparative methods are considered to be major challenges during large-scale production. First, scale-up has mainly been observed to affect the characteristics of nanoparticles, such as particle size, drug encapsulation, process residual materials, colloidal stability, and surface morphology. The scale-up of the emulsion-based method, from the laboratory scale of 60 mL to the industrial scale of 2 L (nearly 33-fold), did not alter the encapsulation efficiency. However, an increase in impeller speed and agitation time decreased the particle sizes. A decrease in the size of the nanoparticles was observed with a decrease in the polymer concentration. Another study by Galindo-Rodriguez et al. (2005) also reported that during scale-up, from a laboratory batch volume of 60 mL to 1.5 L, increasing the stirring rate decreased the particle size in the emulsion-based method. Moreover, they observed that the scale-up process reduced the drug loading of nanoparticles.

Abraxane® (Celgene Corporation, NJ, USA), albumin based nanoparticles is a typical example of the outcome of aggressive research on nanomedicines, and available in the market for the treatment of ovarian cancer and metastatic breast cancer. The launch of new nanomedicine products on the market is preceded by different developmental stages. A forthcoming major challenge among the developmental stages of nanomedicine is industrial scale-up (large-scale industrial production).

1.11.4 Cost considerations for nanomedicine applications

For the major cost-causing disease groups of cancer, cardiovascular, neurodegenerative, and musculoskeletal diseases, technology-dependent costs account for a maximum of 20% of the total costs. Thus, nanotechnology innovations are likely to have a particularly strong impact on healthcare costs if they reduce personnel costs by a reduction of the number of days in hospital. Other savings may be realized in ambulatory care costs, i.e., for diagnostic tests or for pharmaceutical therapies. The major drivers for future healthcare costs were identified: in addition to demographic changes, cancer, cardiovascular, neurodegenerative, and musculoskeletal diseases are expected to be the major cost-causing diseases. Further, personnel-intensive care (e.g., days in hospital) is very cost-intensive.

Taking these issues into account, nanomedicine innovations are likely to reduce future healthcare costs if they are aimed at primary cost-causing diseases, and at the same time, diminish personnel costs, e.g., by reducing the required days of inpatient care, contributing to "healthy aging" by raising the health status of the population. On the other hand, nanomedicine innovations are likely to have no major effect or even increase future healthcare costs if they aim at diseases of minor cost relevance such as infections or diseases with low prevalence and incidence, or come as add-on technology, which offers only a small health effect at significant costs so that the cost–benefit ratio is unfavorable, or result in additional procedures without substantial health effects (e.g., more diagnostic procedures) (Payne and Shabaruddin, 2010).

1.11.5 Addressing the knowledge gaps

Nanomedicine is the application of nanotechnology to the discipline of medicine: the use of nanoscale materials for the diagnosis, monitoring, control, prevention, and treatment of disease. Nanomedicine holds tremendous promise to revolutionize medicine across disciplines and specialties, but this promise has yet to be fully realized. Beyond the typical challenges associated with drug development, the fundamentally different and novel physical and chemical properties of some nanomaterials compared to materials on a larger scale (i.e., their bulk counterparts) can create a unique set of opportunities as well as safety concerns, which have only begun to be explored. As the research community continues to investigate nanomedicines, their efficacy, and the associated safety issues, it is critical to work to close the scientific and regulatory gaps to assure that nanomedicine drives the next generation of biomedical innovation. The variety of innovative and existing nanomedicines is too complex and difficult to address to allow the harmonization of even partially complete guidelines.

A more successful approach could be to harmonize regulatory assessment proposals on follow-on versions of established nanoparticulate therapeutics like most actual representatives of NBCDs (Non-biologic complex drugs), also referred to as nanosimilars. Such NBCDs are defined as synthetic (nonbiologic), large-molecule nanoparticulate complexes in which the entire formulation represents the pharmaceutically active ingredient. Special consideration has to be paid toward the assessment of physicochemical properties and the qualitative and quantitative equivalence between the test and reference product in their relevant components. This includes stoichiometric ratio, iron core characterization, composition and surface properties of the carbohydrate shell, and the labile iron determination under physiologically relevant conditions (Tinkle et al., 2014).

1.12 SUMMARY AND CONCLUSION

Nanotechnology has started to change the scale and methods of medicine formulation technology and drug delivery. In fact, the NIH initiative known as "Nanomedicine Initiatives" reveals that the market of nanotechnology-based nanomedicines is going to be much wider in the coming few years with the advent of novel mechanisms and technologies for the prevention and cure of even more complex diseases. Even the diagnostic and imaging tools have entered into a new era of development with the advent of advanced electron microscopes, microchip devices, nanopore sequencing, siRNA therapeutics, and nanorobotics. However, for the potential in vivo acceptability of nanotechnology in targeted imaging and drug delivery to be realized, nanocarriers have to get smarter.

In this chapter we have explored the employment of these smart nanocarriers with special attention to their diverse characteristics, properties, and applications. Among them, liposomes, dendrimers, polymeric, and metallic nanoparticles (gold and silver) have created a greater difference between conventional therapy and novel approaches, especially to treat complex diseases such as cancer and neurodegenerative disorders where the target specificity is of prime importance. However, these nanocarriers suffer from the well-known hurdle of clinical delivery that is toxicity, which plays an important role for the approval of medicines in the public domain. Therefore, it is essential that fundamental research should be carried out to address these issues if successful efficient application of these technologies is going to be achieved. The future of nanomedicine will depend on rational design of nanotechnology materials and tools based around a detailed and thorough understanding of biological processes rather than forcing applications for some materials currently in vogue. Its nanoscale formulations are a like a science fiction miracle that is exploring novel ways to fight viruses and diseases, destroying cancer cells, and repairing damaged cells and tissues. However, a variety of challenges need to be resolved for the full functionalization and commercialization of nanobased medicines with complete safety and acceptability for the future use.

ACKNOWLEDGMENTS

RKT acknowledges support from the Fundamental Research Grant (FRGS) scheme of the Ministry of Higher Education, Malaysia, for research support on gene delivery. The authors would like to acknowledge International Medical University, Malaysia, for providing research support on cancer and arthritis research. We also acknowledge internal grants to Dr. Tekade from the IMU-JC for providing start up financial support to our research group.

REFERENCES

Arduini, F., Micheli, L., Moscone, D., Palleschi, G., Piermarini, S., Ricci, F., et al., 2016. Electrochemical biosensors based on nanomodified screen-printed electrodes: recent applications in clinical analysis. Trends Anal. Chem. 79, 114–126.
Attari, Z., Bhandari, A., Jagadish, P.C., Lewis, S., 2016. Enhanced ex vivo intestinal absorption of olmesartan medoxomil nanosuspension: preparation by combinative technology. Saudi Pharm. J. 24 (1), 57–63.
Atun, R., Jaffray, D.A., Barton, M.B., Bray, F., Baumann, M., Vikram, B., et al., 2015. Expanding global access to radiotherapy. Lancet Oncol. 16 (10), 1153–1186.
Azimi, B., Nourpanah, P., Rabiee, M., Arbab, S., 2014. Producing gelatin nanoparticles as delivery system for bovine serum albumin. Iranian Biomed. J. 18 (1), 34.

Binns, C., 2010. Introduction to Nanoscience and Nanotechnology, Vol. 14. John Wiley & Sons, New York.

Biswas, S., Torchilin, V.P., 2014. Nanopreparations for organelle-specific delivery in cancer. Adv. Drug Deliv. Rev. 66, 26–41.

Boisseau, P., Loubaton, B., 2011. Nanomedicine, nanotechnology in medicine. C R Phys. 12 (7), 620–636.

Bozzuto, G., Molinari, A., 2015. Liposomes as nanomedical devices. Int. J. Nanomed. 10 (1), 975–999.

Charitidis, C.A., Georgiou, P., Koklioti, M.A., Trompeta, A.F., Markakis, V., 2014. Manufacturing nanomaterials: from research to industry. Manuf. Rev. 1, 11.

Chinen, A.B., Guan, C.M., Ferrer, J.R., Barnaby, S.N., Merkel, T.J., Mirkin, C.A., 2015. Nanoparticle probes for the detection of cancer biomarkers, cells, and tissues by fluorescence. Chem. Rev. 115 (19), 10530–10574.

Chopdey, P.K., Tekade, R.K., Mehra, N.K., Mody, N., Jain, N.K., 2015. Glycyrrhizin conjugated dendrimer and multi-walled carbon nanotubes for liver specific delivery of doxorubicin. J. Nanosci. Nanotechnol. 15 (2), 1088–1100.

Chougule, M.B., Tekade, R.K., Hoffmann, P.R., Bhatia, D., Sutariya, V.B., Pathak, Y., 2014. Nanomaterial-based gene and drug delivery: pulmonary toxicity considerations. Biointeract. Nanomater. 225–248.

Coughlin, A.J., Ananta, J.S., Deng, N., Larina, I.V., Decuzzi, P., West, J.L., 2014. Gadolinium-conjugated gold nanoshells for multimodal diagnostic imaging and photothermal cancer therapy. Small 10 (3), 556–565.

Cunha, C., Panseri, S., Antonini, S., 2011. Emerging nanotechnology approaches in tissue engineering for peripheral nerve regeneration. Nanomed. Nanotechnol. Biol. Med. 7 (1), 50–59.

Desai, N., 2012. Challenges in development of nanoparticle-based therapeutics. AAPS J. 14 (2), 282–295.

Despanie, J., Dhandhukia, J.P., Hamm-Alvarez, S.F., MacKay, J.A., 2015. Elastin-like polypeptides: therapeutic applications for an emerging class of nanomedicines. J. Control. Release 240, 93–108.

Dhakad, R.S., Tekade, R.K., Jain, N.K., 2013. Cancer targeting potential of folate targeted nanocarrier under comparative influence of tretinoin and dexamethasone. Curr. Drug Deliv. 10 (4), 477–491.

Durymanov, M.O., Rosenkranz, A.A., Sobolev, A.S., 2015. Current approaches for improving intratumoral accumulation and distribution of nanomedicines. Theranostics 5 (9), 1007.

Dvir, T., Timko, B.P., Kohane, D.S., Langer, R., 2011. Nanotechnological strategies for engineering complex tissues. Nat. Nanotechnol. 6 (1), 13–22.

Dwivedi, P., Tekade, R.K., Jain, N.K., 2013. Nanoparticulate carrier mediated intranasal delivery of insulin for the restoration of memory signaling in Alzheimer's disease. Curr. Nanosci. 9 (1), 46–55.

Finch, G., Havel, H., Analoui, M., Barton, R.W., Diwan, A.R., Hennessy, M., et al., 2014. Nanomedicine drug development: a scientific symposium entitled "charting a roadmap to commercialization". AAPS J. 16 (4), 698–704.

Gajbhiye, V., Vijayaraj Kumar, P., Tekade, R.K., Jain, N.K., 2007. Pharmaceutical and biomedical potential of PEGylated dendrimers. Curr. Pharm. Design 13 (4), 415–429.

Gajbhiye, V., Palanirajan, V.K., Tekade, R.K., Jain, N.K., 2009. Dendrimers as therapeutic agents: a systematic review. J. Pharm. Pharmacol. 61 (8), 989–1003.

Galindo-Rodríguez, S.A., Puel, F., Briançon, S., Allémann, E., Doelker, E., Fessi, H., 2005. Comparative scale-up of three methods for producing ibuprofen-loaded nanoparticles. European journal of pharmaceutical sciences. 25 (4), 357–367.

Gandhi, N.S., Tekade, R.K., Chougule, M.B., 2014. Nanocarrier mediated delivery of siRNA/miRNA in combination with chemotherapeutic agents for cancer therapy: current progress and advances. J. Control. Release 194, 238–256.

Garg, N.K., Singh, B., Jain, A., Nirbhavane, P., Sharma, R., Tyagi, R.K., et al., 2016. Fucose decorated solid-lipid nanocarriers mediate efficient delivery of methotrexate in breast cancer therapeutics. Colloids Surf. B. Biointerfaces 146, 114–126.

Gener, P., Gouveia, L.P., Sabat, G.R., de Sousa Rafael, D.F., Fort, N.B., Arranja, A., et al., 2015. Fluorescent CSC models evidence that targeted nanomedicines improve treatment sensitivity of breast and colon cancer stem cells. Nanomed. Nanotechnol. Biol. Med. 11 (8), 1883–1892.

Ghanghoria, R., Tekade, R.K., Mishra, A.K., Chuttani, K., Jain, N.K., 2016. Luteinizing hormone-releasing hormone peptide tethered nanoparticulate system for enhanced antitumoral efficacy of paclitaxel. Nanomedicine (Lond) 11 (7), 797–816.

Goldberg, M.S., Hook, S.S., Wang, A.Z., Bulte, J.W., Patri, A.K., Uckun, F.M., et al., 2013. Biotargeted nanomedicines for cancer: six tenets before you begin. Nanomedicine 8 (2), 299–308.

Gong, T., Xie, J., Liao, J., Zhang, T., Lin, S., Lin, Y., 2015. Nanomaterials and bone regeneration. Bone Res. 3, 15029.

Guerrero-Cázares, H., Tzeng, S.Y., Young, N.P., Abutaleb, A.O., Quiñones-Hinojosa, A., Green, J.J., 2014. Biodegradable polymeric nanoparticles show high efficacy and specificity at DNA delivery to human glioblastoma in vitro and in vivo. ACS Nano 8 (5), 5141–5153.

Hafner, A., Lovric, J., Lakos, G.P., Pepic, I., 2014. Nanotherapeutics in the EU: an overview on current state and future directions. Int. J. Nanomed. 9, 1005–1023.

Hainfeld, J.F., Smilowitz, H.M., OConnor, M.J., Dilmanian, F.A., Slatkin, D.N., 2013. Gold nanoparticle imaging and radiotherapy of brain tumors in mice. Nanomedicine 8 (10), 1601–1609.

Heath, J.R., 2015. Nanotechnologies for biomedical science and translational medicine. Proc. Nat. Acad. Sci. 112 (47), 14436–14443.

Hock, S.C., Ying, Y.M., Wah, C.L., 2011. A review of the current scientific and regulatory status of nanomedicines and the challenges ahead. PDA. J. Pharm. Sci. Technol. 65 (2), 177–195.

Hong, J., Li, Y., Xiao, Y., Li, Y., Guo, Y., Kuang, H., et al., 2016. Annonaceous acetogenins (ACGs) nanosuspensions based on a self-assembly stabilizer and the significantly improved anti-tumor efficacy. Colloids Surf. B. Biointerfaces 145, 319–327.

Huang, B.J., Hu, J.C., Athanasiou, K.A., 2016. Cell-based tissue engineering strategies used in the clinical repair of articular cartilage. Biomaterials 98, 1–22.

Huang, C.W., Kearney, V., Moeendarbari, S., Jiang, R.Q., Christensen, P., Tekade, R.K., et al., 2015. Hollow gold nanoparticles as biocompatible radiosensitizer: an in vitro proof of concept study. J. Nano Res. 32, 106–112.

Huang, Y., He, S., Cao, W., Cai, K., Liang, X.J., 2012. Biomedical nanomaterials for imaging-guided cancer therapy. Nanoscale 4 (20), 6135–6149.

Hussain, T., Nguyen, Q.T., 2014. Molecular imaging for cancer diagnosis and surgery. Adv. Drug Deliv. Rev. 66, 90–100.

Jain, N.K., Tekade, R.K., 2013. Dendrimers for enhanced drug solubilization. In: Drug delivery strategies for poorly water-soluble drugs. John Wiley & Sons, New York, NY, pp. 373–409.

Jayakumar, R., Chennazhi, K.P., Muzzarelli, R.A.A., Tamura, H., Nair, S.V., Selvamurugan, N., 2010. Chitosan conjugated DNA nanoparticles in gene therapy. Carbohydr. Polymers 79 (1), 1–8.

Kamaly, N., Yameen, B., Wu, J., Farokhzad, O.C., 2016. Degradable controlled-release polymers and polymeric nanoparticles: mechanisms of controlling drug release. Chem. Rev. 116 (4), 2602–2663.

Karakucuk, A., Celebi, N., Teksin, Z.S., 2016. Preparation of ritonavir nanosuspensions by microfluidization using polymeric stabilizers: I. A Design of Experiment approach. Eur. J. Pharm. Sci. 95, 111–121.

Kayat, J., Gajbhiye, V., Tekade, R.K., Jain, N.K., 2011. Pulmonary toxicity of carbon nanotubes: a systematic report. Nanomed. Nanotechnol. Biol. Med. 7 (1), 40–49.

Kesharwani, P., Tekade, R.K., Jain, N.K., 2015a. Formulation development and in vitro-in vivo assessment of 4.0g PPI dendrimer as cancer targeting vector. Nanomedicine (London) 32, 1438–1450.

Kesharwani, P., Tekade, R.K., Jain, N.K., 2015b.). Dendrimer generational nomenclature: the need to harmonize. Drug Discovery Today 20 (5), 497–499.

Kim, J., Kim, J., Jeong, C., Kim, W.J., 2016. Synergistic nanomedicine by combined gene and photothermal therapy. Adv. Drug Deliv. Rev. 98, 99–112.

Kolk, A., Handschel, J., Drescher, W., Rothamel, D., Kloss, F., Blessmann, M., et al., 2012. Current trends and future perspectives of bone substitute materials–from space holders to innovative biomaterials. J. Cranio-Maxillofacial Surgery 40 (8), 706–718.

Kompella, U.B., Amrite, A.C., Ravi, R.P., Durazo, S.A., 2013. Nanomedicines for back of the eye drug delivery, gene delivery, and imaging. Prog. Retinal Eye Res. 36, 172–198.

Kulkarni, N., Muddapur, U., 2014. Biosynthesis of metal nanoparticles: a review. J. Nanotechnol. 2014 Article ID 510246, 8 pages.

Kumar, L., Verma, S., Prasad, D.N., Bhardwaj, A., Vaidya, B., Jain, A.K., 2015. Nanotechnology: a magic bullet for HIV AIDS treatment. Artificial Cells, Nanomed. Biotechnol. 43 (2), 71–86.

Kumari, A., Yadav, S.K., Yadav, S.C., 2010. Biodegradable polymeric nanoparticles based drug delivery systems. Colloids Surf. B Biointerfaces 75 (1), 1–18.

Kurmi, B.D., Kayat, J., Gajbhiye, V., Tekade, R.K., Jain, N.K., 2010. Micro-and nanocarrier-mediated lung targeting. Exp. Opin. Drug Deliv. 7 (7), 781–794.

Lam, K., Pan, K., Linnekamp, J.F., Medema, J.P., Kandimalla, R., 2016. DNA methylation based biomarkers in colorectal cancer: a systematic review. Biochimica et Biophysica Acta (BBA) 1866 (1), 106–120.

Lee, H., Choi, T.K., Lee, Y.B., Cho, H.R., Ghaffari, R., Wang, L., et al., 2016. A graphene-based electrochemical device with thermoresponsive microneedles for diabetes monitoring and therapy. Nat. Nanotechnol. 11 (6), 566–572.

Lehner, R., Hunziker, P., 2012. Why not just switch on the light?: light and its versatile applications in the field of nanomedicine. Eur. J. Nanomed. 4 (2–4), 73–80.

Li, X., Wu, M., Pan, L., Shi, J., 2015a. Tumor vascular-targeted co-delivery of anti-angiogenesis and chemotherapeutic agents by mesoporous silica nanoparticle-based drug delivery system for synergetic therapy of tumor. Int. J. Nanomed. 11, 93–105.

Li, Z., Liu, Z., Sun, H., Gao, C., 2015b. Superstructured assembly of nanocarbons: fullerenes, nanotubes, and graphene. Chem. Rev. 115 (15), 7046–7117.

Liao, N., Zhuo, Y., Chai, Y.Q., Xiang, Y., Han, J., Yuan, R., 2013. Reagentless electrochemiluminescent detection of protein biomarker using graphene-based magnetic nanoprobes and poly-L-lysine as co-reactant. Biosens. Bioelectron. 45, 189–194.

Liu, K., Feng, J., Kis, A., Radenovic, A., 2014. Atomically thin molybdenum disulfide nanopores with high sensitivity for DNA translocation. ACS Nano 8 (3), 2504–2511.

Lopez-Serrano, A., Olivas, R.M., Landaluze, J.S., Cámara, C., 2014. Nanoparticles: a global vision. Characterization, separation, and quantification methods. Potential environmental and health impact. Anal. Methods 6 (1), 38–56.

Ma, H., Zhang, X., Li, X., Li, R., Du, B., Wei, Q., 2015. Electrochemical immunosensor for detecting typical bladder cancer biomarker based on reduced graphene oxide–tetraethylene pentamine and trimetallic AuPdPt nanoparticles. Talanta 143, 77–82.

Mahesh, K.V., Singh, S.K., Gulati, M., 2014. A comparative study of top-down and bottom-up approaches for the preparation of nanosuspensions of glipizide. Powder Technol. 256, 436–449.

Maheshwari, R., Tekade, R.K., Sharma, P.A., Gajanan, D., Tyagi, A., Patel, R.P., et al., 2012. Ethosomesand ultradeformable liposomes for transdermal delivery of clotrimazole: a comparative assessment. Saudi Pharm. J. 20, 161–170.

Maheshwari, R., Tekade, M., Sharma, P.A., Tekade, R.K., 2015a. Nanocarriers assisted siRNA gene therapy for the management of cardiovascular disorders. Curr. Pharm. Design 21 (30), 4427–4440.

Maheshwari, R., Thakur, S., Singhal, S., Patel, R.P., Tekade, M., Tekade, R.K., 2015b. Chitosan encrusted nonionic surfactant based vesicular formulation for topical administration of ofloxacin. Sci. Adv. Mater. 7 (6), 1163–1176.

Mahmoudi, M., Sant, S., Wang, B., Laurent, S., Sen, T., 2011. Superparamagnetic iron oxide nanoparticles (SPIONs): development, surface modification and applications in chemotherapy. Adv. Drug Deliv. Rev. 63 (1), 24–46.

Malinoski, F.J., 2014. The nanomedicines alliance: an industry perspective on nanomedicines. Nanomed. Nanotechnol. Biol. Med. 10 (8), 1819–1820.

Mansuri, S., Kesharwani, P., Tekade, R.K., Jain, N.K., 2016. Lyophilized mucoadhesive-dendrimer enclosed matrix tablet for extended oral delivery of albendazole. Eur. J. Pharm. Biopharm. 102, 202–213.

Massadeh, S., Al-Aamery, M., Bawazeer, S., AlAhmad, O., AlSubai, R., Barker, S., et al., 2016. Nano-materials for gene therapy: an efficient way in overcoming challenges of gene delivery. J. Biosens. Bioelectron. 7, 195.

Mishra, B., Sahoo, J., Dixit, P.K., 2016. Enhanced bioavailability of cinnarizine nanosuspensions by particle size engineering: optimization and physicochemical investigations. Mater. Sci. Eng. C 63, 62–69.

Mody, N., Tekade, R.K., Mehra, N.K., Chopdey, P., Jain, N.K., 2014. Dendrimer, liposomes, carbon nanotubes and PLGA nanoparticles: one platform assessment of drug delivery potential. Aaps Pharmscitech. 15 (2), 388–399.

Moeendarbari, S., Tekade, R.K., Mulgaonkar, A., Christensen, P., Ramezani, S., Hassan, G., et al., 2016. Theranostic nanoseeds for efficacious internal radiation therapy of unresectable solid tumors. Scientific Reports 6 (20614), 1–9.

Morganti, P., Palombo, M., Tishchenko, G., Yudin, V.E., Guarneri, F., Cardillo, M., et al., 2014. Chitin-hyaluronan nanoparticles: a multifunctional carrier to deliver anti-aging active ingredients through the skin. Cosmetics 1 (3), 140–158.

Mou, D., Chen, H., Wan, J., Xu, H., Yang, X., 2011. Potent dried drug nanosuspensions for oral bioavailability enhancement of poorly soluble drugs with pH-dependent solubility. Int. J. Pharm. 413 (1), 237–244.

Muthu, M.S., Leong, D.T., Mei, L., Feng, S.S., 2014. Nanotheranostics-application and further development of nanomedicine strategies for advanced theranostics. Theranostics 4 (6), 660–677.

Nanomedicine: a matter of rhetoric, 2006. Nat. Mater. 5 (4), 243.

Neacsu, I.A., Nicoara, A.I., Vasile, O.R., Vasile, B.S., 2016. Inorganic micro-and nanostructured implants for tissue engineering. Nanobiomate. Hard Tissue Eng. Applicat. Nanobiomater. 271.

Nguyen, K.T., 2012. Targeted nanoparticles for cancer therapy: promises and challenges. J. Nanomed. Nanotechnol. 2, 103e.

Oberdörster, G., 2010. Safety assessment for nanotechnology and nanomedicine: concepts of nanotoxicology. J. Inter. Med. 267 (1), 89–105.

Ozak, S.T., Ozkan, P., 2013. Nanotechnology and dentistry. Eur. J. Dentistry 7 (1), 145.

Pautler, M., Brenner, S., 2010. Nanomedicine: promises and challenges for the future of public health. Int. J. Nanomed. 5 (8039.2).

Payne, K., Shabaruddin, F.H., 2010. Cost-effectiveness analysis in pharmacogenomics. Pharmacogenomics 11 (5), 643–646.

Perera, J., Weerasekera, M., Kottegoda, N., 2015. Slow release anti-fungal skin formulations based on citric acid intercalated layered double hydroxides nanohybrids. Chem. Central J. 9 (1), 1–7.

Prajapati, R.N., Tekade, R.K., Gupta, U., Gajbhiye, V., Jain, N.K., 2009. Dendimer-mediated solubilization, formulation development and in vitro-in vivo assessment of piroxicam. Mol. Pharm. 6 (3), 940–950.

Rampino, A., Borgogna, M., Blasi, P., Bellich, B., Cesàro, A., 2013. Chitosan nanoparticles: preparation, size evolution and stability. Int. J. Pharm. 455 (1), 219–228.

Ranganathan, R., Madanmohan, S., Kesavan, A., Baskar, G., Krishnamoorthy, Y.R., Santosham, R., et al., 2012. Nanomedicine: towards development of patient-friendly drug-delivery systems for oncological applications. Int. J. Nanomed. 7 (1043), e1060.

Rauwel, P., Rauwel, E., Ferdov, S., Singh, M.P., 2015. Silver nanoparticles: synthesis, properties, and applications. Adv. Mater. Sci. Eng. 2015 Article ID 682749, 9 pages.

Roco, M.C., Mirkin, C.A., Hersam, M.C. (eds) 2010. Nanotechnology research directions for societal needs in 2020: retrospective and outlook. NSF/WTEC report, Springer. http://www.wtec.org/nano2/Nanotechnology_Research_ Directions_to_2020/.

Roco, M.C., 2011. The long view of nanotechnology development: the National Nanotechnology Initiative at 10 years. J. Nanopart. Res. 13 (2), 427–445.

Sainz, V., Conniot, J., Matos, A.I., Peres, C., Zupanǒiǒ, E., Moura, L., et al., 2015. Regulatory aspects on nanomedicines. Biochem. Biophys. Res. Commun. 468 (3), 504–510.

Sanchez, D.A., Schairer, D., Tuckman-Vernon, C., Chouake, J., Kutner, A., Makdisi, J., et al., 2014. Amphotericin B releasing nanoparticle topical treatment of Candida spp. in the setting of a burn wound. Nanomed. Nanotechnol. Biol. Med. 10 (1), 269–277.

Sangshetti, J.N., Deshpande, M., Zaheer, Z., Shinde, D.B., Arote, R., 2014. Quality by design approach: regulatory need. Arabian Journal of Chemistry 2014, 17.

Sanna, V., Pala, N., Sechi, M., 2014. Targeted therapy using nanotechnology: focus on cancer. Int. J. Nanomed. 9, 467.

Sedaghati, T., Yang, S.Y., Mosahebi, A., Alavijeh, M.S., Seifalian, A.M., 2011. Nerve regeneration with aid of nanotechnology and cellular engineering. Biotechnol. Appl. Biochem. 58 (5), 288–300.

Sharma, P., Maheshwari, R., Tekade, M., Kumar Tekade, R., 2015. Nanomaterial based approaches for the diagnosis and therapy of cardiovascular diseases. Curr. Pharm. Design 21 (30), 4465–4478.

Shi, J., Votruba, A.R., Farokhzad, O.C., Langer, R., 2010. Nanotechnology in drug delivery and tissue engineering: from discovery to applications. Nano Lett. 10 (9), 3223–3230.

Singare, D.S., Marella, S., Gowthamrajan, K., Kulkarni, G.T., Vooturi, R., Rao, P.S., 2010. Optimization of formulation and process variable of nanosuspension: an industrial perspective. Int. J. Pharm. 402 (1), 213–220.

Singh, R., Kesharwani, P., Mehra, N.K., Singh, S., Banerjee, S., Jain, N.K., 2015. Development and characterization of folate anchored Saquinavir entrapped PLGA nanoparticles for anti-tumor activity. Drug Develop. Indust. Pharm. 41 (11), 1888–1901.

Soni, N., Soni, N., Pandey, H., Maheshwari, R., Kesharwani, P., Tekade, R.K., 2016. Augmented delivery of gemcitabine in lung cancer cells exploring mannose anchored solid lipid nanoparticles. J. Colloid Interface Sci. 481, 107–116.

Sonneville-Aubrun, O., Simonnet, J.T., L'alloret, F., 2004. Nanoemulsions: a new vehicle for skincare products. Adv. Colloid Interface Sci. 108, 145–149.

Stanley, S.A., Gagner, J.E., Damanpour, S., Yoshida, M., Dordick, J.S., Friedman, J.M., 2012. Radio-wave heating of iron oxide nanoparticles can regulate plasma glucose in mice. Science 336 (6081), 604–608.

Sun, M., Liu, F., Zhu, Y., Wang, W., Hu, J., Liu, J., et al., 2016. Salt-induced aggregation of gold nanoparticles for photoacoustic imaging and photothermal therapy of cancer. Nanoscale 8, 4452–4457.

Sutradhar, K.B., Khatun, S., Luna, I.P., 2013. Increasing possibilities of nanosuspension. J. Nanotechnol. Article 346581, 12 pages.

Tansık, G., Ozkan, A.D., Guler, M.O., Tekinay, A.B., 2016. Nanomaterials for the repair and regeneration of dental tissues. Ther. Nanomater. 153.

Tekade, R.K., 2014. Editorial: contemporary siRNA therapeutics and the current state-of-art. Curr. Pharm. Design 21 (31), 4527–4528.

Tekade, R.K., Chougule, M.B., 2013. Formulation development and evaluation of hybrid nanocarrier for cancer therapy: taguchi orthogonal array based design. BioMed Res. Int. 2013, 712678.

Tekade, R.K., Kumar, P.V., Jain, N.K., 2008a. Dendrimers in oncology: an expanding horizon. Chem. Rev. 109 (1), 49–87.

Tekade, R.K., Dutta, T., Tyagi, A., Bharti, A.C., Das, B.C., Jain, N.K., 2008b. Surface-engineered dendrimers for dual drug delivery: a receptor up-regulation and enhanced cancer targeting strategy. J. Drug Target. 16 (10), 758–772.

Tekade, R.K., Dutta, T., Gajbhiye, V., Jain, N.K., 2009. Exploring dendrimer towards dual drug delivery: pH responsive simultaneous drug-release kinetics. J. Microencapsul. 26 (4), 287–296.

Tekade, R.K., D'Emanuele, A., Elhissi, A., Agrawal, A., Jain, A., Arafat, B.T., et al., 2013. Extraction and RP-HPLC determination of taxol in rat plasma, cell culture and quality control samples. J. Biomed. Res. 27 (5), 394–405.

Tekade, R., Xu, L., Hao, G., Ramezani, S., Silvers, W., Christensen, P., et al., 2014a. A facile preparation of radioactive gold nanoplatforms for potential theranostic agents of cancer. J. Nuclear Med. 55 (1), 1047.

Tekade, R.K., Youngren-Ortiz, S.R., Yang, H., Haware, R., Chougule, M.B., 2014b. Designing hybrid onconase nanocarriers for mesothelioma therapy: a Taguchi orthogonal array and multivariate component driven analysis. Mol. Pharm. 11 (10), 3671–3683.

Tekade, R.K., Youngren-Ortiz, S.R., Yang, H., Haware, R., Chougule, M.B., 2015a. Albumin-chitosan hybrid onconase nanocarriers for mesothelioma therapy. Cancer Res. 75 (15), 3680.

Tekade, R., Maheshwari, Rahul, G.S., Sharma, P., Tekade, M., Singh Chauhan, A., 2015b. siRNA therapy, challenges and underlying perspectives of dendrimer as delivery vector. Curr. Pharm. Design 21 (31), 4614–4636.

Tekade, R.K., Tekade, M., Kumar, M., Chauhan, A.S., 2015c. Dendrimer-stabilized smart-nanoparticle (DSSN) platform for targeted delivery of hydrophobic antitumor therapeutics. Pharm. Res. 32 (3), 910–928.

Tekade, R., Maheshwari, Rahul, G.S., Sharma, A.P., Tekade, M., Singh Chauhan, A., 2015d. siRNA therapy, challenges and underlying perspectives of dendrimer as delivery vector. Curr. Pharm. Design 21 (31), 4614–4636.

Tekade, R.K., Tekade, M., Kesharwani, P., D'Emanuele, A., 2016. RNAi-combined nano-chemotherapeutics to tackle resistant tumors. Drug Discovery Today 21 (11), 1761–1774.

Tinkle, S., McNeil, S.E., Mühlebach, S., Bawa, R., Borchard, G., Barenholz, Y.C., et al., 2014. Nanomedicines: addressing the scientific and regulatory gap. Ann. NY Acad. Sci. 1313 (1), 35–56.

Toy, R., Roy, K., 2016. Engineering nanoparticles to overcome barriers to immunotherapy. Bioeng. Transl. Med. 1, 47–62.

Vela Ramirez, J.E., Roychoudhury, R., Habte, H.H., Cho, M.W., Pohl, N.L.B., Narasimhan, B., 2014. Carbohydrate-functionalized nanovaccines preserve HIV-1 antigen stability and activate antigen presenting cells. J. Biomater. Sci. Polymer Edition 25 (13), 1387–1406.

Wang, Y., Hu, A., 2014. Carbon quantum dots: synthesis, properties and applications. J. Mater. Chem. C 2 (34), 6921–6939.

Weissig, V., Guzman-Villanueva, D., 2015. Nanopharmaceuticals (part 2): products in the pipeline. Int. J. Nanomed. 10, 1245.

Weissig, V., Pettinger, T.K., Murdock, N., 2014. Nanopharmaceuticals (part 1): products on the market. Int. J. Nanomed. 9, 4357–4373.

Widom, J.R., Dhakal, S., Heinicke, L.A., Walter, N.G., 2014. Single-molecule tools for enzymology, structural biology, systems biology and nanotechnology: an update. Arch. Toxicol. 88 (11), 1965–1985.

Wu, L., Qiao, Y., Wang, L., Guo, J., Wang, G., He, W., et al., 2015. A self-microemulsifying drug delivery system (SMEDDS) for a novel medicative compound against depression: a preparation and bioavailability study in rats. AAPS PharmSciTech 16 (5), 1051–1058.

Yameen, B., Choi, W.I., Vilos, C., Swami, A., Shi, J., Farokhzad, O.C., 2014. Insight into nanoparticle cellular uptake and intracellular targeting. J. Control. Release 190, 485–499.

Youngren, S.R., Tekade, R.K., Gustilo, B., Hoffmann, P.R., Chougule, M.B., 2013. STAT6 siRNA matrix-loaded gelatin nanocarriers: formulation, characterization, and ex vivo proof of concept using adenocarcinoma cells. BioMed Res. Int. 2013, 858946.

Zabow, G., Dodd, S.J., Koretsky, A.P., 2015. Shape-changing magnetic assemblies as high-sensitivity NMR-readable nanoprobes. Nature 520 (7545), 73–77.

Zhang, Y., Zhang, J., 2016. Preparation of budesonide nanosuspensions for pulmonary delivery: characterization, in vitro release and in vivo lung distribution studies. Artif. Cells Nanomed. Biotechnol. 44 (1), 285–289.

Zhong, Y., Meng, F., Deng, C., Zhong, Z., 2014. Ligand-directed active tumor-targeting polymeric nanoparticles for cancer chemotherapy. Biomacromolecules 15 (6), 1955–1969.

Zhou, H., Lee, J., 2011. Nanoscale hydroxyapatite particles for bone tissue engineering. Acta Biomaterialia 7 (7), 2769–2781.

Zhou, W., Gao, X., Liu, D., Chen, X., 2015. Gold nanoparticles for in vitro diagnostics. Chem. Rev. 115 (19), 10575–10636.

CHAPTER 2

Current Update on the Role of Enhanced Permeability and Retention Effect in Cancer Nanomedicine

Anfal Jasim*, Sara Abdelghany* and Khaled Greish

Contents

* These authors contributed equally to the manuscript.

Nanotechnology-Based Approaches for Targeting and Delivery of Drugs and Genes.
DOI: http://dx.doi.org/10.1016/B978-0-12-809717-5.00002-6

2.1 INTRODUCTION

In 2012 the World Health Organization (WHO) categorized cancer as the number one cause of morbidity and mortality worldwide, with approximately 14 million new cases and 8.2 million cancer-related deaths annually (Gulland, 2014). However, the great efforts in the pharmaceutical industry in recent years has helped intensely in discovering many cytotoxic drugs that could potentially treat cancer. Still, cancer treatment is usually accompanied by severe side effects and acquired drug resistance. Therefore, current efforts to treat cancer have been focusing on developing targeted therapies that allow higher tumor specificity and less toxicity.

In the 1980s a milestone in anticancer drug field was achieved when Maeda and his colleagues discovered a unique phenomenon called the enhanced permeability and retention (EPR) effect (Matsumura and Maeda, 1986). This is a phenomenon that describes the facilitated extravasation of macromolecular drugs more selectively at tumor tissues due to the distinctive tumor vasculature and to the ability of macromolecules to attain prolonged plasma or local half-lives.

Observations related to this phenomenon were started more than 100 years ago by R. Virchow who made the observation that tumors have a distinct capillary network (Virchow and Chance, 1863). This study was followed by the first systemic tumor vasculature study by E. Goldmann who specifically referred to the vascular system to describe the growth of human malignant tumors. Goldmann concluded that the invading growth disturbs the surrounding blood vessels by forming new vessels mostly in the zone of proliferation and, as tumor grows, the newly formed vessels resulted in a central necrosis in the capsule area of the tumor (Goldmann, 1908). At the beginning of the 20th century, P. Ehrlich theorized an ideal therapy for diseases, a drug that selectively targeted a disease-causing organism. This theory, that became known as "Ehrlich's magic bullet," opened the door toward searching for a distinctive feature of tumor cells to be targeted (Ehrlich, 1913). Later, Judah Folkman demonstrated that solid tumor growth was "angiogenesis-dependent" (Folkman, 1971). He also introduced the concept of "antiangiogenic" therapy based on the idea that cancer can be fought by preventing de novo blood vessel formation and recruitment.

In recent years, these significant findings have inspired the use of macromolecular systems for effective, targeted drug delivery to tumor and inflamed tissues. This chapter will highlight the principle of the EPR effect, and the mechanism and factors contributing to EPR. Moreover, we will discuss the EPR-based nanoconstructs and drugs along with current limitations and challenges of using these systems.

2.2 ENHANCED PERMEABILITY AND RETENTION EFFECT

The EPR effect comprises two main parts: *enhanced permeation* and *enhanced retention*.

When solid tumor cells reach a size of 150–200 μm, they establish their own vasculature system, and start depending on it for their blood, nutrition, and oxygen supply (Wu et al., 1998). This is due to the secretion of VEGF by hypoxic tumor cells, triggering angiogenesis (Bertrand et al., 2013). Blood vessels in tumors have an irregular vascular morphology, with a heterogeneous spatial distribution, abnormally wide interendothelial junctions, transendothelial channels, and an atypical basement membrane (Jain and Stylianopoulos, 2010; Jain, 1987). Owing to their aberrant structure and large fenestrations which may range from 200 to 2000 nm, tumor vasculature becomes leaky and intensely permeable in many areas of the tumor, therefore contributing to the enhanced permeation (Fig. 2.1) (Bertrand et al., 2013).

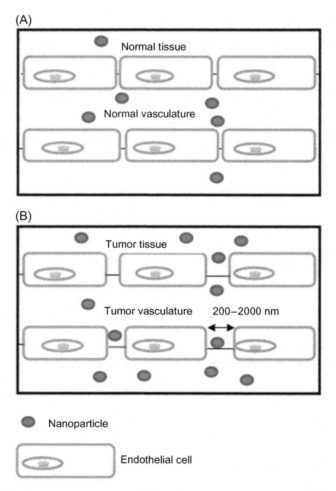

Figure 2.1 (A) Normal tissue vasculature and (B) tumor tissue vasculature with fenestrations ranging from 200 to 2000 nm in size, therefore contributing to the enhanced permeability and retention effect.

In normal tissues excess fluid is constantly drained to the lymphatic system in order to maintain a balance in interstitial fluids (Jain and Stylianopoulos, 2010). Many studies have failed to document the presence of functional lymphatics within the main core of the tumor; functional lymphatic vessels are only found in the periphery of the tumor and belongs to existing lymphatics of neighboring normal tissues (Padera et al., 2002, 2004).

Molecules smaller than 4 nm are not able to maintain and accumulate in the tumor tissue. Meanwhile, macromolecules and nanoparticles (NPs) with considerably large hydrodynamic radii are able to concentrate in the tumor interstitium (Bertrand et al., 2013). Moreover, due to the poor lymphatic drainage and the slow venous return, tumor tissue is capable of retaining macromolecules and nanoparticles. This enhanced retention effect, along with the enhanced permeation, are time-dependent; in which it takes almost 6 h for an anticancer drug to achieve optimum accumulation in the tumor tissue for an extended period of time (Taurin et al., 2012).

2.2.1 Features of enhanced permeability and retention effect

2.2.1.1 Tight junctions in tumor vessels

Endothelial cell junctions constitute tight junctions (TJs) and adherent junctions (Dejana, 2004). TJ proteins are essential for the permeability of endothelial cells, as they are located in the apical side of the cells. TJs consist of two main types of protein, transmembrane adhesion proteins, as well as intracellular proteins. Transmembrane adhesion proteins are made up of the claudin family (claudins 1, 5, and 12), occludin, endothelial selective adhesion molecule, and junctional adhesion molecules (Dejana, 2004). While the intracellular component of the TJs constitutes a complex of proteins, zonaoccludens 1 and 2 (ZO1 and ZO2), partitioning defective 3 and 6 (PAR3 and PAR6), calcium/calmodulin-dependent serine protein kinase (CASK), afadin, and multi-PDZ-domain protein-1 (MUPP1) (Dejana, 2004).

Some studies have shown that several types of inflammatory cytokines are involved in tumor metastasis and proliferation, by decreasing the expression of TJ proteins. For instance, in a study conducted by Wachtel et al. (2001), it was found that tumor necrosis factor alpha (TNF-α) is associated with the downregulation of occludin expression in astrocytes (Wachtel et al., 2001). In another study conducted by Martin et al. (2002), it was revealed that HGF/SF is a potent cytokine with a mitogenic and morphogenic function toward mammary cells, hence stimulating metastasis of MDA-MB-231 breast cancer cell line, by decreasing transendothelial resistance through downregulating the expression of ZO1 and claudin-1 (Martin et al., 2002; Niranjan et al., 1995). Moreover, claudin-5 was found to be highly expressed in vascular endothelial cells, while being downregulated in hepatic cancer (Hewitt et al., 2006; Krause et al., 2008). Therefore it contributes to increased microvascular permeability, and thus acts as a good target for antiangiogenic therapy.

2.2.1.2 Vascular permeability enhancement factors

In addition to the deranged anatomical abnormalities related to the space between endothelial cells, research has shown the presence of a plethora of vasoactive mediators that contributes jointly or individually to the EPR effect.

2.2.1.2.1 Vascular endothelial growth factor

Angiogenesis is controlled by many factors, amongst them, vascular endothelial growth factor (VEGF) is the most important. During pathological angiogenesis, overexpression of vascular growth factors such as VEGF occurs, causing abnormal blood flow as well as irregularities in the metabolic flow and demand (Bates et al., 2002).

VEGF is the main element in both physiological and pathological angiogenesis, which is constituted of four main isoforms: VEGF-121, VEGF-165, VEGF-189, and VEGF-206 (Wei et al., 1996).

It enhances leakiness of endothelial cells, by destabilizing transmembrane adhesion proteins. It also activates protein kinase C (PKC) which in return phosphorylates occludin, hence causing its endocytosis (Murakami et al., 2009).

VEGF induces angiogenesis, by two main mechanisms, one which constitutes enhancing the permeability of venules to circulating fibrinogen, therefore causing the deposition of fibrin, ending with angiogenesis (Dvorak et al., 1995). The second mechanism is by the induction of the vasodilator nitric oxide, which in return increases blood flow and blood vessel generation.

Hypoxia is one of the main stimuli for VEGF production and release, from both tumor and normal cells (Shweiki et al., 1992). Hypoxia stimulates VEGF gene expression which is mediated by the transcription factor, known as hypoxia inducible factor-1 (Ikeda et al., 1995). In addition to hypoxia, VEGF mRNA transcription is stimulated by other factors, such as TNF-α, platelet-derived growth factor, and cytokines (Ferrara and Davis-Smyth, 1997). This stimulatory effect causes an upregulation and activation of growth factor receptors, production of tissue matrix metalloproteinases, and increased vascular permeability (McMahon, 2000). Once released, VEGF has autocrine and paracrine stimulatory effects. The autocrine effect is on tumor cells, while the paracrine effect is for angiogenesis. VEGF binds to two tyrosine kinase receptors, VEGF receptor-2 (VEGFR-2) and VEGF receptor-1 (VEGFR-1) (Dunk and Ahmed, 2001). Once bound to VEGFR-2 found on endothelial cells, VEGF stimulates angiogenesis and mitogenesis. In their study, Dunk et al. have demonstrated the effect of VEGF on the epithelial cancer cell line ECV304 (Dunk and Ahmed, 2001). The team has shown that once VEGF binds to VEGFR-2, it stimulates ECV304 cell proliferation; while on the other hand, once bound to VEGFR-1, VEGF enhances monocyte migration, in order to negatively regulate cell division via the production of nitric oxide (NO) (Dunk and Ahmed, 2001).

2.2.1.2.2 Angiotensin

Along with VEGF, there are other factors that contribute to the EPR effect. One of these factors is angiotensin II (ATII). ATII is a vasoconstrictor which in return increases blood pressure, in normal vasculature. Meanwhile, in tumor blood vessels ATII has no effect, and that is due to the atypical structure of the tumor vasculature, which lacks a smooth layer at the level of precapillary arterioles. However, ATII-induced hypertension, where the systolic blood pressure ranges from 100 to 160 mmHg, causes a passive dilation of the tumor vasculature (Taurin and Greish, 2013). Passive dilation is due to opening of the TJs, hence promoting transvascular leakage, and contributing to the EPR effect (Hornig et al., 1997). Therefore, while normal blood vessels are constricted and tumor vasculature is dilated, macromolecular drugs can concentrate in the tumor with a decreased risk for other normal tissues.

2.2.1.2.3 ACE inhibitors and bradykinin

ACE inhibitors (ACE-I) are one of the classes of antihypertensive drugs. Their mechanism of action comprises inhibition of angiotensin-converting enzyme (ACE), hence limiting the conversion of angiotensin I (ATI), which is a vasodilator, to ATII, a vasoconstrictor; therefore, blocking hypertension.

Due to the amino acid sequence similarity of the C-termini of bradykinin with ATI, as well as the analogous structure of ACE and kininase II, ACE-I is able to increase the levels of bradykinin (Greish et al., 2003; Hornig et al., 1997). Bradykinin is a vasodilating peptide, which exerts its action through binding to the B_2 receptors found in endothelial cells. This then causes the stimulation of the release of prostacyclins (prostaglandin I_2/PGI2), NO and endothelium-derived hyperpolarizing factor; therefore resulting in increased vasodilatation and angiogenesis (Hornig et al., 1997). It should be noted that NO release is blood flow-dependent, an increase in the release of bradykinin will augment the blood flow through vasodilatation.

2.2.1.2.4 Prostacyclin (PGI2) agonists

Although it is considered as a hormone with a very short half-life, PGI2 is one of the most powerful vasodilators and platelet aggregation inhibitors. Vascular endothelial cells are capable of synthesizing PGI2 de novo (Kelton and Blajchman, 1980). The PGI2 platelet aggregation inhibitor effect is mediated by cyclic adenosine monophosphate (cAMP); and its affect is augmented by cAMP degradation inhibitors (Kelton and Blajchman, 1980).

Injection of Beraprost, a PGI2 analogue with a longer in vivo half-life, has stimulated the EPR effect, by loosening the TJs of the vascular endothelial cells, therefore causing leakage of the plasma components (Greish et al., 2003; Hornig et al., 1997). This can be used to selectively accumulate macromolecular cytotoxic drugs or nanoparticles in tumors.

2.2.2 Recent insights into enhanced permeability and retention effect

2.2.2.1 Vascular eruptions and dynamic vents

Although, it has been believed that blood vessels are stagnant, a study has demonstrated that the permeability of the tumor vasculature experiences a dynamic phenomenon controlled by vascular dynamic vents, known as vascular eruptions (Matsumoto et al., 2016).

Dynamic vents are time-limited formations that occur during vascular eruptions, which are neither affected by a pressure spike following the injection of the drug, nor nanoparticle dispersion by time (Matsumoto et al., 2016). In a study, two sizes of fluorescent-labeled polymeric nanoparticles (30 and 70 nm) were tested on human pancreatic BxPC3-GFP tumors implanted in BALB/c nu/nu mice (Matsumoto et al., 2016). The particles were evaluated for their eruption and distribution in the implanted tumor cells, using intravital confocal laser scanning microscopy. During the study, it was shown that the 30 nm nanoparticles were able to travel faster than the 70 nm, which had to face transport resistance through tissues. The 30 nm nanoparticles have shown a more static permeability, as they were capable of penetrating the tumor vasculature via both static and dynamic vents. On the other hand, 70 nm nanoparticles were only able to penetrate the tumor tissue through the dynamic vents. Unlike static vents, dynamic vents do not allow the diffusion of both large and small nanoparticles back into the circulation. This has been detected by the concentration measurement of both particles in the BxPC3 tumors after 8 and 24 h from the time of injection. The 30 nm nanoparticle dose that has been measured at both times was 8% and 11% of the original concentration, while the 70 nm had a concentration of 3% and 4% of the original dose per gram of the tissue (Matsumoto et al., 2016).

The irregularity of the tumor blood vessels allowed rapid extravasation of the nanoparticles, through a pressure-driven mechanism, therefore driving the nanoparticles into the interstitial space of the tumor. In addition, the tumor vent size affects the density and size of the dynamic eruptions. There is a reverse relationship between the density of tumors and their eruption activity. As the tumor's density becomes larger, its ability to resist fluid and particle penetration increases (Matsumoto et al., 2016).

2.3 DESIRED PHYSIOCHEMICAL NANOCONSTRUCTS

To tailor a carrier system that can take advantage of the EPR effect, certain considerations should be accounted for. Accumulated experience with nanoconstructs has shown that subtle changes in the carrier system physicochemical characteristics, such as particle shape, aspect ratio, and plasmon resonance, can significantly alter the drug distribution, uptake by tumor cells, and degradation kinetics. The following points will discuss the different desired properties of nanoparticles, in order to achieve maximum tumor-targeting goals.

2.3.1 Physiochemical properties

Different studies have stated that nanoparticle sizes ranging from 200 nm to less than 1 μm are mechanically filtered into the spleen, whereas, nanoparticles of a size less than 100 nm are able to penetrate the fenestrae of the endothelial lining and accumulate in the blood vessels (Stolnik et al., 1995; Ilium et al., 1982). Champion et al. have noticed that particles that are 500 nm and larger are phagocytosed by macrophages and antigen-presenting cells, therefore allowing for their rapid clearance from the body (Champion et al., 2007).

In another study two sizes of gold colloids, 30 and 50 nm, were tested for protein interaction. Due to its greater surface area, the 30 nm colloid had almost a twofold increase in bound protein mass compared with the 50 nm colloid (Dobrovolskaia et al., 2009).

According to Decuzzi particles have a critical radius size, which is around 150 nm (Decuzzi et al., 2005). For any particle radius smaller than 150 nm, the time required for it to reach the blood vessel walls decreases. Meanwhile, particles with a radius size greater than 150 nm experience van der Waals forces with the blood vessels, hence increasing the time required for them to reach the wall (Decuzzi et al., 2005). Therefore, when designing nanoparticles, it is important to have a single target size. For the current example of GNPs, this is done by slowly growing the seeded particles. If the size of the nanoparticles is small, that will cause the particles to aggregate, thus making it necessary to add a dispersing agent, or increase the size of the particles, by adding a longer alkyl chain (Horikoshi and Serpone, 2013).

Along with the size, particle curvature/shape, aspect ratio, and plasmon resonance greatly affect the margination dynamics of the nanoparticle. Nanoparticles come in various shapes: spherical, rods, filamentous, ellipsoid, etc. (Fig. 2.2). Chan et al.

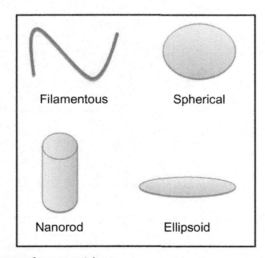

Figure 2.2 Different shapes of nanoparticles.

compared two shapes of gold nanoparticles, one spherical, the other rod-shaped. They reported that the spherical particles had a higher tendency to be internalized by HeLa cells when compared to the rod-shaped cells (Chithrani and Chan, 2007). Hence concluding that cellular uptake of nanoparticles is both size- and shape-dependent.

In another study it was found that long filamentous nanoparticles (i.e., filomicelles), when compared to spherical nanoparticles of the same volume, have a higher potential to be taken in by the tumor's EPR effect (Murphy and Jana, 2002). This is due to the shape deformation that spherical nanoparticles have to go through in order to pass through the tubular pores of the tumor (Fig. 2.3).

The aspect ratio of a shape of a nanoparticle is defined by the length of the major axis divided by the width of the minor axis (Murphy and Jana, 2002). Hence,

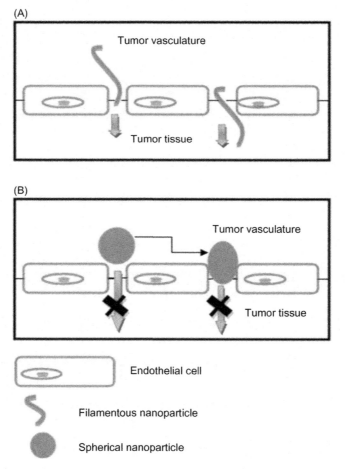

Figure 2.3 (A) Filamentous nanoparticle passing through the wide fenestrae of the tumor vasculature and (B) spherical nanoparticle unable to pass through the fenestrae even after deformation.

nanospheres have an aspect ratio of 1, while nanorods have an aspect ratio greater than 1 but smaller than 20. The visible absorption spectrum depends greatly on the shape of the nanoparticle; the larger the aspect ratio the greater the absorption wavelength (Murphy and Jana, 2002).

There are many techniques when it comes to the characterization of the surface properties of nanoparticles. One of those techniques is the determination of the zeta potential of nanoparticles in an aqueous suspension (Soppimath et al., 2001). An electrical potential is added to the aqueous suspension, and the mobility of the nanoparticles is observed and measured. The zeta potential value may be positive or negative, depending on the polymer used in its construct. Thus the surface charge of the particle may be determined by the measured zeta potential (Soppimath et al., 2001).

When formulating nanoparticles, the surface charge is a critical parameter in the determination of the bioavailability of the drug. It has been recognized that nanoparticles carrying a positive charge, such as an amine group, experience a high rate of phagocytosis, in addition to nonspecific internalization, leading to a short half-life in blood (Alexis et al., 2008). On the other hand, negatively charged or neutral nanoparticles have a lower rate of nonspecific uptake by cells. In a study conducted by Kedmi, the group formulated 1,2-dioleoyl-3-trimethylammoniumpropane (DOTAP)-based positively charged lipid nanoparticles, and administered them via the IV route into mice (Kedmi et al., 2010). The mice showed signs of hepatotoxicity, which was confirmed through an increase in the serum levels of hepatic liver enzymes; in addition to a stimulation of the inflammatory response, indicated by an increase in both Th1 and Th17 cytokines (Kedmi et al., 2010). In another study, it was concluded that cationic lipid-based nanoparticles, as well as positive micelles, once injected into mice will lead to brain lesions (Knudsen et al., 2015). This is mostly due to the attraction of the positive charge of the particles to the negative charge of the cell, therefore enabling the particles to enter the cell, leading to oxidative stress and DNA damage (Knudsen et al., 2015). In contrast, negatively charged particles and micelles have not shown toxic effects. Saying that, this should not be a general rule, every positive and negative nanoparticle should be tested individually for its toxicity.

Another technique which is employed in surface characterization is known as hydrophobic interaction chromatography. This technique is used to identify the hydrophobicity of the surface of nanoparticles (Soppimath et al., 2001). The nanoparticles are passed through a hydrophobic gel matrix, and the interaction of the particles with the matrix is dependent on the surface hydrophobicity of the particles. The eluent is then collected and measured spectrophotometrically at a wavelength of 400 nm to determine the optical density (Soppimath et al., 2001).

Another common technique that is used for surface characterization of nanoparticles is known as X-ray photoelectron spectroscopy. This is performed by irradiating photons on the nanoparticles, therefore ionizing the core-level electrons (Soppimath et al., 2001).

Electrons emitted from the object are specific for the atoms found on the surface of the material.

There are different forms of triggered release of drugs from their nanocarriers; the two most vital mechanisms are physiologically dependent. Nanocarriers with a triggered release pathway are made of materials that are responsive to the pH change or certain enzymes found in the pathological or physiological environment or oxidation potential. For nanocarriers that are triggered by the pH, they are made of anionic or cationic materials with hydrophobic or hydrophilic polymers (Caldorera-Moore et al., 2010). However, pH-triggered release is not disease-specific, hence making it a limitation when designing such nanocarriers. In the case of tumors, certain types of enzymes are upregulated, therefore development of biomaterials that are sensitive to these enzymes allows triggered release of the cytotoxic drug specifically at the tumor tissue.

2.3.2 Biocompatibility and biodegradation

One of the concerns of nanodrugs is biocompatibility. Nonbiodegradable drugs pose a threat to many organs' functions (especially the liver), and may induce malignant transformation. Surface properties of nanoparticles govern the way that the host's body will respond to it. For instance, PEGylation of the nanoparticle surface reduces significantly immunoreactions such as platelet aggregation and protein surface adsorption (Naahidi et al., 2013). As the surface charge of nanoparticles increases (whether negative or positive), reticuloendothelial clearance of the nanoparticles is augmented, therefore leading to a loss in the dose percentage.

Nonbiodegradable nanoparticles have been shown to induce toxicity, therefore different forms of biodegradable polymers have been employed in the design of safe and biocompatible nanoparticles. Biodegradable nanoparticles are made of various materials, such as natural polymers, which include polylactic acid (PLA), and polycaprolactone (PCL), synthetic polymers, and pseudosynthetic polymers (Naahidi et al., 2013). Polymers such as PLA and poly(D,L-lactide-*co*-glycolide) (PLGA) have been highly studied for their drug delivery mechanisms. The drug release rate from PLGA can be adjusted by changing the building blocks or composition of the copolymer (Panyam and Labhasetwar, 2003). After the release of the drug, polymer biodegradation goes through hydrolysis and then the citric acid cycle which eventually removes the polymers from the body (Panyam and Labhasetwar, 2003).

One of the most immunologically safe copolymers used in nanoparticle formulation is *N*-(2-hydroxypropyl) methacrylamide (HPMA). HPMA-bound cytotoxic drugs are not only considered highly effective, but in a study conducted by Rihova and Kovar, HPMA was able to tone down the toxic side effects of a cytotoxic drug, doxorubicin (DOX) (Rihova and Kovar, 2010). Another biocompatible polymer, is the highly branched dendrimers. Dendrimers are made of a hydrophobic core and a hydrophilic outer core, therefore allowing the successful encapsulation and transportation of insoluble drugs. Dendrimers possess intrinsic properties that render them favorable when it comes to choosing nanoparticle polymers. Dendrimers are considered as membrane permeability enhancers, hence stimulating the EPR effect, and antitumor, antibacterial, and antiviral properties (Goldberg et al., 2007).

2.4 ENHANCED PERMEABILITY AND RETENTION-BASED DRUG DELIVERY SYSTEMS

Today, nanotechnologies are increasingly emerging to take an essential place in drug delivery and human therapeutics. The growing demand to construct drug delivery systems capable of binding, adsorbing, and carrying other compounds such as drugs, probes, and proteins as well as having drug targeting competence and reduced toxicity has resulted in a concentration on the development and usage of nanoparticles. To date, the majority of the nanoconstructs designed for drug targeting have relied intensively on the EPR effect. Carrier molecules are derived from biological sources such as cellular components, viral, or bacterial origin: all remain dependent on the EPR effect to drive them to the tumor tissues. In this section we will spotlight the most applicable EPR-based systems of synthetic and biological nature.

2.4.1 Liposomes

Liposomes are the first developed nanoparticles to accumulate drugs via the EPR effect. These are closed spherical vesicles formed of an aqueous core surrounded by a phospholipid bilayer (Bangham et al., 1965). This amphipathic nature allows drug molecules to be entrapped either in the aqueous core or intercalated into the lipid bilayer shell (Massing and Fuxius, 2000). Hence, the application of effective and safer phospholipids for transport of drugs to a specific site has shown a path to the development of a liposomal drug delivery system (Fig. 2.4).

Liposomes can be classified based on their size, number of bilayers, composition, and methods of preparation. However, liposomes are mostly classified in relation to their size, in which they are either multilayer vesicles (MLVs) or unilamellar vesicles (ULVs). MLV consist of two or more layers. These vesicles are 0.5–5 μm in size with a moderate trapped volume. On the other hand, ULV have a single lamellae and are further classified into large unilamellar vesicles (LUVs) and small unilamellar vesicles (SUVs). LUV size is more than 100 nm and has a high trapped volume, whereas SUV size is less than 100 nm with a lower trapped volume.

There is a wide range of natural, semisynthetic, and synthetic phospholipids that can be used for liposome preparation, such as egg and soya lecithin and phosphotidylcholines (Vemuri and Rhodes, 1995). However, liposomes prepared utilizing only phospholipids are usually unstable, which may result in drug leakage. Therefore, other compounds, usually cholesterol or α-tocopherol, are included in the liposomal formula

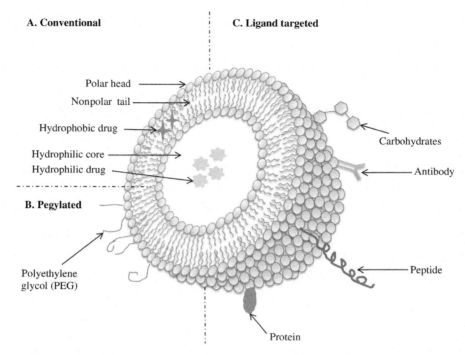

Figure 2.4 Illustration of liposome construction as well as the different drug-loading mechanisms.

to enhance encapsulation efficiency (Sackmann, 1994). Most of liposome preparation methods involve the use of organic solvents or detergents (New, 1990), whose presence has a high number of toxicological side effects. Now, more safe methods have been developed to prepare liposomes such as polyol dilution, bubble method, and heating method (Kikuchi et al., 1994; Talsma et al., 1994; Mozafari et al., 2002). After preparation, liposomal drug formulations are characterized according to their size, percent drug encapsulation, charge, shape, lamellarity, and stability in order to predict these products' in vivo and in vitro behaviors.

In cancer, liposome-encapsulated anticancer drugs have achieved highly efficient anticancer activity with reduced systemic toxicity. Studies have found that the accumulation of liposomes in tumor tissue is governed by the EPR effect (Gabizon and Papahadjopoulos, 1992; Brandl, 2001; Gregoriadis et al., 1971). In order to penetrate malignant tissue, a liposomal system of less than 150nm is required (Harashima et al., 1994). Ultimately, after extravasation into tumor tissue, liposomes are degraded by enzymes and/or attacked by phagocytes to release the loaded cargo that subsequently diffuses to tumor cells.

In general, there are four liposomal drug delivery platforms: conventional liposomes, sterically stabilized liposomes, ligand-targeted liposomes, and a combination of these. Conventional liposomes are the first developed liposomal delivery systems. They consist of a lipid bilayer that can be composed of cationic, anionic, or neutral (phospho)lipids and cholesterol, which enclose an aqueous volume. These formulations were firstly tested in the 1980s where liposomal particles were loaded with DOX and amphotericin and showed enhanced drug delivery and improved therapeutic index (Gabizon et al., 1982).

However, liposomes undergo rapid clearance due to the macrophage uptake of the reticuloendothelial system (RES). Despite that, they are useful in targeting the liver for therapeutic application.

To overcome this limitation, polyethylene glycol (PEG)-coated liposomal nanoparticles have been created to obtain sterically stabilized liposomes, hence providing long circulation and effective targeting (Yu et al., 2013; Klibanov et al., 1990). These PEG-coated liposomes have a circulation half-life greater than 24 h in rats and as high as 45 h in humans (Papahadjopoulos et al., 1991; Allen, 1994). Doxil® was the first PEG-coated liposome that was loaded with DOX, and was the first to be approved by FDA in 1995. While this coating helps liposomes to escape rapid clearance, it reduces the ability to interact with the intended targets (Moghimi and Szebeni, 2003; Forssen and Willis, 1998).

Most liposome nanoparticles that direct molecular targeting of cancer cells are ligand-targeted liposomes. An example of this category is immunoliposomes, where monoclonal-antibody fragments are ligated to liposomes (Ulrich, 2002; Puri et al., 2009). In preclinical studies, those constructs showed superior efficacy in intracellular delivery of encapsulated agents versus all other treatments (Moghimi and Szebeni, 2003). A major obstacle of those immunoliposomes is poor pharmacokinetics and immunogenicity. Therefore, the next-generation liposomes are based on utilizing a combination of those platforms. For example, combining the site-specific feature of immunoliposomes with the stability of PEG-coated liposomes that have been found to be significantly effective (Maruyama, 2002).

Overall, the stable formulation and the capability of targeting specific organs by both passive and active uptake make liposomes one of the preferable nanoparticles. Along with the improved pharmacokinetics and drug safety are increased bioavailability, sustained release, and precise delivery actions.

2.4.2 Polymer nanocarriers

Polymer nanocarriers are either natural or synthetic nanoconstructs that are classified according to their morphology and composition in the core and periphery into around seven platforms including solid polymeric nanoparticles, polymeric micelles, polymer conjugates, dendrimers, polymersomes, polyplexes, and polymer lipid hybrid systems (Fig. 2.5) (Alexis et al., 2008).

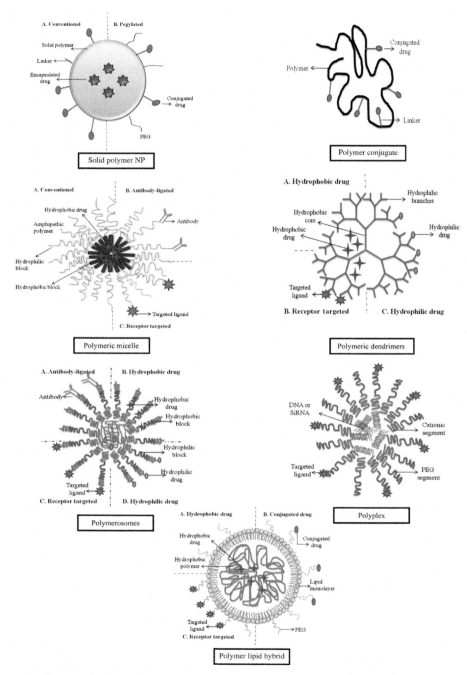

Figure 2.5 Illustration of the different platforms of polymer nanocarriers representing their composition as well as the different drug-loading mechanisms.

Solid polymeric NPs are solid colloidal systems where the therapeutic agent is dissolved, entrapped, encapsulated, or adsorbed into the constituent polymer matrix (Prabhu et al., 2015). Many solid polymeric NPs have been developed for targeted drug delivery. One example is the PEGylated poly (lactide-*co*-glycolide) (PLGA) NP that was loaded with paclitaxel (PTX) and resulted in enhanced survival rate and both in vitro and in vivo improved cytotoxic effect on liver tumor growth when compared with the free drug (Danhier et al., 2009).

Furthermore, polymer conjugates are nanoconstructs that are composed at least of three major components: a soluble polymer backbone, a biodegradable linker, and a covalently linked anticancer drug which is deactivated as a conjugate. Various chemical bonds are used to prepare polymeric conjugates including amide, ester, hydrazide, azide imine, thioether, and urethane. The characteristics of these bonds determine the speed of release of drugs from the polymeric chains. Such variance allows constructing different conjugates to meet different physiological conditions. An example is the clinically tested (HPMA)-Dox conjugate which showed therapeutic responses in both breast and lung cancer (Ruth, 2005; Cassidy, 2000).

On the other hand, polymeric micelle NPs are spherical, amphiphilic, self-assembled core or shell structures that allow drugs encapsulation inside their hydrophobic core. In oncotherapy, polymeric micelles have shown highlighted drug–carrying ability (Oerlemans et al., 2010). An example is research done by Jin et al. to detect the effect of PTX-loaded *N*-octyl-*O*-sulfate chitosan micelles on treating multidrug-resistant cancer. The micelles exhibit twofold higher cellular uptake than the free drugs and high in vivo tumor inhibition (Jin et al., 2014).

Additionally, the dendrimers are synthetic macromolecules with a tree-like structure resulting from numerous extensions protruding from the central core. Therapeutic agents can be either encapsulated or conjugated in the core or on the dendrimer surface (Bharali et al., 2009). One example is the poly(glycol-succinic acid) dendrimers carrying camptothecin. Increased drug uptake and cytotoxicity up to 16- and 7-fold, respectively, were observed in comparison with the free drug when treating human breast adenocarcinoma (MCF-7), colorectal adenocarcinoma (HT-29), nonsmall-cell lung carcinoma (NCI-H460), and glioblastoma (SF-268) (Morgan et al., 2003).

Polymersomes are synthetic copolymers that are self-assembled into discrete hydrophobic and hydrophilic blocks. Despite the fact that they resemble liposomal NPs, polymersomes are more stable, and have more storage capability and longer circulation time (Levine et al., 2008). When compared to liposomal NPs, polymersome vesicles carrying DOX were more capable of inhibiting tumor growth than the liposomal formulation.

Other polymer nanocarriers are the polyplexes. In polyplexes, genes or siRNA are attached to the cationic polymer group through electrostatic interactions. Polyplexes are mainly used for cancer-specific gene therapy. One example is the galactose-modified trimethyl chitosan-cysteine-based polymeric vectors. These polyplexes showed great in vitro and in vivo capability of delivering siRNA when tested in human liver cancer (QGY-7703) cells and human lung cancer (A549) cells, resulting in efficient and persistent gene knockdown (Han et al., 2013).

The last is the polymer lipid hybrid system which is a combination of polymeric NPs and liposomes. This system is composed of a hydrophobic core and a hydrophilic shell with a lipidic monolayer separating them. Such a system was explored by Wong et al. in which a soybean-oil-based polymer was conjugated to DOX and later dispersed with lipid in water. Effective delivery of DOX and enhanced cytotoxicity by eightfold against P-gp-overexpressing human breast cancer cell lines were observed (Wong et al., 2006).

In all these types of polymeric nanocarriers, the therapeutic agent either encapsulated inside the polymeric core or attached to the NP surface (Asati et al., 2010). The EPR-based targeting of those nanoparticles as well as their cytotoxic potential depends on their surface charge (anionic, cationic, or neutral) (Boyer et al., 2013). Because the tumor cell surface is highly negative due to the preferentially expressed phospholipids on tumor endothelial cells, cationic NPs have been found to efficiently interact with tumor cells in vitro, though this proved troublesome in in vivo situations (Beh et al., 2009; Byrne et al., 2008).

Another way of targeting tumor cells by polymer nanocarriers is ligand-based targeting. In this method, ligands such as proteins, nucleic acids, or others are conjugated to the periphery of an NP to facilitate the interaction with specific overexpressed receptors on tumor cells. After efficient delivery of NPs into the tumor site by EPR phenomena, they start binding to the target receptors which are mostly transferrin, folate, and epidermal growth factor (EGF) receptors, and glycoproteins. NPs are then intracellularly endocytosed, followed by lysosomal uptake and degradation promoting drug release and cytotoxicity effect (Cho et al., 2008; Helgason et al., 2009). Currently, the majority of NP-based drug delivery systems being investigated in clinical and preclinical phases belong to the polymeric type.

2.4.3 Solid lipid nanoparticles

Solid lipid nanoparticles (SLNs) were invented at the beginning of the 1990s. They are defined as aqueous colloidal dispersions (50–1000 nm) made of biocompatible and biodegradable lipid solid at room and body temperature (Kaur et al., 2008). SLNs consist of lipids including lipid acids, mono-, di-, or triglycerides, glyceride mixtures or waxes, and are stabilized by biocompatible surfactants (Fig. 2.6). SLNs are believed to combine the advantages of other colloidal carriers as well as avoiding their disadvantages (Tyle, 1987).

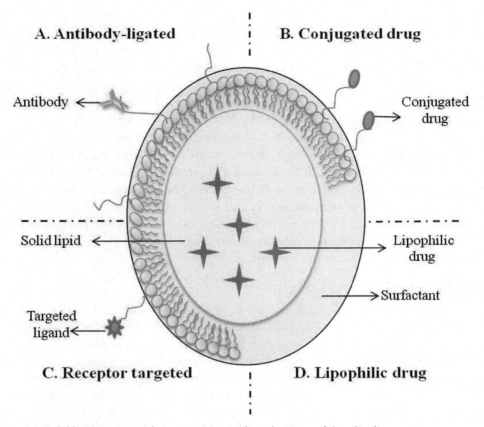

A. Antibody-ligated

B. Conjugated drug

Antibody

Conjugated drug

Solid lipid

Lipophilic drug

Surfactant

Targeted ligand

C. Receptor targeted

D. Lipophilic drug

Figure 2.6 Solid lipid nanoparticle composition and mechanisms of drug loading.

SLNs are prepared from lipid, emulsifier, and water or solvent by different methods such as high-pressure homogenization, ultrasonication, supercritical fluid method, solvent evaporation method, and microemulsion-based method. After preparation, drugs can be incorporated into SLNs by three methods. The first is the homogeneous matrix model where highly lipophilic drug is incorporated into SLNs using hot or cold homogenization resulting in a homogeneous drug–lipid mixture. The second method is drug-enriched shell with lipid core model. This model is performed during SLN production. In this model, drug is fixed in the external shell of a lipid core gradually. The last model is the lipid shell with a drug-enriched core. This model is obtained when dissolving drug in lipid melts. The resulting system is composed of drug-enriched core surrounded by a lipid shell (Pandey et al., 2005).

SLNs encourage high interest because of their advantages, such as physical stability, protection and controlled release of active substance, biocompatibility, selective orientation, absence of organic solvents, and possibility of scaling up. SLNs have been applied and investigated widely in the pharmaceutical industry for various routes of administration including parenteral, oral, rectal, and dermal routes.

In cancer therapy, SLNs were tested and used for various cancer therapies such as liver cancer, breast cancer, colorectal cancer, lung cancer, brain tumors, and others. In breast cancer, SLNs loaded with methotrexate and camptothecin showed prolonged release of drug, reduced systemic toxicity, and improved safety after IV administration (Stuchlík and Žák, 2001; Skog et al., 2008). Moreover, in a study to overcome the limited access of the drug 5-fluro-2′-deoxyuridine to the brain, SLN loaded with 5-fluro-2′-deoxyuridine has revealed effective brain tumor targeting and drug release (Wysoczynski and Ratajczak, 2009). Despite the promising roles of SLNs in targeted drug delivery there are some drawbacks that need to be overcome, for instance polymorphic modifications of the lipid matrix and sterilization and antimicrobial preservation. Although SLNs are in their nascent stage, by overcoming their limitations SLNs could become the most effective delivery system for cancer therapy.

2.5 EXOSOMES

Exosomes are small, 30–140 nm, lipid membrane-bound vesicles that cells secrete into the extracellular environment (Rana et al., 2012). Exosomes contain a variety of molecules from their cell of origin including RNA species, proteins, biofunctional lipids, and occasionally DNA (Szajnik et al., 2010). The main function of exosomes is to provide intercellular communication between cells by carrying information and cargo (Zarovni et al., 2015). Exosomes also participate in waste removal, antigen presentation, and the induction of proinflammatory cytokine release. Interestingly, these significant roles of exosomes are thought to be played in both normal physiological conditions as well as in the pathological processes of many diseases (Zarovni et al., 2015).

In cancer, tumor-derived exosomes are capable of exchanging information between neighboring cancer cells and also can communicate with distant cells and various cell types (Szajnik et al., 2010). Tumor-derived exosomes have been reported to contribute to tumorigenesis, cancer progression and metastasis, angiogenesis, extracellular matrix remodeling, reduced anticancer immune response, and drug resistance (Lazar et al., 2015). These characteristics of exosomes suggest them to be important for both diagnostic and therapeutic purposes.

Current studies focus on targeting tumor exosomes as cancer progenitors as well as utilizing tumor exosomes to identify novel biomarkers and molecular targets to help in early personalized cancer diagnosis and prognosis (Matsumura et al., 2015; Zhou et al., 2015; Chalmin et al., 2010; Phuyal et al., 2014). In addition, a new paradigm is using exosomes as drug delivery devices (Johnsen et al., 2014; Katakowski et al., 2013). The biocompatibility, stability in circulation, and ability to target certain cell types make exosomes ideal vehicles for drug and/or molecule delivery.

Loading exosomes with therapeutic agents can be done endogenously or exogenously. The endogenous loading is performed inside the parent cells prior to exosome isolation by overexpressing RNA species or molecules of interest. In contrast, exogenous loading is performed after exosome isolation and requires either coincubation or electroporation of the exosomes with the drug or molecule of interest (El Andaloussi et al., 2013).

Recently, this approach has been tested in vitro and in vivo, whereby cancer cells were shown to uptake exosomes loaded with different chemotherapeutics and/or molecules, resulting in the killing of cancer cells while also reducing the drug's unfavorable side effects. A recent study by Heikki Saari and his colleagues revealed an enhanced cytotoxicity effect of PTX on autologous prostate cancer cells achieved by using an exosome-mediated drug delivery system (Saari et al., 2015). Furthermore, another study by Katakowski in 2013 showed the efficacy of exosomes derived from marrow stromal cells expressing miR-146b in inhibiting glioma growth. This study indicates the significant role of exosomes as a capable delivery system to carry unstable RNA into a targeted cell site promoting a highly therapeutic effect (Katakowski et al., 2013).

However, exosome delivery system design needs further refinement in order to be validated and used in clinical scale. It is very important to choose immature cell lines from which exosomes are derived to insure that exosomes have no immune-stimulating activity and thus preventing immune resistance in target tissue. Moreover, it has been shown that the interaction between the drug delivery system and the proposed target cell is affected by exosomal surface proteins that differ according to the choice of exosome parent cells. Recently, immature dendritic cells have been favorable choices for exosome isolation and reusage (Yin et al., 2013). Rather, a semisynthetic strategy is now being applied to produce exosomes with targeted peptides via glycosylation sites for enhanced targeted delivery of exosomes for therapeutics (Hung and Leonard, 2015).

2.6 BACTERIAL AND VIRAL DELIVERY SYSTEMS

The intense research work to find ideal delivery systems has fueled using biological NPs as novel drug delivery systems for antigens, nucleic acids, and drugs. So far, virus-based systems are the most dominant biological NPs wherein viruses have naturally evolved to deliver cargos to specific sites. However, many researches and preclinical studies are now shifting toward using bacterial nanoparticles as novel delivery systems.

2.6.1 Viral delivery systems

Many plant, bacteria, insect, and animal viruses have been developed to be used as drug delivery systems (Singh et al., 2006) (Fig. 2.7). Avoiding using human viruses is preferable to prevent subsequent immunological and toxicological responses. In order to use

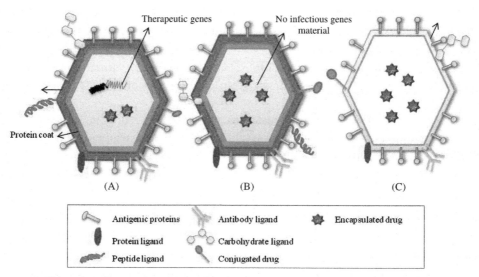

Figure 2.7 Schematic illustration of the different viral nanocarriers composition and mechanism of drug loading. (A) Viral vector, (B) virus-like particle, and (C) virosome.

viruses as drug delivery systems, their intrinsic properties should be considered, such as the ability to self-assemble in the presence and absence of nucleic acids, content of viral coats as targets for different chemical and genetic modifications, and the ability to recognize certain cell types and induce immunological responses in mammalian organisms.

There are many types of virus-based delivery systems, such as the viral vectors that are derived from disease-causing viruses with replaced genetic material. Another type is the virus-like particles (VLPs) which are synthetic multiprotein structures that resemble virus's organization and conformation but lack the viral genome. One more type is virosomes. These are viral envelopes mostly derived from influenza virus, including membrane phospholipids and viral spike glycoproteins but lacking the viral genome (Warfield et al., 2004). Viral-based delivery systems created from plants and bacteria can be produced in gram quantities in plants and bacteriophages. On the other hand, VLPs are often produced in heterologous expression systems such as *Escherichia coli* and yeast. VLPs can also be produced chemically by pH-induced swelling followed by alkaline hydrolysis of the nucleic acids (Schneemann and Young, 2003; Garnier et al., 1994; Mueller et al., 2010). The protein structures of viral nanoparticles can be modified genetically and chemically to insert amino acids, peptides, epitopes, antibodies, carbohydrates, and other ligands to serve as targeting ligands and/or to stimulate the immune response (Miller et al., 2007; Wang et al., 2002; Pokorski and Steinmetz, 2011).

Therapeutic agents can be either bioconjugated to the exterior or interior capsid shell, or encapsulated by the viral system. However, loading viral NPs with different cargos can be accomplished by diffusion and entrapment making use of the swelling mechanisms, or by disassembly and reassembly (Ren et al., 2007; Sikkema et al., 2007; Daniel et al., 2010). Studies have found that the majority of plant viruses require surface modifications such as iron oxide attaching to allow magnetic direction of virus NPs toward specific sites (Rae et al., 2005; Manchester and Singh, 2006; Destito et al., 2009). On the other hand, oncolytic animal viruses have the ability to discriminate between healthy and tumor human cells displaying a cell-selective advantage that is not found in plant viruses (Lundstrom, 2009; Hu and Pathak, 2000; Haviv et al., 2002).

In one experiment, cowpea mosaic virus (CPMV), a viral NP created from a plant virus, was loaded with DOX using different bioconjugate chemistries and tested in vitro against HeLa cells. The results indicated a time-delayed but greatly enhanced efficacy compared to free drug (Aljabali et al., 2013).

Another example is work done by Mukhopadhyay and Dragnea describing the packing of artificial cargoes, including RNA, DNA, proteins, fluorophores, and gold nanoparticles into the capsids of alphaviruses, a type of NP that has the advantage of targeting tumor cells. In a tissue culture experiment, these NPs showed efficient cell targeting and cargo uptake (Steinmetz, 2013).

In a study to monitor virus-based NPs' intracellular fate, the results pointed to a combination of caveolae-dependent endocytosis plus macropinocytosis entry pathway of viral NPs. These are selected by viruses to escape lysosomal degradation. This route is believed to aid targeting specificity and improves the therapeutic effect (Wu et al., 2012; Medina-Kauwe, 2007).

Despite the many advantages of viral NPs such as the ease of attaching different therapeutic molecules and the extreme tumor specificity, bringing these NPs from research to clinic needs more understanding of their behavioral properties and immunological and toxicological impacts.

2.6.2 Bacterial delivery systems

It has been found that several bacterial strains colonize selectively at the hypoxic tumor sites. This property directs working toward using bacteria as a therapeutic as well as a drug delivery system, hence providing a more tumor-targeted therapy (Gardlik and Fruehauf, 2010). Such an effect of live attenuated bacterial infection on tumor growth has been verified in many studies. In a study where the effect of a tumor-targeting *Salmonella typhimurium* strain was tested in various mouse models, bacterial infection significantly eradicated primary tumors, as well as cancer metastases (Hayashi et al., 2009). The unique characteristic of bacterial cells is based on the unique antigenic structures of bacterial membranes where lipopolysaccharides, lipid A, and peptidoglycans are predominantly expressed promoting high immune-stimulating influences.

However, the use of bacterial systems for therapeutic purposes can be enhanced by applying genetic and/or chemical modifications, hence improving bacterial targeting capability and allowing bacterial loading with different therapeutic agents and thus augment the cytotoxicity effect.

For delivery purposes, a novel and progressive approach called the bacterial ghost (BG) system has been created (Fig. 2.8). BG is produced from Gram-negative bacteria wherein expressing a cloned gene (gene A) from bacteriophage ϕX174 results in bacterial cell lysis. This protein-E-mediated lysis results in a bacterial envelope devoid of nucleic acids, ribosomes, and other intracellular contents, although it retains all morphological and structural features of the natural cell. BGs offer high capacity of nucleic acids, antigens, and drug delivery due to the multiple choices of agent's compartmentalization they provide. A therapeutic agent could be prelysis anchored in or a combination of the outer membrane, the inner membrane, the periplasmic space, and the internal lumen of the cytoplasmic space of BGs (Szostak et al., 1996). It has been found that exposure to BGs activates melanoma cells and increases their phagocytosis ability as well as initiating immune responses (Kudela et al., 2010).

BGs can be employed as anticancer therapeutic vehicles by different approaches as follows.

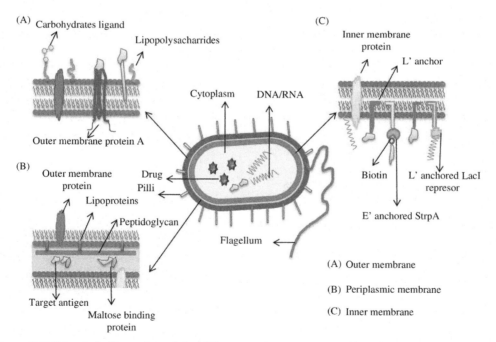

Figure 2.8 Schematic illustration of the different viral nanocarriers composition and sites of therapeutic agent loading.

2.6.2.1 Bactofection

Bactofection is the approach of using BGs to deliver genes into the target organism, organ, or tissue. In cancer therapy, bacterial vehicles usually deliver eukaryotic expression plasmids encoding the therapeutic anticancer gene to the tumor cell cytoplasmic area. After entering the cells, BGs undergo systemic lysis to release the plasmids into the cytoplasm and to prevent the toxicological, long-term survival of BGs inside the organism. Upon release into the cytoplasm, plasmids are localized into the cell nucleus where the therapeutic gene is expressed by the host cell's transcription and translation systems. The resulting protein acts as a cytotoxic molecule toward the target cell as well as surrounding cells, by activating suicide pathways.

In one study, the *Salmonella choleraesuis* strain was engineered to deliver a eukaryotic expression plasmid encoding endostatin gene into tumor cells. A 40–70% inhibition of tumor growth was observed, combined with a significant decrease of the intratumoral microvessel density at the site of infection (Lee et al., 2004).

2.6.2.2 DNA vaccination

This approach resembles the bactofection approach where plasmids are being delivered by BGs into the target tissue, however, in DNA vaccination plasmids are encoding tumor-expressed antigens rather than anticancer genes. Tumor-expressed antigens induce humoral and cellular immune responses in the host, resulting in a tumor growth suppression action.

Experimentally, DNA vaccination studies using *S. typhimurium* strains have exhibited high efficacy in eradicating tumor growth in animal models of malignant melanoma, colorectal carcinoma, glioblastoma, and lung cancer upon oral administration. Those strains fight tumor growth by delivering plasmid-encoding antigens against VEGFR-2 growth factor to targeted tumor sites (Niethammer et al., 2002).

2.6.2.3 Bacterial-mediated protein delivery

Another approach is the use of BGs for delivering proteins into the targeted tumor site. This strategy is based on genetically modifying bacteria to induce therapeutic protein expression inside the host organism. In this approach, the persistence of bacteria in the target tissue is mandatory to achieve sufficient production and ideal local distribution of the therapeutic protein.

A widely researched strategy that uses bacterially mediated protein delivery as an anticancer therapy is the enzyme prodrug approach, in which systemically administered bacteria deliver an enzyme that converts the prodrug into its active compound. One example is using *S. typhimurium* strains to deliver cytosine deaminase in many phase 1 trials (Xiang et al., 2008).

2.6.2.4 *Bacterial delivery of drugs*

In order to avoid the severe side effects, lack of selectivity, and poor solubility resulting from systemic administration of highly aggressive drugs, BGs have been recently developed to act as efficient and safer drug delivery systems. In one study, BGs created from colonic commensal *Mannheimia haemolytica* were used for in vitro delivery of DOX to human colorectal adenocarcinoma (Caco-2) cells. Cytotoxicity test showed twofold higher cytotoxic and antiproliferative activities of BGs loaded with DOX as compared to treating with free DOX (Paukner et al., 2004).

Nowadays, several nanomedicines have reached the market besides many others that have reached preclinical trials. Table 2.1 shows these drugs' statuses and the types of cancer they target.

In Section 2.6.2.3, we highlighted the major synthetic and biological carrier systems taking advantage of the EPR effect by virtue of its size to concentrate in tumor tissues. In the next section we will discuss approaches that have been developed beyond the EPR effect to target tumors at the cellular level rather than the tissue level that is archived through EPR targeting.

2.7 BEYOND ENHANCED PERMEABILITY AND RETENTION

2.7.1 Active targeting

Unlike the enhanced permeability effect, active targeting is achieved by conjugating a targeting element to the nanoparticle, hence increasing the particles' targeting ability toward the cancer tissue. There are two main types of active targeting; receptor- and antigen-based active targeting as well as carbohydrate-based active targeting.

2.7.2 Receptor- and antigen-based active targeting

The optimal receptor- and antigen-based nanoparticle should have a high affinity to the tumor cell surface receptors, and be easily internalized by cells and biodegradable. Two factors that play a major role in the success of the internalization of the drug (Sinha et al., 2006) are the abundance of the receptor and antigens on the tumor cells, and the high affinity of the ligands to the previous molecules.

One example of those nanoparticles is carbon nanotubes. There are two main forms of carbon nanotubes: PEG functionalized single-walled carbon nanotubes (SWNTs), and copolymer-coated multiwalled carbon nanotubes (MWNTs) (Fabbro et al., 2012).

In experimental trials, DOX has been combined with different forms of carbon nanotubes in order to treat various cancers, such as cervical, colon, and breast cancers. In one study, the structural and cytotoxic characteristics of the MWNT–DOX

Table 2.1 Examples of clinically used tumor-targeted nanoparticles

Type of nanomedicine	Name	Therapeutic agent	Status	Cancer type	References/clinical trial number[a]
Liposomes		Doxorubicin	Approved	Ovarian cancer	Gordon et al. (2001)
			Approved	HIV-associated Kaposi's sarcoma	James et al. (1994)
	DaunoXome	Doxorubicin	Phase IV	Breast cancer	Green et al. (2011)
			Approved	HIV-associated Kaposi's sarcoma	FDA approves DaunoXome as first-line therapy for Kaposi's sarcoma (1996)
			Phase III	Acute myeloid leukemia	Latagliata et al. (2008), Fassas and Anagnostopoulos (2005)
	Myocet	Doxorubicin	Approved (Europe/Canada)	Metastatic breast cancer	Swenson et al. (2003)
	Depocyt	Cytarabine	Approved	Malignant lymphomatous meningitis	Chhikara and Parang (2010)
	MEPACT	Muramyl tripeptide phosphatidyl ethanolamine	Approved (Europe)	Osteosarcoma	Ando et al. (2011)
	ThermoDox	Doxorubicin	Phase III	Hepatocellular carcinoma	NCT00617981[a]
			Phase III	Colorectal liver metastases	Yarmolenko et al. (2010)
			Phase II	Breast cancer	NCT00826085[a]
	Allovectin-7	HLA-B7 and beta2-microglobulin complex	Phase III	Melanoma	NCT00395070[a]
			Phase III	Head and neck cancer	NCT00050388[a]
			Phase II	Melanoma	Stopeck et al. (2001)

(*Continued*)

Table 2.1 Examples of clinically used tumor-targeted nanoparticles (Continued)

Type of nanomedicine	Name	Therapeutic agent	Status	Cancer type	References/clinical trial number[a]
	CPX-351 (VYXEOS)	Cytarabine:daunorubicin	Phase IV	Acute myeloid leukemia	NCT0253115[a]
	CPX-1	Irinotecan, floxuridine	Phase II	Colorectal neoplasms	NCT0361842[a]
	SPI-077	Cisplatin	Phase III	Nonsmall-cell lung cancer	Kim et al. (2001), White et al. (2006)
			Phase II	Ovarian cancer	Seetharamu et al. (2010)
	TKM-PLK1 TKM-ApoB (Tekmira)	RNAi targeting polo-like kinase 1	Phase I	Liver tumors	Burnett et al. (2011)
	Oncolipin	Interleukin-2	Phase II	Nonsmall-cell lung cancer	Wang et al. (2008)
	L-Annamycin	Annamycin	Phase II	Breast cancer	Booser et al. (2002)
			Phase I/II	Acute lymphoblastic leukemia	Wetzler et al. (2013)
	Aroplatin	Oxaliplatin	Phase II	Advanced colorectal cancer	Dragovich et al. (2006)
	LE-SN38	SN-38	Phase II	Colorectal cancer	NCT00311610,[a] Bayes et al. (2004)
			Phase I	Neoplasm	NCT00046540[a]
		Thymidylate synthase inhibitor	Phase I	Advanced biliary cancers	Ciuleanu et al. (2007)
			Phase I	Advanced solid tumor	Beutel et al. (2005)
	Endo-Tag-1 (Medigene)	Cationic liposomal paclitaxel	Phase II	Pancreatic cancer	Lohr et al. (2012)
			Phase II	Triple negative breast cancer	Chang and Yeh (2012)
	LEP ETU	Paclitaxel	Phase II	Metastatic breast cancer	NCT01190982[a]
			Phase I	Advanced cancer	NCT00100139,[a] NCT00080418[a]
	LEM-ETU	Mitoxantrone	Phase I	Acute myeloid leukemia, prostate cancer	Immordino et al. (2006), Chang and Yeh (2012)

Onco-TCS	Vincristine	FDA approved	Non-Hodgkin's lymphoma	Allen and Cullis (2013)
OSI-211	Lurtotecan	Phase II	Ovarian cancer	Seiden et al. (2004), Dark et al. (2005)
		Phase II	Head and neck carcinoma	Duffaud et al. (2004)
		Phase II	Small-cell lung cancer	NCT00046787[a]
		Phase I	Advanced leukemia	Giles et al. (2004)
		Phase I	Solid tumors	Gelmon et al. (2004)
ALN-VSP (Alnylam)	siRNA targeting PCSK9 RNAi targeting liver cancer	Phase I	Liver cancer and liver metastases	Tabernero et al. (2013)
Brakiva (Talon)	Topotecan	Phase I/II	Relapsed solid tumors	Seiden et al. (2004), Allen and Cullis (2013)
Alocrest (Talon)	Vinorelbine	Phase I	Newly diagnosed or relapsed solid tumors	Allen and Cullis (2013)
INGN-401	FUS-1	Phase I	Lung cancer	INGN 201 (2007)
MM-302 (Merrimack)	ErbB2/ErbB3-targeted doxorubicin	Phase II	ErbB2-positive breast cancer	NCT02213744[a]
MM-398 (Merrimack)	CPT-11	Phase III	Gastric and pancreatic cancer	NCT01494506[a]
		Phase II	Glioma and colon cancer	Roy et al. (2013)
MCC-465	Doxorubicin	Phase I	Metastatic stomach cancer	Matsumura et al. (2004a)
SGT-53	p53 gene	Phase I	Solid tumors	Heath and Davis (2008)
LErafAON	c-Raf antisense oligonucleotides (AON), cationic	Phase I	Advanced solid tumors	Rudin et al. (2004)

(Continued)

Table 2.1 Examples of clinically used tumor-targeted nanoparticles (Continued)

Type of nanomedicine	Name	Therapeutic agent	Status	Cancer type	References/clinical trial number[a]
	PLD-E1A	Cationic-E1A pDNA	Phase I	Breast and ovarian cancer	Wang and Hung (2000)
	CALAA-01	Small interfering DNA	Phase I	Solid tumors	Davis et al. (2010)
	MBP-426	Oxaliplatin	Phase I	Metastatic solid tumors	NCT00355888[a]
			Phase I/II	Gastric and esophageal adenocarcinoma	NCT00964080[a]
Polymeric micelles	Genexol-PM	Paclitaxel	Approved (South Korea)	Metastatic breast cancer	Oerlemans et al. (2010)
			Phase II	Urothelial cancer	Lee et al. (2011)
			Phase II	Advanced nonsmall-cell lung cancer	Kim et al. (2007)
			Phase II	Advanced ovarian cancer	NCT00886717[a]
			Phase II	Pancreatic cancer	Saif et al. (2010)
	Oncaspar	PEG-L-asparaginase	Approved	Acute lymphoblastic leukemia	Dinndorf et al. (2007)
	PegAsys/PegIntron	IFNα2a/- IFNα2b	Phase III	Myelogenous leukemia	Michallet et al. (2004)
			Phase I–II	Solid tumors	Bukowski et al. (2002)
	SP1049C	Doxorubicin	Phase II	Adenocarcinoma of esophagus and gastroesophageal junction	Valle et al. (2011)
	NK-012	SN-38	Phase II	Small-cell lung cancer	NCT00951613[a]
			Phase II	Metastatic triple negative breast cancer	NCT00951054[a]
	NK105	Paclitaxel	Phase I	Solid tumors	Hamaguchi et al. (2010)
	NC-6004	Cisplatin	Phase II	Gastric cancer	Kato et al. (2011)
			Phase I/II	Pancreatic cancer	NCT00910741[a]
	NK911	Doxorubicin	Phase I	Solid tumors	Plummer et al. (2011)
			Phase I	Solid tumors	Matsumura et al. (2004b)

Polymer–drug conjugates	Smancs	Neocarzinostatin		Hepatocellular carcinoma	Maeda et al. (2009)
	Xyotax/CT-2103/ paclitaxel poliglumex	Paclitaxel	Approved (Japan) Phase III	Nonsmall-cell lung cancer	Paz-Ares et al. (2008)
	Opaxio		Phase II	Esophageal cancer	Dipetrillo et al. (2011)
			Phase II	Ovarian cancer	Sabbatini et al. (2004), Sabbatini et al. (2008)
			Phase II	Metastatic breast cancer	Lin et al. (2007)
	PK1, FCE28068	Doxorubicin	Phase II	Prostate cancer	Beer et al. (2010)
			Phase II	Breast, lung, and colorectal cancer	Seymour et al. (2009)
	IT-101	Camptothecin	Phase II	Ovarian cancer	NCT00753740[a]
			Phase I/II	Advanced solid tumor	NCT00333502[a]
	NKTR–102	Irinotecan	Phase II	Ovarian cancer	NCT00806156[a]
			Phase II	Breast cancer	NCT00802945[a]
			Phase II	Colorectal cancer	NCT00598975[a]
	SN–38	Irinotecan	Phase I/II	Metastatic colorectal cancer	Suenaga et al. (2015)
	CDP–791	Anti–VEGFR–2 Fab	Phase II	Nonsmall-cell lung cancer	Suenaga et al. (2016)
	EZN–2208	SN–38	Phase II	Metastatic breast cancer	NCT01036113[a]
			Phase II	Colorectal carcinoma (NCT00931840)[a]	NCT00931840[a]
			Phase I	Solid tumors and lymphoma	NCT00520637[a]

(Continued)

Table 2.1 Examples of clinically used tumor-targeted nanoparticles (Continued)

Type of nanomedicine	Name	Therapeutic agent	Status	Cancer type	References/clinical trial number[a]
	Prothecan	Camptothecin	Phase II	Metastatic gastric and gastroesophageal junction adenocarcinoma	Scott et al. (2009)
	CT2106	Camptothecin	Phase I	Solid tumors and lymphomas	Posey et al. (2005)
			Phase II	Colorectal cancer	NCT00291785[a]
			Phase II	Ovarian cancer	NCT00291837[a]
			Phase I	Advanced cancer	NCT00059917, [a] Homsi et al. (2007)
	Hepacid/	Arginine deaminase	Phase II	Hepatocellular carcinoma	NCT0005992,[a] Shen and Shen (2006)
	ADI-PEG20		Phase I	Melanoma	NCT00029900,[a] Shen and Shen (2006)
	PK2, FCE28069	Doxorubicin	Phase I/II	Hepatocellular carcinoma	Seymour et al. (2002)
	PNU166945	Paclitaxel	Phase I	Breast cancer	Meerum Terwogt et al. (2001)
	MAG-CPT	Camptothecin	Phase I	Solid tumor	Wachters et al. (2004), Bissett et al. (2004)
	AP5280	Platinum	Phase I/II	Solid tumor	Rademaker-Lakhai et al. (2004)
	AP5346	Platinum	Phase II	Solid tumor	Campone et al. (2007)
	DOX-OXD/AD-70	Doxorubicin	Phase I	Various cancers	Danhauser-Riedl et al. (1993)
	DE-310	Topoisomerase-I-inhibitor, exatecan mesylate	Phase I	Solid tumors	Soepenberg et al. (2005)
	BIND-014	Docetaxel	Phase I	Solid tumors	NCT01300533[a]

Albumin-based nanomedicine		Approved	Metastatic breast cancer	Gradishar (2006)
Abraxane (ABI-007)	Paclitaxel	Approved	Metastatic breast cancer	Gradishar (2006)
		Phase II	Nonsmall-cell lung cancer	Reynolds et al. (2009)
		Phase II	Melanoma	Kottschade et al. (2011), Hersh et al. (2010)
		Phase II	Ovarian cancer	Coleman et al. (2011), Teneriello et al. (2009)
MTX-HSA	Methotrexate	Phase I	Pancreatic cancer	Stinchcombe et al. (2007)
ABI-008	Docetaxel	Phase I	Bladder cancer	McKiernan et al. (2011)
		Phase II	Kidney carcinoma	Vis et al. (2002)
		Phase I/II	Metastatic breast cancer	NCT00531271[a]
ABI-009	Rapamycin	Phase I/II	Prostate cancer	NCT00477529[a]
ABI-010	Tanespimycin (17–AAG)	Phase I	Solid tumors	NCT00635284[a]
ABI-011	Microtubule and topoisomerase inhibitor	Phase I	Solid tumors	NCT00820768[a]
		Phase I	Advanced solid tumors and lymphomas	NCT01163071[a]

[a]ClinicalTrials.gov identifier number.

complex have been studied (Ali–Boucetta et al., 2008). The new complex has shown enhanced cytotoxicity abilities on MCF-7 breast cancer cells, when compared with DOX on its own (Ali–Boucetta et al., 2008).

In another study (Ali–Boucetta et al., 2008) a carbon nanotube was designed by attaching SWNT to an EGF and Quantum dots (Qdots); SWNT-Qdot-EGF. This complex was aimed toward the EGF receptors (EGFRs) that are highly expressed in head and neck squamous carcinoma cells (HNSCCs) (Fig. 2.9). Using Qdot luminescence and confocal microscopy, it was able to detect high-rate internalization of the complex within the HNSCC; which was not the case in noncancerous cells (i.e., cells not containing EGFR). The results indicated that this nanotube complex is a suitable delivery system for cytotoxic drugs, using an active targeting pathway.

In addition to drug delivery, carbon nanotubes are used for the delivery of nucleic acids and immune-therapeutics. In one study, MWNT was constructed and bound to proapoptotic siRNA. The construct was then injected into human lung carcinoma (Calu 6) xenografted in mice (Podesta et al., 2009). The system was observed to prolong the animals' lives and inhibited the growth of the tumor.

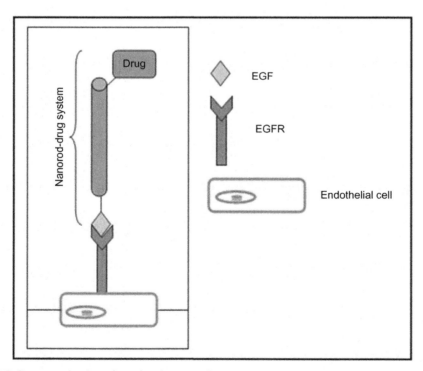

Figure 2.9 Receptor-/antigen-based active targeting.

Another example of receptor-based nanoparticles is gold nanoparticles. Gold nanoparticles pose as a great candidate for loading small drug molecules and large biomolecules, such as DNA, RNA, or proteins. This property makes gold nanoparticles capable of controlled and sustained release of drugs in the system, with a low-toxicity side effect.

In one study to test the efficiency of those nanoparticles, gold nanoparticles were constructed (size of 10 nm) conjugated to folic acid with a PEG spacer to target tumor cells overexpressing folate receptors (Dixit et al., 2006). The particles were tested on two different cell lines, KB cells, which express folate receptors, and originate from a human nasopharyngeal cancer, and W1-38 cells, which are taken from normal human embryonic diploid cells. The W1-38 cells have taken a small amount of the PEG-folic acid-coated gold nanoparticles, when compared to the folate receptor-rich KB cells indicating the efficiency of receptor-based gold nanoparticles in targeting tumor cells.

2.7.3 Carbohydrate-based active targeting

The most famous example of carbohydrate-based active targeting is lectin-carbohydrate. Lectins are proteins that bind specifically to carbohydrates expressed on cells. Lectins facilitate cell-to-cell contact, without the immune system's detection. However, it was found that lectins play a major role in tumor vascularization, tumor cell survival, and adhesion to the endothelium, therefore contributing to tumor apoptosis (Gorelik et al., 2001).

Another frequently used carbohydrate in active targeting is hyaluronic acid (HA). HA is a member of the glycosaminoglycan family, and is found in many parts of the human body, in addition to its involvement in many biological processes, such as cell migration and wound healing (Lapcik et al., 1998). These attributes, which contribute to HA biocompatibility, make HA a good candidate for drug delivery systems. To test its efficacy, HA was bounded to iron oxide-based magnetic nanoparticles, used for active targeting of macrophages (Kamat et al., 2010). Macrophages play an important role in many diseases, such as rheumatoid arthritis and atherosclerosis. Macrophages express CD44 receptors, which are one of the targets of HA (Kamat et al., 2010).

2.8 LIMITATIONS OF THE ENHANCED PERMEABILITY AND RETENTION EFFECT

The discovery of the EPR effect has resulted in a considerable progression in the field of nanomedicine. The EPR effect has solved the main problems that were facing free-drug therapeutic efficacy, which are tumor targeting and systemic toxicity, by permitting the use of macromolecular delivery systems.

However, with the exception of a few clinically used nanomedicines, the transition of nanomedicines from laboratory to effective clinical anticancer drugs has been disappointingly slow. There are several overlooked factors that could account for this slow transition.

2.8.1 Influence of enhanced permeability and retention-based drug design on internalization, release rate, and biocompatibility

To make use of the EPR effect in anticancer therapy it is obligatory to design drug delivery systems with a size more than 7 nm to avoid renal filtration and urinary excretion (Choi et al., 2007). This and other physiochemical characteristics of the nanosystems are important to be considered wherein in many cases the nature of the system itself has been found to limit its transition into clinic.

Below are a few considerations inherent to the design of nanomedicines that may significantly influence the outcome of EPR-based drug targeting.

2.8.1.1 Intercellular internalization

Intercellular internalization was found to be a limiting factor for EPR effect influences wherein sufficient localization of the drug at the tumor site does not always mean efficient internalization into tumor cytoplasm and nucleus. Many factors affect the internalization efficacy. One is the fact that tumor cells exhibit a low endocytosis capacity compared to normal cells. This explains the low drug internalization, bearing in mind that most drug delivery systems are internalized via endocytosis. A second barrier facing the nanosystem's cell uptake is their rapid uptake by the macrophage cells of the RES. RES macrophages have the ability to opsonize nanomedicines through their Toll-like receptor 4 (TLR-4) or through scavenger receptors (Arnida et al., 2011; Nie, 2010). However, hydrophilic nanoparticles are more capable of escaping macrophage opsonization than hydrophobic nanoparticles. Surface modifications such as PEG, folate, or transferrin coating have been applied to minimize nanoparticle uptake by RES and thus permit prolonged drug circulation and higher internalization chance (Klibanov et al., 1990; Papahadjopoulos et al., 1991; Dixit et al., 2006; Bellocq et al., 2003). In addition to the previous barriers, there is a extracellular matrix (EM), which is a highly interconnected network of collagen fibers that can unfavorably interact with the nanoparticle protein coating as well as obstructing convection movements of the nanoconstructs (Haley and Frenkel, 2008). This may prevent the delivery system from reaching leaked capillaries and thus target tumor cells. Nowadays, multiple therapeutic strategies have been applied to improve tumoral cell penetration by nanomedicines (Sugahara et al., 2010; Jain and Stylianopoulos, 2010; Goel et al., 2011).

2.8.2 Release rate

In many drug delivery systems the entrapped drug is attached by covalent, noncovalent, or chemical bonds (Howard et al., 2011; Greish et al., 2003). Therefore, the release of a drug depends on pH, temperature, or enzymatic cleavage (Howard et al., 2011; Torchilin, 2007; Duncan, 2003). Additionally, in many cases it has been established that the stability of these bonds inside the body could influence the drug release rate. For example, an ester bond is very unstable in plasma where the excess of esterases causes rapid breakage of this bond, triggering rapid drug release. Consequently, to ensure an ideal release rate in the site of the tumor, it is important to choose a stable bond in order to avoid drug release in the circulation but at the same time it is essential that this bond breaks rapidly at the site of the tumor to provide adequate therapeutic concentration of the drug. Thus, the choice of a specific linker is critical for a favorable anticancer outcome of EPR-targeted nanosystems.

2.8.3 Biocompatibility and biodegradability

Biocompatibility has been defined as the ability of a material to perform with an appropriate host response in a specific situation (Donaruma, 1987). A high degree of biocompatibility is achieved when the material interacts with the body environment without resulting in unfavorable toxic, immunogenic, thrombogenic, and carcinogenic responses. In nanomedicine, nanocarriers are expected to accomplish a high degree of biocompatibility by ensuring high dose accumulation of a drug at the tumor site and lower accumulation at other body organs. However, it has been found that upon administration the majority of nanocarriers accumulate in the spleen and liver and to a minor extent in the kidneys and lungs (Bae and Park, 2011). Besides lowering the required treatment dose at the targeted site, this accumulation might result in metabolic problems associated with these organs, which eventually can cause organ failure. Therefore, surface modifications of the nanocarriers have been applied to create more stable constructs with low capacity to interact with body tissues and organs. As an example, PEGylated nanocarriers showed increased systemic circulation and better tumor accumulation. Still, in the vast majority of cases, these nanomedicines turned out to be able to reduce systemic toxicity rather than improving the therapeutic efficacy. For instance, DOX-containing liposomes are well known to reduce the cardiotoxicity and the hematologic side effects associated with therapy, but they generally fail to improve therapeutic efficacy. Recent studies are focusing on constructing more advanced and more optimal combination treatments to exploit the biocompatibility and the beneficial biodistribution of tumor-targeted nanomedicines (Bae and Park, 2011).

Another drawback in the use of nanocarriers is their biodegradability. After drug release, nanocarriers should be degraded and eliminated from the body by macrophages. However, the use of nonbiodegradable nanocarriers or the failure of clearing the side products of degradation may induce the formation of foreign body giant cells caused by fusion of multiple macrophages or monocytes and ultimately the formation of lesions resembling granulomas (Anderson et al., 2008; Mukhopadhyay and Gal, 2010).

This may result in the formation of dense fibrous capsules replacing the healthy tissue and/or induce malignancy resulting from prolonged inflammation (O'Neill, 2008). Consequently, to avoid these limitations many biodegradable polymer matrices such as poly-D-L-lactide-*co*-glycolide (PLGA), PLA, and chitosan have been used lately for the preparation of biodegradable and nonimmunological nanocarriers (Kumari et al., 2010; Kim and Lee, 2001; Gan and Wang, 2007).

2.8.4 Translation

Another important pitfall with regard to bringing the nanomedicines into clinic relates to the fact that animal models that are used in preclinical trials are far from being representative of the clinical situation. It has been shown that tumors that take 2–3 weeks to grow into ~1 cm in diameter in rodents may generally take several years to grow to this size in humans. This means that human tumor vasculature is less leaky than that of animal tumors, suggesting an overestimation of the potential usefulness of EPR-based nanomedicines.

Also, tumors growing subcutaneously (SC) are anatomically and physiologically very different from tumors growing in their native environment. The SC implanted tumors in animal models usually take advantage of the extensive cutaneous vascular network to increase their blood supply. Also, animal models developed quickly and thus have marked expression of VEGF and its receptors that can be targeted with ligand-conjugated nanocarriers (Robinson-Smith et al., 2007). This condition is not the same in human tumors where VEGFs are heterogeneously expressed and the vasculature is more complicated.

Moreover, most of the in vivo experiments are done on immunodeficient mice, thereby excluding the positive or negative effect of the immune system on the therapeutic outcome. To overcome these limitations it is mandatory to establish a well-defined panel of animal models in order to test and compare the efficacy of all formulations which are close to being translated to the clinic.

2.8.5 Tumor biology diversity

2.8.5.1 Tumor doubling time

The tumor doubling time (TDT) is the amount of time it takes for a group of cancer cells to double in size. The TDT is different between different tumor types, grades, and stages. TDT is an important factor to consider when designing anticancer nanomedicines. While rapid TDT increases the chance of chemotherapeutics to selectively target tumor cells by targeting processes necessary for mitosis, it may decrease the therapeutic efficacy as the drug amount necessary to kill a given number of cells will double with each tumor doubling (Blankenberg et al., 1995; Malaise et al., 1974). Hence, it is very important to consider TDT variation when designing nanomedicines for different tumors. For example, tumors with low TDT should be targeted by a more stable nanocarrier with a slowly cleaved bond between the backbone and the drug, such as the amide bond. However, fast-releasing nanocarriers, such as micelles, with a rapidly cleaved bond like an ester bond should be used to target tumors with a rapid TDT (Taurin et al., 2012).

2.8.5.2 Tumor extravasation and vascular density

Due to the minimal knowledge about tumor biology and anatomical and physiological properties, there is a tendency to neglect the fact that the EPR effect is a highly heterogeneous phenomenon, which varies from tumor type to tumor type and from patient to patient. Moreover, even within the same tumor there are differences, where some sites show high vascular permeability (~200 nm) but other sites show low permeability (~3 nm) (Jain and Stylianopoulos, 2010). These differences even within the same tumor result from the variation in vascular leakiness that could be compromised by the presence of a dense perivascular lining at some tumor sites, as well as the presence of intact endothelial linings at those sites making them less permeable. The vascular permeability can also be altered by the VEGF and inflammatory mediator expression. High expression promotes increased drug accumulation by the EPR effect and effective penetration of tumor by the nanomedicine.

Moreover, the vascular densities of different human cancers each have distinct structures and functions. Vascular density varies within each tumor type and may also relate to tumor progression. For example, studies showed that renal cell carcinoma is highly vascularized compared to ovarian carcinoma (Ng et al., 2001; Hollingsworth et al., 1995). Also, higher-stage cancer and metastatic cancers have higher vascular density than early and/or nonmetastatic cancers.

Therefore, it is important to consider tumor permeability and vascular density differences between different tumors when constructing drug delivery systems in order to convert the negative impact of these differences into a positive one and to ensure full advantage of the EPR effect is taken.

2.9 FUTURE PROSPECTS

Despite the large investment and success of in vivo studies to image and treat tumors in animal models, the translation of nanomedicines to patient care is still limited due to the several limitations described earlier. Therefore, it is imperative that novel strategies to improve the construction, delivery efficiency, and cytotoxicity of nanomedicines, are designed. Emerging approaches implicate modulating the abnormal tumor microenvironment by vascular normalization, solid stress alleviation, and using tumor-penetrating peptides to enhance tumor penetration and accumulation at the tumor site (Khawar et al., 2015). Moreover, there is a greater emphasis on using bacterial cells to deliver nanoparticles and/or drugs to tumors. Engineering smart nanoparticles that have the ability to exploit different entry pathways into tumors, or being activated specifically at tumor site in response to an external stimulus, are also promising.

However, the key question that should be addressed is whether nanoparticles target diseased tissues or not? This question should be answered individually for each patient. This means that there is need to match the nanodelivery system with specific tumor biological characteristics in a given patient by conducting systemic studies to monitor the basis of interactions not only between nanoparticles and tumors but also between nanoparticles and other organs and tissues that hinder the effectiveness of nanoparticles such as the liver and spleen. New analysis techniques are urgently required to augment what is currently used in testing nanomedicines in different cell, tissue, and animal models. Finally, it is necessary to create a database in order to organize all findings related to the nanoparticles starting from their synthesis and properties and ending with their clinical behavior.

REFERENCES

Alexis, F., et al., 2008. Factors affecting the clearance and biodistribution of polymeric nanoparticles. Mol. Pharm. 5 (4), 505–515.

Ali-Boucetta, H., et al., 2008. Multiwalled carbon nanotube-doxorubicin supramolecular complexes for cancer therapeutics. Chem. Commun. (Camb.) 4, 459–461.

Aljabali, A.A., et al., 2013. CPMV-DOX delivers. Mol. Pharm. 10 (1), 3–10.

Allen, T.M., 1994. The use of glycolipids and hydrophilic polymers in avoiding rapid uptake of liposomes by the mononuclear phagocyte system. Adv. Drug Del. Rev. 13, 285–309.

Allen, T.M., Cullis, P.R., 2013. Liposomal drug delivery systems: from concept to clinical applications. Adv. Drug Deliv. Rev. 65 (1), 36–48.

Anderson, J.M., Rodriguez, A., Chang, D.T., 2008. Foreign body reaction to biomaterials. Semin. Immunol. 20 (2), 86–100.

Ando, K., et al., 2011. Mifamurtide for the treatment of nonmetastatic osteosarcoma. Expert Opin. Pharmacother. 12 (2), 285–292.

Arnida, 2011. Geometry and surface characteristics of gold nanoparticles influence their biodistribution and uptake by macrophages. Eur. J. Pharm. Biopharm. 77 (3), 417–423.

Asati, A., et al., 2010. Surface-charge-dependent cell localization and cytotoxicity of cerium oxide nanoparticles. ACS Nano 4 (9), 5321–5331.

Bae, Y.H., Park, K., 2011. Targeted drug delivery to tumors: myths, reality and possibility. J. Control. Release 153 (3), 198–205.

Bangham, A.D., Standish, M.M., Watkins, J.C., 1965. Diffusion of univalent ions across the lamellae of swollen phospholipids. J. Mol. Biol. 13 (1), 238–252.

Bates, D.O., et al., 2002. Regulation of microvascular permeability by vascular endothelial growth factors. J. Anat. 200 (6), 581–597.

Bayes, M., Rabasseda, X., Prous, J.R., 2004. Gateways to clinical trials. Methods Find Exp. Clin. Pharmacol. 26 (1), 53–84.

Beer, T.M., et al., 2010. A phase II study of paclitaxel poliglumex in combination with transdermal estradiol for the treatment of metastatic castration-resistant prostate cancer after docetaxel chemotherapy. Anticancer Drugs 21 (4), 433–438.

Beh, C.W., et al., 2009. Efficient delivery of Bcl-2-targeted siRNA using cationic polymer nanoparticles: downregulating mRNA expression level and sensitizing cancer cells to anticancer drug. Biomacromolecules 10 (1), 41–48.

Bellocq, N.C., et al., 2003. Transferrin-containing, cyclodextrin polymer-based particles for tumor-targeted gene delivery. Bioconjug. Chem. 14 (6), 1122–1132.

Bertrand, N., et al., 2013. Cancer nanotechnology: the impact of passive and active targeting in the era of modern cancer biology. Adv. Drug Deliv. Rev. 66, 2–25.

Beutel, G., et al., 2005. Phase I study of OSI-7904L, a novel liposomal thymidylate synthase inhibitor in patients with refractory solid tumors. Clin. Cancer Res. 11 (15), 5487–5495.

Bharali, D.J., et al., 2009. Nanoparticles and cancer therapy: a concise review with emphasis on dendrimers. Int. J. Nanomedicine 4, 1–7.

Bissett, D., et al., 2004. Phase I and pharmacokinetic (PK) study of MAG-CPT (PNU 166148): a polymeric derivative of camptothecin (CPT). Br. J. Cancer 91 (1), 50–55.

Blankenberg, F.G., et al., 1995. The influence of volumetric tumor doubling time, DNA ploidy, and histologic grade on the survival of patients with intracranial astrocytomas. Am. J. Neuroradiol. 16 (5), 1001–1012.

Booser, D.J., et al., 2002. Phase II study of liposomal annamycin in the treatment of doxorubicin-resistant breast cancer. Cancer Chemother. Pharmacol. 50 (1), 6–8.

Boyer, C., et al., 2013. Effective delivery of siRNA into cancer cells and tumors using well-defined biodegradable cationic star polymers. Mol. Pharm. 10 (6), 2435–2444.

Brandl, M., 2001. Liposomes as drug carriers: a technological approach. Biotechnol. Annu. Rev. 7, 59–85.

Bukowski, R., et al., 2002. Pegylated interferon alfa-2b treatment for patients with solid tumors: a phase I/II study. J. Clin. Oncol. 20 (18), 3841–3849.

Burnett, J.C., Rossi, J.J., Tiemann, K., 2011. Current progress of siRNA/shRNA therapeutics in clinical trials. Biotechnol. J. 6 (9), 1130–1146.

Byrne, J.D., Betancourt, T., Brannon-Peppas, L., 2008. Active targeting schemes for nanoparticle systems in cancer therapeutics. Adv. Drug Deliv. Rev. 60 (15), 1615–1626.

Caldorera-Moore, M., et al., 2010. Designer nanoparticles: incorporating size, shape and triggered release into nanoscale drug carriers. Expert Opin. Drug Deliv. 7 (4), 479–495.

Campone, M., et al., 2007. Phase I and pharmacokinetic trial of AP5346, a DACH-platinum-polymer conjugate, administered weekly for three out of every 4 weeks to advanced solid tumor patients. Cancer Chemother. Pharmacol. 60 (4), 523–533.

Cassidy, J., 2000. PK1: Results of Phase I studies. Proc 5th Intl Symp on Polymer Therapeutics: From Laboratory to Clinical Practice, Cardiff, UK p. 20.

Chalmin, F., et al., 2010. Membrane-associated Hsp72 from tumor-derived exosomes mediates STAT3-dependent immunosuppressive function of mouse and human myeloid-derived suppressor cells. J. Clin. Invest. 120 (2), 457–471.

Champion, J.A., Katare, Y.K., Mitragotri, S., 2007. Making polymeric micro- and nanoparticles of complex shapes. Proc. Natl. Acad. Sci. U.S.A. 104 (29), 11901–11904.

Chang, H.-I., Yeh, M.-K., 2012. Clinical development of liposome-based drugs: formulation, characterization, and therapeutic efficacy. Int. J. Nanomedicine 7, 49–60.

Chhikara, B.S., Parang, K., 2010. Development of cytarabine prodrugs and delivery systems for leukemia treatment. Expert Opin. Drug Deliv. 7 (12), 1399–1414.

Chithrani, B.D., Chan, W.C., 2007. Elucidating the mechanism of cellular uptake and removal of protein-coated gold nanoparticles of different sizes and shapes. Nano Lett. 7 (6), 1542–1550.

Cho, K., et al., 2008. Therapeutic nanoparticles for drug delivery in cancer. Clin. Cancer Res. 14 (5), 1310–1316.

Choi, H.S., et al., 2007. Renal clearance of quantum dots. Nat. Biotechnol. 25 (10), 1165–1170.

Ciuleanu, T., et al., 2007. A randomised phase II study of OSI-7904L versus 5-fluorouracil (FU)/leucovorin (LV) as first-line treatment in patients with advanced biliary cancers. Invest. New Drugs 25 (4), 385–390.

Coleman, R.L., et al., 2011. A phase II evaluation of nanoparticle, albumin-bound (nab) paclitaxel in the treatment of recurrent or persistent platinum-resistant ovarian, fallopian tube, or primary peritoneal cancer: a Gynecologic Oncology Group study. Gynecol. Oncol. 122 (1), 111–115.

Danhauser-Riedl, S., et al., 1993. Phase I clinical and pharmacokinetic trial of dextran conjugated doxorubicin (AD-70, DOX-OXD). Invest. New Drugs 11 (2–3), 187–195.

Danhier, F., et al., 2009. Paclitaxel-loaded PEGylated PLGA-based nanoparticles: in vitro and in vivo evaluation. J. Control. Release 133 (1), 11–17.

Daniel, M.C., et al., 2010. Role of surface charge density in nanoparticle-templated assembly of bromovirus protein cages. ACS Nano 4 (7), 3853–3860.

Dark, G.G., et al., 2005. Randomized trial of two intravenous schedules of the topoisomerase I inhibitor liposomal lurtotecan in women with relapsed epithelial ovarian cancer: a trial of the national cancer institute of Canada clinical trials group. J. Clin. Oncol. 23 (9), 1859–1866.

Davis, M.E., et al., 2010. Evidence of RNAi in humans from systemically administered siRNA via targeted nanoparticles. Nature 464 (7291), 1067–1070.

Decuzzi, P., et al., 2005. A theoretical model for the margination of particles within blood vessels. Ann. Biomed. Eng. 33 (2), 179–190.

Dejana, E., 2004. Endothelial cell-cell junctions: happy together. Nat. Rev. Mol. Cell Biol. 5 (4), 261–270.

Destito, G., Schneemann, A., Manchester, M., 2009. Biomedical nanotechnology using virus-based nanoparticles. Curr. Top. Microbiol. Immunol. 327, 95–122.

Dinndorf, P.A., et al., 2007. FDA drug approval summary: pegaspargase (oncaspar) for the first-line treatment of children with acute lymphoblastic leukemia (ALL). Oncologist 12 (8), 991–998.

Dipetrillo, T., et al., 2011. Neoadjuvant paclitaxel poliglumex, cisplatin, and radiation for esophageal cancer: a phase 2 trial. Am. J. Clin. Oncol. 35 (1), 64–67.

Dixit, V., et al., 2006. Synthesis and grafting of thioctic acid-PEG-folate conjugates onto Au nanoparticles for selective targeting of folate receptor-positive tumor cells. Bioconjug. Chem. 17 (3), 603–609.

Dobrovolskaia, M.A., et al., 2009. Interaction of colloidal gold nanoparticles with human blood: effects on particle size and analysis of plasma protein binding profiles. Nanomedicine 5 (2), 106–117.

Donaruma, L.G., 1988. Definitions in biomaterials, D. F. Williams, Ed., Elsevier, Amsterdam, 1987, 72 pp. J. Polym. Sci. C Polym. Lett. 26 (9), 414.

Dragovich, T., et al., 2006. A Phase 2 trial of the liposomal DACH platinum L-NDDP in patients with therapy-refractory advanced colorectal cancer. Cancer Chemother. Pharmacol. 58 (6), 759–764.

Duffaud, F., et al., 2004. Phase II study of OSI-211 (liposomal lurtotecan) in patients with metastatic or loco-regional recurrent squamous cell carcinoma of the head and neck. An EORTC New Drug Development Group study. Eur. J. Cancer 40 (18), 2748–2752.

Duncan, R., 2003. The dawning era of polymer therapeutics. Nat. Rev. Drug Discov. 2 (5), 347–360.

Dunk, C., Ahmed, A., 2001. Vascular endothelial growth factor receptor-2-mediated mitogenesis is negatively regulated by vascular endothelial growth factor receptor-1 in tumor epithelial cells. Am. J. Pathol. 158 (1), 265–273.

Dvorak, H.F., et al., 1995. Vascular permeability factor/vascular endothelial growth factor, microvascular hyperpermeability, and angiogenesis. Am. J. Pathol. 146 (5), 1029–1039.

Ehrlich, P., 1913. Address in Pathology, ON CHEMIOTHERAPY: delivered before the Seventeenth International Congress of Medicine. Br. Med. J. 2 (2746), 353–359.

El Andaloussi, S., et al., 2013. Extracellular vesicles: biology and emerging therapeutic opportunities. Nat. Rev. Drug Discov. 12 (5), 347–357.

Fabbro, C., et al., 2012. Targeting carbon nanotubes against cancer. Chem. Commun. (Camb.) 48 (33), 3911–3926.

Fassas, A., Anagnostopoulos, A., 2005. The use of liposomal daunorubicin (DaunoXome) in acute myeloid leukemia. Leuk. Lymphoma 46 (6), 795–802.

FDA approves DaunoXome as first-line therapy for Kaposi's sarcoma. Food and Drug Administration, 1996. J. Int. Assoc. Physicians AIDS Care 2 (5), 50–51.

Ferrara, N., Davis-Smyth, T., 1997. The biology of vascular endothelial growth factor. Endocr. Rev. 18 (1), 4–25.

Folkman, J., 1971. Tumor angiogenesis: therapeutic implications. N. Engl. J. Med. 285 (21), 1182–1186.

Forssen, E., Willis, M., 1998. Ligand-targeted liposomes. Adv. Drug Del. Rev. 29 (3), 249–271.

Gabizon, A., Papahadjopoulos, D., 1992. The role of surface charge and hydrophilic groups on liposome clearance in vivo. Biochim. Biophys. Acta 1103 (1), 94–100.

Gabizon, A., et al., 1982. Liposomes as in vivo carriers of adriamycin: reduced cardiac uptake and preserved antitumor activity in mice. Cancer Res. 42 (11), 4734–4739.

Gan, Q., Wang, T., 2007. Chitosan nanoparticle as protein delivery carrier—systematic examination of fabrication conditions for efficient loading and release. Colloids Surf. B Biointerfaces 59 (1), 24–34.

Gardlik, R., Fruehauf, J.H., 2010. Bacterial vectors and delivery systems in cancer therapy. IDrugs 13 (10), 701–706.

Garnier, A., et al., 1994. Scale-up of the adenovirus expression system for the production of recombinant protein in human 293S cells. Cytotechnology 15 (1–3), 145–155.

Gelmon, K., et al., 2004. A phase 1 study of OSI-211 given as an intravenous infusion days 1, 2, and 3 every three weeks in patients with solid cancers. Invest. New Drugs 22 (3), 263–275.

Giles, F.J., et al., 2004. Phase I and pharmacokinetic study of a low-clearance, unilamellar liposomal formulation of lurtotecan, a topoisomerase 1 inhibitor, in patients with advanced leukemia. Cancer 100 (7), 1449–1458.

Goel, S., et al., 2011. Normalization of the vasculature for treatment of cancer and other diseases. Physiol. Rev. 91 (3), 1071–1121.

Goldberg, M., Langer, R., Jia, X., 2007. Nanostructured materials for applications in drug delivery and tissue engineering. J. Biomater. Sci. Polym. Ed. 18 (3), 241–268.

Goldmann, E., 1908. The growth of malignant disease in man and the lower animals, with special reference to the vascular system. Proc. R. Soc. Med. 1 (Surg Sect), 1–13.

Gordon, A.N., et al., 2001. Recurrent epithelial ovarian carcinoma: a randomized phase III study of pegylated liposomal doxorubicin versus topotecan. J. Clin. Oncol. 19 (14), 3312–3322.

Gorelik, E., Galili, U., Raz, A., 2001. On the role of cell surface carbohydrates and their binding proteins (lectins) in tumor metastasis. Cancer Metastasis Rev. 20 (3–4), 245–277.

Gradishar, W.J., 2006. Albumin-bound paclitaxel: a next-generation taxane. Expert Opin. Pharmacother. 7 (8), 1041–1053.

Green, H., et al., 2011. Pegylated liposomal doxorubicin as first-line monotherapy in elderly women with locally advanced or metastatic breast cancer: novel treatment predictive factors identified. Cancer Lett. 313 (2), 145–153.

Gregoriadis, G., Leathwood, P.D., Ryman, B.E., 1971. Enzyme entrapment in liposomes. FEBS Lett. 14 (2), 95–99.

Greish, K., et al., 2003. Macromolecular therapeutics: advantages and prospects with special emphasis on solid tumour targeting. Clin. Pharmacokinet. 42 (13), 1089–1105.

Gulland, A., 2014. Global cancer prevalence is growing at an "alarming pace," says WHO. BMJ 348, g1338.

Haley, B., Frenkel, E., 2008. Nanoparticles for drug delivery in cancer treatment. Urol. Oncol. 26 (1), 57–64.

Hamaguchi, T., et al., 2010. Phase I study of NK012, a novel SN-38-incorporating micellar nanoparticle, in adult patients with solid tumors. Clin. Cancer Res. 16 (20), 5058–5066.

Han, L., Tang, C., Yin, C., 2013. Effect of binding affinity for siRNA on the in vivo antitumor efficacy of polyplexes. Biomaterials 34 (21), 5317–5327.

Harashima, H., et al., 1994. Enhanced hepatic uptake of liposomes through complement activation depending on the size of liposomes. Pharm. Res. 11 (3), 402–406.

Haviv, Y.S., et al., 2002. Adenoviral gene therapy for renal cancer requires retargeting to alternative cellular receptors. Cancer Res. 62 (15), 4273–4281.

Hayashi, K., et al., 2009. Cancer metastasis directly eradicated by targeted therapy with a modified Salmonella typhimurium. J. Cell. Biochem. 106 (6), 992–998.

Heath, J.R., Davis, M.E., 2008. Nanotechnology and cancer. Annu. Rev. Med. 59, 251–265.

Helgason, T., et al., 2009. Effect of surfactant surface coverage on formation of solid lipid nanoparticles (SLN). J. Colloid. Interface Sci. 334 (1), 75–81.

Hersh, E.M., et al., 2010. A phase 2 clinical trial of nab-paclitaxel in previously treated and chemotherapy-naive patients with metastatic melanoma. Cancer 116 (1), 155–163.

Hewitt, K.J., Agarwal, R., Morin, P.J., 2006. The claudin gene family: expression in normal and neoplastic tissues. BMC Cancer 6, 186.

Hollingsworth, H.C., et al., 1995. Tumor angiogenesis in advanced stage ovarian carcinoma. Am. J. Pathol. 147 (1), 33–41.

Homsi, J., et al., 2007. Phase I trial of poly-L-glutamate camptothecin (CT-2106) administered weekly in patients with advanced solid malignancies. Clin. Cancer Res. 13 (19), 5855–5861.

Horikoshi, S., Serpone, N., 2013. Introduction to nanoparticles Microwaves in Nanoparticle Synthesis. Wiley-VCH Verlag GmbH & Co. KGaA, Weinheim. 1–24

Hornig, B., Kohler, C., Drexler, H., 1997. Role of bradykinin in mediating vascular effects of angiotensin-converting enzyme inhibitors in humans. Circulation 95 (5), 1115–1118.

Howard, M.D., et al., 2011. Polymer micelles with hydrazone-ester dual linkers for tunable release of dexamethasone. Pharm. Res. 28 (10), 2435–2446.

Hu, W.S., Pathak, V.K., 2000. Design of retroviral vectors and helper cells for gene therapy. Pharmacol. Rev. 52 (4), 493–511.

Hung, M.E., Leonard, J.N., 2015. Stabilization of exosome-targeting peptides via engineered glycosylation. J. Biol. Chem. 290 (13), 8166–8172.

Ikeda, E., et al., 1995. Hypoxia-induced transcriptional activation and increased mRNA stability of vascular endothelial growth factor in C6 glioma cells. J. Biol. Chem. 270 (34), 19761–19766.

Ilium, L., et al., 1982. Blood clearance and organ deposition of intravenously administered colloidal particles. The effects of particle size, nature and shape. Int. J. Pharm. 12 (2), 135–146.

Immordino, M.L., Dosio, F., Cattel, L., 2006. Stealth liposomes: review of the basic science, rationale, and clinical applications, existing and potential. Int. J. Nanomedicine 1 (3), 297–315.

INGN 201: Ad-p53, Ad5CMV-p53, adenoviral p53, p53 gene therapy—introgen, RPR/INGN 201, 2007. Drugs R D 8 (3), 176–187.

Jain, R.K., 1987. Transport of molecules across tumor vasculature. Cancer Metastasis Rev. 6 (4), 559–593.

Jain, R.K., Stylianopoulos, T., 2010. Delivering nanomedicine to solid tumors. Nat. Rev. Clin. Oncol. 7 (11), 653–664.

James, N.D., et al., 1994. Liposomal doxorubicin (Doxil): an effective new treatment for Kaposi's sarcoma in AIDS. Clin. Oncol. (R. Coll. Radiol.) 6 (5), 294–296.

Jin, X., et al., 2014. Paclitaxel-loaded N-octyl-O-sulfate chitosan micelles for superior cancer therapeutic efficacy and overcoming drug resistance. Mol. Pharm. 11 (1), 145–157.

Johnsen, K.B., et al., 2014. A comprehensive overview of exosomes as drug delivery vehicles—endogenous nanocarriers for targeted cancer therapy. Biochim. Biophys. Acta 1846 (1), 75–87.

Kamat, M., et al., 2010. Hyaluronic acid immobilized magnetic nanoparticles for active targeting and imaging of macrophages. Bioconjug. Chem. 21 (11), 2128–2135.

Katakowski, M., et al., 2013. Exosomes from marrow stromal cells expressing miR-146b inhibit glioma growth. Cancer Lett. 335 (1), 201–204.

Kato, K., et al., 2011. Phase II study of NK105, a paclitaxel-incorporating micellar nanoparticle, for previously treated advanced or recurrent gastric cancer. Invest. New Drugs 30 (4), 1621–1627.

Kaur, I.P., et al., 2008. Potential of solid lipid nanoparticles in brain targeting. J. Control. Release 127 (2), 97–109.

Kedmi, R., Ben-Arie, N., Peer, D., 2010. The systemic toxicity of positively charged lipid nanoparticles and the role of Toll-like receptor 4 in immune activation. Biomaterials 31 (26), 6867–6875.

Kelton, J.G., Blajchman, M.A., 1980. Prostaglandin I2 (prostacyclin). Can. Med. Assoc. J. 122 (2), 175–179.

Khawar, I.A., Kim, J.H., Kuh, H.J., 2015. Improving drug delivery to solid tumors: priming the tumor microenvironment. J. Control. Release 201, 78–89.

Kikuchi, H., Yamauchi, H., Hirota, S., 1994. A polyol dilution method for mass production of liposomes. J. Liposome Res. 4 (1), 71–91.

Kim, D.W., et al., 2007. Multicenter phase II trial of Genexol-PM, a novel Cremophor-free, polymeric micelle formulation of paclitaxel, with cisplatin in patients with advanced non-small-cell lung cancer. Ann. Oncol. 18 (12), 2009–2014.

Kim, E.S., et al., 2001. A phase II study of STEALTH cisplatin (SPI-77) in patients with advanced non-small cell lung cancer. Lung Cancer 34 (3), 427–432.

Kim, S.Y., Lee, Y.M., 2001. Taxol-loaded block copolymer nanospheres composed of methoxy poly(ethylene glycol) and poly(epsilon-caprolactone) as novel anticancer drug carriers. Biomaterials 22 (13), 1697–1704.

Klibanov, A.L., et al., 1990. Amphipathic polyethyleneglycols effectively prolong the circulation time of liposomes. FEBS Lett. 268 (1), 235–237.

Knudsen, K.B., et al., 2015. In vivo toxicity of cationic micelles and liposomes. Nanomedicine 11 (2), 467–477.

Kottschade, L.A., et al., 2011. A phase II trial of nab-paclitaxel (ABI-007) and carboplatin in patients with unresectable stage IV melanoma: a North Central Cancer Treatment Group Study, N057E(1). Cancer 117 (8), 1704–1710.

Krause, G., et al., 2008. Structure and function of claudins. Biochim. Biophys. Acta 1778 (3), 631–645.

Kudela, P., Koller, V.J., Lubitz, W., 2010. Bacterial ghosts (BGs)—advanced antigen and drug delivery system. Vaccine 28 (36), 5760–5767.

Kumari, A., Yadav, S.K., Yadav, S.C., 2010. Biodegradable polymeric nanoparticles based drug delivery systems. Colloids Surf. B Biointerfaces 75 (1), 1–18.

Lapcik Jr., L., et al., 1998. Hyaluronan: preparation, structure, properties, and applications. Chem. Rev. 98 (8), 2663–2684.

Latagliata, R., et al., 2008. Liposomal daunorubicin versus standard daunorubicin: long term follow-up of the GIMEMA GSI 103 AMLE randomized trial in patients older than 60 years with acute myelogenous leukaemia. Br. J. Haematol. 143 (5), 681–689.

Lazar, I., et al., 2015. Proteome characterization of melanoma exosomes reveals a specific signature for metastatic cell lines. Pigment Cell Melanoma Res. 28 (4), 464–475.

Lee, C.-H., Wu, C.-L., Shiau, A.-L., 2004. Endostatin gene therapy delivered by *Salmonella choleraesuis* in murine tumor models. J. Gene Med. 6 (12), 1382–1393.

Lee, J.L., et al., 2011. Phase II study of a cremophor-free, polymeric micelle formulation of paclitaxel for patients with advanced urothelial cancer previously treated with gemcitabine and platinum. Invest. New Drugs 30 (5), 1984–1990.

Levine, D.H., et al., 2008. Polymersomes: a new multi-functional tool for cancer diagnosis and therapy. Methods 46 (1), 25–32.

Lin, N.U., et al., 2007. Phase II study of CT-2103 as first- or second-line chemotherapy in patients with metastatic breast cancer: unexpected incidence of hypersensitivity reactions. Invest. New Drugs 25 (4), 369–375.

Lohr, J.M., et al., 2012. Cationic liposomal paclitaxel plus gemcitabine or gemcitabine alone in patients with advanced pancreatic cancer: a randomized controlled phase II trial. Ann. Oncol. 23 (5), 1214–1222.

Lundstrom, K., 2009. Alphaviruses in gene therapy. Viruses 1 (1), 13–25.

Maeda, H., Bharate, G.Y., Daruwalla, J., 2009. Polymeric drugs for efficient tumor-targeted drug delivery based on EPR-effect. Eur. J. Pharm. Biopharm. 71 (3), 409–419.

Malaise, E.P., et al., 1974. Relationship between the growth rate of human metastases, survival and pathological type. Eur. J. Cancer 10 (7), 451–459.

Manchester, M., Singh, P., 2006. Virus-based nanoparticles (VNPs): platform technologies for diagnostic imaging. Adv. Drug Deliv. Rev. 58 (14), 1505–1522.

Martin, T.A., Mansel, R.E., Jiang, W.G., 2002. Antagonistic effect of NK4 on HGF/SF induced changes in the transendothelial resistance (TER) and paracellular permeability of human vascular endothelial cells. J. Cell. Physiol. 192 (3), 268–275.

Maruyama, K., 2002. PEG-Immunoliposome. Biosci. Rep. 22 (2), 251–266.

Massing, U., Fuxius, S., 2000. Liposomal formulations of anticancer drugs: selectivity and effectiveness. Drug Resist. Updat. 3 (3), 171–177.

Matsumoto, Y., et al., 2016. Vascular bursts enhance permeability of tumour blood vessels and improve nanoparticle delivery. Nat. Nanotechnol. 11 (6), 533–538.

Matsumura, Y., et al., 2004a. Phase I and pharmacokinetic study of MCC-465, a doxorubicin (DXR) encapsulated in PEG immunoliposome, in patients with metastatic stomach cancer. Ann. Oncol. 15 (3), 517–525.

Matsumura, Y., et al., 2004b. Phase I clinical trial and pharmacokinetic evaluation of NK911, a micelle-encapsulated doxorubicin. Br. J. Cancer 91 (10), 1775–1781.

Matsumura, T., et al., 2015. Exosomal microRNA in serum is a novel biomarker of recurrence in human colorectal cancer. Br. J. Cancer 113 (2), 275–281.

Matsumura, Y., Maeda, H., 1986. A new concept for macromolecular therapeutics in cancer chemotherapy: mechanism of tumoritropic accumulation of proteins and the antitumor agent smancs. Cancer Res. 46 (12 Pt 1), 6387–6392.

McKiernan, J.M., et al., 2011. A phase I trial of intravesical nanoparticle albumin-bound paclitaxel in the treatment of bacillus Calmette-Guerin refractory nonmuscle invasive bladder cancer. J. Urol. 186 (2), 448–451.

McMahon, G., 2000. VEGF receptor signaling in tumor angiogenesis. Oncologist 5 (Suppl. 1), 3–10.

Medina-Kauwe, L.K., 2007. "Alternative" endocytic mechanisms exploited by pathogens: new avenues for therapeutic delivery? Adv. Drug Deliv. Rev. 59 (8), 798–809.

Meerum Terwogt, J.M., et al., 2001. Phase I clinical and pharmacokinetic study of PNU166945, a novel water-soluble polymer-conjugated prodrug of paclitaxel. Anticancer Drugs 12 (4), 315–323.

Michallet, M., et al., 2004. Pegylated recombinant interferon alpha-2b vs recombinant interferon alpha-2b for the initial treatment of chronic-phase chronic myelogenous leukemia: a phase III study. Leukemia 18 (2), 309–315.

Miller, R.A., Presley, A.D., Francis, M.B., 2007. Self-assembling light-harvesting systems from synthetically modified tobacco mosaic virus coat proteins. J. Am. Chem. Soc. 129 (11), 3104–3109.

Moghimi, S.M., Szebeni, J., 2003. Stealth liposomes and long circulating nanoparticles: critical issues in pharmacokinetics, opsonization and protein-binding properties. Prog. Lipid Res. 42 (6), 463–478.

Morgan, M.T., et al., 2003. Dendritic molecular capsules for hydrophobic compounds. J. Am. Chem. Soc. 125 (50), 15485–15489.

Mozafari, M.R., et al., 2002. Construction of stable anionic liposome-plasmid particles using the heating method: a preliminary investigation. Cell. Mol. Biol. Lett. 7 (3), 923–927.

Mueller, A., et al., 2010. In vitro assembly of Tobacco mosaic virus coat protein variants derived from fission yeast expression clones or plants. J. Virol. Methods 166 (1–2), 77–85.

Mukhopadhyay, S., Gal, A.A., 2010. Granulomatous lung disease: an approach to the differential diagnosis. Arch. Pathol. Lab. Med. 134 (5), 667–690.

Murakami, T., Felinski, E.A., Antonetti, D.A., 2009. Occludin phosphorylation and ubiquitination regulate tight junction trafficking and vascular endothelial growth factor-induced permeability. J. Biol. Chem. 284 (31), 21036–21046.

Murphy, C.J., Jana, N.R., 2002. Controlling the aspect ratio of inorganic nanorods and nanowires. Adv. Mater. 14 (1), 80–82.

Naahidi, S., et al., 2013. Biocompatibility of engineered nanoparticles for drug delivery. J. Control. Release 166 (2), 182–194.

New, R.R.C., 1990. Preparation of liposomes. In: New, R.R.C. (Ed.), Liposomes A Practical Approach. Oxford Press Ltd, Oxford, pp. 33–104.

Ng, I.O., et al., 2001. Microvessel density, vascular endothelial growth factor and its receptors Flt-1 and Flk-1/KDR in hepatocellular carcinoma. Am. J. Clin. Pathol. 116 (6), 838–845.

Nie, S., 2010. Understanding and overcoming major barriers in cancer nanomedicine. Nanomedicine (London, England) 5 (4), 523–528.

Niethammer, A.G., et al., 2002. A DNA vaccine against VEGF receptor 2 prevents effective angiogenesis and inhibits tumor growth. Nat. Med. 8 (12), 1369–1375.

Niranjan, B., et al., 1995. HGF/SF: a potent cytokine for mammary growth, morphogenesis and development. Development 121 (9), 2897–2908.

Oerlemans, C., et al., 2010. Polymeric micelles in anticancer therapy: targeting, imaging and triggered release. Pharm. Res. 27 (12), 2569–2589.

O'Neill, L.A.J., 2008. How frustration leads to inflammation. Science 320 (5876), 619–620.

Padera, T.P., et al., 2002. Lymphatic metastasis in the absence of functional intratumor lymphatics. Science 296 (5574), 1883–1886.

Padera, T.P., et al., 2004. Pathology: cancer cells compress intratumour vessels. Nature 427 (6976), 695.

Pandey, R., Sharma, S., Khuller, G.K., 2005. Oral solid lipid nanoparticle-based antitubercular chemotherapy. Tuberculosis 85 (5–6), 415–420.

Panyam, J., Labhasetwar, V., 2003. Biodegradable nanoparticles for drug and gene delivery to cells and tissue. Adv. Drug Deliv. Rev. 55 (3), 329–347.

Papahadjopoulos, D., et al., 1991. Sterically stabilized liposomes: improvements in pharmacokinetics and antitumor therapeutic efficacy. Proc. Natl. Acad. Sci. U.S.A. 88 (24), 11460–11464.

Paukner, S., Kohl, G., Lubitz, W., 2004. Bacterial ghosts as novel advanced drug delivery systems: antiproliferative activity of loaded doxorubicin in human Caco-2 cells. J. Control. Release 94 (1), 63–74.

Paz-Ares, L., et al., 2008. Phase III trial comparing paclitaxel poliglumex vs docetaxel in the second-line treatment of non-small-cell lung cancer. Br. J. Cancer 98 (10), 1608–1613.

Phuyal, S., et al., 2014. Regulation of exosome release by glycosphingolipids and flotillins. FEBS J. 281 (9), 2214–2227.

Plummer, R., et al., 2011. A Phase I clinical study of cisplatin-incorporated polymeric micelles (NC-6004) in patients with solid tumours. Br. J. Cancer 104 (4), 593–598.

Podesta, J.E., et al., 2009. Antitumor activity and prolonged survival by carbon-nanotube-mediated therapeutic siRNA silencing in a human lung xenograft model. Small 5 (10), 1176–1185.

Pokorski, J.K., Steinmetz, N.F., 2011. The art of engineering viral nanoparticles. Mol. Pharm. 8 (1), 29–43.

Posey, J.A., et al., 2005. Phase 1 study of weekly polyethylene glycol-camptothecin in patients with advanced solid tumors and lymphomas. Clin. Cancer Res. 11 (21), 7866–7871.

Prabhu, R.H., Patravale, V.B., Joshi, M.D., 2015. Polymeric nanoparticles for targeted treatment in oncology: current insights. Int. J. Nanomedicine 10, 1001–1018.

Puri, A., et al., 2009. Lipid-based nanoparticles as pharmaceutical drug carriers: from concepts to clinic. Crit. Rev. Ther. Drug Carrier Syst. 26 (6), 523–580.

Rademaker-Lakhai, J.M., et al., 2004. A Phase I and pharmacological study of the platinum polymer AP5280 given as an intravenous infusion once every 3 weeks in patients with solid tumors. Clin. Cancer Res. 10 (10), 3386–3395.

Rae, C.S., et al., 2005. Systemic trafficking of plant virus nanoparticles in mice via the oral route. Virology 343 (2), 224–235.

Rana, S., et al., 2012. Toward tailored exosomes: the exosomal tetraspanin web contributes to target cell selection. Int. J. Biochem. Cell Biol. 44 (9), 1574–1584.

Ren, Y., Wong, S.M., Lim, L.Y., 2007. Folic acid-conjugated protein cages of a plant virus: a novel delivery platform for doxorubicin. Bioconjug. Chem. 18 (3), 836–843.

Reynolds, C., et al., 2009. Phase II trial of nanoparticle albumin-bound paclitaxel, carboplatin, and bevacizumab in first-line patients with advanced nonsquamous non-small cell lung cancer. J. Thorac. Oncol. 4 (12), 1537–1543.

Rihova, B., Kovar, M., 2010. Immunogenicity and immunomodulatory properties of HPMA-based polymers. Adv. Drug Deliv. Rev. 62 (2), 184–191.

Robinson-Smith, T.M., et al., 2007. Macrophages mediate inflammation-enhanced metastasis of ovarian tumors in mice. Cancer Res. 67 (12), 5708–5716.

Roy, A.C., et al., 2013. A randomized phase II study of PEP02 (MM-398), irinotecan or docetaxel as a second-line therapy in patients with locally advanced or metastatic gastric or gastro-oesophageal junction adenocarcinoma. Ann Oncol 24 (6), 1567–1573.

Rudin, C.M., et al., 2004. Delivery of a liposomal c-raf-1 antisense oligonucleotide by weekly bolus dosing in patients with advanced solid tumors: a phase I study. Clin. Cancer Res. 10 (21), 7244–7251.

Ruth, D., 2005. N-(2-Hydroxypropyl) methacrylamide Copolymer Conjugates Polymeric Drug Delivery Systems. Informa Healthcare, New York, NY. 1–92.

Saari, H., et al., 2015. Microvesicle- and exosome-mediated drug delivery enhances the cytotoxicity of Paclitaxel in autologous prostate cancer cells. J. Control. Release 220 (Part B), 727–737.

Sabbatini, P., et al., 2004. Phase II study of CT-2103 in patients with recurrent epithelial ovarian, fallopian tube, or primary peritoneal carcinoma. J. Clin. Oncol. 22 (22), 4523–4531.

Sabbatini, P., et al., 2008. A phase II trial of paclitaxel poliglumex in recurrent or persistent ovarian or primary peritoneal cancer (EOC): a Gynecologic Oncology Group Study. Gynecol. Oncol. 111 (3), 455–460.

Sackmann, E., 1994. Membrane bending energy concept of vesicle- and cell-shapes and shape-transitions. FEBS Lett. 346 (1), 3–16.

Saif, M.W., et al., 2010. Phase II clinical trial of paclitaxel loaded polymeric micelle in patients with advanced pancreatic cancer. Cancer Invest. 28 (2), 186–194.

Schneemann, A., Young, M.J., 2003. Viral assembly using heterologous expression systems and cell extracts. Adv. Protein Chem. 64, 1–36.

Scott, L.C., et al., 2009. A phase II study of pegylated-camptothecin (pegamotecan) in the treatment of locally advanced and metastatic gastric and gastro-oesophageal junction adenocarcinoma. Cancer Chemother. Pharmacol. 63 (2), 363–370.

Seetharamu, N., et al., 2010. Phase II study of liposomal cisplatin (SPI-77) in platinum-sensitive recurrences of ovarian cancer. Anticancer Res. 30 (2), 541–545.

Seiden, M.V., et al., 2004. A phase II study of liposomal lurtotecan (OSI-211) in patients with topotecan resistant ovarian cancer. Gynecol. Oncol. 93 (1), 229–232.

Seymour, L.W., et al., 2002. Hepatic drug targeting: phase I evaluation of polymer-bound doxorubicin. J. Clin. Oncol. 20 (6), 1668–1676.

Seymour, L.W., et al., 2009. Phase II studies of polymer-doxorubicin (PK1, FCE28068) in the treatment of breast, lung and colorectal cancer. Int. J. Oncol. 34 (6), 1629–1636.

Shen, L.J., Shen, W.C., 2006. Drug evaluation: ADI-PEG-20—a PEGylated arginine deiminase for arginine-auxotrophic cancers. Curr. Opin. Mol. Ther. 8 (3), 240–248.

Shweiki, D., et al., 1992. Vascular endothelial growth factor induced by hypoxia may mediate hypoxia-initiated angiogenesis. Nature 359 (6398), 843–845.

Sikkema, F.D., et al., 2007. Monodisperse polymer-virus hybrid nanoparticles. Org. Biomol. Chem. 5 (1), 54–57.

Singh, P., et al., 2006. Canine parvovirus-like particles, a novel nanomaterial for tumor targeting. J. Nanobiotechnol. 4 (1), 1–11.

Sinha, R., et al., 2006. Nanotechnology in cancer therapeutics: bioconjugated nanoparticles for drug delivery. Mol. Cancer Ther. 5 (8), 1909–1917.

Skog, J., et al., 2008. Glioblastoma microvesicles transport RNA and proteins that promote tumour growth and provide diagnostic biomarkers. Nat. Cell Biol. 10 (12), 1470–1476.

Soepenberg, O., et al., 2005. Phase I and pharmacokinetic study of DE-310 in patients with advanced solid tumors. Clin. Cancer Res. 11 (2 Pt 1), 703–711.

Soppimath, K.S., et al., 2001. Biodegradable polymeric nanoparticles as drug delivery devices. J. Control. Release 70 (1–2), 1–20.

Steinmetz, N.F., 2013. Viral nanoparticles in drug delivery and imaging. Mol. Pharm. 10 (1), 1–2.

Stinchcombe, T.E., et al., 2007. Phase I and pharmacokinetic trial of carboplatin and albumin-bound paclitaxel, ABI-007 (Abraxane) on three treatment schedules in patients with solid tumors. Cancer Chemother. Pharmacol. 60 (5), 759–766.

Stolnik, S., Illum, L., Davis, S.S., 1995. Long circulating microparticulate drug carriers. Adv. Drug Del. Rev. 16, 195–214.

Stopeck, A.T., et al., 2001. Phase II study of direct intralesional gene transfer of allovectin-7, an HLA-B7/beta2-microglobulin DNA-liposome complex, in patients with metastatic melanoma. Clin. Cancer Res. 7 (8), 2285–2291.

Stuchlík, M., Žák, S., 2001. Lipid-based vehicle for oral drug delivery. Biomed. Pap. 145 (2), 17–26.

Suenaga, M., et al., 2015. Phase II study of reintroduction of oxaliplatin for advanced colorectal cancer in patients previously treated with oxaliplatin and irinotecan: RE-OPEN study. Drug Design Dev. Ther. 9, 3099–3108.

Suenaga, M., et al., 2016. Serum VEGF-A and CCL5 levels as candidate biomarkers for efficacy and toxicity of regorafenib in patients with metastatic colorectal cancer. Oncotarget 7 (23), 34811–34823.

Sugahara, K.N., et al., 2010. Coadministration of a tumor-penetrating peptide enhances the efficacy of cancer drugs. Science 328 (5981), 1031–1035.

Swenson, C.E., et al., 2003. Pharmacokinetics of doxorubicin administered i.v. as Myocet (TLC D-99; liposome-encapsulated doxorubicin citrate) compared with conventional doxorubicin when given in combination with cyclophosphamide in patients with metastatic breast cancer. Anticancer Drugs 14 (3), 239–246.

Szajnik, M., et al., 2010. Tumor-derived microvesicles induce, expand and up-regulate biological activities of human regulatory T cells (Treg). PLoS One 5 (7), e11469.

Szostak, M.P., et al., 1996. Bacterial ghosts: non-living candidate vaccines. J. Biotechnol. 44 (1–3), 161–170.

Tabernero, J., et al., 2013. First-in-humans trial of an RNA interference therapeutic targeting VEGF and KSP in cancer patients with liver involvement. Cancer Discov. 3 (4), 406–417.

Talsma, H., et al., 1994. A novel technique for the one-step preparation of liposomes and nonionic surfactant vesicles without the use of organic solvents. Liposome formation in a continuous gas stream: the 'bubble'™ method. J. Pharm. Sci. 83 (3), 276–280.

Taurin, S., Greish, K., 2013. Enhanced vascular permeability in solid tumors: a promise for anticancer nanomedicine. In: Martin, A.T., Jiang, G.W. (Eds.), Tight Junctions in Cancer Metastasis. Springer Netherlands, Dordrecht, pp. 81–118.

Taurin, S., Nehoff, H., Greish, K., 2012. Anticancer nanomedicine and tumor vascular permeability; where is the missing link? J. Control. Release 164 (3), 265–275.

Teneriello, M.G., et al., 2009. Phase II evaluation of nanoparticle albumin-bound paclitaxel in platinum-sensitive patients with recurrent ovarian, peritoneal, or fallopian tube cancer. J. Clin. Oncol. 27 (9), 1426–1431.

Torchilin, V.P., 2007. Micellar nanocarriers: pharmaceutical perspectives. Pharm. Res. 24 (1), 1–16.

Tyle, P., 1988. Controlled drug delivery: fundamentals and applications. Edited by Joseph R. Robinson and Vincent H. L. Lee. Marcel Dekker, Inc., New York. 1987. 739 pp. 16 × 23.5 cm. ISBN 0-8247-7588-0. $125.00. J. Pharm. Sci. 77 (1), 94.

Ulrich, A.S., 2002. Biophysical aspects of using liposomes as delivery vehicles. Biosci. Rep. 22 (2), 129–150.

Valle, J.W., et al., 2011. A phase 2 study of SP1049C, doxorubicin in P-glycoprotein-targeting pluronics, in patients with advanced adenocarcinoma of the esophagus and gastroesophageal junction. Invest. New Drugs 29 (5), 1029–1037.

Vemuri, S., Rhodes, C.T., 1995. Preparation and characterization of liposomes as therapeutic delivery systems: a review. Pharm. Acta Helv. 70 (2), 95–111.

Virchow, R., Chance, F., 1863. Cellular Pathology as Based Upon Physiological and Pathological Histology. J.B. Lippincott, Philadelphia, PA.

Vis, A.N., et al., 2002. A phase II trial of methotrexate-human serum albumin (MTX-HSA) in patients with metastatic renal cell carcinoma who progressed under immunotherapy. Cancer Chemother. Pharmacol. 49 (4), 342–345.

Wachtel, M., et al., 2001. Down-regulation of occludin expression in astrocytes by tumour necrosis factor (TNF) is mediated via TNF type-1 receptor and nuclear factor-kappaB activation. J. Neurochem. 78 (1), 155–162.

Wachters, F.M., et al., 2004. A phase I study with MAG-camptothecin intravenously administered weekly for 3 weeks in a 4-week cycle in adult patients with solid tumours. Br. J. Cancer 90 (12), 2261–2267.

Wang, Q., et al., 2002. Natural supramolecular building blocks. Cysteine-added mutants of cowpea mosaic virus. Chem. Biol. 9 (7), 813–819.

Wang, S.C., Hung, M.C., 2000. Transcriptional targeting of the HER-2/neu oncogene. Drugs Today (Barc.) 36 (12), 835–843.

Wang, X., et al., 2008. Application of nanotechnology in cancer therapy and imaging. CA Cancer J. Clin. 58 (2), 97–110.

Warfield, K.L., et al., 2004. Marburg virus-like particles protect guinea pigs from lethal Marburg virus infection. Vaccine 22 (25–26), 3495–3502.

Wei, M.H., et al., 1996. Localization of the human vascular endothelial growth factor gene, VEGF, at chromosome 6p12. Hum. Genet. 97 (6), 794–797.

Wetzler, M., et al., 2013. Phase I/II trial of nanomolecular liposomal annamycin in adult patients with relapsed/refractory acute lymphoblastic leukemia. Clin. Lymphoma Myeloma Leuk. 13 (4), 430–434.

White, S.C., et al., 2006. Phase II study of SPI-77 (sterically stabilised liposomal cisplatin) in advanced non-small-cell lung cancer. Br. J. Cancer 95 (7), 822–828.

Wong, H.L., et al., 2006. A new polymer-lipid hybrid nanoparticle system increases cytotoxicity of doxorubicin against multidrug-resistant human breast cancer cells. Pharm. Res. 23 (7), 1574–1585.

Wu, J., Akaike, T., Maeda, H., 1998. Modulation of enhanced vascular permeability in tumors by a bradykinin antagonist, a cyclooxygenase inhibitor, and a nitric oxide scavenger. Cancer Res. 58 (1), 159–165.

Wu, Z., et al., 2012. Development of viral nanoparticles for efficient intracellular delivery. Nanoscale 4 (11), 3567–3576.

Wysoczynski, M., Ratajczak, M.Z., 2009. Lung cancer secreted microvesicles: underappreciated modulators of microenvironment in expanding tumors. Int. J. Cancer 125 (7), 1595–1603.

Xiang, R., et al., 2008. Oral DNA vaccines target the tumor vasculature and microenvironment and suppress tumor growth and metastasis. Immunol. Rev. 222, 117–128.

Yarmolenko, P.S., et al., 2010. Comparative effects of thermosensitive doxorubicin-containing liposomes and hyperthermia in human and murine tumours. Int. J. Hyperthermia 26 (5), 485–498.

Yin, W., et al., 2013. Immature dendritic cell-derived exosomes: a promise subcellular vaccine for autoimmunity. Inflammation 36 (1), 232–240.

Yu, K.F., et al., 2013. The antitumor activity of a doxorubicin loaded, iRGD-modified sterically-stabilized liposome on B16-F10 melanoma cells: in vitro and in vivo evaluation. Int. J. Nanomedicine 8, 2473–2485.

Zarovni, N., et al., 2015. Integrated isolation and quantitative analysis of exosome shuttled proteins and nucleic acids using immunocapture approaches. Methods 87, 46–58.

Zhou, J., et al., 2015. Urinary microRNA-30a-5p is a potential biomarker for ovarian serous adenocarcinoma. Oncol. Rep. 33 (6), 2915–2923.

CHAPTER 3

Systematic Development of Nanocarriers Employing Quality by Design Paradigms

Bhupinder Singh, Sumant Saini, Shikha Lohan and Sarwar Beg

Contents

Nanotechnology-Based Approaches for Targeting and Delivery of Drugs and Genes.
DOI: http://dx.doi.org/10.1016/B978-0-12-809717-5.00003-8

Disclosures: There is no conflict of interest and disclosures associated with the manuscript.

3.1 INTRODUCTION

In the therapeutic archives, the approach to treat a disease has remained restricted to exploration of the *right* medicine for a particular ailment. Today, modern therapeutics aims at developing the *right* therapeutic systems capable of delivering the *right* therapeutic agent at the *right* time in minimal dose at the *right* target site. The present strategies aim at effective and targeted drug delivery (Svenson, 2004). After remaining shrouded for several years, nanomedicine is fast evolving to yield enormous benefits to patients as well as society. Verily, the nanostructured drug delivery systems (DDSs) have proved their incredible potential for revolutionizing the efficacy of therapeutics and diagnostics (i.e., theranostics) by several orders of magnitude. As their clinical applications require wide acceptability among patients and approval from regulatory agencies, scientists need to develop such DDS that are markedly, but not just marginally, better than the existing ones. This calls for harmonized efforts among scientists and adoption of systematic approaches for developing such DDS. Particularly, this is a fastidious exercise for advanced supramolecular nanostructured systems, wherein the interplay of molecules amid the milieu of intricate cellular systems plays a critical role.

Development of such impeccable DDSs, however, involves rational blending of diverse functional excipients and precise control of process(es). Such delivery systems not only require sound knowledge and understanding of the critical attributes of the drug substance, diverse functional and nonfunctional excipients, but also on their influence on the process(es). Further, these specialized delivery systems more often call for precise process control.

Optimizing the formulation and the manufacturing process to achieve the drug product with desired attributes is a cumbersome process. Traditionally, optimization involved optimizing one variable at a time (OVAT). The solution was somehow achieved, but the attainment of the true optimum formulation was never guaranteed using this approach. Such inadequacies and limitations were the result of factor–factor interaction. The final product may be satisfactory but not precisely the true optimum formulation of interest (Singh et al., 2005a). Further, the OVAT methodology results only in "just satisfactory" solutions, as detailed study of all variables is prohibitive. As one cannot establish "cause–effect relationships" using OVAT, the technique becomes futile when all variables are changed simultaneously.

Several drug product inconsistencies tend to prevail in an OVAT approach as the result of inadequate knowledge of the causal factor–response relationship. Of late, systematic optimization employing the multivariate techniques has evolved, which is being widely practiced to alleviate such inconsistencies. The rational approach involves the application of appropriate experimental designs coupled with the generation of mathematical models which can also be expressed in a graphical format. A graphical format gives the pictorial representation of the mathematical model.

A Design of Experiments (DoE) approach has been far more successful, as it overcomes all the shortcomings associated with one factor at a time (OFAT). DoE aids in achieving optimal formulation with minimal experimentation. The screening techniques are employed to separate out important input variables from less significant variables. The product or the process behavior can be simulated by using a model mathematical equation(s), which can significantly save energy, time, and resources.

Lately, a holistic DoE-based philosophy of Quality by Design (QbD) has been permeating into the mindset and practice of the researchers, especially in industrial environs. This popularity of QbD in pharma circles is largely attributable to the recent impetus provided by the International Council for Harmonisation (ICH), US Food Drug and Administration (USFDA), European Medicines Agency (EMA), Medicines & Healthcare products Regulatory Agency (MHRA), Therapeutics Good Administration (TGA), World Health Organization (WHO), and several other key regulatory agencies through their own respective federal guidelines, or endorsement of ICH/USFDA Q8–Q11 and PAT guidance. With the growing pressure from these federal statutes, the pharmaceutical industry has been thoroughly coerced to prepare themselves by reorienting its strategies. Regulatory agencies, today, require neither "Quality by Chance" nor "Quality by Testing and Inspection," rather only require "Quality by Design" (QbD). As DoE has much a wider domain of application, beyond even the pharma sector, therefore, on the heels of QbD paradigm, a terser jargon, namely "Formulation by Design" (FbD), has recently been proposed, applicable to the integrated use of DoE and other QbD principles, specifically in the development of pharmaceutical dosage forms (Singh et al., 2011a).

One of the merits associated with the use of FbD is that it can detect the factor–factor interactions with a high degree of precision. It gives a deeper insight into the product, making it easy to trace and rectify any "problem" in a systematic manner. The use of DoE ensures a robust, economical, and federal compliant product. Table 3.1 highlights the various merits of FbD over OFAT methodology.

3.1.1 FbD terminology

Specific terminology, both technical and otherwise, is employed usually during FbD practice. To facilitate better understanding of the precepts of FbD of DDS, important terms have been compiled in Table 3.2.

As a prelude to the application of FbD, it is essential to have awareness of the FbD terminology and prior multidisciplinary knowledge on various possible product and process variables ahead. A "knowledge space," i.e., entire worth-exploring realm, therefore, has to be identified from the possible vast ocean of scientific information based upon prior knowledge. A "knowledge space," thereby, encompasses information on all those product and process variables that may even minutely affect the overall product quality. A "design space" has to be demarcated as a subset construct of "knowledge space," ensuring optimal product quality or process performance involving "selected

Table 3.1 Comparison of OFAT and FbD methodology (Singh et al., 2011a,b)

Attribute	OFAT	FbD
Choice of optimum formulation	Usually results only in suboptimal solutions	Yields the best possible formulation
Resource-economics	Highly resource-intensive, as it leads to unnecessary runs and batches	Economical, as it furnishes information on product/ process performance using minimal trials
Time-economics	Highly time-consuming, as each product is individually evaluated for its performance	Can simulate the product or process behavior using model equations
Interaction among the variables	Inept to reveal possible interactions	Estimates any synergistic or antagonistic interaction among constituents
Scale-up and postapproval changes	Very difficult to design formulation slightly differing from the desired formulation, especially beyond Level II	Changes in the optimized formulation can easily be incorporated, as all response variables are quantitatively governed by a set of input variables

OFAT, one factor at a time; *FbD*, Formulation by Design.

Table 3.2 Essential terminology employed during Formulation by Design (FbD) optimization of drug delivery systems

Term	Definition
Optimize	Make as perfect, effective, or functional as possible
Optimization	Implementation of systematic approaches to achieve "*the best*" combination of product and/or process characteristics under a given set of conditions using FbD and computers
Independent variables	Input variables, which are directly under the control of the product development scientist
Categorical variables	Qualitative variables which cannot be quantified
Runs or trials	Experiments conducted according to the selected experimental design
Factors	Independent variables, which tend to influence the product/process characteristics or output of the process
Design matrix	Layout of experimental runs in matrix form, as per experimental design
Explorable space	Dimensional space defined by coded variables for the factors being investigated
Knowledge space	Scientific elements to be considered and explored on the basis of previous knowledge as product attributes and process parameters
Design space	Multidimensional combination and interaction of input variables and process parameters, demonstrated to provide quality assurance

(*Continued*)

Table 3.2 Essential terminology employed during Formulation by Design (FbD) optimization of drug delivery systems (Continued)

Term	Definition
Control space	Domain of design space selected for detailed controlled strategy
Levels	Values assigned to a factor
Constraints	Restrictions imposed on the factor levels
Response variables	Characteristics of the finished drug product or the in-process material
Quality target product profile	Prospective and dynamic summary of the quality characteristics of a drug product that would ideally be achieved to ensure its quality, safety, and efficacy
Critical quality attributes	Parameters ranging within appropriate limits, which ensure the desired product quality
Critical process parameters	Independent process parameters most likely to affect the quality attributes of a product or intermediates
Critical formulation attributes	Formulation parameters affecting critical quality attributes
Effect	Magnitude of the change in response caused by varying the factor level(s)
Main effect	The effect of a factor averaged over all the levels of other factors
Interaction	Lack of additivity of factor effects during their simultaneous validation
Antagonism	Overall, negative change due to interaction among factors
Synergism	Overall, positive change due to interaction between factors
Nuisance factors	Uncontrollable factors which complicate the estimation of main effect or interactions
Orthogonality	The estimated effects are due to the main factor of interest and independent of interactions
Confounding	Lack of orthogonality
Resolution	Measure of the degree of confounding
Coding (or normalization)	Process of transforming a natural variable into a nondimensional coded variable
Factor space	Dimensional space defined by the coded variables
Experimental domain	Part of the factor space, investigated experimentally for optimization
Blocks	Sets of relatively homogeneous experimental conditions, wherein every level of the primary factor occurs the same number of times with each level of nuisance factor
Response surface	Graphical depiction of the mathematical relationship
Empirical model	Mathematical model describing factor–response relation using polynomial equations
Response surface plot	3D graphical representation of a response plotted between two independent variables and one response variable
Contour plot	Geometric illustration of a response obtained by plotting one independent variable against another, while holding the magnitude of response and other variables as constant

(Continued)

Table 3.2 Essential terminology employed during Formulation by Design (FbD) optimization of drug delivery systems (Continued)

Term	Definition
Failure mode effect analysis (FMEA)	Systematic method to enhance safety and customer satisfaction by identifying failure modes or potential risks based upon severity, likelihood, and/or detectability of the plausible failures
Risk assessment	Process to identify and mitigate risks, find root causes of process failure, prevent problems to improve quality and reliability of product
Quality risk management	Systematic process for identification, assessment, and control of risk to the product quality across its lifecycle
Control strategy	Comprehensive plan to ensure that the final product meets critical requirements
Continuous improvement	Monitoring of process capability to reproduce the product quality

few" influential variables. "Control space" is further deduced from this "design space" as the experimental domain earmarked for the detailed in-house studies, especially in an industrial set-up within the refined ranges of input variables. "Design space" applies a systematic approach on archival data to convert the "knowledge space" into "control space" (Singh et al., 2011a). Extensive experimentation may be necessary for relatively intricate DDSs in order to reduce uncertainty and justify a design space than that required for conventional formulation systems like tablets. As working within the design space is not considered as a "change," it would not initiate any postapproval changes as per the federal guidelines by ICH and USFDA. The enhanced understanding on the formula components and process(es) enables scientists to strategize rationally to control the entire formulation system. The entire plan, largely, is at times referred to as the "control strategy."

3.1.2 Formulation by Design methodology

FbD hits the bull's eye using five vital strengths, namely, meticulous drug product development, apt choice of experimental designs, identification of critical quality attributes (CQAs), critical formulation attributes (CFAs) and critical process parameters (CPPs), precise definition of design and control space, and accurate computer-aided optimization. Fig. 3.1 illustrates the typical five-step FbD optimization methodology employed for the development of drug products with complete understanding of the product(s) and process(es) (Singh et al., 2005a).

- The FbD study begins with *Step I*, where an endeavor is made to explicitly ascertain the drug delivery objective(s). A quality target product profile (QTPP) is defined encompassing the basic attributes of the product to be prepared or aspired

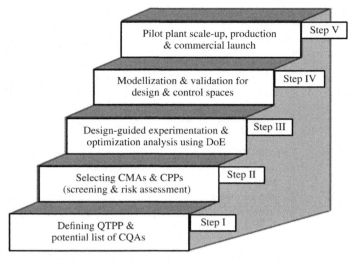

Figure 3.1 Five-step Formulation by Design methodology for developing optimized drug delivery systems.

as a "goal-setting" exercise and its desired drug delivery characteristics. Various CQAs or response variables, which pragmatically epitomize the objective(s), are earmarked for the purpose. All the independent product/process variables are also listed likewise.

- In *Step II*, the critical material attributes (CMAs) or CFAs and CPPs which directly influence the response variables (CQAs) and represent the product quality (e.g., size, emulsification time) are prioritized through an Ishikawa–Fish bone diagram and prioritization studies employing principles of quality risk management (QRM) approach. For that, the factors with moderate to high risk are chosen from patient perspectives through brainstorming among the team members using techniques like comparison matrix, risk estimation matrix (REM), fault tree analysis, hazard operability analysis and, above all, failure mode effects analysis (FMEA). These techniques help in identifying and sorting the potential risk associated with each CMA or CPP, as applicable to the identified CQAs. Selection of the critical factors out of the many possible factors is termed *factor screening*. Based on the experimental results, factor levels are demarcated. Apt use of screening designs, in this regard, helps to manage and reduce the number of potential CMAs and/or CPPs and identification of CMAs and/or CPPs actually affecting the CQAs.
- In *Step III*, a suitable experimental design is worked out to study the responses based on study objective(s), responses (CQAs) being explored, number and the type of factors, and factor levels, namely, high, medium, or low. A *design matrix*,

i.e., matrix type arrangement of experimental runs as per experimental design, serves as a guide to the pharmaceutical scientist. The drug delivery formulations are experimentally prepared according to the design matrix, and the chosen response variables are evaluated meticulously.

- In *Step IV*, numeric and/or graphical optimization is carried out and statistical significance is determined. The relation between the various factors and response variables within the desired conditions and constraints is studied using RSM polynomials. Design space is represented graphically as a 2D-contour or 3D-response surface plot. Optimum formulation compositions are searched within this design space. Usually in industrial milieu, a narrower domain of an apt control space is further constructed from the design space for further implicit and explicit studies.

- *Step V* is the ultimate stage in the FbD treatment, which is carried out in an industrial set-up, involving validation of response predictive ability of the proposed design model. The *optimum formulation* is scaled-up through intermediate pilot-plant and commercial levels, set forth ultimately for the final production scale. For these complex DDS, a set of experimentation at the pilot-scale level, preferably using DoE, confirms not only the scale-up principles, but also verifies extrapolation of the formulation understanding developed at small scale to higher scales and establishes process robustness in the proposed ranges. Finally, demonstration of the successful implementation of the proposed parameters at the commercial scale wraps up the experimentation strategy prior to control strategy proposal. Not only does the final product become available in the "optimized" form befitting product excellence and federal compliance, but the whole exercise leads to comprehensive product and process understanding too. A "control strategy" is implemented, to achieve the goal of "continuous improvement" of drug delivery even postapproval. Furthermore, the principle of Scale-up Postapproval Changes–QbD approach is based on the product risk assessment which can help to improve the quality and robustness of the final drug product.

3.2 EXPERIMENTAL DESIGNS EMPLOYED DURING FORMULATION BY DESIGN OF DRUG DELIVERY SYSTEM

An experimental design constitutes the gist of the entire FbD exercise. Systematic FbD optimization of a DDS includes a careful "screening" of influential variables and subsequent RSM analysis using experimental designs. Fig. 3.2 provides a bird's eye view of key experimental designs employed for systematic optimization of various carrier-based DDS. Out of all the experimental designs, the factorial, composite, and mixture designs have been employed most extensively and frequently to optimize various DDS. Table 3.3 provides a comparative account of key experimental designs employed for optimization of DDS, listing their advantages and disadvantages.

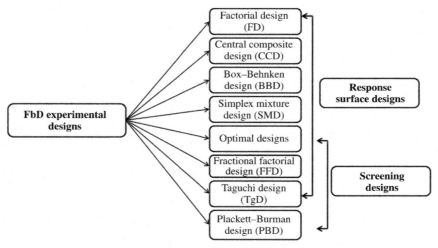

Figure 3.2 Experimental designs usually embarked upon during Formulation by Design optimization of drug delivery systems.

3.2.1 Selection of experimental designs

The low-resolution designs like full or fractional factorial designs (FD or FFD), Plackett–Burman design (PBD), or Taguchi design (TgD) suffice for the purpose of simpler screening of a large number of experimental parameters. Screening designs support only the linear responses, hence considered as low-resolution designs (Singh et al., 2005a). If a nonlinear response is detected, high-resolution experimental design like response surface is required. Such exercise is crucial when the investigator is interested in estimating interactions and/or quadratic effects, or intends to have an idea of the local shape of the response surface, the response surface designs, capable of mapping nonlinear responses, are used. The compilation in Table 3.4 acts as a guide to help while selecting an experimental design, based upon the desired motive of the study.

In a nutshell, the major consideration while selecting the experimental design can be summarized as:

- All designs can be applied for optimization of product characteristics, but simplex mixture design should not be employed for process optimization.
- Any design out of 2^k FD, x^k FD, FFD, PBD, D-OD, or TgD can be employed for screening studies.
- For estimation of main effects, all two-level designs except PBD can be employed. However, for a higher number of factors (>6), screening should first be employed using FFD, PBD, or TgD.

Table 3.3 Experimental designs usually employed during Formulation by Design of drug delivery systems

Design	Description	Diagrammatic representation
(A) Response surface designs		

Factorial designs (FD)	A factorial experiment is one wherein all levels (x) of a given factor (k) are combined with all levels of every other factor in the experiment, with the total number of experiments being x^k
	
	Merits:
	  • Efficient in estimating main effects and interactions
	  • Maximum usage of data
	
	Demerits:
	  • Reflection of curvature not possible in a two-level design
	  • More experimental runs are required
		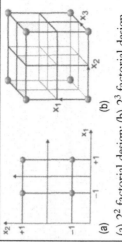
		(a) 2^2 factorial design; (b) 2^3 factorial design

Central composite designs (CCD) or Box–Wilson design	For nonlinear responses requiring second-order models, CCDs are most frequently employed. The total number of factor combinations in a CCD is given by $2^k + 2k + 1$
	
	Merits:
	  • Allows the work to proceed in stages while augmenting the FD by adding a center point
	  • Requires fewer experiments
	
	Demerits:
	  • Difficult to practice with fractional values of rotability α
		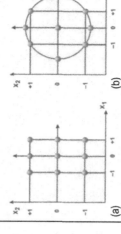
		(a) CCD (rectangular domain) with $\alpha = 1$;
		(b) CCD (spherical domain) with $\alpha = 1.414$

Box–Behnken designs (BBD)	A specially made design requires only three levels for each factor, i.e., -1, 0, and $+1$. A BBD is an economical alternative to CCD	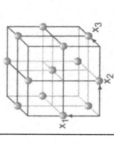 BBD for three factors

(Continued)

Table 3.3 Experimental designs usually employed during Formulation by Design of drug delivery systems (Continued)

Design	Description	Diagrammatic representation
Equiradial designs (EqD)	EqDs are first-degree response surface designs, consisting of N points on a circle around the center of interest in the form of a regular polygon	(a) (b) Two-factor EqD: (a) triangular four-run design; (b) square five-run design, hexagonal design
Mixture designs	In drug delivery system with multiple excipients, the characteristics of the finished drug product usually depend not so much on the quantity of each substance present but on their proportions. Simplex mixture designs (SMDs) are highly recommended in such cases. In a two-component mixture, only one factor level can be independently varied, while in a three-component mixture, only two factor levels can be independently varied Merits: • Suitable for formulations wherein a constraint is imposed on some combination of factor levels Demerits: • Difficulty in comprehending the polynomials generated • Interactions and quadratic effects are not estimated	(a) (b) SMDs: (a) linear model; (b) quadratic model
Optimal designs	When the domain is irregular in shape, optimal designs can be used. These are the nonclassic *custom designs* generated by exchange algorithm using computer. In general, such custom designs are generated based on a specific *optimality criterion* such as D-, A-, G-, I-, and V-optimality criteria Merits: • Can be employed even if the value of experimental domain is either irregular or unknown Demerits: • Involves a relatively complex model	

(B) Screening designs

Fractional factorial designs (FFD)	The erstwhile high number of experiments in an FD can be significantly reduced in a systematic way in an FFD. An FFD is a finite fraction $(1/x^r)$ of a complete or full FD, where r is the degree of fractionation and x^{k-r} is the total number of experiments required Merits: • Suitable for large number of factors or factor levels Demerits: • Difficult to construct • Effects are confounded with interaction terms.	 (a) (b) (a) 2^{3-1} FFD with design points as spheres, (b) 2^{3-1} FFD with added center point
Plackett–Burman designs (PBD)	PBDs are special two-level FFDs used generally for screening of K factors, i.e., $N-1$ factors, where N is a multiple of 4. Also known as *Hadamard designs* or *symmetrically reduced* 2^{k-r} *FDs*, the designs can easily be constructed employing a minimum number of trials Merits: • Suitable for very large number of factors, where even FFDs require a large number of experiments Demerits: • Design structure is complex • Results in confounding of effects, as number of experiments is quite less	
Taguchi designs (TgD)	Employed to develop the products or processes as robust amidst natural variability. Also referred to as "offline quality control," experimental designs, the TgDs ensure good performance in the development of products or processes. Based upon the magnitude of signal-to-noise ratio, the TgDs can be used especially in factor screening for maximization or minimization of responses, or matching to a targeted value	 Inner 2^3 and outer 2^2 arrays of TgD

Table 3.4 Application of important experimental designs depending upon the nature of factor, models, and strategies

Design → Trait ↓	2^k FD	X^k FD	FFD	PBD	CCD	BBD	SMD	TgD	D-OD	EqD
Factor type										
Formulation	✓	✓	✓	✓	✓	✓	✓	✓	✓	✓
Process	✓	✓	✓	✓	✓	✓	—	✓	✓	✓
Both	✓	✓	✓	✓	✓	✓	—	✓	✓	✓
Number of factors										
≤3	✓	✓	✓	—	✓	✓	✓	✓	✓	✓
4–6	✓	✓	✓	✓	✓	✓	—	✓	✓	—
>6	—	—	✓	✓	—	—	—	✓	—	—
Factor level										
2	✓	—	✓	✓	—	—	✓	✓	✓	—
≥3	—	✓	—	—	✓	✓	✓	✓	✓	✓
Model proposed										
Linear model	✓	✓	✓	✓	✓	✓	—	✓	—	✓
Interaction model	✓	✓	✓	✓	✓	✓	✓	✓	—	✓
Quadratic model	—	—	—	—	—	—	—	—	—	—
Mixture model	—	—	—	—	—	—	✓	—	—	—
Custom made model	—	—	—	—	—	—	✓	—	✓	—
Screening studies	✓	✓	✓	✓	—	—	✓	✓	✓	—
Response surface modeling	—	✓	✓	—	✓	✓	✓	✓	✓	✓

BBD, Box–Behnken design; *CCD*, central composite design; *D-OD*, D-optimal design; *EqD*, equiradial design; *FD*, fractional design; *FFD*, fractional factorial design; *PBD*, Plackett–Burman design; *SMD*, simplex mixture design; *TgD*, Taguchi design.

- If there are only two factor levels, any design out of 2^k FD, FFD, or mixture design can be employed. However, in case of >3 factor levels, CCD, BBD, Equiradial design, mixture, simplex, and optimal designs are preferred.
- For quadratic models, x^k FD, CCD, BBD, or EqD are preferred for process optimization.

3.2.2 Formulation by Design model development

A model is a mathematical or graphical expression defining the quantitative dependence of a response variable on the independent variables. Numeric models can be either empirical or theoretical. An empirical model provides a way to describe the factor–response relationship. Usually, it is a set of polynomials of a given order or degree. The models are mostly employed to describe responses as first-, second-, and very rarely as third-order polynomials (Box et al., 1978). A first-order model is used in the first instance. In the case that a simple model is found to be inadequate for describing the phenomenon, the higher-order models are followed.

The coefficients for quantitative factors can be estimated using regression analysis. However in the case of qualitative factors, as interpolation between discrete (i.e., categorical) factor values is meaningless, regression analysis is not employed. For more factors, interactions and higher order terms, multiple linear regression analysis (MLRA) is usually preferred. For a combination of categorical and continuous variables, polynomial equations between response and the continuous variable are generated for each level of categorical variable. Multiple nonlinear regression analysis (MNLRA) should be preferred when the factor–response relationship is nonlinear.

When a large number of variables are to be computed, multivariate methods of analysis using chemometric techniques can be employed like partial least squares (PLSs) and/or principal component analysis (PCA) can also be employed for regression (Box and Draper, 1987). PLS is an extension of MLRA, which is used when there are a lower number of observations than predictor variables. In general, model analysis is conducted considering statistical tools like analysis of variance (ANOVA), multiple analysis of variance (MANOVA), Student's t-test, predicted residual sum of squares (PRESS), and Pearsonian regression coefficient (r^2).

3.2.3 Model diagnostic plots in Formulation by Design optimization

One or more diagnostic plots are usually employed during FbD optimization as tools for analyzing the factor sparsity and at times to investigate the goodness of fit of the proposed FbD model through half-normal plots, Cook's distance plot, leverage plot, Pareto charts, outlier-T plot, Box–Cox plot, actual versus predicted plot, etc. Fig. 3.3 illustrates the typical instances on various model diagnostic plots such as half-normal plot, actual versus predicted plot, outlier-T plot, and Box–Cox plot used in the FbD optimization of the DDS.

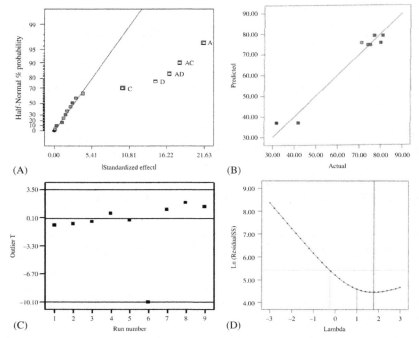

Figure 3.3 Typical Formulation by Design model diagnostic plots: (A) half-normal plot; (B) actual versus predicted plot; (C) outlier-T plot; (D) Box–Cox plot.

3.2.4 Validation of Formulation by Design models

The major graphical tools employed for testing and revising an FbD model encompass linear correlation plots, normal probability plot of residuals: response versus prediction plot, residual lag plot, residuals histogram, etc. These model validation plots furnish information on the location, spread, skewness, outliers, and multiple nodes of the data.

3.3 OPTIMUM SEARCH

From the models thus selected, optimization of one response or the simultaneous optimization of multiple responses needs to be accomplished graphically, numerically, or using artificial neural networks (ANNs) (Schwartz et al., 1973). A pictographic account on various modeling approaches used for the optimum search in the FbD methodology is depicted in Fig. 3.4.

3.3.1 Graphical optimization

Graphical optimization deals with selecting the best possible formulation out of a feasible factor space region (Lewis et al., 1999). The desirable limits of response variables

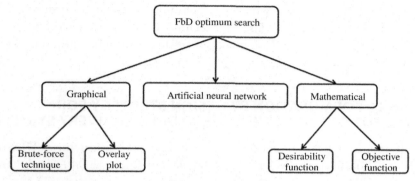

Figure 3.4 Various modeling approaches employed for optimum search during Formulation by Design.

are set and the factor levels are screened accordingly. Graphical optimization can be accomplished through one or more of the following methodologies *"brute-force method," "overlay plots,"* and *"canonical analysis."* The brute-force method is also known as an exhaustive search. It is the simplest and most accurate of all possible optimization search methods, as it implies checking every single point in the function space. The overlay plots include bidimensional response contour plots that are superimposed over each other to search for the best compromise visually. Most often, this overlay plot is considered as the best option to present the design space for assuring desired quality levels and regulatory purpose. Minimum and maximum boundaries are set for acceptable objective values. The region is highlighted wherein all the responses are acceptable. Within this area, an optimum is located, "trading off" different responses. Canonical analysis indicates the predictability of each of the extracted components of the criterion set of variables from the corresponding components, extracted from the predictor set of variables. The technique can only be employed for single response optimization.

3.3.2 Mathematical optimization

Graphical analysis is usually considered adequate in the case of a single response. However, in cases of multiple responses, it is usually advisable to conduct mathematical or numerical optimization first to uncover a feasible region. It can be accomplished using techniques like *desirability function*, objective function, sequential unconstrained minimization technique, and ANNs.

Other optimum search tools which are less commonly employed include steepest ascent (or descent) methods involving direct optimization methods for first-order designs, especially when the optimum is outside the domain and is to be arrived at rapidly (Lewis et al., 1999). The optimum path method is analogous to the steepest ascent method, but is employed where the optimum is searched outside the experimental domain by extrapolation. Also, the technique of evolutionary operations uses the production procedure (formulation and process) to locate the optimum solution by careful planning and constant repetition in industrial processes.

3.4 OVERALL FORMULATION BY DESIGN STRATEGY FOR DEVELOPMENT OF NANOSTRUCTURED DRUG DELIVERY SYSTEM

The overall approach for conduct of an FbD study in nanostructured DDS can be described by a holistic plan (Myers, 2003; Singh and Ahuja, 2004). The salient steps involved in the FbD strategy include the following.

3.4.1 Problem definition

The FbD problem is comprehended clearly, QTPP vividly defined, and possible responses (CQAs) ascertained.

3.4.2 Selection of factors and factor levels

The independent factors (i.e., CMAs, CFAs, and/or CPPs) are identified amongst the score of quantifiable and easily controllable variables.

3.4.3 Design of experimental protocol

Based on the choice of independent factors and the response variables (i.e., CQAs), a suitable experimental design is selected and the number of experimental runs is determined.

3.4.4 Formulating and evaluating the dosage form

As per the experimental design, the different formulations are developed and evaluated for their response(s).

3.4.5 Selection of design space and optimum formulation

A robust mathematical model is generated using the experimental data using suitable graphical and/or numeric methods.

3.4.6 Validation of Formulation by Design optimization

The predicted optimal formulation is prepared and the responses evaluated. Results, if validated, are carried further to the production cycle via pilot plant operations and scale-up techniques.

3.4.7 Control strategy and continuous improvement

At the end of FbD-based product development, the control strategy is needed to be formulated for continuous improvement after regulatory approval too.

3.5 SOFTWARE USAGES DURING FORMULATION BY DESIGN

The merits of QbD techniques are numerous and they have good acceptability. However, a colossal amount of mathematical and statistical intricacies is required for the same. With the availability of economical and powerful hardware, coupled with comprehensive QbD software, the computational hiccups have been greatly surmounted and the process has been streamlined. Fig. 3.5 lists selected software commercially available for carrying out QbD studies. Some important software employed for QbD and optimization includes Design-Expert, MODDE, Unscrambler, JMP, Statistica, Minitab, etc. There are a few software that are employed for multivariate analysis using various chemometric tools like MNLRA, PCA, PLS, ANN, etc. encompassing CODDESA, IBM-SPSS, Matlab, MODDE, SIMCA, and Unscrambler. For QRM studies using Fish bone diagrams, REM and FMEA matrices during risk assessment studies, etc., software like FMEA-Pro, iGrafx, Minitab, Risk, Statgraphics, etc., can be used.

3.6 FORMULATION BY DESIGN OPTIMIZATION OF NANOSTRUCTURED DRUG DELIVERY SYSTEM: LITERATURE UPDATE

Today, FbD has successfully been employed for wide-ranging objectives in pharmaceutical sciences including preformulation studies, stability studies, development of drug products, performance evaluation of formulation or process, etc.

Figure 3.5 Select computer software used during Quality by Design implementation for product and process optimization.

Albeit the experimental designs have long been employed to optimize various industrial products and/or processes since 1926 (Box and Wilson, 1951; Plackett and Burman, 1946; Scheffe, 1958), their use permeated into the domain of pharmaceutical product/process development around four decades ago (Singh et al., 2005b). The first literature report on the rational use of optimization appeared in 1967, when a tablet of sodium salicylate was optimized using an FD (Marlowe and Shangraw, 1967). Since then, these systematic approaches have been put into practice in the development of drug formulations at a relatively steady pace. Nanostructured DDS primarily includes a diverse variety of vesicular and nonvesicular systems. Instances of vesicular systems include liposomes, niosomes, ethosomes, and transferosomes, whereas the nonvesicular systems include nanoparticulate systems, nanomaterials, and nanocomposites too.

3.6.1 Vesicular systems

Vesicular systems are a novel means of drug delivery that can enhance the bioavailability of an encapsulated drug and provide therapeutic activity in a controlled manner for a prolonged period of time. Such systems invariably contain phospholipids as the integral part of the carrier along with cholesterol and surfactants for vesicle stabilization (Mosley et al., 2012). Liposomes are the most promisingly used vesicular carriers made of lipid bilayer. Liposomes can be prepared by disrupting biological membranes, e.g., naturally derived phospholipids with mixed lipid chains (e.g., egg phosphatidyl ethanolamine) or other surfactants by sonication. These can be filled with drugs to deliver them at the desired site of action for treatment of cancer and other diseases. Niosomes are vesicles composed of nonionic surfactants, which are a biodegradable, relatively nontoxic, more stable, and inexpensive alternative to liposomes. Transferosomes are the flexible-membrane vesicles containing a high amount of edge activator and provide enhanced permeation characteristics for topical and transdermal drug delivery. Table 3.5 provides an account of various vesicular systems optimized using FbD.

3.6.2 Nonvesicular systems

These include various nanostructured systems like polymeric nanoparticles, metallic nanoparticles, magnetic nanoparticles, nanospheres, nanocapsules, nanofibers, nanofluids, nanomicelles, nanoemulsions, solid lipid nanoparticles (SLNs), nanostructured lipid carriers (NLCs), self-nanoemulsifying systems, nanoorganogels, in situ gelling systems, carbon nanotubes (CNTs), and dendrimers. The FbD approach has been used meticulously in developing such optimized systems by critically analyzing the material attributes influencing the product performance. A detailed account on the individual systems is given below.

Table 3.5 Select literature instances on Formulation by Design optimization of vesicular DDS

Drug	Factors/lipidic carriers	Design	Year
Liposomes			
Clodronate	Phospholipid to cholesterol ratio, drug to lipid ratio, sonication time	BBD	Ailiesei et al. (2016)
Paclitaxel	Concentration of phospholipid and cholesterol	CCD	Barbosa et al. (2015)
Hyaluronidase	Phosphatidylcholine and speed of rotation	CCD	Kasinathan et al. (2014)
Sirolimus	Molar ratio of DPPC/cholesterol and molar ratio of DOPE/DPPC	CCD	Ghanbarzadeh et al. (2013)
Dithranol	Concentration of Compritol 888 ATO, Phospholipon 90G	CCD	Raza et al. (2013)
Paeonol	DC-Chol concentration, molar ratio of lipid/drug, polymer concentration	BBD	Shi et al. (2012)
Plasmid DNA	DC-Chol/DOPE, amount of total lipid contents, charge ratio of cationic lipid to pDNA	BBD	Yang et al. (2012)
Epimedium polysaccharide	Drug:lipid ratio, soybean phospholipid: cholesterol ratio, ultrasonic time, temperature	TgD	Gao et al. (2012)
Sinomenine	Proportion of phospholipid, cholesterol	SMD	Wang et al. (2009)
Ciprofloxacin	Molar concentration of ciprofloxacin and cholesterol	FD	Mehanna et al. (2009)
Diclofenac	Concentration of cholesterol, Span 80, phospholipid	D-OMD	Bhatia et al. (2009)
Benzocaine	Surfactant concentration, volume of hydration phase, vesicle lipid phase, percent of ethanol	D-OD	Mura et al. (2008)
Lidocaine hydrochloride	Concentration of CH coating solution, stirring rate	FFD	Gonzalez-Rodriguez et al. (2007)
Glipizide	Amount of paraffin wax, proportion of stearic acid in the wax	FD	Shivakumar et al. (2007)
Leuprolide	Volume of aqueous phase, HSPC/DSPG and HSPC/cholesterol	ANN, FD	Arulsudar et al. (2005)
Nimesulide	Concentration of phospholipid and cholesterol	CCD	Singh et al. (2005a,b,c)
Protamine-DNA complex	Weight ratio of protamine/DNA, Chems/DNA, molar ratio of Chems/DOPE	CCD	Zhong et al. (2007)
Gadolinium	Phospholipid type, amount of cholesterol, drug/lipid ratio	FFD	Glogard et al. (2002)

(Continued)

Table 3.5 Select literature instances on Formulation by Design optimization of vesicular DDS (Continued)

Drug	Factors/lipidic carriers	Design	Year
Niosomes			
Clarithromycin	Surfactant:cholesterol ratio, amount of solvent	FD	Asthana et al. (2016)
Methotrexate	Concentration of hydration medium, total weight of niosomes, surfactant: cholesterol ratio	BBD	Abdelbary and AbouGhaly (2015)
Valsartan	Phospholipids 90G, ethanol, valsartan, sonication time	BBD	Ahad et al. (2013)
Itraconazole	Surfactant:cholesterol ratio, quantity of ethanol	FD	Wagh and Deshmukh (2012)
Clotrimazole	Concentration of cholesterol, surfactant	FD	Nasr et al. (2008)
Piroxicam	Molar ratio of Span 60: cholesterol, amount of drug	BBD	Solanki et al. (2007)

ANN, artificial neural network; *BBD*, Box–Behnken design; *CCD*, central composite design; *D-OMD*, D-optimal mixture design; *DOPE*, dioleyl phosphatidyl ethanolamine; *DPPC*, distearoyl phosphatidylcholine; *FD*, fractional design; *FFD*, fractional factorial design; *HSPC*, hydrogenated stearyl phosphtaidylcholine; *OD*, orthogonal design; *SMD*, simplex mixture design; *TgD*, Taguchi design.

3.6.2.1 Nanoparticles and nanomaterials

Pharmaceutical nanotechnology has emerged as a booming area in drug delivery because of the several merits associated with it, making it superior in comparison to other delivery systems. Their size allows them to be administered intravenously via injection for both temporal and spatial drug delivery application with enhanced therapeutic benefits. Owing to their small size, tuneable properties, and enormous surface area, they can escape from phagocytotic uptake by the reticuloendothelial system to target the drug and therapeutic biomolecules more efficiently. Examples of nanoparticles used in drug delivery include polymeric nanoparticles, metallic nanoparticles, magnetic nanoparticles, SLNs, and NLCs (Singh et al., 2009b). Table 3.6 provides a terse account on key instances of various nanoparticulate systems, optimized using FbD. Important CMAs/CPPs used in the preparation of such nanoparticulate systems include concentration of monomer, concentration of surfactant, concentration of polymer, molecular weight of the polymer, initiator concentration, stabilizer concentration, stirring speed, and temperature, which tend to affect the CQAs like yield, particle size, drug loading, entrapment, in vitro drug release, polydispersity index, and zeta potential. The key literature instances on other nanostructured systems like nanocapsules, nanofibers, and nanofluids have also been listed.

Table 3.6 Select literature instances on Formulation by Design optimization of nanostructured systems

Drug	Factors/polymers/carriers	Design	References
Polymeric nanoparticles			
Indometacin	Concentrations of tripolyphosphate, chitosan, and stirring time	BBD	Kalam et al. (2016)
Quercetin	Concentration of PLGA, PVA, and stirring speed	BBD	Tefas et al. (2015)
Hydrocortisone butyrate	PLGA, PVA, stirring speed, sonication time	PBD	Yang et al. (2014)
Glibenclamide	Amount of Poloxamer 188, amount of PVP S 630 D, solvent to antisolvent volume ratio	PBD	Shah et al. (2013a,b)
Pyridostigmine bromide	Concentration of poly(lactic acid) and PVA	CCD	Tan et al. (2012)
Sildenafil	Concentration of polymer, stabilizer, pH	FD	Beck-Broichsitter et al. (2011)
Insulin	pH, polymer:insulin ratio, polymer type	D-OD	Mahjub et al. (2011)
Acyclovir	Amount of PLGA, Pluronic F68, Polycarbophil	FD	Bhosale et al. (2011)
Paclitaxel	Amount of polymer, ultrasonication time	RCCD	Kollipara et al. (2010)
Gentamycin	Molecular weight of PLLA	OD	He et al. (2009)
Diclofenac	Inlet air temperature, spray rate feed	FD	Beck et al. (2006)
Insulin	Ratio of polymers (PCL/RS ratio), volume, pH of the aqueous solution of polyvinyl alcohol	CCD	Attivi et al. (2005)
5-Fluorouracil	Type of surfactant, amount of acetone, molecular weight of the polymer	OD	Bozkir and Saka (2005)
Metallic nanoparticles			
Streptomyces viridochromogenes	$AgNO_3$ concentration, incubation period, pH level, inoculum size	FD	El-Naggar Nel and Abdelwahed (2014)
Gold	Stirring rate, sodium citrate concentration, and ionic strength	BBD	Honary et al. (2013)
ZnO	Surfactant concentration, time, and temperature	BBD	Ismaila et al. (2005)
Silver	Amount of precursor ($AgNO_3$), stabilizer (Daxad), reducing agent (Asc), and pH controller (HNO_3)	FFD	Lim and Lee (2008)

(Continued)

Table 3.6 Select literature instances on Formulation by Design optimization of nanostructured systems (Continued)

Drug	Factors/polymers/carriers	Design	References
Magnetic nanoparticles			
Tacrine	Amount of Span 80, glutaraldehyde saturated toluene, and NaCl	BBD	Elmizadeh et al. (2013)
Iron oxide	pH of solution, amount of extractant, and amount of nanoparticles	BBD	Khajeh (2009)
Colloidal silica	Inlet temperature, spray flow rate, liquid feed rate, concentration of nanocapsules, and silicon dioxide	FD	Tewa-Tagne et al. (2007)
Nanocapsules			
Food	Concentration of PCL, PEG	RSM	Esmaeili and Gholami (2015)
Diclofenac	Inlet temperature and spray feed rate	FD	Beck et al. (2006)
Cyclosporin A	Amount of lipid (Miglitol 840), amount of poly-ε-caprolactone, volume of organic solvent (acetone)	FD	Calvo et al. (1996)
Nanospheres			
Atorvastatin	Polymer:drug ratio, pH, stirring time	FFD	Hashem et al. (2015)
Cadmium sulfate	Complexing agent, pH buffer, temperature, and time	TD	Leal-Cruz et al. (2013)
Amoxicillin	Concentration of Carbopol, inlet temperature, and flow rate	CCD	Harsha et al. (2013)
Solid lipid nanoparticles			
Voriconazole	Concentration of lecithin, Tween 80, cetyltriethyl ammonium bromide	FD	Sousa and Pessine (2015)
Lopinavir	Solid lipid concentration, stirring speed and time	BBD	Ravi et al. (2014)
Ondansetron	Concentration of lipid, surfactant, and cosurfactant	FD	Joshi et al. (2012)
Isotretinoin	Percentage of Compritol and Phospholipon 90G	CCD	Raza et al. (2012)
Doxorubicin	Ratio between the lipid phase and aqueous phase, ratio of surfactant (Poloxamer) and cosurfactant (cetylpyridinium chloride)	FD	Taveira et al. (2012)
Chloramphenicol	Amount of lipid, surfactant, and drug/lipid ratio	BBD	Hao et al. (2011)
Quercetin	Amount of Compritol and Tween 80	CCD	Dhawan et al. (2011)

Flurbiprofen	Concentration of liquid lipid, drug, surfactant	CCD	Gonzalez-Mira et al. (2010)
Vitamin K1	Concentrations of the surfactants, i.e., Myverol, Pluronic	CCD	Liu et al. (2010)
Amikacin	Amount of lipid phase, drug to lipid ratio, volume of aqueous phase	CCD	Varshosaz and Keihanfar (2001)
Simvastatin	Amount of glycerol monostearate, concentration of poloxamer, volume of isopropyl alcohol	FD	Shah et al. (2010a,b)
Buspirone–HCl	Surfactant percentage, homogenization speed, acetone:dichloromethane ratio	BBD	Varshosaz et al. (2010)
Allopurinol	Drug:wax ratio, homogenization speed	FD	El-Gibaly and Abdel-Ghaffar (2005)

Nanostructured lipid carriers

Pioglitazone	Tween 80, lipid, carbopol	BBD	Prasad et al. (2016)
Lamotrigine	% lipid, ratio of liquid:solid lipid, pioglitazone	BBD	Alam et al. (2016)
	Amount of solid lipid, liquid lipid, surfactant, sonication time	BBD	Alam et al. (2015)
5-Fluorouracil	Concentrations of lipid and surfactant	FD	Andalib et al. (2012)
Econazole nitrate	Concentrations of drug, lipid, and emulsifier	CCD	Keshri et al. (2012)
Flurbiprofen	Ratio of liquid lipid versus total lipid, concentration of the surfactant and drug	FCCD	Gonzalez-Mira et al. (2011)
Oxybenzone	Concentrations of liquid lipid, oxybenzone	FD	Sanad et al. (2010)

Nanofluids

Titanium dioxide	Ultrasonic power, ultrasonication time, TiO$_2$ volume	BBD	LotfizadehDehkordi et al. (2013)

Nanofibers

Cellulose acetate	Potential difference, distance between tip-to-collector and feed rate	BBD	Konwarh et al. (2013)
Polyacrylamide-MWCNTs	PAAm concentration, MWCNTs content, flow rate, and applied voltage	Taguchi	Amini et al. (2013)
Polyurethane	Voltage applied and distance from the tip to collector	FD	Yanilmaz et al. (2013)

BBD, Box–Behnken design; *CCD*, central composite design; *DFD*, double factorial design; *D-OD*, D-optimal design; *FD*, fractional design; *OD*, orthogonal design; *RD*, robust design; *SMD*, simplex mixture design; *TgD*, Taguchi design.

3.6.2.2 Self-nanoemulsifying drug delivery systems

Self-nanoemulsifying drug delivery systems (SNEDDSs) due to their impeccable potential, have emerged as a boon for improving oral bioavailability of poorly water-soluble drugs. Following their oral administration, these systems rapidly disperse in gastrointestinal (GI) fluids, yielding micro- or nanoemulsions containing the solubilized drug (Singh et al., 2009a). Owing to their miniscule globule size, the micro/nanoemulsified drug can easily be absorbed through lymphatic pathways, bypassing the hepatic first-pass effect and P-gp efflux. Concentrations of lipids, surfactants, cosurfactants, cosolvents, polymeric precipitation inhibitors, and cationic charge inducer are primarily considered as the CMAs, whereas emulsification time, globule size, zeta potential, in vitro drug release, and ex vivo permeation are taken as CQAs for optimization of SNEDDS. Table 3.7 provides an account of self-emulsifying systems optimized using FbD.

Table 3.7 Select literature instances on Formulation by Design optimization of self-nanoemulsifying systems

Drug	Factors/lipids/surfactants/cosurfactants	Design	Year
Lopinavir	Amount of Maisine, Tween 80, Transcutol HP	D-OD	Garg et al. (2016)
Candesartan	Amount of Lauroglycol 90, Tween 40, Transcutol HP	D-OD	Sharma et al. (2015)
Olmesartan	Amount of oleic acid, Tween 40, Transcutol HP	D-OD	Beg et al. (2015)
Tamoxifen	Amount of Capmul MCM, Cremophor RH 40, Labrafil M1944CS	D-OD	Jain et al. (2014)
Valsartan	Amount of Capmul MCM, Acrysol K-140, Transcutol HP	BBD	Shah et al. (2013a,b)
Carvedilol	Amount of Capmul MCM, Nikkol HCO 50	CCD	Singh et al. (2012)
Ezetimibe	Amount of lipid and surfactant	CCD	Bandyopadhyay et al. (2012)
Simvastatin	Amount of Ethyl oleate, Labrasol, Tween 80	MD	Patil et al. (2007)
Astilbin	Amount of lipid and surfactant	CCD	Mezghrani et al. (2011)
Carvedilol	Amount of Cremophor EL, Transcutol HP	CCD	Singh et al. (2011a,b)
Patchoulic alcohol	Amount of Cremophor EL, Tween 80, PEG 400-isopropyl myristate-patchoulic alcohol ratio	CCD	You et al. (2010)

(Continued)

Table 3.7 Select literature instances on Formulation by Design optimization of self-nanoemulsifying systems (Continued)

Drug	Factors/lipids/surfactants/ cosurfactants	Design	Year
Candesartan	Amount of Lauroglycol 90, Labrasol, Transcutol	SMD	Singh et al. (2010a,b)
Lacidipine	Amount of oil, surfactant, cosurfactant	D-OD	Basalious et al. (2010)
GBE50	Amount of isopropyl myristate, Cremophor RH 40	OD	Xiong et al. (2009)
Genistein	Amount of Maisine 35-1, Labrafac, Cremophor, Labrasol, Transcutol	BBD	Zhu et al. (2009)
Curcumin	Concentration of oil, surfactant, cosurfactant	SLD	Cui et al. (2009)
Oridonin	Oil percentage, surfactant:cosurfactant ratio	CCD	Liu et al. (2009)
Probucol	Amount of surfactant and cosurfactant	BBD	Zaghloul et al. (2008)
QHN17	Oil content, weight ratio of surfactant and cosurfactant	CCD	Zhang et al. (2007)
Cyclosporine	Amounts of Emulphor EL-620, Capmul MCM-C8	BBD	Zidan et al. (2007)
Ketoprofen	Conc. of cosurfactant and gelling agent	FD	Patil et al. (2004)
Celecoxib	Concentration of Capmul PG8, Acconon MC-82, Tween 80	MD	Subramanian et al. (2004)
Coenzyme Q10	Amount of R-(+)-limonene, surfactant, cosurfactant	BBD	Palamakula et al. (2004)

BBD, Box–Behnken design; *CCD*, central composite design; *D-OD*, D-optimal design; *FD*, fractional design; *OD*, orthogonal design; *PEG*, polyethylene glycol; *SLD*, simplex lattice design; *SMD*, simplex mixture design.

3.6.2.3 Nanoemulsions

Nanoemulsions are clear, stable, isotropic liquid mixtures of oil, water, and surfactant, frequently in combination with a cosurfactant (Hassan and Elshafeey, 2010). Such systems contain an aqueous phase along with the oil which is a complex mixture of different hydrocarbons (saturated as well as unsaturated). In contrast to ordinary emulsions, microemulsions form upon simple mixing of the components and do not require the high shear conditions generally used in the formation of ordinary emulsions. Based on the composition, there can be two types of emulsions, namely, oil-in-water (o/w) and water-in-oil (w/o, reversed). The two immiscible oil and water phases are present with the surfactant forming a monolayer structure at the oil–water interface. The surfactants have their polar heads toward the aqueous portion

Table 3.8 Select literature instances on Formulation by Design optimization of microemulsion systems

Drug	Factors/excipients	Design	Year
Lidocaine Prilocaine	Concentration of IPM, Tween 80, Labrasol/Lauroglycol 90	D-OD	Negi et al. (2016)
	Emulsifier concentration, homogenization speed	BBD	Negi et al. (2015)
Curcumin	Amount of oil, surfactant, cosurfactant	BBD	Sood et al. (2014)
Diazepam	Oil content, lecithin type, and the presence of diazepam	FD	Đordević et al. (2013)
Tanshinone IIA	Content of Tween 80, lecithin, and water	FD	Chang et al. (2011)
β-Carotene	Time and shear rate of homogenization	FD	Silva et al. (2011)
β-Carotene	Emulsifier concentration, homogenization pressure, temperature	FD	Yuan et al. (2008)
Carbamazepine	Type of oil and type of lipophilic emulsifier	FD	Kelmann et al. (2007)

FD, factorial design; *IPM*, isopropyl myristate.

and hydrophobic tails directed toward the oil phase. The amount of oils, surfactants, cosurfactants, and cosolvents present in the nanoemulsion formulations are major CMAs, which invariably affect the CQAs like permeation, flux, globule size, zeta potential, and polydispersity index. Table 3.8 provides an account of microemulsions optimized using FbD.

3.6.2.4 Nanoorganogels

Nanoorganogels are novel supramolecular structured systems, which contain organic solvents to form a three-dimensional polymeric network. Examples of gelable organic solvents include aliphatic and aromatic hydrocarbons, alcohols, silicone oil, and vegetable oils. Such systems have advantages over conventional hydrogels owing to their high permeation and flux, and better drug entrapment (Murdan, 2005). In the past decade, nanoorganogel-based DDS have been quite meticulously explored for topical drug delivery. Literature reports suggested that the nanoorganogels primarily contain phospholipids, L-alanine, and sorbitan monostearate in an organic solvent like ethanol, dimethyl sulfoxide, etc. The particle size, entrapment efficiency, permeation, and flux are the major CQAs of nanoorganogel formulations which are dependent on the CMAs like concentration of phospholipid, cholesterol, amount of organic phase in the formulation, etc. Table 3.9 lists an account on literature reports on FbD optimization of the organogel systems for various drug delivery applications.

Table 3.9 Select literature instances on Formulation by Design optimization of nanoorganogels

Drug	Factors/excipients	Design	Year
Methotrexate	Amount of lipid, surfactant, stabilizer	FD	Avasatthi et al. (2016)
Novicidin	Flow rate, Novicidin ratio, OSA-HA concentration	BBD	Water et al. (2015)
Tamoxifen	Concentration of lecithin, Span 80, water	D-OD	Bhatia et al. (2013)
Estradiol	Concentration of lecithin, N-methyl-2-pyrrolidinone, N-lauroyl l-lysine methyl ester	CCD	Yang et al. (2012)
Propranolol HCl	Percentages of drug, lecithin, and water	CCD	Hadidi et al. (2009)
Cyclosporin A	Concentration of lecithin, Span 80, water	D-OD	

CCD, central composite design; *D-OD*, D-optimal design.

3.6.2.5 Nanomicelles

Nanomicelles are self-assembling amphiphilic macromolecules. They are particularly useful for drugs exhibiting poor oral bioavailability by virtue of their poor aqueous solubility and/or limited drug absorption through the GI mucosa (Aliabadi and Lavasanifar, 2006). Such systems primarily contain a blend of hydrophilic and/or hydrophobic polymers prepared by dissolving both the polymer and drug in an aqueous solvent. Important applications of polymeric micelles include drug solubilization, controlled-release drug delivery, and drug targeting. Table 3.10 illustrates an account of polymeric nanomicellar systems optimized using FbD, including the list of various CMAs/CPPs like concentrations of drug, diblock–copolymers, amount of surfactant, surfactant type, stirring speed, rate, hydration temperature, and time, which have a direct influence on the entrapment efficiency, drug loading, and particle size of the formulation as CQAs.

3.6.2.6 Nanosuspensions

Nanosuspensions are the promising drug delivery technologies containing submicron colloidal dispersion of drug particles stabilized by surfactants. These are particularly suitable for BCS class II drugs exhibiting low oral bioavailability owing to their poor aqueous solubility only (Rabinow, 2004). Nanosuspensions have average particle sizes ranging between 200 and 600 nm. They are usually prepared by wet milling of the drug particles in an aqueous medium containing surfactant. Such systems have been increasingly explored in the bioavailability enhancement of hydrophobic drugs like antimalarials, and anticancer and cardiovascular agents. Table 3.11 illustrates an account of nanosuspension systems optimized using FbD along with various CMAs/CPPs such as drug:stabilizer ratio, size of the bead, mill speed, and milling time, which tend to affect the CQAs like particle size and size distribution, zeta potential, etc.

Table 3.10 Select literature instances on Formulation by Design optimization of polymeric nanomicelles

Drug	Factors/carriers	Design	Year
Dexamethasone	Polymer amount and dexamethasone amount	RSD	Vaishya et al. (2014)
Celastrol	Concentrations of drug and block-copolymers	CCD	Peng et al. (2012)
Curcumin	Concentrations of drug, amount of mixture of Pluronics P123 and F68	CCD	Chaudhary et al. (2011)
Paclitaxel	Pluronics P123 mass fraction, amount of water, feeding of drug and hydration temperature	ED	Wei et al. (2009)
Methoxy-PEG conjugates	Amount of copolymer, type of fatty acid, and micelle formulation	D-OD	Alexis et al. (2008)

CCD, central composite design; D-OD, D-optimal design; ED, equiradial design; PEG, polyethylene glycol; RSD, response surface design.

Table 3.11 Select literature instances on Formulation by Design optimization of nanosuspensions

Drug	Factors/polymers	Design	Year
Ritonavir	Concentration of HPMC/PVP, drug to polymer ratio, number of passes	FD	Karakucuk et al. (2016)
Resveratrol	Concentration of drug, stabilizer, surfactants	BBD	Hao et al. (2015)
Indometacin	Inlet temperature, flow rate, aspiration	FD	Kumar et al. (2014)
Carvedilol	Concentration of drug and alpha-tocopherol succinate, level of sodium lauryl sulfate	CCD	Liu et al. (2012)
Rebamipide	Concentration of drug, Lutrol F127, and Kollidon 90F	CCD	Shi et al. (2012)
Simvastatin	Amount of Compritol, Poloxamer, and volume of acetone	FD	Shah et al. (2010a,b)
Pyrimethamine	Amount of Lutrol F68 and stirring speed	FD	Dhapte and Pokharkar (2011)
Itraconazole	Concentration of stabilizer and amount of milling media	FD	Nakarani et al. (2012)
Cyclosporine A	Concentration of stabilizer and amount of media milling	FD	Nakarani et al. (2010)

CCD, central composite design; FD, factorial design.

3.6.2.7 In situ gelling systems

In situ gelling systems encompass a polymeric blend of smart (or intelligent) polymers, which are bioresponsive to various stimuli like change in pH, temperature, and/or ionic strength. Such systems have been used in oral, periodontal, ocular, and nasal drug delivery owing to their property of providing high residence time at the point of administration (Ajazuddin et al., 2012). The biodegradable polymers include polylactic acid (PLA), polycaprolactone (PCL), polylactic glycolic acid (PLGA), and polyglycolic acid (PGA). Invariably, the polymers used in preparing these systems have a high degree of influence on the biopharmaceutical performance of the formulations. Rational optimization of the concentration of polymers may provide robust products. Instances of CMAs used for the optimization of such systems include concentration of gelling polymers, freezing agent, stirring speed, and stirring time, which were optimized for CAQs like gelling temperature, gel strength, bioadhesion strength, and drug release. Table 3.12 gives an account of in situ gelling systems optimized using FbD for various drug delivery applications.

Table 3.12 Select literature instances on Formulation by Design optimization of in situ gelling systems

Drug	Formulation and route	Factors/gelling polymers	Design	Year
Dexamethasone sodium	Gel/ocular	Concentration of Poloxamer 407, HPMC K4M	CCD	Patel et al. (2016)
Acyclovir	Gel/oral	Concentration of sodium alginate, gellan	CCD	Singh et al. (2016)
Bromfenac	Gel/ocular	Concentration of gelrite, HPMC E15LV	FD	Shahi et al. (2016)
Meloxicam	Implant/ intramuscular	Concentration of PLGA, N-methyl pyrrolidone, PLGA activity	BBD	Ibrahim et al. (2014)
Chlorhexidine hydrochloride	Gel/ocular	Concentration of Poloxamer 407, Carbopol 934P	CCD	Garala et al. (2013)
Ornidazole	Nanoparticles/ periodontal	Concentration of Lutrol F68, gellan	CCD	Lohit et al. (2012)
Ofloxacin, Ornidazole	Gels/ periodontal	Concentration of gellan, calcium carbonate	CCD	Nandi et al. (2012)
Sodium cromoglycate	Gel/nasal	Concentration of Carbopol 940, HPMC K4M	3^2 FD	Shah et al. (2011)
Acyclovir	Gastroretentive gels/oral	Concentration of sodium alginate, gellan	CCD	Singh et al. (2010a,b)

CCD, central composite design; *FD*: factorial design.

3.6.2.8 Carbon nanotubes

CNTs are a subfamily of fullerenes (or Buckminter fullerenes or bucky balls), which are pseudo-one-dimensional carbon allotropes with high aspect ratio, high surface area, and excellent material attributes in terms of electrical and thermal conductivities and mechanical strength. CNTs are considered as an ideal nanocarrier in the field of drug delivery for therapeutic molecules. Of the diverse applications of CNTs, the vital ones encompass controlled drug delivery and targeting of the drug molecules to specific sites like the lymphatic system, brain, ocular system, and cancerous tissue. Such properties can be attributed to their nanosized hollow tube-shaped structure, favoring encapsulation of drug molecules or by possible attachment of drug molecules on the nanotube walls. Further, CNTs find their application in delivery of biotechnology products like vaccines, hormones, enzymes, and also genes. Table 3.13 illustrates an updated literature on the application of FbD approach of optimizing CNTs for drug delivery. Major CMAs/CPPs involved during the preparation of CNTs include concentration of the starting material, reaction time, and reaction temperature, which have a direct influence on the yield of the functionalized CNTs, solubility, drug loading, etc.

3.6.2.9 Dendrimers

Dendrimers are repetitively branched molecules, which constitute a core and spherical three-dimensional morphology. A single functional entity of the dendrimer is called a dendron, where the drug molecules can be tethered on the surface of dendrimer and its sprouting branches. On the basis of growth process of the dendrimers, these can be

Table 3.13 Updated literature instances on Formulation by Design optimization of dendrimers and carbon nanotubes

Drug/CNT	Factors/independent variables	Design	Year
Dendrimers			
Aceclofenac	Concentration of dendrimer and temperature	FD	Garala et al. (2009)
	Concentration of dendrimer D12 and methyl methacrylate	FD	
Carbon nanotubes			
	Concentration of Nafion and pH of the stripping buffer solution	FFD	Lien et al. (2013)
SWNT	Type of phospholipid-polyethylene glycol 200 and 500	D-OD	Hadidi et al. (2011)
	PL-PEG/SWCNT weight ratio, sonication time, Pl-PEG type	D-OD	Hadidi et al. (2011)
	Fe:MgO ratio, reaction temperature, Ar volumetric flow rate	BBD	Kukovecz et al. (2005)

BBD, Box–Behnken design; D-OD, D-optimal design; FD, factorial design; FFD, fractional factorial design; PBD, Plackett–Burman design.

classified according to their generation number, such as $G_{0.5}$, G_1, G_2, G_3, G_4, and G_5. Dendrimers can be synthesized by both divergent and convergent methods. Examples of several types of dendrimers include those which have immense applications in drug delivery including poly(amidoamine) (PAMAM) dendrimers, poly(propylene imine) (PPI) dendrimers, polyether-copolyester (PEPE) dendrimers, PEGylated dendrimers, peptide dendrimers, etc. Important factors used for FbD optimization of dendrimers include concentration of reactant, catalyst as the CMAs to achieve the desired yield of a particular generation of dendrimer. Table 3.12 illustrates an updated literature on the application of the FbD approach for optimization of the dendrimer-mediated DDS.

3.7 CONCLUSIONS

A formulation scientist can derive unique benefits of FbD employing multivariate DOE approaches in rational development and formulation optimization of various nanostructured DDS and processes associated with them. FbD merits extend beyond the selection of the optimum formulation. It helps in improving the product characteristics as a function of change in process or product parameter(s). Today, the federal agencies in terms of QbD need "in-built product quality" rather than testing the quality of the finished product. Comprehending the formulation or process variables rationally using FbD not only would help in attaining product development excellence with phenomenal ease at low cost but also aid in federal compliance too. As a rule, when finding the correct solution is not simple, a pharmaceutical scientist should mandatorily consider the use of FbD for developing novel and nanostructured DDS, wherein the variability and vulnerability of the systems make them ultrasensitive to diverse formulation factors and processes these systems tend to undergo. FbD is a quality-centric approach, which provides enormous benefits to meet the unmet needs of patients as well as pharmaceutical manufacturers for development of efficacious, cost-effective, safe, and robust drug products.

Notwithstanding the immense vitality of this FbD philosophy in the industrial milieu, its importance is enormous in diverse fields beyond the formulation development such as in analytical method development, pharmaceutical manufacturing, and stability studies. This approach not only helps in developing optimal solutions, but also in permeating rational research mindsets toward evolving "out-of-the-box" strategies. Apt implementation of the FbD paradigm, accordingly, would be pivotal in achieving a "win–win situation" not only for patients, drug industry, and regulators, but for the whole gamut of drug delivery scientists too.

REFERENCES

Abdelbary, A.A., AbouGhaly, M.H., 2015. Design and optimization of topical methotrexate loaded niosomes for enhanced management of psoriasis: application of Box-Behnken design, in-vitro evaluation and in-vivo skin deposition study. Int. J. Pharm. 485, 235–243.

Ahad, A., Aqil, M., Kohli, K., Sultana, Y., Mujeeb, M., 2013. Enhanced transdermal delivery of an antihypertensive agent via nanoethosomes: statistical optimization, characterization and pharmacokinetic assessment. Int. J. Pharm. 443, 26–38.

Ailiesei, I., Anuta, V., Mircioiu, C., Cojocaru, V., Orbesteanu, A.M., Cinteza, L.O., 2016. Application of statistical design of experiments for the optimization of clodronate loaded liposomes for oral administration. Rev. Chim. 67, 1566–1570.

Ajazuddin, Alexander, A., Khan, J., Giri, T.K., Tripathi, D.K., Saraf, S., 2012. Advancement in stimuli triggered in situ gelling delivery for local and systemic route. Expert Opin. Drug Deliv. 9, 1573–1592.

Alam, S., Aslam, M., Khan, A., Imam, S.S., Aqil, M., Sultana, Y., et al., 2016. Nanostructured lipid carriers of pioglitazone for transdermal application: From experimental design to bioactivity detail. Drug Deliv. 23, 601–609.

Alam, T., Pandit, J., Vohora, D., Aqil, M., Ali, A., Sultana, Y., 2015. Optimization of nanostructured lipid carriers of lamotrigine for brain delivery: in vitro characterization and in vivo efficacy in epilepsy. Expert Opin. Drug Deliv. 12, 181–194.

Alexis, F., Pridgen, E., Molnar, L.K., Farokhzad, O.C., 2008. Factors affecting the clearance and biodistribution of polymeric nanoparticles. Mol. Pharm. 5, 505–515.

Aliabadi, H.M., Lavasanifar, A., 2006. Polymeric micelles for drug delivery. Expert Opin. Drug Deliv. 3, 139–162.

Amini, N., Kalaee, M., Mazinani, S., Pilevar, S., Ranaei-Siadat, S.-O., 2013. Morphological optimization of electrospun polyacrylamide/MWCNTs nanocomposite nanofibers using Taguchi experimental design. Int. J. Adv. Mfg. Technol. 69, 139–146.

Andalib, S., Varshosaz, J., Hassanzadeh, F., Sadeghi, H., 2012. Optimization of LDL targeted nanostructured lipid carriers of 5-FU by a full factorial design. Adv. Biomed. Res. 1, 45.

Arulsudar, N., Subramanian, N., Muthy, R.S., 2005. Comparison of artificial neural network and multiple linear regression in the optimization of formulation parameters of leuprolide acetate loaded liposomes. J Pharm. Pharm. Sci. 8, 243–258.

Asthana, G.S., Sharma, P.K., Asthana, A., 2016. In vitro and in vivo evaluation of niosomal formulation for controlled delivery of clarithromycin. Scientifica 2016, 1–10.

Attivi, D., Wehrle, P., Ubrich, N., Damge, C., Hoffman, M., Maincent, P., 2005. Formulation of insulin-loaded polymeric nanoparticles using response surface methodology. Drug Dev. Ind. Pharm. 31, 179–189.

Avasatthi, V., Pawar, H., Dora, C.P., Bansod, P., Gill, M.S., Suresh, S., 2016. A novel nanogel formulation of methotrexate for topical treatment of psoriasis: optimization, in vitro and in vivo evaluation. Pharm. Dev. Technol. 21, 554–562.

Bandyopadhyay, S., Katare, O.P., Singh, B., 2012. Optimized self nano-emulsifying systems of ezetimibe with enhanced bioavailability potential using long chain and medium chain triglycerides. Colloids Surf. B Biointerfaces 100, 50–61.

Barbosa, M.V., Monteiro, L.O., Carneiro, G., Malagutti, A.R., Vilela, J.M., Andrade, M.S., et al., 2015. Experimental design of a liposomal lipid system: a potential strategy for paclitaxel-based breast cancer treatment. Colloids Surf. B Biointerfaces 136, 553–561.

Basalious, E.B., Shawky, N., Badr-Eldin, S.M., 2010. SNEDDS containing bioenhancers for improvement of dissolution and oral absorption of lacidipine. I: development and optimization. Int. J. Pharm. 391, 203–211.

Beck, R.S.R., Haas, S.A., Guterres, S.S., Rã, M.I., Benvenutti, E.V., Pohlmann, A.R., 2006. Nanoparticle-coated organic-inorganic microparticles: experimental design and gastrointestinal tolerance evaluation. Quím Nova 29, 1–10.

Beck-Broichsitter, M., Schmehl, T., Gessler, T., Seeger, W., Kissel, T., 2011. Development of a biodegradable nanoparticle platform for sildenafil: formulation optimization by factorial design analysis combined with application of charge-modified branched polyesters. J. Control. Release 157, 469–477.

Beg, S., Sharma, G., Thanki, K., Jain, S., Katare, O.P., Singh, B., 2015. Positively charged self-nanoemulsifying oily formulations of olmesartan medoxomil: Systematic development, in vitro, ex vivo and in vivo evaluation. Int. J. Pharm. 493, 466–482.

Bhatia, A., Amarji, B., Singh, B., Raza, K., Katare, O.P., 2009. Development, optimization and evaluation of diclofenac loaded flexible liposomes, Canadian society for pharmaceutical sciences. In: 12th Annual Meeting on Drug Development to Regulatory Approval, Toronto, Ontario, Canada.

Bhatia, A., Singh, B., Raza, K., Wadhwa, S., Katare, O.P., 2013. Tamoxifen-loaded lecithin organogel (LO) for topical application: development, optimization and characterization. Int. J. Pharm. 444, 47–59.

Bhosale, U.V., Devi, V.K., Jain, N., 2011. Formulation and optimization of mucoadhesive nanodrug delivery system of acyclovir. J. Young Pharm. 3, 275–283.

Box, G.E.P., Wilson, K.B., 1951. On the experimental attainment of optimum conditions. J. R. Stat. Soc. Ser. B 13, 1–45.

Box, G.E.P., Draper, N.R., 1987. Empirical Model-Building and Response Surfaces, first ed. Wiley, New York, NY.

Box, G.E.P., Hunter, W.G., Hunter, J.S., 1978. Statistics for Experimenters. Wiley, New York, NY.

Bozkir, A., Saka, O.M., 2005. Formulation and investigation of 5-FU nanoparticles with factorial design-based studies. Farmaco 60, 840–846.

Calvo, P., Sanchez, A., Martinez, J.S., Lapez, M.A., Calonge, M., Pastor, J.C., et al., 1996. Polyester nanocapsules as new topical ocular delivery systems for cyclosporin A. Pharm. Res. 13, 311–315.

Chang, L.-C., Wu, C.-L., Liu, C.-W., Chuo, W.-H., Li, P.-C., Tsai, T.-R., 2011. Preparation, L. Keshri, K. Pathak, (2013). Development of thermodynamically stable nanostructured lipid carrier system using central composite design for zero order permeation of Econazole nitrate through epidermis. Pharm. Dev. Technol. 18, 634–644.

Chaudhary, H., Kohli, K., Amin, S., Rathee, P., Kumar, V., 2011. Optimization and formulation design of gels of diclofenac and curcumin for transdermal drug delivery by Box-Behnken statistical design. J. Pharm. Sci. 100, 580–593.

Cui, J., Yu, B., Zhao, Y., Zhu, W., Li, H., Lou, H., et al., 2009. Enhancement of oral absorption of curcumin by self-microemulsifying drug delivery systems. Int. J. Pharm. 371, 148–155.

Dhapte, V., Pokharkar, V., 2011. Polyelectrolyte stabilized antimalarial nanosuspension using factorial design approach. J. Biomed. Nanotechnol. 7, 139–141.

Dhawan, S., Kapil, R., Singh, B., 2011. Formulation development and systematic optimization of solid lipid nanoparticles of quercetin for improved brain delivery. J. Pharm. Pharmacol. 63, 342–351.

Đorđević, S.M., Radulović, T.S., Cekić, N.D., Ranđelović, D.V., Savić, M.M., Krajišnik, D.R., et al., 2013. Experimental design in formulation of diazepam nanoemulsions: physicochemical and pharmacokinetic performances. J. Pharm. Sci. 102, 4159–4172.

El-Gibaly, I., Abdel-Ghaffar, S.K., 2005. Effect of hexacosanol on the characteristics of novel sustained-release allopurinol solid liposheres (SLS): factorial design application and product evaluation. Int. J. Pharm. 294, 33–51.

Elmizadeh, H., Khanmohammadi, M., Ghasemi, K., Hassanzadeh, G., Nassiri-Asl, M., Garmarudi, A.B., 2013. Preparation and optimization of chitosan nanoparticles and magnetic chitosan nanoparticles as delivery systems using Box-Behnken statistical design. J. Pharm. Biomed. Anal. 80, 141–146.

El-Naggar Nel, A., Abdelwahed, N.A., 2014. Application of statistical experimental design for optimization of silver nanoparticles biosynthesis by a nanofactory Streptomyces viridochromogenes. J. Microbiol. 52, 53–63.

Esmaeili, A., Gholami, M., 2015. Optimization and preparation of nanocapsules for food applications using two methodologies. Food Chem. 179, 26–34.

Gao, H., Fan, Y., Wang, D., Hu, Y., Liu, J., Zhao, X., et al., 2012. Optimization on preparation condition of epimedium polysaccharide liposome and evaluation of its adjuvant activity. Int. J. Biol. Macromol. 50, 207–213.

Garala, K., Joshi, P., Shah, M., Ramkishan, A., Patel, J., 2013. Formulation and evaluation of periodontal in situ gel. Int. J. Pharm. Investig. 3, 29–41.

Garala, K.C., Shinde, A.J., More, H.N., 2009. Solubility enhancement of aceclofenac using dendrimer. Res. J. Pharma Dosage Forms Tech. 1, 94–96.

Garg, B., Katare, O.P., Beg, S., Lohan, S., Singh, B., 2016. Systematic development of solid self-nanoemulsifying oily formulations (S-SNEOFs) for enhancing the oral bioavailability and intestinal lymphatic uptake of lopinavir. Colloids Surf. B Biointerfaces 141, 611–622.

Ghanbarzadeh, S., Valizadeh, H., Zakeri-Milani, P., 2013. Application of response surface methodology in development of sirolimus liposomes prepared by thin film hydration technique. Bioimpacts 3, 75–81.

Glogard, C., Stensrud, G., Hovland, R., Fossheim, S.L., Klaveness, J., 2002. Liposomes as carriers of amphiphilic gadolinium chelates: the effect of membrane composition on incorporation efficacy and in vitro relaxivity. Int. J. Pharm. 233, 131–140.

Gonzalez-Mira, E., Egea, M.A., Souto, E.B., Calpena, A.C., García, M.L., 2011. Optimizing flurbiprofen-loaded NLC by central composite factorial design for ocular delivery. Nanotechnology 22, 4.

Gonzalez-Rodriguez, M.L., Barros, L.B., Palma, J., Gonzalez-Rodriguez, P.L., Rabasco, A.M., 2007. Application of statistical experimental design to study the formulation variables influencing the coating process of lidocaine liposomes. Int. J. Pharm. 337, 336–345.

Hadidi, N., Nazari, N., Aboofazeli, R., 2009. Formulation and optimization of microemulsion-based organogels containing propranolol hydrochloride using experimental design methods. DARU 17, 217–224.

Hadidi, N., Kobarfard, F., Nafissi-Varcheh, N., Aboofazeli, R., 2011. Optimization of single-walled carbon nanotube solubility by noncovalent PEGylation using experimental design methods. Int. J. Nanomedicine 6, 737–746.

Hao, J., Fang, X., Zhou, Y., Wang, J., Guo, F., Li, F., et al., 2011. Development and optimization of solid lipid nanoparticle formulation for ophthalmic delivery of chloramphenicol using a Box-Behnken design. Int. J. Nanomedicine 6, 683–692.

Hao, J., Gao, Y., Zhao, J., Zhang, J., Li, Q., Zhao, Z., et al., 2015. Preparation and optimization of resveratrol nano-suspensions by antisolvent precipitation using Box-Behnken design. AAPS PharmSciTech. 16, 118–128.

Harsha, S., Attimard, M., Khan, T.A., Nair, A.B., Aldhubiab, B.E., Sangi, S., et al., 2013. Design and formulation of mucoadhesive microspheres of sitagliptin. J. Microencap. 30, 257–264.

Hashem, F.M., Al-Sawahli, M.M., Nasr, M., Ahmed, O.A., 2015. Optimized zein nanospheres for improved oral bioavailability of atorvastatin. Int. J. Nanomedicine 10, 4059–4069.

Hassan, A.O., Elshafeey, A.H., 2010. Nanosized particulate systems for dermal and transdermal delivery. J. Biomed. Nanotechnol. 6, 621–633.

He, Z., Xing, J., Kong, H., Xu, H., 2009. Optimization of preparation parameters for gentamicin-loaded PLLA nanoparticles and the drug release behavior in vitro. Sheng Wu Yi Xue Gong Cheng Xue Za Zhi 26, 351–355.

Honary, S., Ebrahimi, P., Ghasemitabar, M., 2013. Preparation of gold nanoparticles for biomedical applications using chemometric technique. Trop. J. Pharm. Res. 12, 295–298.

Ibrahim, H.M., Ahmed, T.A., Hussain, M.D., Rahman, Z., Samy, A.M., Kaseem, A.A., et al., 2014. Development of meloxicam in situ implant formulation by quality by design principle. Drug Dev. Ind. Pharm. 40, 66–73.

Ismaila, A.A., El-Midanyb, T.A., Abdel-Aala, E.A., El-Shall, H., 2005. Application of statistical design to optimize the preparation of ZnO nanoparticles via hydrothermal technique. Mat. Lett., 1924–1928.

Jain, A.K., Thanki, K., Jain, S., 2014. Solidified self-nanoemulsifying formulation for oral delivery of combinatorial therapeutic regimen: part I. Formulation development, statistical optimization, and in vitro characterization. Pharm. Res. 31, 923–945.

Joshi, A.S., Patel, H.S., Belgamwar, V.S., Agrawal, A., Tekade, A.R., 2012. Solid lipid nanoparticles of ondansetron HCl for intranasal delivery: development, optimization and evaluation. J. Mater. Sci. Mater. Med. 23, 2163–2175.

Kalam, M.A., Khan, A.A., Khan, S., Almalik, A., Alshamsan, A., 2016. Optimizing indomethacin-loaded chitosan nanoparticle size, encapsulation, and release using Box-Behnken experimental design. Int. J. Biol. Macromol. 87, 329–340.

Karakucuk, A., Celebi, N., Teksin, Z.S., 2016. Preparation of ritonavir nanosuspensions by microfluidization using polymeric stabilizers: I. A Design of Experiment approach. Eur. J. Pharm. Sci. 95, 111–121.

Kasinathan, N., Volety, S.M., Josyula, V.R., 2014. Application of experimental design in preparation of nanoliposomes containing hyaluronidase. J. Drug Deliv. 2014, 1–7.

Kelmann, R.G., Kuminek, G., Teixeira, H.F., Koester, L.C.S., 2007. Carbamazepine parenteral nanoemulsions prepared by spontaneous emulsification process. Int. J. Pharm. 342, 231–239.

Keshri, L., Pathak, K., 2012. Development of thermodynamically stable nanostructured lipid carrier system using central composite design for zero order permeation of econazole nitrate through epidermis. Pharm. Dev. Technol. 18 (3), 634–644.

Khajeh, M., 2009. Application of Box-Behnken design in the optimization of a magnetic nanoparticle procedure for zinc determination in analytical samples by inductively coupled plasma optical emission spectrometry. J. Hazard. Mater. 172, 385–389.

Kollipara, S., Bende, G., Movva, S., Saha, R., 2010. Application of rotatable central composite design in the preparation and optimization of poly(lactic-co-glycolic acid) nanoparticles for controlled delivery of paclitaxel. Drug Dev. Ind. Pharm. 36, 1377–1387.

Konwarh, R., Misra, M., Mohanty, A.K., Karak, N., 2013. Diameter-tuning of electrospun cellulose acetate fibers: a Box-Behnken design (BBD) study. Carbohydr. Polym. 92, 1100–1106.

Kukovecz, A., Méhn, D., Nemes-Nagy, E., Szabó, R., Kiricsi, I., 2005. Optimization of CCVD synthesis conditions for single-wall carbon nanotubes by statistical design of experiments (DoE). Carbon 43, 2842–2849.

Kumar, S., Gokhale, R., Burgess, D.J., 2014. Quality by Design approach to spray drying processing of crystalline nanosuspensions. Int. J. Pharm. 464, 234–242.

Leal-Cruz, A.L., Berman-Mendoza, D., Vera-Marquina, A., García-Juárez, A., Villa Velazquez-Mendoza, C., Zaldívar-Huerta, I.E., 2013. Synthesis and characterization of CBD-CdS Nanospheres. IEEE 978, 4673–6155.

Lewis, G.A., Mathieu, D., Phan-Tan-Luu, R., 1999. Pharmaceutical Experimental Design, first ed. Marcel Dekker, New York, NY, Singh, B., Ahuja, N. (2000). Book Review on Pharmaceutical Experimental Design. Int. J. Pharm. 195, 247–248.

Lien, C.-H., Chang, K.-H., Hu, C.C., Wang, D.S.-H., 2013. Optimizing bismuth-modified graphene-carbon nanotube composite-coated screen printed electrode for lead-ion sensing through the experimental design strategy. J. Elect. Chem. Soc. 160, B107–B112.

Lim, J.H., Lee, J.S., 2008. A statistical design and analysis illustrating the interactions between key experimental factors for the synthesis of silver nanoparticles. Colloids Surf. A 322, 155–163.

Liu, C.H., Wu, C.T., Fang, J.Y., 2010. Characterization and formulation optimization of solid lipid nanoparticles in vitamin K1 delivery. Drug 36, 751–761.

Liu, D., Xu, H., Tian, B., Yuan, K., Pan, H., Ma, S., et al., 2012. Fabrication of carvedilol nanosuspensions through the anti-solvent precipitation-ultrasonication method for the improvement of dissolution rate and oral bioavailability. AAPS PharmSciTech. 13, 295–304.

Liu, Y., Zhang, P., Feng, N., Zhang, X., Wu, S., Zhao, J., 2009. Optimization and in situ intestinal absorption of self-microemulsifying drug delivery system of oridonin. Int. J. Pharm. 365, 136–142.

Lohit, R. Kaur, B. Garg, B. Singh B, 2012. Development of "optimized" nanoparticles in stimuli-bioresponsive in situ gelling periodontal systems of ornidazole. In: 64nd Indian Pharmaceutical Congress (IPC). Abstract No. AX-277, Chennai, India.

LotfizadehDehkordi, B., Ghadimi, A., Metselaar, H.C., 2013. Box-Behnken experimental design for investigation of stability and thermal conductivity of TiO_2 nanofluids. J. Nanopart. Res. 1369 (15), 1–9.

Mahjub, R., Dorkoosh, F.A., Amini, M., Khoshayand, M.R., Rafiee-Tehrani, M., 2011. Preparation, statistical optimization, and in vitro characterization of insulin nanoparticles composed of quaternized aromatic derivatives of chitosan. AAPS PharmSciTech. 12, 1407–1419.

Marlowe, E., Shangraw, R.F., 1967. Dissolution of sodium salicylate from tablet matrices prepared by wet granulation and direct compression. J. Pharm. Sci. 56, 498–504.

Mehanna, M.M., Elmaradny, H.A., Samaha, M.W., 2009. Ciprofloxacin liposomes as vesicular reservoirs for ocular delivery: formulation, optimization, and in vitro characterization. Drug Dev. Ind. Pharm. 35, 583–593.

Mezghrani, O., Ke, X., Bourkaib, N., Xu, B.H., 2011. Optimized self-microemulsifying drug delivery systems (SMEDDS) for enhanced oral bioavailability of astilbin. Pharmazie 66, 754–760.

Mosley, G.L., Yamanishi, C.D., Kamei, D.T., 2012. Mathematical modeling of vesicle drug delivery systems 1: vesicle formation and stability along with drug loading and release. J. Lab. Autom. 18, 34–45.

Mura, P., Capasso, G., Maestrelli, F., Furlanetto, S., 2008. Optimization of formulation variables of benzocaine liposomes using experimental design. J. Liposome Res. 18, 113–125.

Murdan, S., 2005. Organogels in drug delivery. Expert Opin. Drug Deliv. 2, 489–505.

Myers, W.R., 2003. Response surface methodology. In: Chow, S.C. (Ed.), Encyclopedia of Biopharmaceutical Statistics. Marcel Dekker, New York, NY.

Nakarani, M., Misra, A.K., Patel, J.K., Vaghani, S.S., 2012. Itraconazole nanosuspension for oral delivery: formulation, characterization and in vitro comparison with marketed formulation. Daru 18, 84–90.

Nakarani, M., Patel, P., Patel, J., Murthy, R.S., Vaghani, S.S., 2010. Cyclosporine a-nanosuspension: formulation, characterization and in vivo comparison with a marketed formulation. Sci. Pharm. 78, 345–361.

Nandi M., Kaur, R., Garg, B., Singh, B., 2012. Systematic development of optimized periodontal drug delivery systems of ofloxacin and ornidazole using stimuli bioresponsive smart polymers. In: International Conference on "Frontiers in Nanoscience, Nanotechnology and their Applications" NanoSciTech 2012, Chandigarh. Abstract D11.

Nasr, M., Mansour, S., Mortada, N.D., El Shamy, A.A., 2008. Liposomes as carriers for topical delivery of aceclofenac: preparation, characterization and in vivo evaluation. AAPS PharmSciTech. 9, 154–162.

Negi, P., Singh, B., Sharma, G., Beg, S., Katare, O.P., 2015. Biocompatible lidocaine and prilocaine loaded-nanoemulsion system for enhanced percutaneous absorption: QbD-based optimisation, dermatokinetics and in vivo evaluation. J. Microencap. 32, 419–431.

Negi, P., Singh, B., Sharma, G., Beg, S., Raza, K., Katare, O.P., 2016. Phospholipid microemulsion-based hydrogel for enhanced topical delivery of lidocaine and prilocaine: QbD-based development and evaluation. Drug Deliv. 23, 941–957.

Palamakula, A., Nutan, M.T., Khan, M.A., 2004. Response surface methodology for optimization and characterization of limonene-based coenzyme Q10 self-nanoemulsified capsule dosage form. AAPS PharmSciTech. 5, e66.

Patel, N., Thakkar, V., Metalia, V., Baldaniya, L., Gandhi, T., Gohel, M., 2016. Formulation and development of ophthalmic in situ gel for the treatment ocular inflammation and infection using application of quality by design concept. Drug Dev. Ind. Pharm. 42, 1406–1423.

Patil, P., Joshi, P., Paradkar, A., 2004. Effect of formulation variables on preparation and evaluation of gelled self-emulsifying drug delivery system (SEDDS) of ketoprofen. AAPS PharmSciTech. 5, e42.

Patil, P., Patil, V., Paradkar, A., 2007. Formulation of a self-emulsifying system for oral delivery of simvastatin: in vitro and in vivo evaluation. Acta Pharm. 57, 111–122.

Peng, X., Wang, J., Song, H., Cui, D., Li, L., Li, J., et al., 2012. Optimized preparation of celastrol-loaded polymeric nanomicelles using rotatable central composite design and response surface methodology. J. Biomed. Nanotechnol. 8, 491–499.

Plackett, R.L., Burman, J.P., 1946. The design of optimum multifactorial experiments. Biometrica 33, 305–325.

Prasad, P.S., Imam, S.S., Aqil, M., Sultana, Y., Ali, A., 2016. QbD-based carbopol transgel formulation: characterization, pharmacokinetic assessment and therapeutic efficacy in diabetes. Drug Deliv. 23, 1057–1066.

Rabinow, B.E., 2004. Nanosuspensions in drug delivery. Nat. Rev. Drug Discov. 3, 785–796.

Ravi, P.R., Vats, R., Dalal, V., Murthy, A.N., 2014. A hybrid design to optimize preparation of lopinavir loaded solid lipid nanoparticles and comparative pharmacokinetic evaluation with marketed lopinavir/ritonavir coformulation. J. Pharm. Pharmacol. 66, 912–926.

Raza, K., Singh, B., Singal, P., Wadhwa, S., Katare, O.P., 2012. Systematically optimized biocompatible isotretinoin-loaded solid lipid nanoparticles (SLNs) for topical treatment of acne. Colloids Surf. B Biointerfaces 105C, 67–74.

Raza, K., Katare, O.P., Setia, A., Bhatia, A., Singh, B., 2013. Improved therapeutic performance of dithranol against psoriasis employing systematically optimized nanoemulsomes. J. Microencapsul. 30, 225–236.

Sanad, R.A., AbdelMalak, N.S., elBayoomy, T.S., Badawi, A.A., 2010. Formulation of a novel oxybenzone-loaded nanostructured lipid carriers (NLCs). AAPS PharmSciTech. 11, 1684–1694.

Scheffe, H., 1958. Experiments with mixtures. J. R. Stat. Soc. Ser. B. 20, 344–360.

Schwartz, J.B., Flamholz, J.R., Press, R.H., 1973. Computer optimization of pharmaceutical formulations. I. General procedure. J. Pharm. Sci. 62, 1165–1170.

Shah, M., Chuttani, K., Mishra, A.K., Pathak, K., 2010a. Oral solid compritol 888 ATO nanosuspension of simvastatin: optimization and biodistribution studies. Drug Dev. Ind. Pharm. 37, 526–537.

Shah, M., Pathak, K., 2010b. Development and statistical optimization of solid lipid nanoparticles of simvastatin by using 2(3) full-factorial design. AAPS PharmSciTech. 11, 489–496.

Shah, R.A., Mehta, M.R., Patel, D.M., Patel, C.N., 2011. Design and optimization of mucoadhesive nasal in situ gel containing sodium cromoglycate using factorial design. Asian J. Pharm. 5, 65–74.

Shah, M.K., Madan, P., Lin, S., 2013a. Preparation, in vitro evaluation and statistical optimization of carvedilol-loaded solid lipid nanoparticles for lymphatic absorption via oral administration. Pharm. Res. 19, 475–485.

Shah, S.R., Parikh, R.H., Chavda, J.R., Sheth, N.R., 2013b. Self-nanoemulsifying drug delivery system of glimepiride: design, development, and optimization. PDA J. Pharm. Sci. 67, 201–213.

Shahi, S.R., Dabir, P.D., Nawale, R.B., Wagh, D.G., Deore, S.V., 2016. QbD approach in formulation and evaluation of gelrite based in situ ophthalmic gel of bromfenac sodium sesquihydrate. World J. Pharm. Pharm. Sci. 5, 1790–1805.

Sharma, G., Beg, S., Thanki, K., Katare, O.P., Jain, S., Kohli, K., et al., 2015. Systematic development of novel cationic self-nanoemulsifying drug delivery systems of candesartan cilexetil with enhanced biopharmaceutical performance. RSC Adv. 5, 71500–71513.

Shi, J., Ma, F., Wang, X., Wang, F., Liao, H., 2012. Formulation of liposomes gels of paeonol for transdermal drug delivery by Box-Behnken statistical design. J. Microencapsul. 22, 270–278.

Shivakumar, H.N., Patel, P.B., Desai, B.G., Ashok, P., Arulmozhi, S., 2007. Design and statistical optimization of glipizide loaded lipospheres using response surface methodology. Acta Pharm. 57, 269–285.

Silva, H.D., Cerqueira, M.A., Souza, B.W.S., Ribeiro, C., Avides, M.C., Quintas, M.A.C., et al., 2011. Effects of small and large molecule emulsifiers on the characteristics of β-carotene nanoemulsions prepared by high pressure homogenization. J. Food Eng. 102, 130–135.

Singh, B., Ahuja, N., 2004. Response surface optimization of drug delivery systems. In: Jain, N.K. (Ed.), Progress in Controlled and Novel Drug Delivery Systems, first ed. CBS Publishers and Distributors, New Delhi, pp. 470–509.

Singh, B., Kumar, R., Ahuja, N., 2005a. Optimizing drug delivery systems using systematic "design of experiments." Part I: fundamental aspects. Crit. Rev. Ther. Drug Carrier Syst. 22, 27–105.

Singh, B., Dahiya, M., Saharan, V., Ahuja, N., 2005b. Optimizing drug delivery systems using "Design of Experiments" Part II: Retrospect and prospects. Crit. Rev. Ther. Drug Carrier Syst. 22, 215–292.

Singh, B., Mehta, G., Kumar, R., Bhatia, A., Ahuja, N., Katare, O.P., 2005c. Design, development and optimization of nimesulide-loaded liposomal systems for topical application. Curr. Drug Deliv. 2, 143–153.

Singh, B., Bandopadhyay, S., Kapil, R., Singh, R., Katare, O.P., 2009a. Self-emulsifying drug delivery systems (SEDDS): formulation development, characterization, and applications. Crit. Rev. Ther. Drug Carrier Syst. 26, 427–521.

Singh, B., Bandyopadhyay, S., Kapil, R., Katare, O.P., 2009b. Novel nanostructured lipidic drug delivery systems. Pharma Rev. 118–122.

Singh, B., Kaur, A., Nandi, M., Bandyopadhyay, S., 2010a. Formulation development and optimization of novel stimuli-responsive gastroretentive drug delivery systems of acyclovir using smart polymer blends. In: 62nd Indian Pharmaceutical Congress (IPC). Manipal, 17-19 December 2010: Abstract No. A-191, IPC-2010. Manipal, Karnataka.

Singh, B., Sharma, G., Tripathi, C.B., Bandyopadhyay, S., 2010b. Formulation and Optimization Positively Charged Self Nanoemulsifying System of Candesartan with Enhanced Bioavailability Potential. Biotechnica Chandigarh, Panjab University, Chandigarh.

Singh, B., Kapil, R., Nandi, M., Ahuja, N., 2011a. Developing oral drug delivery systems using formulation by design: vital precepts, retrospect and prospects. Expert Opin. Drug Deliv. 8, 1341–1360.

Singh, B., Khurana, L., Bandyopadhyay, S., Kapil, R., Katare, O.O., 2011b. Development of optimized self-nano-emulsifying drug delivery systems (SNEDDS) of carvedilol with enhanced bioavailability potential. Drug Deliv. 18, 599–612.

Singh, B., Singh, R., Bandyopadhyay, S., Kapil, R., Garg, B., 2012. Optimized nanoemulsifying systems with enhanced bioavailability of carvedilol. Colloids Surf. B Biointerfaces 101C, 465–474.

Singh, B., Kaur, A., Dhiman, S., Garg, B., Khurana, R.K., Beg, S., 2016. QbD-enabled development of novel stimuli-responsive gastroretentive systems of acyclovir for improved patient compliance and biopharmaceutical performance. AAPS PharmSciTech. 17, 454–465.

Solanki, A.B., Parikh, J.R., Parikh, R.H., 2007. Formulation and optimization of piroxicam proniosomes by 3-factor, 3-level Box-Behnken design. AAPS PharmSciTech. 8, E86.

Sood, S., Jain, K., Gowthamarajan, K., 2014. Optimization of curcumin nanoemulsion for intranasal delivery using design of experiment and its toxicity assessment. Colloids Surf. B Biointerfaces 113, 330–337.

Sousa, M.D., Pessine, F.B.T., 2015. Production of mannosylated solid lipid nanoparticles by using experimental design: application to saquinavir. J. Pharm. Sci. Pharmacol. 2, 64–72.

Subramanian, N., Ray, S., Ghosal, S.K., Bhadra, R., Moulik, S.P., 2004. Formulation design of self-microemulsifying drug delivery systems for improved oral bioavailability of celecoxib. Biol. Pharm. Bull. 27, 1993–1999.

Svenson, S., 2004. Carrier-Based Drug Delivery. American Chemical Society, New York, NY.

Tan, Q.Y., Xu, M.L., Wu, J.Y., Yin, H.F., Zhang, J.Q., 2012. Preparation and characterization of poly(lactic acid) nanoparticles for sustained release of pyridostigmine bromide. Pharmazie 67, 311–318.

Taveira, S.N.F., Araújo, L.M., de Santana, D.C.A.S., Nomizo, A., de Freitas, L.A.P., Lopez, R.F.V., 2012. Development of cationic solid lipid nanoparticles with factorial design-based studies for topical administration of doxorubicin. J. Biomed. Nanotechnol. 8, 219–228.

Tefas, L.R., Tomuta, I., Achim, M., Vlase, L., 2015. Development and optimization of quercetin-loaded PLGA nanoparticles by experimental design. Clujul Med. 88, 214–223.

Tewa-Tagne, P., Degobert, G., Briançon, S.P., Bordes, C., Gauvrit, J.-Y., Lanteri, P., et al., 2007. Spray-drying nanocapsules in presence of colloidal silica as drying auxiliary agent: formulation and process variables optimization using experimental designs. Pharm. Res. 24, 650–661.

Vaishya, R.D., Gokulgandhi, M., Patel, S., Minocha, M., Mitra, A.K., 2014. Novel dexamethasone-loaded nanomicelles for the intermediate and posterior segment uveitis. AAPS PharmSciTech. 15, 1238–1251.

Varshosaz, J., Keihanfar, M., 2001. Development and evaluation of sustained-release propranolol wax microspheres. J. Microencapsul. 18, 277–284.

Varshosaz, J., Tabbakhian, M., Mohammadi, M.Y., 2010. Formulation and optimization of solid lipid nanoparticles of buspirone HCl for enhancement of its oral bioavailability. J. Liposome Res. 20, 286–296.

Wagh, V.D., Deshmukh, O.J., 2012. Itraconazole niosomes drug delivery system and its antimycotic activity against *Candida albicans*. ISRN Pharm. 2012, 653465.

Wang, Y., Cong, Z., Liu, Q., Ling, J., Zhou, L., 2009. Study on optimization of formulation and preparation process of sinomenine liposomes. Zhongguo Zhong Yao Za Zhi 34, 275–278.

Water, J.J., Kim, Y., Maltesen, M.J., Franzyk, H., Foged, C., Nielsen, H.M., 2015. Hyaluronic acid-based nanogels produced by microfluidics-facilitated self-assembly improves the safety profile of the cationic host defense peptide novicidin. Pharm. Res. 32, 2727–2735.

Wei, Z., Hao, J., Yuan, S., Li, Y., Juan, W., Sha, X., et al., 2009. Paclitaxel-loaded Pluronic P123/F127 mixed polymeric micelles: formulation, optimization and in vitro characterization. Int. J. Pharm. 376, 176–185.

Xiong, Y., Liu, Q.D., Lai, L., Chen, J.H., 2009. Preparation of the oral self-microemulsifying drug delivery system of GBE50. Yao Xue Xue Bao 44, 803–808.

Yang, X., Patel, S., Sheng, Y., Pal, D., Mitra, A.K., 2014. Statistical design for formulation optimization of hydrocortisone butyrate-loaded PLGA nanoparticles. AAPS PharmSciTech. 15, 569–587.

Yang, Y., Xu, L., Gao, Y., Wang, Q., Che, X., Li, S., 2012. Improved initial burst of estradiol organogel as long-term in situ drug delivery implant: formulation, in vitro and in vivo characterization. Drug Dev. Ind. Pharm. 38, 550–556.

Yanilmaz, M., Kalaoglu, F., Karakas, H., 2013. Investigation on the effect of process variables on polyurethane nanofibre diameter using a factorial design. Fibres Text. East. Eur. 21, 19–21.

You, X., Wang, R., Tang, W., Li, Y., He, Z., Hu, H., et al., 2010. Self-microemulsifying drug delivery system of patchoulic alcohol to improve oral bioavailability in rats. Zhongguo 35, 694–698.

Yuan, Y., Gao, Y., Maoa, L., Zhao, J., 2008. Optimisation of conditions for the preparation of b-carotene nanoemulsions using response surface methodology. Food Chem. 107, 1300–1306.

Zaghloul, A., Khattab, I., Nada, A., Al-Saidan, S., 2008. Preparation, characterization and optimization of probucol self-emulsified drug delivery system to enhance solubility and dissolution. Pharmazie 63, 654–660.

Zhang, J.Y., Gan, Y., Gan, L., Zhu, C.L., Pan, W.S., 2007. Optimization and evaluation of a new antischistosomal drug QH917 self-microemulsifying drug delivery system. Yao Xue Xue Bao 42, 434–439.

Zhong, Z.R., Liu, J., Deng, Y., Zhang, Z.R., Song, Q.G., Wei, Y.X., et al., 2007. Preparation and characterization of a novel nonviral gene transfer system: procationic-liposome-protamine-DNA complexes. Drug Deliv. 14, 177–183.

Zhu, S., Hong, M., Liu, C., Pei, Y., 2009. Application of Box-Behnken design in understanding the quality of genistein self-nanoemulsified drug delivery systems and optimizing its formulation. Pharm. Dev. Technol. 14, 642–649.

Zidan, A.S., Sammour, O.A., Hammad, M.A., Megrab, N.A., Habib, M.J., Khan, M.A., 2007. Quality by design: understanding the formulation variables of a cyclosporine A self-nanoemulsified drug delivery systems by Box-Behnken design and desirability function. Int. J. Pharm. 332, 55–63.

Current Technologies in Nanomedicine

CHAPTER 4

Liposomal-Based Therapeutic Carriers for Vaccine and Gene Delivery

Mahfoozur Rahman, Sarwar Beg, Amita Verma, Firoz Anwar, Abdus Samad and Vikas Kumar

Contents

4.1 INTRODUCTION

Vaccines are those agents which are mainly used for immunization against infectious diseases. They play a keen role in the eradication of diseases born from viruses, bacteria, and actinomycetes. A vaccine mainly provides active immunity to the body by premature development of specific antibodies for pathogens, which may attack the human body during the life span. Moreover they build up cellular immunity (Moingeon et al., 2002). Basically, vaccines containing inactivated strains or heat-killed strains of disease-causing microorganisms, which is endogenous protein. Conventional-based vaccine delivery has poor targeting ability and reports higher hypersensitive reactions (Moingeon et al., 2002). Other significant problems associated with them are higher toxicity. To minimize such problems, liposomal-based drug delivery systems have gained immense potential and popularity in the application of vaccine drug delivery. There are numerous literatures available on liposomal drug delivery systems which have revealed their safety and effective delivery of vaccine. Such systems can be

Nanotechnology-Based Approaches for Targeting and Delivery of Drugs and Genes.
DOI: http://dx.doi.org/10.1016/B978-0-12-809717-5.00005-1

delivered through parenteral, oral, transmucosal, and/or transcutaneous immunization (Gupta and Vyas, 2011). Thus liposomal carriers provide newer paradigms for noninvasive delivery of vaccines with more immune recognition potential. This enormously explains the mechanism of immune recognition by liposomal carriers in the vaccine delivery (Beg et al., 2013). In the 21st century, gene therapy has emerged as a novel therapy because it aims to eradicate the symptoms of diseases by incorporating a normal function of a gene on a mutated gene into the nucleus of cells (Naldini, 2015; Anderson 1998). However, efficient delivery of gene at physiological conditions into the body is always easier said than done, because the naked DNA and cells possess negative charge and thus entry of gene or DNA into the cell is not likely to be an efficient process (Hacein-Bey-Abina et al., 2002). The successful delivery of genes is critically dependent on the development or use of efficient and safer gene carriers, which are called transfection vectors (Hollon, 2000). In gene therapy, there are two types of vectors commonly used, viral and nonviral. Viral vectors are highly efficient and superior to nonviral carriers (Hollon, 2000). But thereafter they have a significant adverse immunogenic reaction associated with the use of viral carriers and concurrently therefore they increase the demand for nonviral-based gene delivery as the vector of choice (Check, 2003). Development of cationic liposomes have several merits (Gao and Huang, 1991). They have several merits, such as robust manufacture, ease in handling and preparation, ability to inject large lipid:DNA complexes, and poor immunogenic response (Vigneron et al., 1996). Therefore, this chapter highlights the major achievements in the area of liposomal-based vaccine development and gene-mediated targeted delivery has been addressed. Moreover, the challenging issues and future promises of liposomes and their modified forms, such as cationic liposomes, and transfersomes for vaccines and gene delivery are discussed.

4.2 INTRODUCTION TO LIPOSOMES AND THEIR THERAPEUTIC APPLICATIONS IN VACCINE DELIVERY

Liposomes are lipid-based bilayer vesicles with a lamellar structure, which are composed of phospholipids and cholesterol. They provide both closely packed structure and impart vesicle integrity (Moingeon et al., 2002). Due to their biocompatibility and biodegradable characteristics, liposomes have gained wider acceptance in the drug delivery area including vaccine delivery. The antigen may entrap in the hydrophilic and lipophilic structure of the liposomes, which ultimately depends on their hydrophilic/hydrophobic nature (Alving, 1991). The liposome composition of lipid is similar to natural lipid in the cell membrane; this resemblance allows it to permeate higher into the cell and allows direct entry into the reticuloendothelial system via endocytic uptake and further it provides targeted immune responsive action (Beg et al., 2013). Liposomal systems provide highly effective action for inducing the systemic, mucosal, and trans-cutaneous immunization which reflects its versatility in the vaccine delivery system. Liposomes act as a vehicle carrying antigens to the antigen-presenting cells (APCs) and effectively delivering them into the cytoplasm of APCs, these attributes are due to their membrane fusion properties (Moingeon et al., 2002). Intraperitoneal administration of liposomal-based vaccines provides an effective immune response by rapidly reaching the spleen and lymph nodes (Gregoriadis and Allison, 1974; Alving, 1991). Literatures have revealed the encapsulation of water-soluble antigens in the liposomal systems which provide controlled-release profile of delivery for prolonged periods of time. Liposomes also have adjuvant-like action which is similar to lipid A (Gram-negative bacterial polysaccharide) and muramyl peptidase, which assists in augmentation of the immune response (Alving and Rao, 2008). Liposomes are mainly classified as unilamellar vesi-cles (ULVs) and multilamellar vesicles. ULVs are further classified into small unilamellar vesicles, large unilamellar vesicles, and oligolamellar vesicles. There are several common methods employed for the preparation of liposomes such as solvent evaporation, thin film hydration, and the ethanol and ether injection method (Griebel and Hein, 1996). However, these methods have limited application for industrial-scale manufacturing because of the use of organic solvents in high concentration, which may cause severe toxicity. Apart from these, other smarter techniques have been applied like extrusion and supercritical reverse phase evaporation in the preparation of liposomes (Mozafari, 2005; Frederiksen et al., 1997). By applying these methods vesicular size of less than 20 nm was obtained, so they were named nanoliposomes. Moreover, it enhanced the targeting ability to the macrophage organs. Furthermore, such nanoliposomal-based vaccine delivery has been extensively explored for immunization of several diseases like prostate cancer, colorectal carcinoma, influenza, malaria, hepatitis A, and HIV (Zolnik et al., 2010). In this regard, Stimuvax emerged as a nanoliposomal-based vaccine for radical immunization in lung cancer. It acts by targeting the cell surface protein, mucin-1, which is overexpressed in nonsmall-cell lung carcinoma, currently it has taken part in phase III clinical trials (Zolnik et al., 2010). PEGylation of liposome resulted in better attachment of antigens onto the surface of T cells, which provides better tar-geting ability to the class I major histocompatibility complex (MHC) to induce cell-mediated immunity (Mellman and Steinman, 2001). In the same fashion, the presence of cationic charge on the liposome results in enhanced permeability into the targeted APCs as compared to conventional liposomal vaccines (Nakanishi et al., 1999). As for monosylated liposomes containing tetanus toxoid, they show enhanced uptake into the dendritic cells to induce T-cell proliferation over neutral liposomes (Copland et al., 2003). Table 4.1 summarizes the various liposomal vaccines available in the market and many are at various stages of clinical trial. Liposomal vaccine has also proved its effi-cacy against pathogenic organisms like tetanus, diphtheria, pertussis, tuberculosis, and

Table 4.1 Various liposomal vaccines with their clinical developments (Beg et al., 2013)

Name of disease	Liposomal-based vaccines	Targeted antigen (s)	Clinical stage development	Manufacturer
Influenza	Trivalent influenza vaccine (i.m.)	H1N1, H2N2	Under development	Swiss serum, Switzerland
Hepatitis A and B	HAV–HB-IRIV combined vaccine (i.m.)	Hepatitis B conjugated HAV	Under development	Swiss serum, Switzerland
Hepatitis A	Epaxal Berna vaccine (i.m.)	Inactivated hepatitis A	Developed	Swiss serum, Switzerland
Diphtheria, tetanus and hepatitis A	Diphtheria, tetanus/hepatitis A combined vaccine (i.m.)	Diphtheria, tetanus	Developed	Swiss serum, Switzerland
Hepatitis A and B, diphtheria, tetanus, and influenza	Hepatitis A7 B/ diphtheria/ tetanus/influenza supercombined vaccine (i.m.)	Inactivated HAV virions, diphtheria, tetanus, influenza	Under development	Swiss serum, Switzerland
Shigella flexneri 2A infection	*Shigella flexneri* 2A vaccine (oral)	*Shigella flexneri* 2A (killed)	Under development	Novavax
Escherichia coli 015/infection	*Escherichia coli* 0157:H7 vaccine (oral)	*Escherichia coli* 0157:H7 vaccine (killed)	Under development	Novavax
Influenza	IRIV liposomes (trivalent influenza vaccine)	Hemagglutinin/ neuraminidase from influenza strains	Phase III	Swiss serum, Switzerland
Hepatitis A and B, diphtheria, tetanus, and influenza	IRIV liposomes (hepatitis A, B, diphtheria, tetanus)	Inactivated hepatitis A virions; diphtheria and tetanus toxoids	Phase I	Swiss serum, Switzerland
Hepatitis A	IRIV liposomes (Epaxal Berna)	Inactivated hepatitis A virions	Approved in Switzerland	Berna Biologics
Diphtheria, tetanus, and hepatitis A	IRIV liposomes (diphtheria, tetanus, hepatitis A combined vaccine)	Diphtheria and tetanus toxoids; inactivated HAV virions	Phase I	Swiss serum, Switzerland
Hepatitis A and B	IRIV liposomes (combined HAV/HBV)	Genetically engineered hepatitis B	Phase I	Swiss serum, Switzerland

HAV, hepatitis A vaccine; *HBV*, hepatitis B vaccine; *IRIVs*, immunopotentiating reconstituted influenza virosomal carriers.

many more. Lipovaxin is a multiple unit vaccine which was developed by Lipotek Inc. It delivers the antigens into the dendritic cells to mediate the cellular immunity. This multiple unit vaccine therefore has tremendous potential for immunization against malaria and tuberculosis. Moreover, it is also administered through nasal route against *Yersinia pestis* infection (plague) (Henriksen-Lacey et al., 2011). Polymer-coated liposomes have been developed with the help of chitosan, which provides optimal mucosal therapeutic action against diphtheria toxoid (Martin et al., 2005). Furthermore, they also provide gastric protection (Rescia et al., 2011). In another way, addition of immunomodulators into liposomal vaccines loaded with tetanus, diphtheria antigen, results in enhanced immune response. As for tetanus toxoid, encapsulated liposomes with coadjuvants augment the systemic and mucosal immune response (Tafaghodi et al., 2006). Liposomal-based vaccine delivery has also shown promising action against cancer by inducing antibody formation and immune response (Saupe et al., 2006).

4.2.1 Mechanism involved in antigen loading in liposomal vaccines

Antigen is carried by the liposomal system on its surface through an adsorption mechanism or entrapment in its central core. Surface-adsorbed loaded antigen on liposome releases antigen into cells by a lysosmotic mechanism and further facilitates endocytosis uptake via fusion with the lipid bilayers of immune-compromised cells (Reddy et al., 1991; Van Rooijen and van Nieuwmegen, 1980). Moreover, they can easily attach with the T cells by covalent linkages which results in better expression of class I and II MHC to provoke cell-mediated immunity (Raphael and Tom, 1984; Gregoriadis et al., 1987).

4.2.2 Adjuvant characteristic-dependent factors for liposomal vaccines

There are several factors which affect the adjuvant capacity of liposomes, including phospholipid-to-protein ratio, phase transition temperature (Pt) of phospholipids, surface charge, and particle size. Lipid-to-protein ratio refers to the dry weight of lipids as compared to solid proteins. A higher lipid-to-protein ratio destabilizes the antigen by rupture of the vesicular structure (Hedlund et al., 1988; Alpar et al., 2001). Phase transition temperature (Pt) is the final temperature required to change the physical forms of lipids from the sol to gel state. Alteration in this temperature leads to change in the nature of antigenicity in the liposomal vaccines. Moreover, the antigenicity also depends on chain length and degree of saturation of the hydrocarbon chains of the phospholipids and electrostatic properties of the head group. Higher value of Pt makes the liposomes more elastic and helps in better permeation through biological barriers to augment the immune response (Martin et al., 2005). Literature has reported the phospholipid Pt in the range of $-32°C$ to $41.5°C$ shows a good antigen release profile. Enhancement in vesicle rigidity renders poor permeation characteristics and also lower antigen release (Martin et al., 2005). Apart from these, the zeta potential is also a causal element for physical stability and higher permeability of liposomes.

4.3 ARCHAEOSOMES

These are self-assembled lipid-based vesicular carriers, primarily composed of polar phospholipid which is derived from the archaebacteriae *Sulfolobus acidocaldarius*, and is formed by aggregation at below the critical micelle concentration. This carrier is primarily useful in the delivery of antigens through the oral route to induce the systemic immune response (Patel and Sprott, 1999; Krishnan and Dennis Sprott, 2003). The archaeal lipids mainly contain 20–40 carbon atoms with a uniformly branched phytanyl chain which has saturated isoprenoid chains attached through the ether bonds between sn-2, 3 carbon atoms. The archaeal lipid has an absence of the glycerol backbone. This carrier has numerous merits, such as its unique structure, excellent physicochemical stability at high temperature, alkaline pH, and serum proteins. Moreover, archaeosomes have better self-adjuvant property and evoke immune response over conventional liposomes (Patel and Chen, 2005). In the case of mucosal immunization, antigen-loaded archaeosomes are given by the oral route. Further, they are readily taken up by macrophages and release the loaded antigens to initiate the immune cascade or antibody formation by activation of MHC class I and II pathways (Patel et al., 2004). It also provides robust antigen-specific cell-mediated immunity by CD8+ CTL responses for a sustained period and permits the body to read out the immune response in the memory T cells. Therefore, archaeosomes induce both humoral and cell-mediated immunity against intracellular and extracellular pathogens. Patel et al. found better systemic and oral mucosal immunization by parenteral and oral administration of ovalbumin (Patel et al., 2007; Kunisawa et al., 2001).

4.4 TRANSFERSOMES

These are also called elastic liposomes or deformable vesicles, which are to deliver high-molecular-weight antigens noninvasively through the skin for transcutaneous immunization (Vinod et al., 2012). It is primarily composed of phospholipid, cholesterol, along with an edge activator such as ethanol, which reveals its flexibility. Their deformable property makes better penetration through the stratum corneum (Vinod et al., 2012). Dermal application of transfersomes can effectively deliver the antigen materials of >5000 Da to induce the antibody formation and develop a cell-mediated immune response. Thus, these systems are mainly suitable for the diseases which are mainly associated with lymphoid organs such as gut-associated lymphoid tissue, Payer's patches, and Langerhans cells. Gupta et al. (2005) found that transfersomes loaded with tetanus toxoid application after percutaneous immunization received superior efficacy over conventional delivery of tetanus toxoid delivered in alum. Possessing of gap junction protein in transfersomes provides them with better epicutaneous immunization through the skin and gives higher skin permeation over conventional liposomes

(Paul et al., 1998). Other than the oral route, transfersomes also induce immunization through transdermal drug delivery. Therefore this approach is very useful in the delivery of vaccines for treatment of various disease including cancer, HIV, and other infectious diseases (Rai et al., 2008). Surface modification in terms of cationic charge leads to developing cationic transfersomes which are capable of topical delivery of a genetic DNA vaccine against hepatitis B to induce antihepatitis B antibody titer as compared to the naked DNA alone (Mahor et al., 2007).

4.5 IMMUNOPOTENTIATING RECONSTITUTED INFLUENZA VIROSOMAL CARRIERS

Immunopotentiating reconstituted influenza virosomal carriers (IRIVs) are virosomal carriers, which are made up of spherical hexagonal ULVs with a diameter of 150 nm. They are composed of two phospholipids: phosphatidylcholine (PC) and phosphatidylethanolamine (PE) as a backbone (Gluck et al., 1992). Moreover, they provide an immunopotentiating effect. These deliver vaccine antigens and antigenic macromolecular proteins for elicit effective immunization. Unlike other novel carriers they are comparatively safe, having no toxicity or adverse reactions. There are several important features which makes them more efficient colloidal carriers, including induction of innate immunity, adjuvant-like property, and lack of immunogenicity (Moser et al., 2011). They are mainly employed in oral mucosal immunization for the treatment of influenza, AIDS, hepatitis A, B, polyvalent hepatitis A, diphtheria, and tetanus. IRIVs' immune recognition is due to the binding with HA1 globular subunits which are expressed on the sialic acid residue of macrophages and immunocompetent cells, resulting in inducing endocytic uptake into the APCs to produce an immune response (Jackson et al., 1991). However, at acidic pH the IRIVs are anchored on the endosomal surface of HA2 subunits to undergo conformational changes, and further expose the fusogenic peptides to APCs for immune recognition (Poltl-Frank et al., 1999). IRIVs have gained immense potential in the delivery of peptide-based malaria vaccine (SPf66). The vaccine contains reduced SPf66 peptide molecules containing terminal cysteine residues, which covalently attached to phosphatidylethanolamine with a hetero bifunctional crosslinker, i.e., γ-maleimidobutyric acid N-hydroxysuccinimide ester. SPf66-phosphatidylethanolamine-loaded IRIVs have been injected intramuscularly into BALB/c mice, which effectively induced the immune response and revealed better efficacy of IRIV-based vaccine over marketed products (Tamborrini et al., 2011). Tamborrini et al. (2011) developed the virosomal preparations containing *Plasmodium falciparum* GLURP-MSP3 chimeric proteins which provides an enhanced immune response against malaria and concurrently provides higher stability too. Lately, IRIVs have been utilized in protection against influenza which is derived from influenza viruses. Thus, it provides effective immunization against influenza (Calcagnile and

Zuccotti, 2010; Daemen et al., 2000). Besides, IRIVs also provide better immunization against H5N1 virus through the sublingual route as compared to the intranasal and intramuscular routes (Pedersen et al., 2011). IRIV applications are also listed in Table 4.1.

4.6 LIPOSOMAL-BASED GENE DELIVERY

As discussed earlier, the versatility of liposome preparation and their characteristic features are made to incorporate a variety of structures for delivery to mammalian cells (Gregoriadis and Florence, 1993). A number of key factors such as size, surface charge, composition and bilayer fluidity make it a promising therapeutic agent for treating various diseases. DNA entrapments by liposomes have been especially categorized as positive and anionic containing charge or so-called cationic and anionic liposomes (Shim et al., 2013). Thus, both of these have been applied for DNA transfer. Cationic liposomes entrapped the DNA in the peripheral positive charge region, whereas anionic liposomes hold the DNA in the aqueous region (shown in Fig. 4.1). Cationic amphiphiles are used for efficient transfection (Remy et al., 1995). *N*-[1–(2,3-dioleyloxy) propyl]-*N,N,N*-trimethyl-ammonium chloride (DOTMA) has been utilized as a transfectant (Life Technologies Gaithersburg, Maryland, USA). It forms

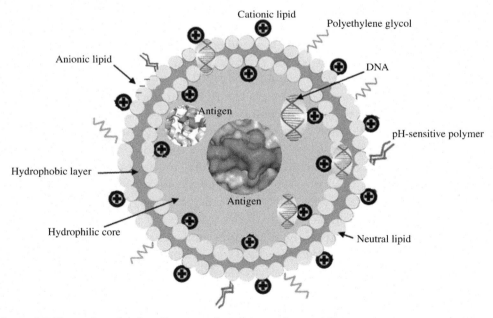

Figure 4.1 Illustration of various liposomes in vaccine and gene delivery.

lipid–DNA complexes resulting in a higher loading efficiency. Apart from this, they are more attractive by virtue of possessing their degradation after transfection and lower toxicity (Tirrell et al., 1985). In this regard, some liposomes destabilize the liposomal membrane at low pH, these are known as pH-sensitive liposomes. The sensitivity of pH liposomes is dependent on the combination of PE with acidic amphiphiles and titrable polymers (Tirrell et al., 1985; Uster and Deamer, 1985) or titrable synthetic peptide (Parente et al., 1988). Besides applications of pH-sensitive liposomes in the delivery of anticancer drugs, antiinflammatory drugs, etc., they also have been utilized to carry exogenous thymidine kinase (*TK*) gene for transfection of mouse L-tk cells, which lack *TK* (Wang and Huang, 1987). The plasmid pPCTK-64 which contains *TK* gene is delivered by liposome to controlled expression of the enzyme in a cAMP-dependent manner. The said system has been widely utilized for in vivo targeted delivery of a bacterial-containing toxin gene to ascites (Wang and Huang, 1987). Liposomal gene delivery is also summarized in Table 4.2.

4.6.1 Mechanism of gene delivery

Cationic liposomes have gained spectacular potential in delivery of nucleic acid for transferring the gene at the in vitro and in vivo level (Legendre and Szoka, 1992). The mechanism involved is an electrostatic interaction between the cationic charge of lipid and negative-charged nucleic acid, as overall this has a positive charge that facilitates binding to the negative-charged cell surface (Felgner and Ringold, 1989). Therefore the delivery of DNA into the target cell is totally dependent upon characteristics of cationic liposomes. Lipopolyamines act as cationic amphiphiles due to positive head on it and self-aggregating nature too. It is easily able to bind DNA or gene (Behr, 1986). Literatures have suggested that cationic liposomes may enter into the cells via two mechanisms, the first is membrane fusion and the second is spontaneous endocytosis (Haywood, 1975). In another way, pH-sensitive liposomes have entered into the target cell by endocytosis, via binding to the cell surface. Once it has entered, it proceeds for lysosome degradation. This depends on time duration involvement in endosomes, which have been categorized as early or late endosomes. At an early phase, pH is more acidic than the external region and in late phase it becomes even more acidic in endosomes (Mellman et al., 1986; Schmid et al., 1988). Liposomal encapsulated antigens are processed in lysosome, recycled, and presented to T cells, whereas pH-sensitive liposomes gives cytoplasmic delivery of loaded material. Cytoplasmic delivery has occurred in three ways: (1) rupture of endosomal membranes; (2) leakage into endosomal lumen; and (3) fusion with endosome membrane. Endocytosis pathways have been utilized for membrane-based receptors in targeted delivery of asialofetuin-labeled liposomes (AF-liposomes) loaded plasmid pCMV beta DNA to rat hepatocytes. Another is the presence of amphiphiles like DOPE, making more efficient delivery (Hara et al., 1996).

Table 4.2 Various liposomal-based gene deliveries with their clinical developments (Martin and Boulikas, 1998)

Nature of disease	Liposomal carrier	In vitro/in vivo study	Route of administration	Clinical stage
Cancer/melanoma/ adenocarcinoma	Cationic liposome complex/DCChol/ HLA-B7/beta-2 microglobulin cDNA	In vivo	Intratumoral/direct Injection/catheter delivery to pulmonary nodules	Phase I
Cancer	Cationic liposome complex/HLA-B7/ beta-2 microglobulin cDNA	In vivo	Direct injection/ catheter delivery to pulmonary nodules	Phase I
Glioblastoma	Liposome complex/ lipofectin (Gibco BRL)/insulin-like growth factor antisense	In vitro	Subcutaneous injection	Phase I
Small cell lung cancer	Cationic liposome complex/lipofectin (Gibco BRL)/ cytokine/interleukin-2 cDNA/neomycin phosphotransferase cDNA	In vitro	Subcutaneous injection	Phase I
Metastatic malignancies (breast adenocarcinoma, renal cell carcinoma, melanoma, colorectal adenocarcinoma, non-Hodgkin's lymphoma)	Cationic liposome complex/DMRIE-DOPE Vical VCL-1005/HLA-B7/beta-2 microglobulin cDNA	In vivo	Direct intratumoral injection	Phase II
Melanoma	Cationic liposome complex/DMRIE-DOPE Vical VCL-1005/HLA-B7/beta-2 microglobulin cDNA	In vivo	Direct intratumoral injection	Phase II
Head and neck squamous cell carcinoma	Cationic liposome complex/DMRIE-DOPE/Vical VCL-1005/HLA-B7/beta-2 microglobulin cDNA	In vivo	Direct intratumoral injection	Phase II

cDNA, complementary DNA; *DMRIE*, *N*-[1-(2,3-dimyristyloxy)propyl]-*N*,*N*-dimethyl-*N*-(2-hydroxyethyl) ammonium bromide; *DOPE*, dioleyl phosphatidylethanolamine (neutral fusogenic lipid).

Cristiano and Curiel (1996) revealed that synthetic vectors have been utilized for gene delivery via the receptor-mediated endocytic pathway. These vectors have a two-functional domain, one is a ligand domain and the other is a DNA-binding domain. The results of binding achieved specific targeting on receptor-targeted cells by incorporation of the transgene into the vector (Cristiano and Curiel, 1996). There is a common agent used for DNA binding which is polylysine, and this is used for ligand targeting including transferrin (Tf), epidermal growth factor, surfactant-associated protein B, etc. (Cristiano and Curiel, 1996).

4.6.2 Liposome-mediated nucleic acid transfer

Liposomal applicability has been used for gene transfer at in vitro and in vivo levels. Among liposomes, cationic liposomes gained immense potential in this regard (Canonico et al., 1994a,b). Cationic liposome-loaded DNA is used to deliver a gene for cystic fibrosis. Thus, the gene is responsible for regulating the *CFTR* gene in cystic fibrosis (Hyde et al., 1993). There has been a great deal of literature supporting the applications of cationic liposome-based gene transfer in gene therapy of cystic fibrosis in humans (Logan et al., 1995; Stribling et al., 1992). A recombinant prostaglandin *GHI (PGII)* gene encapsulated cationic liposomes has been investigated in rabbit lungs for protection from endotoxin injury and endotoxin-induced pulmonary hypertension (Conary et al., 1994). Injection of liposome-loaded DNA into tumor easily facilitates gene delivery and develops a cellular-based immune response (Plautz et al., 1993). Human leukocyte antigen gene and *HLA-B7* were delivered by liposomes and found to be effective against melanoma (Nabel et al., 1993b). In cancer gene therapy, the degree of growth inhibition of malignant cells is dependent on the lipid to DNA and total DNA complex which is administered to the cells (Hofland and Huang, 1995). Besides these, liposome-based gene transfer also targeted various tissues, such as endothelial tissue and the gastrointestinal tract (GIT). Endothelial cell targeting in catheter-mediated cationic lipid–DNA complexes has been shown in in vivo transduction and expression into arterial walls (Nabel et al., 1992, 1993a). In the case of the GIT, the liposome-lacZ marker gene has revealed an opportunity to study gastrointestinal physiology and the possibility of gene therapy for GIT disorders (Schmid et al., 1994). Literatures also suggested that liposomes might be used for gene delivery to the liver (Leibiger et al., 1991). Cationic liposomes showed efficient transfection into cultured murine hepatocytes over conventional methods (Watanabe et al., 1994). One of the important uses of liposomes for the controlling of viral or cellular gene expression is by delivery of the antisense oligonucleotides and ribozymes into infected cells. Antisense technology faces many problems like degradation of oligonucleotides by nuclease enzymes and poor membrane permeability. There are a number of factors that affect the cellular availability of antisense nucleotides to achieve antisense activity. It has been revealed in a great deal of the literature that the liposomal carrier can effectively deliver antisense nucleotides into the target cell without any degradation (Lappalainen et al., 1994; Thierry and Dritschilo, 1992). Development of immunoliposomes has been effective in targeted delivery and to control HIV-1 replication in chronically infected CEM cells (Zelphati et al., 1994). Another is ribozyme technology, liposome as liposome–ribozyme complex has gained potential application to minimize intracellular degradation. Anti-HIV-1 drugs are delivered into the virus-producing tissues such as lymphoid tissues by liposomes loaded with antireverse transcriptase enzyme inhibitor 2′,3′-dideoxyinosine (Harvie et al., 1995). Apart from DNA delivery, the use of processed mRNA gained a potential role for its delivery and has emerged as gene therapeutic approaches (Lu et al., 1994). Certain modifications made in the liposomal structure resulted in enhanced transfer and targeting capacity (Puyal et al., 1995; Zhou and Huang 1994). Hemagglutinating viruses of Japan (HVJ) or Sendai virus also mediate the delivery of gene. Moreover the liposomes have the characteristics of DNA carriers and viral (HVJ virus) fusion properties (Kaneda et al., 1989; Tomita et al., 1994). This modification results in enhancing the therapeutic gene delivery into the host cell nucleus by avoiding the shuttling of the liposome–DNA complex through endosomes (Tomita et al., 1994). To enhance nuclear translocation and expression of the gene, involves mixing of DNA with a nonhistone, this is a chromosomal high-mobility group-1 protein (HMG-l) (Kato et al., 1991).

4.7 CONCLUSION

Liposomal drug delivery systems are very effective in the delivery of vaccines and genes due to their adjuvant property and targeting ability which elicit the immune response of the body through antibody formation and corrected gene inputs. Along with this, the mechanism of vaccine and gene delivery is also explored. There have been several recent investigations into vaccine delivery by liposomes approved by the USFDA and many are in the development stage. In gene delivery, various factors like liposomal preparation, size, and various types of liposomes such as cationic and anionic are responsible for the efficiency of transfection of the gene. First-generation liposomes in gene delivery suffered from several limitations such as poor encapsulation efficiency, poor release, and lower in vivo targetibility. Among second-generation liposomes, cationic liposomes have found better efficiency and good targeting ability for DNA delivery as compared with conventional liposomes. In the area of liposomal-based vaccine and gene delivery, transfection efficiency, toxicity, cellular, and gene delivery need to be studied in future to make it more efficient in this regard.

REFERENCES

Alpar, H.O., Eyles, J.E., Williamson, E.D., Somavarapu, S., 2001. Intranasal vaccination against plague, tetanus and diphtheria. Adv. Drug Deliv. Rev. 51 (1–3), 173–201.
Alving, C.R., 1991. Liposomes as carriers of antigens and adjuvants. J. Immunol. Methods 140 (1), 1–13.

Alving, C.R., Rao, M., 2008. Lipid A and liposomes containing lipid A as antigens and adjuvants. Vaccine 26 (24), 3036–3045.

Anderson, W.F., 1998. Human gene therapy. Nature 392, 25–30.

Beg, S., Samad, A., Nazish, I., Sultana, R., Rahman, M., Ahmad, M., et al., 2013. Colloidal drug delivery systems in vaccine delivery. Curr. Drug Targets 14 (1), 123–137.

Behr, J.P., 1986. DNA strongly binds to micelles and vesicles containing lipopolyamines or lipointercalatants. Tet. Lett. 27, 5861–5864.

Calcagnile, S., Zuccotti, G.V., 2010. The virosomal adjuvanted influenza vaccine. Expert Rev. Vaccines 10 (2), 191–200.

Canonico, A.E., Conary, J.T., Meyrick, B., Brigham, K.L., 1994a. Aerosol and intravenous transfection of human alpha-1 antitrypsin gene to lungs of rabbits. Am. J. Respir. Cell Mol. Biol. 10, 24–29.

Canonico, A.E., Plitman, J.D., Conary, J.T., Meyrick, B.O., Brigham, K.L., 1994b. No lung toxicity after repeated aerosol or intravenous delivery of plasmid-cationic liposome complexes. J. Appl. Physiol. 77, 415–419.

Check, E., 2003. Harmful potential of viral vectors fuels doubt over gene therapy. Nature 423, 573–574.

Conary, J.T., Parker, R.E., Christman, B.W., Faulks, R.D., King, G.A., Meyrick, B.O., et al., 1994. Protection of rabbit lungs from endotoxin injury by in vivo hyperexpression of the prostaglandin GIH synthase gene. J. Clin. Invest. 93, 1834–1840.

Copland, M.J., Baird, M.A., Rades, T., et al., 2003. Liposomal delivery of antigen to human dendritic cells. Vaccine 21 (9–10), 883–890.

Cristiano, R.J., Curiel, D.T., 1996. Strategies to accomplish gene delivery via the receptor-mediated endocytosis pathway. Cancer Gene Ther. 3, 49–57.

Daemen, T., Bungener, L., Huckriede, A., et al., 2000. Virosomes as an antigen delivery system. J. Liposome Res. 10 (4), 329–338.

Felgner, P.L., Ringold, G.M., 1989. Cationic liposome-mediated transfection. Nature 337, 387–388.

Frederiksen, L., Anton, K., van Hoogevest, P., et al., 1997. Preparation of liposomes encapsulating water-soluble compounds using supercritical carbon dioxide. J. Pharm. Sci. 86 (8), 921–928.

Gao, X., Huang, L., 1991. A novel cationic liposome reagent for efficient transfection of mammalian cells. Biochem. Biophys. Res. Commun. 179, 280–285.

Gluck, R., Mischler, R., Brantschen, S., et al., 1992. Immuno-potentiating reconstituted influenza virus virosome vaccine delivery system for immunization against hepatitis A. J. Clin. Invest. 90 (6), 2491–2495.

Gregoriadis, G., Davis, D., Davies, A., 1987. Liposomes as immunological adjuvants: antigen incorporation studies. Vaccine 5 (2), 145–151.

Gregoriadis, G., Allison, A.C., 1974. Entrapment of proteins in liposomes prevents allergic reactions in pre-immunised mice. FEBS Lett. 45 (1), 71–74.

Gregoriadis, G., Florence, A.T., 1993. Liposome Technology, second ed. RCR Presses Inc, Boca Raton, FL.

Griebel, P.J., Hein, W.R., 1996. Expanding the role of Peyer's patches in B-cell ontogeny. Immunol. Today 17 (1), 30–39.

Gupta, P.N., Vyas, S.P., 2011. Colloidal carrier systems for transcutaneous immunization. Curr. Drug Targets 12 (4), 579–597.

Gupta, P.N., Mishra, V., Singh, P., et al., 2005. Tetanus toxoid-loaded transfersomes for topical immunization. J. Pharm. Pharmacol. 57 (3), 295–301.

Hacein-Bey-Abina, S., Le Deist, F., Carlier, F., Bouneaud, C., Hue, C., De Villartay, J.P., et al., 2002. Sustained correction of X-linked severe combined immunodeficiency by ex vivo gene therapy. N. Engl. J. Med. 346, 1185–1193.

Hara, T., Kuwasawa, H., Aramak, Y., Takada, S., Koike, K., Ishidate, K., et al., 1996. Effects of fusogenic and DNA-binding amphiphilic compounds on the receptor mediated gene transfer into hepatic cells by asialofetuin-labeled liposomes. Biochem. Biophys. Acta 1278 (1), 51–58.

Harvie, P., Desormeaux, A., Gagne, N., Tremblay, M., Poulin, L., Beauchamp, D., et al., 1995. Lymphoid tissues targeting of liposome-encapsulated 2′, 3′dideoxyinosine. AIDS 9, 701–707.

Haywood, A.M., 1975. Phagocytosis of Sendai virus by model membranes. Gen. Virol. 29, 63–68.

Hedlund, G., Jansson, B., Brodin, T., et al., 1988. In vivo use of liposome incorporated membrane antigens. In: Gregoriadis, G. (Ed.), Liposomes as Drug Carriers. *John Wiley and Sons*, New York, NY, pp. 167–182.

Henriksen-Lacey, M., Korsholm, K.S., Andersen, P., et al., 2011. Liposomal vaccine delivery systems. Expert Opin. Drug Deliv. 8 (4), 505–519.

Hofland, H., Huang, L., 1995. Inhibition of human ovarian-carcinoma cell proliferation by liposome-plasmid DNA complex. Biochem. Biophys. Res. Commun. 207, 492–496.

Hollon, T., 2000. Researchers and regulators reflect on first gene therapy death. Nature Med. 6, 6.

Hyde, S.C., Gill, D.R., Higgins, C.E., Trezise, A.E.O., Macvinish, L.J., Cuthbert, A.W., et al., 1993. Correction of the ion-transport defect in cystic fibrosis transgenic mice by gene therapy. Nature 362, 250–255.

Jackson, D.C., Crabb, B.S., Poumbourios, P., et al., 1991. Three antibody molecules can bind simultaneously to each monomer of the tetramer of influenza virus neuraminidase and the trimer of influenza virus hemagglutinin. Arch. Virol. 116 (1–4), 45–56.

Kaneda, Y., Iwai, K., Uchida, T., 1989. Increased expression of DNA co-introduced with nuclear protein in adult rat liver. Science 243, 375–378.

Kato, K., Nakanishi, M., Kaneda, Y., Uchida, T., Okada, Y., 1991. Expression of hepatitis B virus surface anti-gen in adult rat liver: co-introduction of DNA and nuclear protein by a simplified liposome method. J. Biol. Chem. 266, 3361–3364.

Krishnan, L., Dennis Sprott, G., 2003. Archaeosomes as self-adjuvanting delivery systems for cancer vaccines. J. Drug Target. 11 (8–10), 515–524.

Kunisawa, J., Nakagawa, S., Mayumi, T., 2001. Pharmacotherapy by intracellular delivery of drugs using fusogenic liposomes: application to vaccine development. Adv. Drug Deliv. Rev. 52 (3), 177–186.

Lappalainen, K., Urtti, A., Jaaskelainen, I., Syrjanen, K., 1994. Cationic liposomes mediated delivery of antisense oligonucleotides targeted to HPV 16 E7 messenger RNA in Caski cells. Antivir. Res. 23, 119–130.

Legendre, J.Y., Szoka Jr., F.C., 1992. Delivery of plasmid DNA into mammalian cell lines using pH-sensitive liposomes: comparison with cationic liposomes. Pharm. Res. 9 (10), 1235–1242.

Leibiger, B., Leibiger, B., Sarrach, D., Zuhlke, H., 1991. Expression of exogenous DNA in rat liver cells after liposome mediated transfection in vivo. Biochem. Biophys. Res. Commun. 174, 1223–1231.

Logan, J.J., Bebok, Z., Walker, L.C., Peng, S.Y., Felgner, P.L., Siegal, G.P., et al., 1995. Cationic lipids for reporter gene and CFTR transfer to pulmonary epithelium. Gene Ther. 2, 38–49.

Lu, D., Benjamin, R., Kim, M., Conry, R.M., Curiel, D.T., 1994. Optimization of methods to achieve mRNA-mediated transfection of tumor cells in vitro and in vivo employing cationic liposome vectors. Cancer Gene Ther. 1, 245–252.

Mahor, S., Rawat, A., Dubey, P.K., et al., 2007. Cationic transfersomes based topical genetic vaccine against hepatitis B. Int. J. Pharm. 340 (1–2), 13–19.

Martin, C., Somavarapu, S., Alpar, H.O., 2005. Mucosal delivery of diphtheria toxoid using polymer-coated bioadhesive liposomes as vaccine carriers. J. Drug Del. Sci. Tech. 15 (4), 301–306.

Martin, F., Boulikas, T., 1998. The challenge of liposomes in gene therapy. Gene Ther. Mol. Biol. 1, 173–214.

Mellman, I., Steinman, R.M., 2001. Dendritic cells: specialized and regulated antigen processing machines. Cell 106 (3), 255–258.

Mellman, I., Fuchs, R., Helenius, A., 1986. Acidification of the endocytic and exocytic pathways. Ann. Rev. Biochem. 55, 663–700.

Moingeon, P., Taisne, Cd, Almond, J., 2002. Delivery technologies for human vaccines. Br. Med. Bull. 62, 29–44.

Moser, C., Amacker, M., Zurbriggen, R., 2011. Influenza virosomes as a vaccine adjuvant and carrier system. Expert Rev. Vaccines 10 (4), 437–446.

Mozafari, M.R., 2005. Liposomes: an overview of manufacturing techniques. Cell. Mol. Biol. Lett. 10 (4), 711–719.

Nabel, E.G., Plautz, G., Nabel, G.J., 1992. Transduction of a foreign histocompatibility gene into the arterial wall induces vasculitis. Proc. Natl. Acad. Sci. U.S.A. 89, 5157–5161.

Nabel, E.G., Yang, Z.Y., Plautz, G., Forough, R., Zhan, X., Haudenschild, C.C., et al., 1993a. Recombinant fibroblast growth factor 1 promotes intimal hyperplasia and angiogenesis in arteries in vivo. Nature 362, 844–846.

Nabel, G.J., Nabel, E.G., Yang, Z.Y., Fox, B.A., Plautz, G.E., Gao, X., et al., 1993b. Direct gene transfer with DNA liposome complexes in melanoma: expression, biologic activity and lack of toxicity in humans. Proc. Natl. Acad. Sci. U.S.A. 90, 11307–11311.

Nakanishi, T., Kunisawa, J., Hayashi, A., et al., 1999. Positively charged liposome functions as an efficient immunoadjuvant in inducing cell-mediated immune response to soluble proteins. J. Control. Release 61 (1–2), 233–240.

Naldini, L., 2015. Gene therapy returns to centre stage. Nature 526, 351–360.

Parente, R.A., Nir, S., Szoka, F.C., 1988. pH-dependent fusion of phosphatidylcholine small vesicles: induction by a synthetic amphiphilic peptide. J. Biol. Chem. 263, 4724.

Patel, G.B., Sprott, G.D., 1999. Archaeobacterial ether lipid liposomes (archaeosomes) as novel vaccine and drug delivery systems. Crit. Rev. Biotechnol. 19 (4), 317–357.

Patel, G.B., Chen, W., 2005. Archaeosome immunostimulatory vaccine delivery system. Curr Drug Deliv. 2 (4), 407–421.

Patel, G.B., Zhou, H., KuoLee, R., et al., 2004. Archaeosomes as adjuvants for combination vaccines. J. Liposome Res. 14 (3–4), 191–202.

Patel, G.B., Zhou, H., Ponce, A., et al., 2007. Mucosal and systemic immune responses by intranasal immunization using archaeal lipid adjuvanted vaccines. Vaccine 25 (51), 8622–8636.

Paul, A., Cevc, G., Bachhawat, B.K., 1998. Transdermal immunisation with an integral membrane component, gap junction protein, by means of ultra-deformable drug carriers, transfersomes. Vaccine 16 (2–3), 188–195.

Pedersen, G.K., Ebensen, T., Gjeraker, I.H., et al., 2011. Evaluation of the sublingual route for administration of influenza H5N1 virosomes in combination with the bacterial second messenger c-di-GMP. PLoS One 6 (11), e26973.

Plautz, G.E., Yang, Z.Y., Wu, B.Y., Gao, X., Huang, L., Nabel, G.J., 1993. Immunotherapy of malignancy by in vivo gene transfer into tumors. Proc. Natl. Acad. Sci. U.S.A. 90, 4645–4649.

Poltl-Frank, F., Zurbriggen, R., Helg, A., et al., 1999. Use of reconstituted influenza virus virosomes as an immuno-potentiating delivery system for a peptide-based vaccine. Clin. Exp. Immunol. 117 (3), 496–503.

Puyal, C., Milhaud, P., Bienvenue, A., Philippot, J.R., 1995. A new cationic liposome encapsulating genetic material: a potential delivery system for polynucleotides. Eur. J. Biochem. 228, 697–703.

Rai, K., Gupta, Y., Jain, A., et al., 2008. Transfersomes: self-optimizing carriers for bioactives. PDA J. Pharm. Sci. Technol. 62 (5), 362–379.

Raphael, L., Tom, B.H., 1984. Liposome facilitated xenogeneic approach for studying human colon cancer immunity: carrier and adjuvant effect of liposomes. Clin. Exp. Immunol. 55 (1), 1–13.

Reddy, R., Zhou, F., Huang, L., et al., 1991. pH sensitive liposomes provide an efficient means of sensitizing target cells to class I restricted CTL recognition of a soluble protein. J. Immunol. Methods 141 (2), 157–163.

Remy, J.S., Sirlin, C., Behr, J.P., 1995. Gene transfer with cationic amphiphiles The Liposomes. *CRC Press Inc, Boca Raton, FL.* pp. 159–170.

Rescia, V.C., Takata, C.S., de Araujo, P.S., et al., 2011. Dressing liposomal particles with chitosan and poly(vinylic alcohol) for oral vaccine delivery. J. Mar. Res. 21 (1), 38–45.

Saupe, A., McBurney, W., Rades, T., et al., 2006. Immuno-stimulatory colloidal delivery systems for cancer vaccines. Expert Opin. Drug Deliv. 3 (3), 345–354.

Schmid, R.M., Weidenbach, H., Draenert, G.F., Lerch, M.M., Liptay, S., Schorr, J., et al., 1994. Liposome-mediated in vivo gene-transfer into different tissues of the gastrointestinal tract. Z. Gastroenterol. 32, 665–670.

Schmid, S.L., Fuchs, R., Male, P., Mellman, I., 1988. Two distinct subpopulations of endosomes involved in membrane recycling and transport to lysosomes. Cell 52, 73–83.

Shim, G., Kim, M., Park, J.Y., Oh, Y.K., 2013. Application of cationic liposomes for delivery of nucleic acids. Asian J. Pharm. Sci. 8 (2), 72–80.

Stribling, R., Brunette, E., Liggitt, D., Gaensler, K., Debs, R., 1992. Aerosol gene delivery in vivo. Proc. Natl. Acad. Sci. U.S.A. 89, 11277–11281.

Tafaghodi, M., Jaafari, M.R., Sajadi Tabassi, S.A., 2006. Nasal immunization studies using liposomes loaded with tetanus toxoid and CpG-ODN. Eur. J. Pharm. Biopharm. 64 (2), 138–145.

Tamborrini, M., Stoffel, S.A., Westerfeld, N., et al., 2011. Immunogenicity of a virosomally-formulated *Plasmodium falciparum* GLURP-MSP3 chimeric protein-based malaria vaccine candidate in comparison to adjuvanted formulations. Malar. J. 10, 359.

Thierry, A.R., Dritschilo, A., 1992. Intracellular availability of unmodified, phosphorothioated and liposomally encapsulated oligodeoxynucleotides for antisense activity. Nucl. Acid Res. 20, 5691–5698.

Tirrell, D.A., Takigawa, D.Y., Seki, K., 1985. pH-sensitization of phospholipid vesicles via complexation with synthetic poly (carboxylic acids). Ann. N.Y. Acad. Sci. 446, 237.

Tomita, N., Higaki, J., Ogihara, T., Kondo, T., Kaneda, Y., 1994. A novel gene-transfer technique mediated by HVJ (Sendai virus), nuclear protein and liposomes. Cancer Detect. Prevent. 18, 485–491.

Uster, P.S., Deamer, D., 1985. pH-dependent fusion of liposomes using titrable polycations. Biochemistry 24, 1–8.

Van Rooijen, N., van Nieuwmegen, R., 1980. Liposomes in immunology: evidence that their adjuvant effect results from surface exposition of the antigens. Cell Immunol. 49 (2), 402–407.

Vigneron, J.P., Oudrhiri, N., Fauquet, M., et al., 1996. Guanidium-cholesterol cationic lipids: efficient vectors for the transfection of eukaryotic cells. Proc. Natl. Acad. Sci. 93, 9682–9986.

Vinod, K.R., Kumar, M.S., Anbazhagan, S., et al., 2012. Critical issues related to transfersomes—novel vesicular system. Acta Sci. Pol. Technol. Aliment 11 (1), 67–82.

Wang, C.Y., Huang, L., 1987. pH-sensitive immunoliposomes mediate target-cell-specific delivery and controlled expression of a foreign gene in mouse. Proc. Natl. Acad. Sci. U.S.A. 84, 7851–7855.

Watanabe, Y., Nomoto, H., Takezawa, R., Miyoshi, N., Akaike, T., 1994. Highly efficient transfection into primary cultured mouse hepatocytes by use of cation-liposomes: an application for immunization. J. Biochem. 116, 1220–1226.

Zelphati, O., Imbach, J.L., Signoret, N., Zon, G., Rayner, B., Leserman, L., 1994. Antisense oligonucleotides in solution or encapsulated in immuno-liposomes inhibit replication of HIV-1 by several different mechanisms. Nucl. Acid Res. 22, 4307–4314.

Zhou, X.H., Huang, L., 1994. DNA transfection mediated by cationic liposomes containing lipopolylysine-characterization and mechanism of action. Biochim. Biophys. Acta 1189, 195–203.

Zolnik, B.S., Gonzalez-Fernandez, A., Sadrieh, N., et al., 2010. Nanoparticles and the immune system. Endocrinology 151, 458–465.

CHAPTER 5

Polymeric Micelles for Drug Targeting and Delivery

Mohd Cairul Iqbal Mohd Amin, Adeel M. Butt, Muhammad W. Amjad and Prashant Kesharwani

Contents

5.1 INTRODUCTION

Polymeric micelles (PMs) have been extensively explored in order to facilitate the delivery of hydrophobic drugs. PMs are spontaneously formed from amphiphilic copolymers in aqueous media, when the critical micelle concentration (CMC) is achieved, as shown in Fig. 5.1. The differences between the solubility of the hydrophilic and hydrophobic blocks of an amphiphilic copolymer in an aqueous solution drive the formation of distinctive and unique PMs with a core–shell architecture (Fig. 5.2), as described in later sections. PM diameters vary from a few tens of nanometers to several hundred nanometers.

Nanotechnology-Based Approaches for Targeting and Delivery of Drugs and Genes.
DOI: http://dx.doi.org/10.1016/B978-0-12-809717-5.00006-3

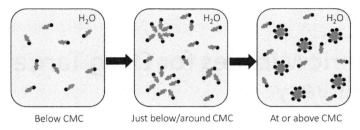

Figure 5.1 Schematic of micelle formation.

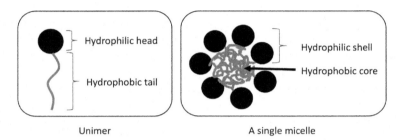

Figure 5.2 Graphical representation of a unimer (amphiphile) and aggregate-forming micelle.

PMs have advantages over surfactant micelle systems because of their enhanced stability, low CMCs, retarded dissociation, controlled drug release, and longer retention times. Their mesoscopic size range and narrow size distribution are an important and critical factor for the enhanced permeability and retention (EPR) effect. The inner core serves as a nanochamber for hydrophobic drugs, while the shell shields it from the surrounding aqueous environment. Shell-forming polymers have been engineered to produce PMs with prolonged blood circulation times; this helps to evade the immune system and facilitates specific and active drug targeting. In addition, PMs have a number of other advantages over its major nanocompetitor liposomes; these include the simple methods used in their preparation, the lack of any special requirements for parent drug modification in order to achieve adequate drug loading, and a controlled drug release profile.

5.2 MECHANISM OF MICELLE FORMATION

A dispersion consists of two phases; the dispersed phase and the continuous phase. Similarly, micelles could be considered as a colloidal dispersion, with micelles forming the dispersed phase and the media constituting the continuous phase (Torchilin, 2001a). The process of PM formation follows thermodynamic principles whereby the hydrophobic and hydrophilic blocks are arranged to minimize the free energy of the system (Jones and Leroux, 1999). The addition of water results in collapse of the hydrophobic blocks, which results in the formation of the dispersed phase. The formation of hydrophobic block particles in water reduces the entropy, leading to an increased structuring of the water molecules. This whole process results in the formation of a cavity that drives aggregation of the hydrophobic blocks; this in turn restores the structure of water and thus increases the entropy of the system (Nagarajan, 1996).

Following the aggregation of the hydrophobic blocks, they relax to a favorable conformation, which results in increased entropy. This has an overall negative Gibbs free energy, which confirms that PM formation is a spontaneous process. The main force driving PM formation is the overall reduction in the free energy of the system, which is caused by the collapse of the hydrophobic blocks from the aqueous phase. On the other hand, the hydrophilic blocks help to stabilize the PMs by maintaining contact with the water molecules at the particle surface. The formation of PMs at the CMC also depends upon the free energy of the system. When the concentration of amphiphilic copolymer in the solvent is high enough, it becomes favorable for PMs to form, in order to reduce the free energy of the system. As the PMs begin to form, at a concentration around the CMC, some solvent molecules remain inside the cores and these result in the formation of swollen cores. This explains why the PM size is slightly larger at or around CMC than the size observed at a higher copolymer concentration (Jones and Leroux, 1999).

5.3 MORPHOLOGY OF POLYMERIC MICELLES

Several PM morphologies have been reported. These include, but are not limited to, star shapes (Wei et al., 2008), flower-like (Hameed et al., 2015), spherical supramolecular assemblies (Butt et al., 2012), worm shapes (Schmelz et al., 2012), vesicles, toroids, and helices (Liu et al., 2011). Some commonly reported morphologies are shown in Fig. 5.3. However, spherical PMs are observed most frequently. The application of PMs has not been limited to cancer chemotherapy or the health sciences. PMs have also been used by material scientists, bioengineers, and by scientists in the microelectronics field; this indicates their versatility and benefits.

5.4 TYPES OF POLYMERIC MICELLES

PMs can be divided into two general types, based on the choice of materials used.

5.4.1 Block copolymer micelles

Block copolymer micelles are one of the most promising classes of nanoparticles used in drug delivery. Block copolymers are formed by one homopolymer attached to one or more other homopolymers. These blocks are repeated throughout the structure of the copolymer and lend specific properties to it. These copolymers allow the engineering of one or more blocks, providing the potential to combine desirable characteristics.

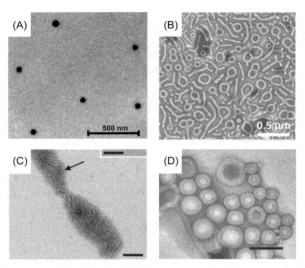

Figure 5.3 Transmission electron micrographs of some commonly reported micelle morphologies: (A) spheres, (B) toroids, (C) worm-like, and (D) vesicles. *Reproduced and adapted from Butt, A.M., Amin, M.C.I.M., Katas, H., Sarisuta, N., Witoonsaridsilp, W., Benjakul, R., 2012. In vitro characterization of pluronic F127 and D-α-tocopheryl polyethylene glycol 1000 succinate mixed micelles as nanocarriers for targeted anticancer-drug delivery. J. Nanomaterials 2012, 112; Schmelz, J., Schedl, A.E., Steinlein, C., Manners, I., Schmalz, H., 2012. Length control and block-type architectures in worm-like micelles with polyethylene cores. J. Am. Chem. Soc. 134 (34), 14217–14225; Liu, C., Chen, G., Sun, H., Xu, J., Feng, Y., Zhang, Z., et al., 2011. Toroidal micelles of polystyrene-block-poly (acrylic acid). Small 7 (19), 2721–2726.*

For example, a block copolymer would have some of the properties of its constituent blocks provided that the length of blocks is not too short. PMs self-assemble from amphiphilic block copolymers in aqueous solutions (Fig. 5.2). The hydrophobic block forms the inner core, while the hydrophilic block forms the outer shell, which is also referred to as the corona. Hydrophobic blocks come together to produce a solid core that is shielded by the corona. This core can be used to encapsulate hydrophobic drugs, or the hydrophobic block can be chemically conjugated with the hydrophobic drugs prior to PM formation (Cammas–Marion et al., 1999; Lavasanifar et al., 2002).

5.4.2 Polyion complex micelles

Polyion complex micelles (PICMs) are potential carriers for charged molecules such as small interfering RNA (siRNA) or DNA. PICMs are formed by the electrostatic interaction between oppositely charged polymer macromolecules and block copolymers, forming a core–shell structure. Although several attempts have been made to use PICMS for siRNA and gene delivery, there are still several problems associated with their clinical and therapeutic application.

5.5 POLYMERS COMMONLY USED TO FABRICATE POLYMERIC MICELLES

The synthesis of PMs mainly involves the use of amphiphilic diblock copolymers, although graft and triblock copolymers are also used. These three copolymer types have exclusive benefits for the delivery of drugs, such as the prolongation of drug circulation time, control of the drug-release profile, or the ability to add targeting ligands. The outer hydrophilic shell usually consists of polyethers such as polyethylene oxide (PEO) or polyethylene glycol (PEG). Examples of other shell polymers include poly(trimethylene carbonate), poly(acryloylmorpholine) (Zhang et al., 2006), and poly(vinylpyrrolidone) (Torchilin et al., 1995). Occasionally, polymers are combined to form this hydrophilic PM shell (Štepánek et al., 2001). These polymers endow PMs with stealth properties, facilitating their escape from the reticuloendothelial system (RES); this is vital for the extension of their circulation half-life. The length of the PEG chains employed is generally 1–15 kDa; the use of longer PEG chains results in a heavier hydrophilic corona, which enhances the stealth properties and thus the in vivo circulation half-life of the PMs. The properties of core-forming blocks can be improved by using block copolymers with functional groups (Li and Kwon, 2000).

The commercial Pluronic range comprises triblock copolymers of PEOm/2-b and poly(propylene oxide) PPOn-b-PEOm/2, where n and m represent the numbers of repeating PPO and PEO units, respectively, and b represents the "block." The partitioning of hydrophobic moieties and the PM CMC is affected by the size of the PPO block (Kozlov et al., 2000). An exceptional characteristic of Pluronic copolymers is their ability to inhibit P-glycoprotein (Pgp)-mediated efflux, a process associated with multiple drug resistance (MDR) (Alakhov et al., 1996). Pgp functionality largely depends on sustained ATP levels. The use of Pluronic increased membrane fluidity and altered intracellular ATP levels in endothelial and MDR cells (Batrakova et al., 2001; Kabanov et al., 2001), indicating that this may underlie its inhibition of Pgp (Hrycyna et al., 1998). As shown in Fig. 5.4, MDR is a major obstacle to the successful delivery of biomolecules.

Poly (L-glutamate and aspartate) are the most frequently used poly(L-amino acids) (PAAs) in drug delivery systems, while poly(ε-caprolactone) (PCL), poly(D,L-lactide) (PDLLA), and poly(glycolide) (PGA) are commonly employed polyesters. PAAs are nontoxic, economic, biocompatible, and contain thiol, amino, carboxyl, and hydroxyl functional groups. The alterations in the PM core to facilitate drug conjugation rely on the versatility of these functional groups. Regular PAAs are linked via peptide bonds between the α-carboxylic and α-amine groups of the L-amino acids, and may either act as hydrophilic or hydrophobic copolymer blocks. Poly(L-glutamate/lysine/aspartate) are frequently used hydrophilic parts. A random derivatization of the PAA amino or carboxyl groups yields a hydrophobic block, thus rendering them similar to the graft-type copolymer. PCL, PGA, and PDLLA polyesters are frequently used to

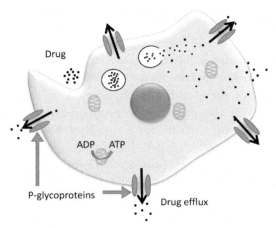

Figure 5.4 Schematic of drug efflux from the cells. Drug is transported out of the cells by P-glycoproteins, which act as drug efflux pumps.

synthesize nanoparticles and PMs. Their molecular weight (MW) and polymer crystallinity influence the degree of hydrolysis of the ester bonds, with smaller and more hydrophilic (PGA > PDLLA > PCL) chains degrading more rapidly (Park, 1995; Belbella et al., 1996; Husmann et al., 2002). Polyesters are often used for the targeted delivery of small hydrophilic molecules and nucleic acids (Brannon-Peppas et al., 2007).

Hence, PMs possess the ability to suppress MDR. It is facilitated mainly by monomers, which constitute a particular polymer that self-assembles to form micelles. Their distribution and temporal control of drug release may be modulated using the block copolymers described above. Features that affect PM distribution can alter drug release and thus control its delivery to the target site (Rösler et al., 2012). Temporal control provides the ability to activate drug release at a specific time-point during treatment. PM modifications that are frequently employed include surface functionalization, the addition of supporting moieties, and crosslinking within the PM shell or core (Lavasanifar et al., 2001; O'Reilly et al., 2006).

5.5.1 Selection of hydrophilic and hydrophobic blocks

Although all block copolymer designs follow the same basic pattern by including hydrophobic and hydrophilic blocks, different chemotherapeutics may have distinctive requirements. Hence, the application needs to be considered when optimizing a carrier for a particular chemotherapeutic. The recent advances in chemistry and synthetic techniques could be instrumental in moving toward a major breakthrough in nanocarrier-based drug delivery. These include controlled/living ring opening polymerization (ROP) or controlled/living radical polymerization and reversible addition–fragmentation chain transfer (RAFT). Organocatalytic ROP has been used for biomedical applications, while ROP and RAFT have both been used for the synthesis and modification of biodegradable copolymers.

The selection of the shell- and core-forming materials is an important step in the preparation of copolymers to be used as drug delivery carriers. The shell directly affects PM disposition, distribution, and ultimately, fate, while the core drives the micellization process. The segregation of core-forming segments from the aqueous milieu drives micelle formation by a combination of hydrophobic and electrostatic interactions, hydrogen bonding, metal complexation, and intermolecular forces. Core-forming materials can be engineered to incorporate genes, siRNA, small peptides, or proteins. Core engineering can also produce stronger interactions with the PM cargo, while shell functionalization or surface modification of PMs is particularly important for receptor or active targeting. Several strategies can be employed to achieve this, such as the inclusion of end-functionalized block copolymers.

The PM corona plays an important role in its protection from immune system-mediated removal from the circulation. This protects the PM and its cargo from environmental factors such as pH and a plethora of enzymes, thus enabling them to reach their target sites. A number of shell-forming materials have therefore been tested, with PEG and PEO being the most widely used. For example, it was shown that doxorubicin (DOX)-loaded mixed PMs based on Pluronic F127 and D-α-tocopheryl polyethylene glycol 1000 (TPGS) remained stable in DMEM containing fetal bovine serum (Fig. 5.5). Pluronic F127 and TPGS have PEO and PEG in their structures, respectively, which reduced the aggregation of PMs and their interactions with serum proteins (Butt et al., 2012).

On the other hand, the core plays an important role in PM formation and the material employed must induce the hydrophobic interactions necessary to form a PM. Some examples of such materials are PPO, PDLLA, PCL, and poly(L-aspartate) (Sutton et al., 2007).

Surface properties in particular are key features of smart PM drug delivery systems. Targeted drug delivery using colloidal nanocarriers has encountered problems due to nonspecificity, such as RES uptake. Evasion of the RES is thus essential in order to achieve active targeting and to enhance PM blood circulation times.

5.5.2 Effect of drug loading on polymeric micelle stability

PMs have lower CMCs than surfactant micelles and this feature makes them more suitable for drug delivery. The lower CMC increases PM stability, even after very large dilutions. A question that could arise relates to the effects of the drugs within the hydrophobic cores. If these significantly alter the copolymer properties, this could reduce their usefulness for drug delivery. However, it has been shown that although the entrapment of small molecules in the PM core produced a slight increase in

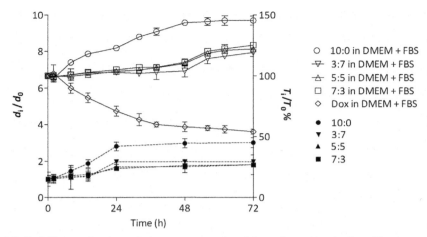

Figure 5.5 Stability evaluation of drugs encapsulated in polymeric micelles. The transmittance and diameter of mixed micelles (of Pluronic F127 and TPGS) are represented as T_i/T_0 (%) (right y-axis) and d_i/d_0 (left y-axis). *Reproduced from Butt, A.M., Amin, M.C.I.M., Katas, H., Sarisuta, N., Witoonsaridsilp, W., Benjakul, R., 2012. In vitro characterization of pluronic F127 and D-α-tocopheryl polyethylene glycol 1000 succinate mixed micelles as nanocarriers for targeted anticancer-drug delivery. J. Nanomaterials 2012, 112.*

size (Xing and Mattice, 1998), they did not significantly affect the CMC (Jones and Leroux, 1999).

PM composition and CMC is one determinant of micelle stability, while PM dissociation is another important aspect. In surfactant micelles, dissociation kinetics are relatively simple and are similar to the association kinetics. In contrast, block copolymers show different association and dissociation kinetics (Tuzar, 1996). As described earlier, the formation of solid hydrophobic cores increases PM stability. The presence of crystallized or rigid cores provides stability by avoiding rapid degradation in serum or in vivo. Although made up of hydrophobic blocks, an anisotropic water distribution exists in the PM cores. The highest concentration of water is at the interface with the hydrophilic block, near the surface, and this decreases toward the center of the core (Torchilin, 2001a). Drug distribution within the PM thus depends upon its polarity.

5.5.3 Effects of hydrophobic block properties on polymeric micelle characteristics

The length of the hydrophobic chain affects the properties of PMs, including their drug-loading capacity and size. If the hydrophobic chain length is increased, drug loading or encapsulation is increased, while increasing the MW of the hydrophobic block will increase the PM size (Govender et al., 2000). This also reduces the Laplace pressure, which in turn slightly enhances drug loading (Torchilin, 2001a).

5.6 ADVANTAGES OF POLYMERIC MICELLES AS DRUG DELIVERY CARRIERS

The properties of PMs provide a number of benefits for drug delivery carriers. Their small size and hydrophilic shells render PMs invisible to macrophages and the RES. This increases the circulation times of their drug cargos. The bioavailability of poorly soluble drugs is increased by encapsulation inside PM cores, which can reduce the required dose and thereby minimize side effects. The small size and enhanced blood circulation times of PMs also enhance passive targeting and the EPR effect (described in Section 5.7.1). Moreover, the hydrophilic block that forms the corona or shell can be modified by attaching ligands to enhance specific targeting and therapeutic efficacy, while reducing side effects (described in Section 5.7.2). In addition, PMs can be readily prepared in large quantities owing to their spontaneous formation; this facilitates the scaling-up process. Despite these advantages, PMs also have the disadvantage of increased instability following dilution by body fluids (Miyata et al., 2011; Solomatin et al., 2003). However, this problem can be overcome by using polymers with a low CMC, which enhances stability.

Furthermore, PMs have advantages over dendrimers or drug–polymer conjugates because no chemical modifications or covalent conjugations between the cargo and carrier molecules are required (Gillies and Frechet, 2005a). This generally makes it easier to use PMs than dendrimers and drug–polymer conjugates, which require the presence of a functional moiety to form chemical conjugates. In addition, this covalent linkage necessitates a specific drug-release mechanism, such as enzymatic degradation or hydrolytic cleavage, e.g., acid-catalyzed hydrolysis (Gopin et al., 2006). Dendrimers are also particularly small, typically around 10 nm, and are easily cleared via the renal glomeruli; this results in a very short blood half-life (Greenwald et al., 2003; Peer et al., 2007).

5.7 POLYMERIC MICELLE TARGETING APPROACHES

There are two main types of tumor-targeting routes: passive and active targeting, as depicted in Fig. 5.6.

5.7.1 Passive targeting

One of the most promising approaches for the delivery of drugs and genes to tumor sites is based on the phenomenon of passive targeting, which exploits the anatomical and pathophysiological abnormalities of the tumor vasculature and utilizes the EPR effect (Iyer et al., 2006; Kesharwani and Iyer, 2015). Nanosized particles such as dendrimers, PMs, and liposomes, as well as macromolecules that are larger than the renal excretion threshold (typically >40 kDa), have a tendency to accumulate more in tumor

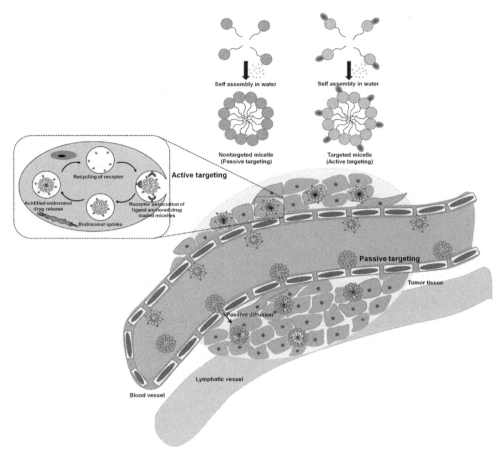

Figure 5.6 The mechanisms underlying drug delivery by micelles. The extravasation of micelles via ineffective lymphatic drainage, the enhanced permeability and retention effect, and the increased permeability of the tumor vasculature drive passive tissue targeting. Functionalizing the surface of micelles with ligands that enhance cell-specific recognition and binding is referred to as active targeting. The micelles can release the drugs in close vicinity to the target sites, attach to the cell membrane to act as a sustained extracellular drug release depot, or enter the cell.

tissues than in normal tissues. This is because tumor blood vessels are poorly aligned, defective, and highly leaky. These blood vessels thus allow the extravasation of plasma components, including macromolecules and nanoparticles, into the tumor interstitium. Furthermore, tumor tissues have poor lymphatic clearance and this prevents clearance of the accumulated nanoparticles. These features allow nanoparticles to remain in the tumor microenvironment for long periods of time, thus facilitating the sustained release of their drug or gene cargos into the tumor tissues.

For example, Sun et al. (2013) used PAA-g-PEG micelles for delivery of DOX. They showed that encapsulation of DOX in these PMs increased its accumulation in tumors, decreased metastatic organ damage, and reduced cardiac toxicity as a result of passive targeting and the EPR effect. Similarly, enhanced transvascular permeability and retention of passively targeted paclitaxel (PTX)-loaded PMs were observed (Danhier et al., 2015).

Using cationic block polymers, negatively charged pDNA and siRNA can be entrapped within the core via electrostatic interactions, forming stable complexes (polyplex micelles) at a physiological pH (Peer and Lieberman, 2011; Maeda, 2015; Mishra et al., 2014). In a reported study, Kim et al. developed stable siRNA-PEG/ polyethyleneimine (PEI) PMs (polyelectrolyte complex) for siRNA delivery by conjugating vascular endothelial growth factor (VEGF) siRNA with PEG. Administration of these PMs to PC-3 prostate cancer xenograft-bearing mice led to downregulation of tumor VEGF expression at both the mRNA and protein levels; this reduced the tumor microvessel density and suppressed tumor growth more effectively that the naked siRNA or the siRNA/PEI complex. The enhanced gene silencing by siRNA-PEG/ PEI PMs was due to the extended siRNA circulation time and an enhanced tumor accumulation via the EPR effect (Kim et al., 2008d). Another study of atelocollagen (a positively charged peptide) complexed with negatively charged siRNA found that these particles showed passive accumulation within the targeted tumor due to the EPR effect (Mishra et al., 2014; Rettig and Behlke, 2012).

5.7.2 Active targeting

Active targeting can be employed to enhance the interaction between PMs and tumor cells, to extend half-life, and to penetrate cells via receptor-mediated endocytosis. The development of ligand-coupled and stimuli-responsive micelles for active targeting is described below.

5.7.2.1 Ligand-coupled polymeric micelles

PMs can be conjugated with a range of other molecules: small molecules including folic acid (FA), tripolyphosphate (TPP), and 2-[3-[5-amino-1-carboxypentyl]-ureido]- pentanedioic acid (ACUPA) (Xiao et al., 2012; Werner et al., 2011; Hrkach et al., 2012); proteins including transferrin, ankyrin repeat proteins, and affibodies (Bartlett et al., 2007; Winkler et al., 2009; Alexis et al., 2008); peptides such as CGNKRTRGC (LyP-1), F3 peptide, iRGD, KLWVLPKGGGC, KLWVLPK, and aptides (Karmali et al., 2009; Park et al., 2008; Graf et al., 2012; Kamaly et al., 2013; Saw et al., 2013); antibodies and antibody fragments (F(ab)/2, F(ab/), and scFv) (Gao et al., 2004; Park et al., 2002; Kirpotin et al., 2006); nucleic acid-based ligands such as the A10 aptamer and A9 CGA aptamer (Farokhzad et al., 2006; Cheng et al., 2007; Kim et al., 2010).

5.7.2.1.1 FA-coupled polymeric micelles

FA-coupled PMs are primarily used to transport drugs or siRNAs into cells. FA has a strong affinity for its binding proteins and receptors, which are selectively overexpressed by cancer cells (Gao et al., 2002; Hobbs et al., 1998). For example, tumors of the brain, mammary gland, ovary, prostate, lung, and colon epithelial tissue have been shown to overexpress FA receptors. Therefore, appropriately designed FA-conjugated PMs can be targeted to and internalized by tumor cells via receptor-mediated endocytosis. Moreover, FA-conjugated micelles may subsequently be exported by the drug efflux pumps expressed by cancer cells (Torchilin, 2005) and diffuse into the cytosol or remain present in recycled endosomes (pH 5–6). These features provide important means to improve the cellular uptake of the PM cargo and to evade potential lysosomal degradation of the PMs.

It was shown that uptake of DOX was enhanced significantly by functionalization with FA, resulting in increased cytotoxicity in human epidermoid carcinoma (KB) cells. Furthermore, FA functionalization also increased the antitumor efficacy by twofold in vivo (Yoo and Park, 2004). In another study by Butt and colleagues (2015), FA-conjugated micelles were shown to enhance the uptake and cytotoxicity of DOX in an ovarian carcinoma cell line (SKOV3), as compared to free DOX or nontargeting micelles (Fig. 5.7). This study showed that FA conjugation did not affect the PM drug release properties.

Human pharyngeal cancer cells were exposed to fluorescent FA-conjugated and FA-unconjugated pH-sensitive PMs for 3 and 24 h in vitro, resulting in intracellular accumulation of fluorescence; this indicated that FA conjugation significantly improved PM uptake by tumor cells (Campbell et al., 1991). Another study used PMs formed from FA-conjugated pH-sensitive PEO-*b*-PDLLA and FA-PEG-*b*-poly (L-histidine) P(His) block copolymers. These PMs also exhibited effective drug delivery to tumor cells in vitro.

5.7.2.1.2 Hyaluronic acid-coupled polymeric micelles

Amiji and coworkers assessed the potential of hyaluronic acid-PEI/hyaluronic acid-PEG for the delivery of siRNA targeting the MDR gene and also explored the efficacy of these PMs coloaded with siRNA and PTX for the suppression of ovarian cancer growth (Yang et al., 2014, 2015). They found that the hyaluronic acid-PEI/hyaluronic acid-PEG PMs delivered MDR siRNA efficiently to ovarian cancer cells showing MDR, leading to downregulation of the MDR gene and reduced expression of Pgp. Administration of siRNA-complexed hyaluronic acid-PEI/hyaluronic acid-PEG PMs, followed by treatment with PTX, produced significant tumor growth inhibition, reduced Pgp expression, and enhanced apoptosis in mice with ovarian cancer showing MDR. These results indicated that these PMs carrying CD44 gene-targeted siRNA could serve as a therapeutic tool to evade MDR in ovarian cancer.

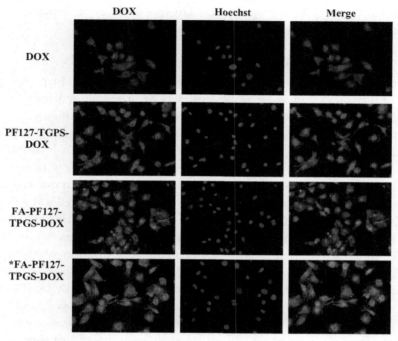

Figure 5.7 Intracellular doxorubicin (*DOX*) uptake in ovarian carcinoma cells (SKOV3) mediated by folic acid-conjugated micelles. *Reproduced from Butt, A.M., Amin, M.C.I.M., Katas, H., 2015. Synergistic effect of pH-responsive folate-functionalized poloxamer 407-TPGS-mixed micelles on targeted delivery of anticancer drugs. Int. J. Nanomedicine 10, 1321–1334.*

In a separate study, Amiji and coworkers designed and screened a range of hyaluronic acid-based self-assembling nanoparticulate systems for the targeted delivery of CD44 siRNA (Ganesh et al., 2013). To assess siRNA encapsulation, hyaluronic acid polymer was functionalized with lipids of varying nitrogen content/carbon chain lengths and polyamines. Dynamic light-scattering analyses and gel retardation assays found that numerous hyaluronic acid derivatives could efficiently complex siRNAs and form self-assembled nanoparticulate systems. Several derivatives of hyaluronic acid could be used to transfect siRNAs into CD44 receptor-overexpressing cancer cells. Fascinatingly, blockade of cellular CD44 receptors using free soluble hyaluronic acid prior to incubation with hyaluronic acid nanoparticles loaded with cy3-labeled siRNA resulted in >90% inhibition of this receptor-mediated uptake, confirming target specificity. Moreover, hyaluronic acid-PEI/PEG nanosystems complexed with Sjögren syndrome antigen B (SSB)/Polo-like kinase-1 siRNA produced target-specific and concentration-dependent gene knockdown in both resistant and sensitive A549 lung cancer cells that overexpressed CD44 receptors. Furthermore, these siRNA-loaded

PMs exhibited tumor-selective uptake and target-specific gene knockdown in solid and metastatic tumors in vivo.

5.7.2.1.3 Transferrin-coupled polymeric micelles

Since certain cancer cells overexpress the transferrin receptor (depending on the degree of malignancy), tumor targeting may also be accomplished using transferrin-conjugated PMs. Torchilin and coworkers used spheroid ovarian cancer cells that showed MDR and in vivo tumor models to assess codelivery of curcumin (CUR) and PTX (Sarisozen et al., 2014). CUR and PTX were coloaded into PMs composed of PEG-phosphatidyl ethanolamine functionalized with transferrin. Confocal imaging and flow cytometry verified a significantly higher penetration of transferrin-functionalized PMs into the spheroids, as compared to nontargeted micelles. The cytotoxicity of PTX to both spheroids and monolayers was significantly increased when it was delivered as a single agent in transferrin-targeted micelles or when it was codelivered with CUR in nontargeted micelles; transferrin functionalization of these coloaded micelles did not produce any further increase in cytotoxicity.

Ren et al. (2010) fabricated transferrin-functionalized PEG-poly(lactic acid) (PLA) PMs for potential targeted in vivo brain glioma drug delivery. Flow cytometry exhibited the in vitro targeting of tumor cells by these PMs, and fluorescence microscopy of brain sections from rats bearing C6 glioma tumors demonstrated that they were able to penetrate the tumor in vivo.

5.7.2.1.4 Luteinizing hormone-releasing hormone-coupled polymeric micelles

Luteinizing hormone-releasing hormone (LHRH) receptors are prevalent on the plasma membranes of normal cells but are also overexpressed by cancer cells, including ovarian cancer cells. Dharap et al. (2003) used synthetic peptides analogous to the BCL-2 homology 3 (BH3) peptide to suppress cellular antiapoptotic defense systems and LHRH as a targeting moiety. Three types of camptothecin (CPT) conjugates, (CPT)-PEG, CPT-PEG-LHRH, and CPT-PEG-BH3 were fabricated and evaluated in A2780 human ovarian cancer cells. The expression of genes encoding caspases 3 and 9, apoptotic peptidase activating factor 1 (APAF-1) proteins, second mitochondria-derived activator of caspase (SMAC), B-cell lymphoma-extra large (BCL-XL), B-cell lymphoma 2 (BCL-2) was investigated, along with cytotoxicity, apoptosis, and the activities of caspases 3 and 9. When compared to free CPT, PEG-CPT conjugates produced much greater apoptosis and cytotoxicity.

Park and coworkers modified polyelectrolyte complex micelles with cancer cell targeting moieties for intracellular delivery of VEGF siRNA (Kim et al., 2008c). LHRH peptide was coupled to the distal end of the PEG–siRNA conjugate as a targeting ligand. These PMs showed increased cellular uptake in A2780 cancer cells, in comparison to those without LHRH, leading to enhanced VEGF gene silencing through receptor-mediated endocytosis.

5.7.2.1.5 Epidermal growth factor-coupled polymeric micelles

Conjugation of PMs to Epidermal growth factor (EGF) provides an alternative strategy for tumor targeting. Zeng et al. (2006) synthesized methoxy PEG-*b*-poly (δ-valerolactone) (MePEG-*b*-PVL) and conjugated these with EGF to selectively target the overexpressed EGF receptors (EGFRs) on the plasma membrane of breast cancer cells. CM-DiI, a hydrophobic fluorescent probe, was loaded into the MePEG-*b*-PVL and EGF-PEG-*b*-PVL PMs. Confocal laser scanning microscopy (CLSM) revealed that the CM-DiI-functionalized EGF-PEG-*b*-PVL micelles accumulated intracellularly in MDA-MB-468 breast cancer cells after a 2-h incubation, whereas no noticeable cellular uptake of the CM-DiI-functionalized MePEG-*b*-PVL micelles was observed. These findings were verified by calculating the intracellular CM-DiI fluorescence in cell lysates. Moreover, the presence of free EGF was found to reduce the uptake of EGF-PEG-*b*-PVL PMs. Nuclear staining using Hoechst 33258 showed that EGF-PEG-*b*-PVL micelles primarily accumulated in the perinuclear area, while some were observed in the nucleus.

5.7.2.1.6 α2-Glycoprotein-coupled polymeric micelles

Cerebral glial cells may be targeted by α2-glycoprotein-conjugated micelles. Kabanov et al. (1992) prepared PMs for drug targeting using Pluronic. Drug molecules were solubilized in the hydrophobic core of Pluronic PMs. The solubilization of drug and fluorescein isothiocyanate (FITC) in Pluronic micelles was investigated using ultracentrifugation and fluorescence measurements. It was observed that FITC solubilization in Pluronic micelles significantly influenced its distribution in mouse tissues, leading to pronounced enhancement of FITC fluorescence in lungs. A specific targeting of FITC to the brain was witnessed when α2-glycoprotein was incorporated into these PMs. Under these circumstances, significant enhancement of FITC fluorescence in the brain and reduction in lungs was observed.

5.7.2.1.7 Oligopeptide-based targeted polymeric micelles

Cai et al. used the RGD peptide for targeted DOX delivery, which resulted in enhanced cytotoxicity. They also showed that RGD-modified PMs selectively accumulated in an integrin–overexpressing human hepatocellular carcinoma cell line (BEL-7402), and did not accumulate in a human epithelial carcinoma cell line (HeLa) (Cai et al., 2011).

For siRNA-based cancer therapy, Kataoka and coworkers developed PICMs to enhance siRNA accumulation in tumor cells after systemic administration (Oe et al., 2014). The polymer was designed to incorporate disulfide crosslinking and cyclic RGD peptide ligands in order to target cancer cells, while the use of cholesterol-modified siRNA (chol-siRNA) provided further hydrophobic stabilization of the micelles. Functionalization using cyclic RGD efficiently facilitated siRNA accumulation in an in vivo model of subcutaneous cervical cancer. These cyclic RGD/

chol–siRNA micelles showed significant in vivo gene silencing, possibly due to their targeting capability and the stability derived from disulfide crosslinking and hydrophobic interactions with cholesterol.

Asn-Gly-Arg (NGR)-containing peptides can be used as ligands for the targeted delivery of PMs to target tumor angiogenesis. For example, Zhao and colleagues prepared NGR-conjugated DSPE-PEG PMs containing PTX. These PMs had a lower IC_{50} in murine brain microvascular endothelial cells than did free drug or PTX-loaded DSPE-PEG PMs without NGR. These targeted PMs also produced a greater reduction of C6 glioma tumors than did the commercial formulation, Taxol (Zhao et al., 2011). Son and coworkers (2010) synthesized a polymer based on branched PEI that was thiolated with propylene sulfide and mixed this with alpha-maleimide-omega-N-hydroxysuccinimide ester polyethylene glycol (MW: 5000 Da) and a cyclic NGR peptide. This gene nanocarrier exhibited effective tumor targeting by the cyclic NGR peptide. Wang et al. also evaluated the antitumor efficacy of NGR peptide-decorated PEG-b-PLA PMs (Wang et al., 2009). HT1080 cells were selected as a positive tumor cell model, whereas HUVECs were used as the tumor endothelial cell model. This study reported that actively targeted PMs showed better uptake and stronger adhesion than did undecorated PMs.

Angiopep-2 was used to functionalize poly(lactic-co-glycolic acid) (PLGA) nanocarriers, which encapsulated both EGFR siRNA and DOX (Wang et al., 2015). This PM efficiently delivered siRNA and DOX into U87MG cells, resulting in significant EGFR silencing, apoptosis, and cell inhibition in vitro. An in vivo experiment using the brain orthotopic U87MG glioma xenograft model showed that these PMs not only prolonged the lifespan of glioma-bearing mice, but also produced obvious cell apoptosis in glioma tissue.

A multilayered PM system consisting of two components: (1) a micelles system-based poly-L-lactide (PLLA) core and PEGylated TAT shell and (2) a pH-sensitive diblock copolymer of poly(methacryloyl sulfadimethoxine), and PEG was prepared for targeted delivery of anticancer drugs. Confocal microscopy indicated pH-responsive uptake in tumors, with enhanced accumulation in the nucleus (Sethuraman and Bae, 2007). Kanazawa et al. (2013) designed nose-to-brain siRNA delivery PMs consisting of PEG-PCL copolymers conjugated to a cell-penetrating peptide, TAT (MPEG-PCL-TAT). Dextran (MW: 10,000 Da) was used as a model (for siRNA) in this study. As compared to intravenous delivery of dextran, with or without MPEG-PCL-TAT, intranasal dextran delivery using MPEG-PCL-TAT produced greater delivery to the brain. Moreover, the use of MPEG-PCL-TAT enhanced transport along the trigeminal and olfactory nerve pathways, owing to the high permeation of these PMs across the nasal mucosa.

5.7.2.1.8 Antibody-coupled polymeric micelles

Palanca–Wessels and colleagues (2011) designed a PM delivery system consisting of (1) a streptavidin-conjugated monoclonal antibody raised against CD22 and (2) a diblock copolymer consisting of a pH-responsive block, to facilitate endosome release, and a positively charged siRNA condensing block. Improved uptake of siRNA was exhibited in transduced HeLa-R (R: resistant) and DoHH2 lymphoma cells expressing CD22, but not in CD22-negative HeLa-R cells. As compared to nontargeted PMs, gene knockdown was significantly enhanced by these CD22-targeted PMs. CD22-targeted PMs containing 15 nmol/L siRNA produced a 70% reduction of gene expression in DoHH2 cells.

It was reported that diacyllipid-PEG PMs conjugated to a monoclonal antibody (2C5 or 2G4) increased PTX accumulation in lung tumors fourfold and also enhanced cytotoxicity, as compared to nonfunctionalized control PMs (Torchilin et al., 2003b).

Dou et al. (2012) designed a nanocarrier comprising an anti-Her2 antibody fragment fusion protein and a positively charged protamine (F5-P) for the delivery of siRNA targeting DNA methyltransferases 1 and/or 3b (siDNMTs) into Her2-expressing BT474 breast cancer cells. The F5-P carrier efficiently bound to the siRNA and delivered it to these cells, but not to MDA-MB-231 breast cancer cells, which do not express Her2. siDNMTs delivery to BT474 cells efficiently silenced target gene expression and enhanced demethylation of the Ras association domain family 1 isoform A (RASSF1A) tumor suppressor gene promoter, resulting in suppression of tumor cell proliferation.

5.7.2.1.9 Carbohydrate-coupled polymeric micelles

Wang and colleagues prepared PTX-loaded PMs based on diblock copolymers of poly(ethyl ethylene phosphate) and PCL, with surface conjugation of galactosamine in order to target the asialoglycoprotein receptor in HepG2 cells. This study found that these PMs exhibited comparable cytotoxic activity to that of free PTX (Wang et al., 2008).

Wang and coworkers (2013) designed N-acetylgalactosamine-functionalized mixed micellar nanoparticles (Gal-MNP), capable of delivering siRNA to hepatocytes and silencing target gene expression upon systemic administration. The hepatocyte-targeting effect of Gal-MNP was exhibited by efficient accumulation of fluorescent siRNA in primary hepatocytes in vitro and in vivo. Following the intravenous administration of Gal-MNP loaded with siRNA targeting apolipoprotein B to BALB/c mice, efficient downregulation of liver apolipoprotein B mRNA and protein expression was achieved. Innate immunity or positive hepatotoxicity was not induced by systemic delivery of siRNA-loaded Gal-MNP.

Zhu and Mahato (2010) conjugated siRNA to mannose 6-phosphate PEG (M6P-PEG) and galactosylated-PEG for targeted delivery to hepatic stellate cells and hepatocytes, respectively. Without transfection reagents, both M6P-PEG-siRNA and Gal-PEG-siRNA conjugates reduced luciferase gene expression by about 40%, whereas 98% gene silencing was observed using a similar concentration of cationic nanocarriers. The conjugation of M6P-PEG and Gal-PEG to TGF-β1 siRNA also silenced endogenous TGF-β1 gene expression.

5.7.2.1.10 Aptamer-coupled polymeric micelles

Aptamers are single-stranded RNA or DNA oligonucleotides with the capacity to recognize and bind to target molecules with high affinity and specificity. Aptamers have been used by Farokhzad et al. (2006) for the targeted delivery of PEG-PLGA PMs loaded with docetaxel, for the treatment of prostate cancer. These authors reported total regression of LNCap tumor xenografts in BALB/c nude mice treated with aptamer-conjugated PMs.

5.7.2.2 Stimuli-responsive polymeric micelles
5.7.2.2.1 pH-responsive polymeric micelles

One approach to the design of the pH-sensitive PMs is to simply combine positively charged drugs with negatively charged entities. There are two types of pH-sensitive PMs, unimolecular (e.g., amphiphilic hyperbranched block copolymers) and multi-molecular (e.g., Pluronic F127 conjugated with acrylic acid) (Jones et al., 2003; Sant et al., 2004; Kim et al., 2008b). The unimolecular PMs have no CMC and are stable upon dilution. These respond to pH changes by modulation of the core polarity, which results in an increase or decrease in drug cargo release. On the other hand, multimolecular PMs dissociate when there is an increase in media pH, thus releasing their cargo (Jones et al., 2003). pH-sensitive PMs are triggered to release their cargo by dissociation or destabilization in response to the acidic pH of tumors, endosomes, or lysosomes.

PICMs are pH-sensitive PMs prepared by encapsulation of a hydrophobic drug in the core, which does not carry a charge, while the shell carries a negative charge (polyanion) (Taillefer et al., 2001). In this case, drug release occurs when there is protonation of the shell, resulting in a disruption of the core–shell structure. PMs carrying a positive charge (polycations) can be used for the delivery of nucleic acids. In this case, an additional polycationic molecule is introduced, which keeps the PM system intact. The addition of this polycationic nature has an additional advantage because it can enhance transfection efficiency, if properly selected, as these cationic polymers help to disrupt lysosomes and release the encapsulated drugs (Stiriba et al., 2002; Felber et al., 2012). These types of polyion complex PMs are called ternary PICMs. For example, methyl methacrylate has been used as a cationic block along with a number of other polyanionic blocks such as PEG-*b*-poly(aminoethyl methacrylate) or PEG-*b*-poly(propyl methacrylate-*co*-methacrylic acid) and poly(amido amine) dendrimers (Felber et al., 2012;Yessine et al., 2007).

PICMs dissociate in the acidic pH present inside endosomes, resulting in the release of the incorporated nucleic acid. As the polyanionic block becomes protonated, the polycationic block interacts with the endosome membrane, resulting in endosomal escape (Elsabahy et al., 2009). Furthermore, these PICMs can be engineered to incorporate surface functional targeting moieties, which are taken up by endocytosis. For example, PICMs carrying siRNA were effective in silencing their target genes (Felber et al., 2011).

The extracellular pH (pHe) of most solid tumors is distinctly different from the physiological pH (7.4) of the adjacent normal tissue (Tannock and Rotin, 1989; Stubbs et al., 2000). For example, Engin and coworkers (1995) demonstrated this by measuring the pHe in tumor nodules of patients using needle microelectrodes. The average tumor pHe is around 7.06, and it commonly ranges from pH 5.7 to 7.8. Noninvasive imaging of the in vivo tumor pHe has been made possible by magnetic resonance spectroscopy, which detects pH-dependent resonance frequency differences (van Sluis et al., 1999; Ojugo et al., 1999). The local pH of the endosomal and lysosomal compartments ranges from 5 to 6 in the majority of cell types. Therefore, a weak acidity is believed to be one of the functional triggers for the specific release of chemotherapeutics or siRNA at tumor tissue sites and/or inside tumor cells (Manchun et al., 2012; Liu et al., 2013). In the early 1980s, Yatvin and coworkers designed the first pH-responsive drug carriers (Yatvin et al., 1980).

Titratable moieties, such as weak acids or weak bases, are extensively applied to micelle-forming copolymers to cause disruption or interior structural alterations of PMs in acidic biological environments (Na et al., 2003; Felber et al., 2012). In particular, P(His) is generally used as a pH-responsive element in PMs (Asayama et al., 2004; Yang et al., 2006). The imidazole of the histidine residue is freely protonated, yielding positively charged micelles.

A few studies have reported pH-responsive PMs comprised of PEG-*b*-P(His) block copolymers (Lee et al., 2003; Gao et al., 2005). A PEG-*b*-P(His) micelle produced a significant increase in the in vivo and in vitro intracellular accumulation of drugs in mice bearing human ovarian carcinoma (A2780) xenografts, due to the tumor's acidic pHe (Gao et al., 2005). Nevertheless, the P(His)-based cores exhibited a marginally unstable structure at physiological pH. Hence, Lee et al. designed block copolymer micelles using PEG-*b*-P(His) and PEG-*b*-PDLLA to improve their stability at a physiological pH (Lee et al., 2003). Additionally, pH-responsive, FA-conjugated block copolymer micelles [FA-PEG-*b*-P(His)/PEG-*b*-PDLLA] were also designed and showed efficient intracellular localization and cytotoxicity against tumor cells in vitro.

Kim and coworkers reported a method to stabilize pH-responsive P(His)-based cores by the addition of hydrophobic L-phenylalanine into the P(His) main chains

(P(His–*co*–Phe)) (Kim et al., 2005). Kim et al. also designed pH-responsive PMs to overcome MDR in tumors; these consisted of a block copolymer blend of PEG-*b*-P(His–*co*–Phe) and FA-PEG-*b*-PDLLA in order to optimize the response to early endosome pH (Kim et al., 2008a). Thus, a fast intracellular siRNA release could be attained by combining His-induced disruption of the endosomal membrane and the endosomal pH response. The linkage of poly(2-(diisopropylamino)ethyl methacrylate) with the biocompatible poly(2-methacryloyloxyethyl phosphorylcholine) block to form pH-responsive PMs has also been described (Licciardi et al., 2006).

Using a carboxylic acid-based system, Leroux and coworkers designed pH-responsive PMs containing a ternary arbitrary coblock of methacrylic acid, octadecylacrylate, and *N*-isopropylacrylamide as a pH-responsive unit (Taillefer et al., 2000; Leroux et al., 2001). These PMs showed a pH-dependent phase transition at around pH 5.8, as evidenced by the altered structure of their inner core.

Biodegradable polyesters have been used as hydrophobic core-forming polymers in the design of block copolymer PMs. Biodegradable polyesters are commonly used as hydrophobic blocks in the hydrolysis-induced disruption of PMs, although the rate of polyester hydrolysis is comparatively slow under slightly acidic conditions (pH 5–6), necessitating multiple days to attain complete disruption of the PM structure (Akimoto et al., 2008). Heller and coworkers (2002) described PMs consisting of PEG-*b*-poly(ortho ester) block copolymers, which are pH-sensitive. These PMs had a nanoscopic size (50–80 nm) and low CMC. Acid-labile chemical bonds such as those present in acetyl (Fife and Jao, 1965; Gillies et al., 2004a), hydrazine (Greenfield et al., 1990; Kaneko et al., 1991), *cis*-aconityl (Shen and Ryser, 1981; Hudecz et al., 1990), and oxime (Jin et al., 2011) moieties are commonly used either to modify the structure of micelle-forming polymers or to conjugate drugs to polymer backbones; this facilitates the design of efficient drug delivery systems that are sensitive to the acidic pH of endosomes or lysosomes. For example, PEG-*b*-(PDLLA) with a *cis*-aconityl or hydrazine spacer was designed for acid-induced release (Yoo et al., 2002). Fréchet et al. designed PEG PMs with pH-sensitive disruption of the hydrophobic core by using an acid-labile core-forming P(Asp) derivative to increase pH-induced drug release (Gillies and Fréchet, 2003). In their innovative system, a PEG–dendrimer hybrid was used for the design of these pH-sensitive PMs. The addition of hydrophobic aromatic groups to neighboring sites that are involved in the core formation was accomplished via cyclic acetyl linkage (Gillies et al., 2004b). The features of the designed PMs (CMC, disruption rate, and size) could be controlled by fine-tuning the dendrimer generation, chemical structure, corona-forming PEG length, and the type of hydrophobic acetyl linkage (Gillies and Fréchet, 2005b).

Chen et al. also reported pH-responsive PMs with biodegradable and acid-labile polycarbonate cores (Chen et al., 2009). Kataoka and coworkers designed pH-sensitive

PMs that released anticancer drugs at the tumor endosomal pH and/or pHe (Bae et al., 2003, 2005b). These were designed by the multiassembly of PEG-b-P(Asp) block copolymers, using hydrazone spacers. Recently, NanoCarrier (Chiba, Japan) designed a pH-sensitive PM carrier system (Takahashi et al., 2013) for breast cancer treatment that is currently in phase I clinical trials in Japan.

5.7.2.2.2 Temperature-responsive polymeric micelles

The temperature change during cooling/heating processes can be used as a trigger to alter the functional and structural characteristics of stimuli-responsive PMs. In particularly, mild heating (up to 41–43°C) of the body may be used as a potential cancer treatment (Ponce et al., 2006) because of the elevated temperature-sensitivity of cancer cells; mild heating specifically affects the biological processes within these cells, including altered receptor expression, microtubule disruption, and inhibited DNA repair and synthesis (Jain, 1987).

Temperature-responsive polymers exhibit a characteristic conformational change in reaction to the environmental temperature. These polymers are divided into two types on the basis of the temperature dependency of their solubility. One type of polymer shrinks or precipitates below a certain temperature, which is defined as the upper critical solution temperature (Katono et al., 1991; Shimada et al., 2011), whereas the other type shrinks and becomes insoluble in water above a specific temperature, defined as the lower critical solution temperature (LCST) (Fujishige et al., 1989). LCST-type polymers have been widely used in drug delivery systems and as biomaterials. In particular, poly(N-isopropylacrylamide) (PIPAAm) has fascinated researchers working on intelligent biomedical applications (Heskins and Guillet, 1968; Schild, 1992). Developments in polymer chemistry facilitate the selection of several types of polymers that are temperature-responsive, such as those based on poly(2-alkyl-2-oxazoline) (Park and Kataoka, 2006), poly(oligo(ethylene glycol) methacrylate) (Lutz, 2008), and polypeptides such as elastin-like polypeptide (Urry, 1992). Generally, the LCST control of poly(N-substituted acrylamide) derivatives is achieved by random radical copolymerization with different ratios of comonomers with hydrophobic or hydrophilic properties (Feil et al., 1993; Takei et al., 1993).

Okano et al. were the first to study PMs with temperature-responsive coronas using AB-type block copolymers containing hydrophobic and PIPAAm-based blocks (Chung et al., 1999). An aqueous solution of PIPAAm-b-poly(butyl methacrylate) (PBMA) PMs showed the same thermal phase transition as that of those formed with IPAAm homopolymers, despite the incorporation of the hydrophobic PBMA block (Chung et al., 2000). This physicochemical behavior reflects the phase-separated PBMA core and PIPAAm corona structure. Therefore, covalently linked hydrophobic PBMA segments barely affect the LCST of the PIPAAm corona and their thermo-responsive behavior thus differs from that of random copolymers, which reduce the LCST.

Recently, outer-surface functionalization of temperature-responsive PMs has been achieved using block copolymers such as end-functional thermo-responsive blocks in order to add exclusive features, including bioimaging and active targeting functions (Nakayama and Okano, 2005). Hydrophobic groups placed at the PM periphery significantly stimulated the dehydration of the corona-forming PIPAAm derivatives and led to a significant LCST shift to a lower temperature, as compared to PMs with hydrophilic surface moieties. Additionally, the amplitude of the LCST shifts depended on the MW of the temperature-responsive chains (Nakayama and Okano, 2005). The findings of earlier experiments showed that hydrophobically terminated linear PIPAAm systems had a lower LCST value than nonmodified pure PIPAAm because of the increased dehydration of the proximal IPAAm units through the freely movable hydrophobic end-groups (Chung et al., 1998; Duan et al., 2006).

A PIPAAm-corona PM formed by mixing hydroxyl- and phenyl-based block copolymers showed a sharp phase transition at a temperature between the specific LCST values of the individual homogeneous micelles (Nakayama and Okano, 2008). This distinctive feature indicates that micellar LCST values can be closely controlled by modifying surface chemistry using certain stimuli such as bio-related interactions, redox reactions, light, and pH. A thermo-responsive PM that was surface-derivatized with sulfadimethoxine, which is pH-responsive, showed a phase shift at acidic pH values (Nakayama et al., 2012).

Okano and coworkers initially investigated PMs with PIPAAm-based coronas as temperature-triggered release systems (Chung et al., 2000). In PIPAAm-*b*-PBMA micelles, a clear variation in temperature-triggered drug release was observed over the LCST of the PIPAAm corona. In comparison, temperature-triggered drug release was limited in temperature-responsive PMs designed using PIPAAm-*b*-polystyrene. On the other hand, Wei and coworkers designed a thermo-responsive PM possessing a poly(methyl methacrylate) core with an elevated glass transition temperature (Wei et al., 2006). Additionally, biodegradable polyesters such as PCL, PLA, and their copolymers have been used to construct hydrophobic cores. For example, Wei et al. engineered a star-shaped block copolymer of PLLA-*sb*-P(IPAAm-*co*HMAAm) to design temperature-responsive PMs for use in combination with mild hyperthermic therapy (Wei et al., 2008). An in vitro release study revealed that these PMs were responsive to temperature changes in the range of 37–40°C. Alternatively, Liu et al. developed a binary temperature- and pH-responsive PM using a block copolymer of PIPAAm-*co*-N,N-dimethylacrylamide-*co*-2-aminoethyl methacrylate)-*b*-poly(10-undecenoic acid) (Liu et al., 2007). The LCST of these PMs varied from 38°C (pH 7.4) to 36.2°C (pH 6.6) because of the protonation/deprotonation of the carboxylic acids in the poly(10-undecenoic acid) block.

5.7.2.2.3 Light-responsive polymeric micelles

UV, visible, or near-infrared (NIR) light provides a practical activation signal as it can be applied to the body externally, under temporal and spatial control. Several additional factors, including the light intensity and wavelength, can also be controlled during photo-induced reactions. PMs with light-responsive properties are categorized into two types on the basis of their disruption mechanism; this can involve light-induced cleavage of chemical bonds or light-switchable property changes, such as photo-isomerization. Photochromic compounds are well-characterized molecules that display reversible isomeric transformations in response to light at a particular wavelength (Rau, 1990). Photochromic compounds are widely used to prepare light-responsive materials (Ercole et al., 2010).

In the past five decades, azobenzene derivatives have been some of the most extensively studied photochromic compounds (Orihara et al., 2001; Lee et al., 2006). Zhao et al. reported PMs that altered their morphology via photo-isomerization of the azobenzene units present in the block copolymer (Wang et al., 2004). These morphological changes are largely accredited to the adaptable polarity of the hydrophobic blocks when exposed to visible and UV light.

Some years ago, the photo-isomerization of spiropyran (SP) derivatives was exploited to form core-corona PMs that exhibited reversible light-induced formation. SP compounds undergo reversible isomerization between the hydrophilic zwitterionic and the hydrophobic merocyanine forms upon exposure to visible and UV light (Berkovic et al., 2000; Sanchez et al., 2003). Hence, the significant change in the polarity of the SP units, present in the core-forming chain, results in reversible block copolymer assembly. PMs containing a PEG-*b*-poly(SP)-methacrylate block copolymer are disrupted by UV radiation at a wavelength of 365 nm, and are recovered at the visible light wavelength of 620 nm (Lee et al., 2007). Therefore, light-induced disruption of the PM structure facilitates release of the drug cargo. Zhao and colleagues (Lee et al., 2007) and Yan et al. (2011) designed PMs exhibiting an irreversible light-responsive dissociation profile. Poly(1-pyrenylmethyl methacrylate) was used for the hydrophobic core-forming chains in the first report of these PMs (Jiang et al., 2005, 2006).

Micelles comprised of poly(2-nitrobenzylmethyl methacrylate) (Jiang et al., 2006) are another example of light-responsive PMs, which show light-triggered drug release from the core. Increasing the intensity of the UV light accelerated the drug release rate. A drug release system that is sensitive to NIR irradiation can also be achieved by including 2-nitrobenzyl, although the drug release rate is far slower than that induced by UV irradiation because of the low effectiveness of two-photon absorption by the 2-nitrobenzyl unit. This issue was addressed by introducing 7-diethylamino-4-(hydroxymethyl) coumarinyl, which showed two-photon absorption at 794 nm, as a substitute chromophore in the hydrophobic block of poly ([7-(diethylamino)coumarin-4-yl]methyl methacrylate) (Babin et al., 2009).

5.7.2.2.4 Ultrasound-based tumor targeting

Ultrasound-based tumor targeting uses focused ultrasound to induce drug release. Ultrasound activates PM drug release, along with disruption of the plasma membrane (Kaur et al., 2015) (Fig. 5.8), thereby improving cellular drug uptake (Rapoport et al., 2002; Marin et al., 2002). In this approach, the transducer is positioned to interact with a layer of water or water-based gel applied to the skin, avoiding the need for an invasive or surgical administration route (Rapaport et al., 2003a).

Ultrasound may improve the cellular uptake of both released and entrapped drugs at the target site, enhance PM drug release into the tumor interstitium, increase the drug diffusion through the tumor interstitium, improve the accumulation of drug-loaded PMs at the tumor site, and produce thermal effects. All of these properties improve the accumulation of drugs inside the tumor.

This technique can be further improved by optimizing several features including the physicochemical properties of the PMs, the time between PM administration and ultrasound application, the sonication frequency, and the type of ultrasound waves applied (Husseini et al., 2000). The tunable physicochemical properties of the PMs include the state of the inner core and the PM hydrophobicity, as this affects their accumulation in the tumor (Rapoport et al., 2000, 2003b). Ultrasound at a frequency of 69 kHz, applied to PMs at $3.2 W/cm^2$ for 10 min, caused the death of 66% of cells that showed MDR, in contrast to 53% cell death in the absence of ultrasound (Rapoport, 2004).

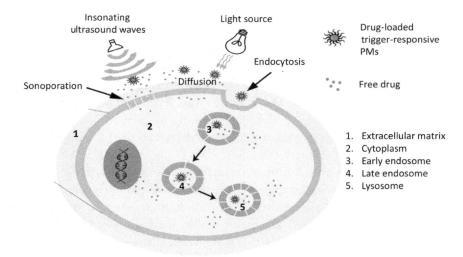

Figure 5.8 External and internal triggers exploited for cancer therapy. Drug release can be triggered by exposure to ultrasound waves or to a light source. Internal triggers include the pH of the extracellular matrix, cytoplasm, endosome, or lysosome. *Adapted from Kaur, S., Prasad, C., Balakrishnan, B., Banerjee, R., 2015. Trigger responsive polymeric nanocarriers for cancer therapy. Biomater. Sci. 3 (7), 955–987.*

5.7.2.3 Immuno polymeric micelles

Immuno PMs have specific antibodies attached to their surface, which target the PM cargo by specific interactions and may be adapted to a variety of targets. It has been shown that specific monoclonal, antinuclear nucleosome-restricted autoantibodies identify the exterior of several tumors through surface-attached nucleosomes, but do not interact with healthy cells (Iakoubov and Torchilin, 1997; Torchilin et al., 2003a). An improved in vivo accumulation and greater inhibition of tumor growth was observed after intravenous injection of PMs with the monoclonal antibody, 2C5, into mice with Lewis lung carcinoma.

A modified design method that uses PEG-poly(ethylene) with a *p*-nitrophenylcarboxyl-activated free PEG end to link monoclonal antibodies to PMs has been introduced (Torchilin et al., 2001c). This novel PEG-poly(ethylene) linker retains its stability at a pH lower than 6; once the pH rises above 7.5, the linker reacts with the amino groups of several peptides and proteins, including monoclonal antibodies, yielding a stable carbamate linkage. The immuno PMs can be quantified using sodium dodecyl sulfate-polyacrylamide gel electrophoresis or fluorescently labeled antibody techniques. Several researchers have reported that coupling of different antibodies to PMs did not significantly alter their size (Choi et al., 2010; Torchilin et al., 2001b). For example, the in vitro interactions between immune PMs decorated with the monoclonal antibody, 2C5, and tumor cells were studied by fluorescence microscopy (Gao et al., 2003).

5.7.2.4 Multifunctional polymeric micelles

Given all their advantages and unique properties, there has been a focus on the development of multifunctional PMs. These could incorporate the following features in a single PM: a targeting ligand for active targeting; a drug for therapeutic purposes; a dye for imaging; and stimulus sensitivity for triggered drug release. Various such PM systems have been developed. For example, FA-conjugated PEG-p(Asp-Hyd-Adr) PMs that show pH sensitivity and active targeting were developed by Bae et al. (2005a). This multifunctional system was more effective than the micelles prepared by individual polymers against KB cells. Similarly, a multifunctional PM system based on PEG-PLA has been developed; this included a cyclic RGD targeting ligand, pH-stimulated drug release, and a magnetic resonance imaging agent. A 50% reduction in tumor size was observed after a 10-day treatment with multifunctional PMs, which also facilitated contrast magnetic resonance imaging (Sutton et al., 2007).

5.7.3 Tumor-targeted codelivery of siRNA and anticancer drugs

RNAi-based therapeutics have been investigated extensively since the late 1990s. Tumor-targeted delivery of siRNA is challenging but the use of PMs has produced promising results. For example, a mixture of cationic copolymer micellar nanoparticles was complexed with siRNA targeting hypoxia inducible factor 1α expression. These

PMs efficiently inhibited tumor growth by ~60% and induced sensitivity to DOX (Liu et al., 2012).

Amjad and colleagues (2015) prepared PMs from cholic acid and PEI diblock copolymers and conjugated these with FA for selective and targeted codelivery of the anticancer drug, DOX, and siRNA targeting MDR-1.

Polo-like kinase-1 siRNA and PTX were administered in a mouse breast cancer xenograft model. The PMs used to codeliver this siRNA and anticancer drug produced very potent effects and the authors suggested that the dose of PTX could be reduced by 1000-fold if coadministered with this siRNA (Sun et al., 2011). Synergistic effects were also observed using other siRNA and chemotherapeutics codelivery systems. For example, hypoxia inducible factor 1α siRNA and gemcitabine were delivered using a polymer–lipid hybrid nanoparticle system (Zhao et al., 2015), MDR-1 siRNA and DOX were delivered using cationic PLGA (Misra et al., 2014), and Pgp siRNA and DOX were delivered by a FA/m-PEG-b-P(LG-Hyd)-b-PDMAPMA) triblock copolymer system (Xu et al., 2015).

Yin et al. (2015) developed hyaluronic acid-based amphiphilic conjugate (HSOP) redox-sensitive PMs for tumor-targeted codelivery of siRNA targeting aurora A kinase and PTX. HSOP showed efficient loading capacities for both siRNA and PTX, with desirable redox-sensitivity and adjustable dosing ratios, as confirmed by PM morphology changes and the in vitro release of both drugs in different reducing environments. Furthermore, confocal microscopy and flow cytometry analyses confirmed that HSOP micelles were able to codeliver siRNA and PTX simultaneously into MDA-MB-231 breast cancer cells by hyaluronic acid receptor-mediated endocytosis, followed by the rapid transport of siRNA and PTX into the cytoplasm. This successful transport and delivery enhanced the synergistic effects of these molecules, resulting in greater antitumor efficacy than nonsensitive coloaded micelles or single molecule-loaded micelles. An in vivo study demonstrated that HSOP micelles successfully accumulated in tumors and produced greater antitumor efficacy than redox-sensitive single-drug controls or the nonsensitive codelivery control.

5.8 POLYMERIC MICELLES FOR DRUG DELIVERY: FUTURE PROSPECTS

Small-molecule drugs have shown disappointing results, particularly in tumor therapy, due to the heterogeneous nature of cancer tissues (Saunders et al., 2012). There are multiple mutations in cancer cells and this leads to the emergence of mutant cells that are not affected by therapeutic approaches targeting specific molecules. The American Society of Clinical Oncology reported drugs acting via specific molecular targets only had a 4%–5% response rate. This was well below expectations, given their very high development costs. However, traditional chemotherapy will remain a key therapeutic approach for the foreseeable future. Most anticancer drugs with low MWs have the ability to move in and out of blood vessels (Maeda et al., 2009). This problem can be solved by forming a drug–polymer conjugate or by attaching a ligand to target the desired pathological site. An alternative and less cumbersome approach is to incorporate these drugs into a PM core.

PMs have been widely explored for anticancer drug delivery applications, particularly for hydrophobic drugs. Block copolymers have been, and are still, under intense scrutiny by drug delivery researchers due to their uniform and narrow size range and ease of use. However, there is a need to develop more advanced multifunctional delivery approaches, such as the codelivery of drugs and nucleic acids. Inclusion of siRNA sequences that target-specific genes could enhance the therapeutic efficacies of anticancer drugs several fold and could also help to overcome the longstanding problem of MDR. PICMs encapsulating anticancer drugs have the potential to provide such a delivery system. Moreover, covalent attachment of active targeting ligands on PMs helps to reduce the off-target side effects of chemotherapeutic agents. The addition of polymers, which act to enhance the activity of anticancer drugs, to micellar systems affords stability and also offers additional selective cytotoxicity against cancer cells. A PM system with all of these characteristics could be ideal for the enhancement of cancer chemotherapeutic efficacy and reduction of side effects.

Despite these advantages, there are some problems that need to be addressed; these include the low drug loading capacity of PMs due to their small size, as well as the need to increase the rate of cargo release in vivo. Increasing the drug loading may promote aggregation and premature drug release, which would in turn result in subtherapeutic levels of drug reaching the target site. Furthermore, delivery of siRNA still represents the major barrier to its therapeutic use. There is a need to develop safer and more effective drug delivery carriers or approaches to translate RNAi-based therapeutics from the laboratory to the clinical setting.

A working collaboration between chemists and drug delivery investigators is required to develop the appropriate chemical approaches to the synthesis of new copolymers with optimal characteristics; these include the availability of functional moieties, biodegradable and biocompatible chemistries, high drug-loading capacities, and active targeting of the desired tissues.

REFERENCES

Akimoto, J., Nakayama, M., Sakai, K., Okano, T., 2008. Molecular design of outermost surface functionalized thermoresponsive polymeric micelles with biodegradable cores. J. Polym. Sci. Part A Polym. Chem. 46 (21), 7127–7137.

Alakhov, V.Y., Moskaleva, E.Y., Batrakova, E.V., Kabanov, A.V., 1996. Hypersensitization of multidrug resistant human ovarian carcinoma cells by pluronic P85 block copolymer. Bioconjug. Chem. 7 (2), 209–216.

Alexis, F., Basto, P., Levy-Nissenbaum, E., Radovic-Moreno, A.F., Zhang, L., Pridgen, E., et al., 2008. HER-2-targeted nanoparticle-affibody bioconjugates for cancer therapy. ChemMedChem 3 (12), 1839–1843.

Amjad, M.W., Amin, M.C.I.M., Katas, H., Butt, A.M., Kesharwani, P., Iyer, A.K., 2015. In vivo antitumor activity of folate-conjugated cholic acid-polyethylenimine micelles for the codelivery of doxorubicin and siRNA to colorectal adenocarcinomas. Mol. Pharm. 12 (12), 4247–4258.

Asayama, S., Kawakami, H., Nagaoka, S., 2004. Design of a poly (L-histidine)-carbohydrate conjugate for a new pH-sensitive drug carrier. Polym. Adv. Technol. 15 (8), 439–444.

Babin, J., Pelletier, M., Lepage, M., Allard, J.F., Morris, D., Zhao, Y., 2009. A new two-photon-sensitive block copolymer nanocarrier. Angew. Chem. Int. Ed. 48 (18), 3329–3332.

Bae, Y., Fukushima, S., Harada, A., Kataoka, K., 2003. Design of environment-sensitive supramolecular assemblies for intracellular drug delivery: polymeric micelles that are responsive to intracellular pH change. Angew. Chem. Int. Ed. 42 (38), 4640–4643.

Bae, Y., Jang, W.-D., Nishiyama, N., Fukushima, S., Kataoka, K., 2005a. Multifunctional polymeric micelles with folate-mediated cancer cell targeting and pH-triggered drug releasing properties for active intracellular drug delivery. Mol. BioSyst. 1 (3), 242–250.

Bae, Y., Nishiyama, N., Fukushima, S., Koyama, H., Yasuhiro, M., Kataoka, K., 2005b. Preparation and biological characterization of polymeric micelle drug carriers with intracellular pH-triggered drug release property: tumor permeability, controlled subcellular drug distribution, and enhanced in vivo antitumor efficacy. Bioconjug. Chem. 16 (1), 122–130.

Bartlett, D.W., Su, H., Hildebrandt, I.J., Weber, W.A., Davis, M.E., 2007. Impact of tumor-specific targeting on the biodistribution and efficacy of siRNA nanoparticles measured by multimodality in vivo imaging. Proc. Natl. Acad. Sci. 104 (39), 15549–15554.

Batrakova, E.V., Li, S., Vinogradov, S.V., Alakhov, V.Y., Miller, D.W., Kabanov, A.V., 2001. Mechanism of pluronic effect on P-glycoprotein efflux system in blood-brain barrier: contributions of energy depletion and membrane fluidization. J. Pharmacol. Exp. Ther. 299 (2), 483–493.

Belbella, A., Vauthier, C., Fessi, H., Devissaguet, J.-P., Puisieux, F., 1996. In vitro degradation of nanospheres from poly (D, L-lactides) of different molecular weights and polydispersities. Int. J. Pharm. 129 (1), 95–102.

Berkovic, G., Krongauz, V., Weiss, V., 2000. Spiropyrans and spirooxazines for memories and switches. Chem. Rev. 100 (5), 1741–1754.

Brannon-Peppas, L., Ghosn, B., Roy, K., Cornetta, K., 2007. Encapsulation of nucleic acids and opportunities for cancer treatment. Pharm. Res. 24 (4), 618–627.

Butt, A.M., Amin, M.C.I.M., Katas, H., Sarisuta, N., Witoonsaridsilp, W., Benjakul, R., 2012. In vitro characterization of pluronic F127 and D-α-tocopheryl polyethylene glycol 1000 succinate mixed micelles as nanocarriers for targeted anticancer-drug delivery. J. Nanomaterials 2012, 112.

Butt, A.M., Amin, M.C.I.M., Katas, H., 2015. Synergistic effect of pH-responsive folate-functionalized poloxamer 407-TPGS-mixed micelles on targeted delivery of anticancer drugs. Int. J. Nanomedicine 10, 1321–1334.

Cai, L.-L., Liu, P., Li, X., Huang, X., Ye, Y.-Q., Chen, F.-Y., et al., 2011. RGD peptide-mediated chitosan-based polymeric micelles targeting delivery for integrin-overexpressing tumor cells. Int. J. Nanomedicine 6, 3499–3508.

Cammas-Marion, S., Okano, T., Kataoka, K., 1999. Functional and site-specific macromolecular micelles as high potential drug carriers. Colloids Surf. B Biointerfaces 16 (1), 207–215.

Campbell, I.G., Jones, T.A., Foulkes, W.D., Trowsdale, J., 1991. Folate-binding protein is a marker for ovarian cancer. Cancer Res. 51 (19), 5329–5338.

Chen, W., Meng, F., Li, F., Ji, S.-J., Zhong, Z., 2009. pH-responsive biodegradable micelles based on acid-labile polycarbonate hydrophobe: synthesis and triggered drug release. Biomacromolecules 10 (7), 1727–1735.

Cheng, J., Teply, B.A., Sherifi, I., Sung, J., Luther, G., Gu, F.X., et al., 2007. Formulation of functionalized PLGA-PEG nanoparticles for in vivo targeted drug delivery. Biomaterials 28 (5), 869–876.

Choi, C.H., Alabi, C.A., Webster, P., Davis, M.E., 2010. Mechanism of active targeting in solid tumors with transferrin-containing gold nanoparticles. Proc. Natl. Acad. Sci. 107 (3), 1235–1240.

Chung, J., Yokoyama, M., Aoyagi, T., Sakurai, Y., Okano, T., 1998. Effect of molecular architecture of hydrophobically modified poly(N-isopropylacrylamide) on the formation of thermoresponsive core-shell micellar drug carriers. J. Control. Release 53 (1), 119–130.

Chung, J., Yokoyama, M., Yamato, M., Aoyagi, T., Sakurai, Y., Okano, T., 1999. Thermo-responsive drug delivery from polymeric micelles constructed using block copolymers of poly (N-isopropylacrylamide) and poly (butylmethacrylate). J. Control. Release 62 (1), 115–127.

Chung, J.E., Yokoyama, M., Okano, T., 2000. Inner core segment design for drug delivery control of thermo-responsive polymeric micelles. J. Control. Release 65 (1), 93–103.

Danhier, F., Danhier, P., De Saedeleer, C.J., Fruytier, A.-C., Schleich, N., Rieux, Ad, et al., 2015. Paclitaxel-loaded micelles enhance transvascular permeability and retention of nanomedicines in tumors. Int. J. Pharm. 479 (2), 399–407.

Dharap, S., Qiu, B., Williams, G., Sinko, P., Stein, S., Minko, T., 2003. Molecular targeting of drug delivery systems to ovarian cancer by BH3 and LHRH peptides. J. Control. Release 91 (1), 61–73.

Dou, S., Yao, Y.-D., Yang, X.-Z., Sun, T.-M., Mao, C.-Q., Song, E.-W., et al., 2012. Anti-Her2 single-chain antibody mediated DNMTs-siRNA delivery for targeted breast cancer therapy. J. Control. Release 161 (3), 875–883.

Duan, Q., Miura, Y., Narumi, A., Shen, X., Sato, S.I., Satoh, T., et al., 2006. Synthesis and thermoresponsive property of end-functionalized poly (N-isopropylacrylamide) with pyrenyl group. J. Polym. Sci. Part A Polym. Chem. 44 (3), 1117–1124.

Elsabahy, M., Wazen, N., Bayó-Puxan, N., Deleavey, G., Servant, M., Damha, M.J., et al., 2009. Delivery of nucleic acids through the controlled disassembly of multifunctional nanocomplexes. Adv. Funct. Mater. 19 (24), 3862–3867.

Engin, K., Leeper, D., Cater, J., Thistlethwaite, A., Tupchong, L., McFarlane, J., 1995. Extracellular pH distribution in human tumours. Int. J. Hyperthermia 11 (2), 211–216.

Ercole, F., Davis, T.P., Evans, R.A., 2010. Photo-responsive systems and biomaterials: photochromic polymers, light-triggered self-assembly, surface modification, fluorescence modulation and beyond. Polym. Chem. 1 (1), 37–54.

Farokhzad, O.C., Cheng, J., Teply, B.A., Sherifi, I., Jon, S., Kantoff, P.W., et al., 2006. Targeted nanoparticle-aptamer bioconjugates for cancer chemotherapy in vivo. Proc. Natl. Acad. Sci. 103 (16), 6315–6320.

Feil, H., Bae, Y.H., Feijen, J., Kim, S.W., 1993. Effect of comonomer hydrophilicity and ionization on the lower critical solution temperature of N-isopropylacrylamide copolymers. Macromolecules 26 (10), 2496–2500.

Felber, A.E., Castagner, B., Elsabahy, M., Deleavey, G.F., Damha, M.J., Leroux, J.-C., 2011. siRNA nanocarriers based on methacrylic acid copolymers. J. Control. Release 152 (1), 159–167.

Felber, A.E., Dufresne, M.-H., Leroux, J.-C., 2012. pH-sensitive vesicles, polymeric micelles, and nanospheres prepared with polycarboxylates. Adv. Drug Deliv. Rev. 64 (11), 979–992.

Fife, T.H., Jao, L., 1965. Substituent effects in acetal hydrolysis. J. Org. Chem. 30 (5), 1492–1495.

Fujishige, S., Kubota, K., Ando, I., 1989. Phase transition of aqueous solutions of poly (N-isopropylacrylamide) and poly (N-isopropylmethacrylamide). J. Phys. Chem. 93 (8), 3311–3313.

Ganesh, S., Iyer, A.K., Morrissey, D.V., Amiji, M.M., 2013. Hyaluronic acid based self-assembling nanosystems for CD44 target mediated siRNA delivery to solid tumors. Biomaterials 34 (13), 3489–3502.

Gao, X., Cui, Y., Levenson, R.M., Chung, L.W., Nie, S., 2004. In vivo cancer targeting and imaging with semiconductor quantum dots. Nat. Biotechnol. 22 (8), 969–976.

Gao, Z., Lukyanov, A.N., Singhal, A., Torchilin, V.P., 2002. Diacyllipid-polymer micelles as nanocarriers for poorly soluble anticancer drugs. Nano Lett. 2 (9), 979–982.

Gao, Z., Lukyanov, A., Chakilam, A., Torchilin, V., 2003. PEG-PE/phosphatidylcholine mixed immunomicelles specifically deliver encapsulated taxol to tumor cells of different origin and promote their efficient killing. J. Drug Target. 11 (2), 87–92.

Gao, Z., Lee, D., Kim, D., Bae, Y., 2005. Doxorubicin loaded pH-sensitive micelle targeting acidic extracellular pH of human ovarian A2780 tumor in mice. J. Drug Target. 13 (7), 391–397.

Gillies, E.R., Fréchet, J.M., 2003. A new approach towards acid sensitive copolymer micelles for drug delivery. Chem. Commun. 14, 1640–1641.

Gillies, E.R., Frechet, J.M., 2005a. Dendrimers and dendritic polymers in drug delivery. Drug Discov. Today 10 (1), 35–43.

Gillies, E.R., Fréchet, J.M., 2005b. pH-responsive copolymer assemblies for controlled release of doxorubicin. Bioconjug. Chem. 16 (2), 361–368.

Gillies, E.R., Goodwin, A.P., Fréchet, J.M., 2004a. Acetals as pH-sensitive linkages for drug delivery. Bioconjug. Chem. 15 (6), 1254–1263.

Gillies, E.R., Jonsson, T.B., Fréchet, J.M., 2004b. Stimuli-responsive supramolecular assemblies of linear-dendritic copolymers. J. Am. Chem. Soc. 126 (38), 11936–11943.

Gopin, A., Ebner, S., Attali, B., Shabat, D., 2006. Enzymatic activation of second-generation dendritic prodrugs: conjugation of self-immolative dendrimers with poly (ethylene glycol) via click chemistry. Bioconjug. Chem. 17 (6), 1432–1440.

Govender, T., Riley, T., Ehtezazi, T., Garnett, M.C., Stolnik, S., Illum, L., et al., 2000. Defining the drug incorporation properties of PLA–PEG nanoparticles. Int. J. Pharm. 199 (1), 95–110.

Graf, N., Bielenberg, D.R., Kolishetti, N., Muus, C., Banyard, J., Farokhzad, O.C., et al., 2012. alpha(V) beta(3) integrin-targeted PLGA-PEG nanoparticles for enhanced anti-tumor efficacy of a Pt(IV) pro-drug. ACS Nano 6 (5), 4530–4539.

Greenfield, R.S., Kaneko, T., Daues, A., Edson, M.A., Fitzgerald, K.A., Olech, L.J., et al., 1990. Evaluation in vitro of adriamycin immunoconjugates synthesized using an acid-sensitive hydrazone linker. Cancer Res. 50 (20), 6600–6607.

Greenwald, R.B., Choe, Y.H., McGuire, J., Conover, C.D., 2003. Effective drug delivery by PEGylated drug conjugates. Adv. Drug Deliv. Rev. 55 (2), 217–250.

Hameed, N., Salim, N.V., Parameswaranpillai, J., Fox, B.L., 2015. Flower like micellar assemblies in poly (styrene)-block-poly (4-vinyl pyridine)/poly (acrylic acid) complexes. Mater. Lett. 147, 92–96.

Heller, J., Barr, J., Ng, S.Y., Abdellauoi, K.S., Gurny, R., 2002. Poly (ortho esters): synthesis, characterization, properties and uses. Adv. Drug Deliv. Rev. 54 (7), 1015–1039.

Heskins, M., Guillet, J.E., 1968. Solution properties of poly (N-isopropylacrylamide). J. Macromol. Sci. Chem. 2 (8), 1441–1455.

Hobbs, S.K., Monsky, W.L., Yuan, F., Roberts, W.G., Griffith, L., Torchilin, V.P., et al., 1998. Regulation of transport pathways in tumor vessels: role of tumor type and microenvironment. Proc. Natl. Acad. Sci. 95 (8), 4607–4612.

Hrkach, J., Von Hoff, D., Mukkaram Ali, M., Andrianova, E., Auer, J., Campbell, T., et al., 2012. Preclinical development and clinical translation of a PSMA-targeted docetaxel nanoparticle with a differentiated pharmacological profile. Sci. Transl. Med. 4 (128), 3003651.

Hrycyna, C.A., Ramachandra, M., Ambudkar, S.V., Ko, Y.H., Pedersen, P.L., Pastan, I., et al., 1998. Mechanism of action of human p-glycoprotein atpase activity photochemical cleavage during a cata-lytic transition state using orthovanadate reveals cross-talk between the two ATP sites. J. Biol. Chem. 273 (27), 16631–16634.

Hudecz, F., Ross, H., Price, M.R., Baldwin, R.W., 1990. Immunoconjugate design: a predictive approach for coupling of daunomycin to monoclonal antibodies. Bioconjug. Chem. 1 (3), 197–204.

Husmann, M., Schenderlein, S., Luck, M., Lindner, H., Kleinebudde, P., 2002. Polymer erosion in PLGA microparticles produced by phase separation method. Int. J. Pharm. 242 (1–2), 277–280.

Husseini, G.A., Myrup, G.D., Pitt, W.G., Christensen, D.A., Rapoport, N.Y., 2000. Factors affecting acousti-cally triggered release of drugs from polymeric micelles. J. Control. Release 69 (1), 43–52.

Iakoubov, L.Z., Torchilin, V.P., 1997. A novel class of antitumor antibodies: nucleosome-restricted antinuclear autoantibodies (ANA) from healthy aged nonautoimmune mice. Oncol. Res. 9 (8), 439–446.

Iyer, A.K., Khaled, G., Fang, J., Maeda, H., 2006. Exploiting the enhanced permeability and retention effect for tumor targeting. Drug Discov. Today 11 (17), 812–818.

Jain, R.K., 1987. Transport of molecules across tumor vasculature. Cancer Metastasis Rev. 6 (4), 559–593.

Jiang, J., Tong, X., Zhao, Y., 2005. A new design for light-breakable polymer micelles. J. Am. Chem. Soc. 127 (23), 8290–8291.

Jiang, J., Tong, X., Morris, D., Zhao, Y., 2006. Toward photocontrolled release using light-dissociable block copolymer micelles. Macromolecules 39 (13), 4633–4640.

Jin, Y., Song, L., Su, Y., Zhu, L., Pang, Y., Qiu, F., et al., 2011. Oxime linkage: a robust tool for the design of pH-sensitive polymeric drug carriers. Biomacromolecules 12 (10), 3460–3468.

Jones, M.-C., Leroux, J.-C., 1999. Polymeric micelles–a new generation of colloidal drug carriers. Eur. J. Pharm. Biopharm. 48 (2), 101–111.

Jones, M.-C., Ranger, M., Leroux, J.-C., 2003. pH-sensitive unimolecular polymeric micelles: synthesis of a novel drug carrier. Bioconjug. Chem. 14 (4), 774–781.

Kabanov, A.V., Batrakova, E.V., Melik-Nubarov, N.S., Fedoseev, N.A., Dorodnich, T.Y., Alakhov, V.Y., et al., 1992. A new class of drug carriers: micelles of poly (oxyethylene)-poly (oxypropylene) block copolymers as microcontainers for drug targeting from blood in brain. J. Control. Release 22 (2), 141–157.

Kabanov, A.V., Batrakova, E.V., Li, S., Alakhov, V.Y., 2001. Selective energy depletion and sensitization of multiple drug-resistant cancer cells by pluronic block copolymer. Macromol. Symp. 172 (1), 103–112.

Kamaly, N., Fredman, G., Subramanian, M., Gadde, S., Pesic, A., Cheung, L., et al., 2013. Development and in vivo efficacy of targeted polymeric inflammation-resolving nanoparticles. Proc. Natl. Acad. Sci. 110 (16), 6506–6511.

Kanazawa, T., Akiyama, F., Kakizaki, S., Takashima, Y., Seta, Y., 2013. Delivery of siRNA to the brain using a combination of nose-to-brain delivery and cell-penetrating peptide-modified nano-micelles. Biomaterials 34 (36), 9220–9226.

Kaneko, T., Willner, D., Monkovic, I., Knipe, J.O., Braslawsky, G.R., Greenfield, R.S., et al., 1991. New hydrazone derivatives of adriamycin and their immunoconjugates-a correlation between acid stability and cytotoxicity. Bioconjug. Chem. 2 (3), 133–141.

Karmali, P.P., Kotamraju, V.R., Kastantin, M., Black, M., Missirlis, D., Tirrell, M., et al., 2009. Targeting of albumin-embedded paclitaxel nanoparticles to tumors. Nanomedicine 5 (1), 73–82.

Katono, H., Maruyama, A., Sanui, K., Ogata, N., Okano, T., Sakurai, Y., 1991. Thermo-responsive swelling and drug release switching of interpenetrating polymer networks composed of poly (acrylamide-co-butyl methacrylate) and poly (acrylic acid). J. Control Release 16 (1), 215–227.

Kaur, S., Prasad, C., Balakrishnan, B., Banerjee, R., 2015. Trigger responsive polymeric nanocarriers for cancer therapy. Biomater. Sci. 3 (7), 955–987.

Kesharwani, P., Iyer, A.K., 2015. Recent advances in dendrimer-based nanovectors for tumor-targeted drug and gene delivery. Drug Discov. Today 20 (5), 536–547.

Kim, G.M., Bae, Y.H., Jo, W.H., 2005. pH-induced micelle formation of poly (histidine-co-phenylalanine)-block-poly (ethylene glycol) in aqueous media. Macromol. Biosci. 5 (11), 1118–1124.

Kim, D., Lee, E.S., Oh, K.T., Gao, Z.G., Bae, Y.H., 2008a. Doxorubicin-loaded polymeric micelle overcomes multidrug resistance of cancer by double-targeting folate receptor and early endosomal pH. Small 4 (11), 2043–2050.

Kim, S., Kim, J.Y., Huh, K.M., Acharya, G., Park, K., 2008b. Hydrotropic polymer micelles containing acrylic acid moieties for oral delivery of paclitaxel. J. Control. Release 132 (3), 222–229.

Kim, S.H., Jeong, J.H., Lee, S.H., Kim, S.W., Park, T.G., 2008c. LHRH receptor-mediated delivery of siRNA using polyelectrolyte complex micelles self-assembled from siRNA-PEG-LHRH conjugate and PEI. Bioconjug. Chem. 19 (11), 2156–2162.

Kim, S.H., Jeong, J.H., Lee, S.H., Kim, S.W., Park, T.G., 2008d. Local and systemic delivery of VEGF siRNA using polyelectrolyte complex micelles for effective treatment of cancer. J. Control. Release 129 (2), 107–116.

Kim, D., Jeong, Y.Y., Jon, S., 2010. A drug-loaded aptamer-gold nanoparticle bioconjugate for combined CT imaging and therapy of prostate cancer. ACS Nano 4 (7), 3689–3696.

Kirpotin, D.B., Drummond, D.C., Shao, Y., Shalaby, M.R., Hong, K., Nielsen, U.B., et al., 2006. Antibody targeting of long-circulating lipidic nanoparticles does not increase tumor localization but does increase internalization in animal models. Cancer Res. 66 (13), 6732–6740.

Kozlov, M.Y., Melik-Nubarov, N.S., Batrakova, E.V., Kabanov, A.V., 2000. Relationship between pluronic block copolymer structure, critical micellization concentration and partitioning coefficients of low molecular mass solutes. Macromolecules 33 (9), 3305–3313.

Lavasanifar, A., Samuel, J., Kwon, G.S., 2001. The effect of alkyl core structure on micellar properties of poly(ethylene oxide)-block-poly(L-aspartamide) derivatives. Colloids Surf. B Biointerfaces 22 (2), 115–126. Epub 2001/07/14.

Lavasanifar, A., Samuel, J., Kwon, G.S., 2002. Poly (ethylene oxide)-block-poly (L-amino acid) micelles for drug delivery. Adv. Drug Deliv. Rev. 54 (2), 169–190.

Lee, E.S., Shin, H.J., Na, K., Bae, Y.H., 2003. Poly (l-histidine)–PEG block copolymer micelles and pH-induced destabilization. J. Control. Release 90 (3), 363–374.

Lee, H.-I., Pietrasik, J., Matyjaszewski, K., 2006. Phototunable temperature-responsive molecular brushes prepared by ATRP. Macromolecules 39 (11), 3914–3920.

Lee, H.I., Wu, W., Oh, J.K., Mueller, L., Sherwood, G., Peteanu, L., et al., 2007. Light-induced reversible formation of polymeric micelles. Angew. Chem. 119 (14), 2505–2509.

Leroux, J.-C., Roux, E., Le Garrec, D., Hong, K., Drummond, D.C., 2001. N-isopropylacrylamide copolymers for the preparation of pH-sensitive liposomes and polymeric micelles. J. Control. Release 72 (1–3), 71–84.

Li, Y., Kwon, G.S., 2000. Methotrexate esters of poly (ethylene oxide)-block-poly (2-hydroxyethyl-L-aspartamide). Part I: Effects of the level of methotrexate conjugation on the stability of micelles and on drug release. Pharm. Res. 17 (5), 607–611.

Licciardi, M., Giammona, G., Du, J., Armes, S.P., Tang, Y., Lewis, A.L., 2006. New folate-functionalized biocompatible block copolymer micelles as potential anti-cancer drug delivery systems. Polymer 47 (9), 2946–2955.

Liu, C., Chen, G., Sun, H., Xu, J., Feng, Y., Zhang, Z., et al., 2011. Toroidal micelles of polystyrene-block-poly (acrylic acid). Small 7 (19), 2721–2726.

Liu, S.-Q., Wiradharma, N., Gao, S.-J., Tong, Y.W., Yang, Y.-Y., 2007. Bio-functional micelles self-assembled from a folate-conjugated block copolymer for targeted intracellular delivery of anticancer drugs. Biomaterials 28 (7), 1423–1433.

Liu, X.-Q., Xiong, M.-H., Shu, X.-T., Tang, R.-Z., Wang, J., 2012. Therapeutic delivery of siRNA silencing HIF-1 alpha with micellar nanoparticles inhibits hypoxic tumor growth. Mol. Pharm. 9 (10), 2863–2874.

Liu, Y., Wang, W., Yang, J., Zhou, C., Sun, J., 2013. pH-sensitive polymeric micelles triggered drug release for extracellular and intracellular drug targeting delivery. Asian J. Pharm. Sci. 8 (3), 159–167.

Lutz, J.F., 2008. Polymerization of oligo (ethylene glycol)(meth) acrylates: toward new generations of smart biocompatible materials. J. Polym. Sci. Part A Polym. Chem. 46 (11), 3459–3470.

Maeda, H., 2015. Toward a full understanding of the EPR effect in primary and metastatic tumors as well as issues related to its heterogeneity. Adv. Drug Deliv. Rev. 91, 3–6.

Maeda, H., Bharate, G., Daruwalla, J., 2009. Polymeric drugs for efficient tumor-targeted drug delivery based on EPR-effect. Eur. J. Pharm. Biopharm. 71 (3), 409–419.

Manchun, S., Dass, C.R., Sriamornsak, P., 2012. Targeted therapy for cancer using pH-responsive nanocarrier systems. Life Sci. 90 (11), 381–387.

Marin, A., Sun, H., Husseini, G.A., Pitt, W.G., Christensen, D.A., Rapoport, N.Y., 2002. Drug delivery in pluronic micelles: effect of high-frequency ultrasound on drug release from micelles and intracellular uptake. J. Control. Release 84 (1), 39–47.

Mishra, V., Kesharwani, P., Jain, N.K., 2014. siRNA nanotherapeutics: a Trojan horse approach against HIV. Drug Discov. Today 19 (12), 1913–1920.

Misra, R., Das, M., Sahoo, B.S., Sahoo, S.K., 2014. Reversal of multidrug resistance in vitro by co-delivery of MDR1 targeting siRNA and doxorubicin using a novel cationic poly (lactide-co-glycolide) nanoformulation. Int. J. Pharm. 475 (1), 372–384.

Miyata, K., Christie, R.J., Kataoka, K., 2011. Polymeric micelles for nano-scale drug delivery. React. Funct. Polym. 71 (3), 227–234.

Na, K., Seong Lee, E., Bae, Y.H., 2003. Adriamycin loaded pullulan acetate/sulfonamide conjugate nanoparticles responding to tumor pH: pH-dependent cell interaction, internalization and cytotoxicity in vitro. J. Control. Release 87 (1), 3–13.

Nagarajan, R., 1996. Solubilization of hydrophobic substances by block copolymer micelles in aqueous solutions. In: Webber, S., Munk, P., Tuzar, Z. (Eds.), Solvents and Self-Organization of Polymers. Springer, Netherlands, pp. 121–165.

Nakayama, M., Okano, T., 2005. Polymer terminal group effects on properties of thermoresponsive polymeric micelles with controlled outer-shell chain lengths. Biomacromolecules 6 (4), 2320–2327.

Nakayama, M., Okano, T., 2008. Unique thermoresponsive polymeric micelle behavior via cooperative polymer corona phase transitions. Macromolecules 41 (3), 504–507.

Nakayama, M., Kawahara, Y., Akimoto, J., Kanazawa, H., Okano, T., 2012. pH-induced phase transition control of thermoresponsive nano-micelles possessing outermost surface sulfonamide moieties. Colloids Surf. B Biointerfaces 99, 12–19.

Oe, Y., Christie, R.J., Naito, M., Low, S.A., Fukushima, S., Toh, K., et al., 2014. Actively-targeted polyion complex micelles stabilized by cholesterol and disulfide cross-linking for systemic delivery of siRNA to solid tumors. Biomaterials 35 (27), 7887–7895.

Ojugo, A.S., McSheehy, P.M., McIntyre, D.J., McCoy, C., Stubbs, M., Leach, M.O., et al., 1999. Measurement of the extracellular pH of solid tumours in mice by magnetic resonance spectroscopy: a comparison of exogenous 19F and 31P probes. NMR Biomed. 12 (8), 495–504.

O'Reilly, R.K., Hawker, C.J., Wooley, K.L., 2006. Cross-linked block copolymer micelles: functional nanostructures of great potential and versatility. Chem. Soc. Rev. 35 (11), 1068–1083. Epub 2006/10/24.

Orihara, Y., Matsumura, A., Saito, Y., Ogawa, N., Saji, T., Yamaguchi, A., et al., 2001. Reversible release control of an oily substance using photoresponsive micelles. Langmuir 17 (20), 6072–6076.

Palanca-Wessels, M.C., Convertine, A.J., Cutler-Strom, R., Booth, G.C., Lee, F., Berguig, G.Y., et al., 2011. Anti-CD22 antibody targeting of pH-responsive micelles enhances small interfering RNA delivery and gene silencing in lymphoma cells. Mol. Ther. 19 (8), 1529–1537.

Park, J.W., Hong, K., Kirpotin, D.B., Colbern, G., Shalaby, R., Baselga, J., et al., 2002. Anti-HER2 immunoliposomes: enhanced efficacy attributable to targeted delivery. Clin. Cancer Res. 8 (4), 1172–1181.

Park, J.-S., Kataoka, K., 2006. Precise control of lower critical solution temperature of thermosensitive poly (2-isopropyl-2-oxazoline) via gradient copolymerization with 2-ethyl-2-oxazoline as a hydrophilic comonomer. Macromolecules 39 (19), 6622–6630.

Park, J.H., von Maltzahn, G., Zhang, L., Schwartz, M.P., Ruoslahti, E., Bhatia, S.N., et al., 2008. Magnetic iron oxide nanoworms for tumor targeting and imaging. Adv. Mater. 20 (9), 1630–1635.

Park, T.G., 1995. Degradation of poly (lactic-co-glycolic acid) microspheres: effect of copolymer composition. Biomaterials 16 (15), 1123–1130.

Peer, D., Lieberman, J., 2011. Special delivery: targeted therapy with small RNAs. Gene Ther. 18 (12), 1127–1133.

Peer, D., Karp, J.M., Hong, S., Farokhzad, O.C., Margalit, R., Langer, R., 2007. Nanocarriers as an emerging platform for cancer therapy. Nat. Nanotechnol. 2 (12), 751–760.

Ponce, A.M., Vujaskovic, Z., Yuan, F., Needham, D., Dewhirst, M.W., 2006. Hyperthermia mediated liposomal drug delivery. Int. J. Hyperthermia 22 (3), 205–213.

Rapoport, N., 2004. Combined cancer therapy by micellar-encapsulated drug and ultrasound. Int. J. Pharm. 277 (1), 155–162.

Rapoport, N., Marin, A.P., Timoshin, A.A., 2000. Effect of a polymeric surfactant on electron transport in HL-60 cells. Arch. Biochem. Biophys. 384 (1), 100–108.

Rapoport, N., Marin, A., Christensen, D., 2002. Ultrasound activated micellar drug delivery. Drug Deliv. Syst. Sci. 2, 37–46.

Rapaport, N., Marin, A., Muniruzzaman, M., Christensen, D., 2003a. Controlled drug delivery to drug-sensitive and multidrug resistant cells: effects of Pluronic micelles and ultrasound. In: Dinh, S.M., Liu, P. (Eds.), Advances in Controlled Drug Delivery: Science, Technology and Products. American Chemical Society, Washington, DC, pp. 85–101.

Rapoport, N., Pitt, W.G., Sun, H., Nelson, J.L., 2003b. Drug delivery in polymeric micelles: from in vitro to in vivo. J Control. Release 91 (1), 85–95.

Rau, H., 1990. Photochromism: Molecules and Systems. Elsevier, Amsterdam, NY, pp. 165–192.

Ren, W.-H., Chang, J., Yan, C.-h, Qian, X.-m, Long, L.-x, He, B., et al., 2010. Development of transferrin functionalized poly (ethylene glycol)/poly (lactic acid) amphiphilic block copolymeric micelles as a potential delivery system targeting brain glioma. J. Mater. Sci. Mater. Med. 21 (9), 2673–2681.

Rettig, G.R., Behlke, M.A., 2012. Progress toward in vivo use of siRNAs-II. Mol. Ther. 20 (3), 483–512.

Rösler, A., Vandermeulen, G.W., Klok, H.-A., 2012. Advanced drug delivery devices via self-assembly of amphiphilic block copolymers. Adv. Drug Deliv. Rev. 64, 270–279.

Sanchez, C., Lebeau, B., Chaput, F., Boilot, J.P., 2003. Optical properties of functional hybrid organic–inorganic nanocomposites. Adv. Mater. 15 (23), 1969–1994.

Sant, V.P., Smith, D., Leroux, J.-C., 2004. Novel pH-sensitive supramolecular assemblies for oral delivery of poorly water soluble drugs: preparation and characterization. J. Control. Release 97 (2), 301–312.

Sarisozen, C., Abouzeid, A.H., Torchilin, V.P., 2014. The effect of co-delivery of paclitaxel and curcumin by transferrin-targeted PEG-PE-based mixed micelles on resistant ovarian cancer in 3-D spheroids and in vivo tumors. Eur. J. Pharm. Biopharm. 88 (2), 539–550.

Saunders, N.A., Simpson, F., Thompson, E.W., Hill, M.M., Endo-Munoz, L., Leggatt, G., et al., 2012. Role of intratumoural heterogeneity in cancer drug resistance: molecular and clinical perspectives. EMBO Mol. Med. 4 (8), 675–684.

Saw, P.E., Kim, S., Lee, I., Park, J., Yu, M., Lee, J., et al., 2013. Aptide-conjugated liposome targeting tumor-associated fibronectin for glioma therapy. J. Mat. Chem. B 1 (37), 4723–4726.

Schild, H.G., 1992. Poly(N-isopropylacrylamide): experiment, theory and application. Prog. Polym. Sci. 17 (2), 163–249.

Schmelz, J., Schedl, A.E., Steinlein, C., Manners, I., Schmalz, H., 2012. Length control and block-type architectures in worm-like micelles with polyethylene cores. J. Am. Chem. Soc. 134 (34), 14217–14225.

Sethuraman, V.A., Bae, Y.H., 2007. TAT peptide-based micelle system for potential active targeting of anti-cancer agents to acidic solid tumors. J. Control. Release 118 (2), 216–224.

Shen, W.-C., Ryser, H.J.-P., 1981. Cis-Aconityl spacer between daunomycin and macromolecular carriers: a model of pH-sensitive linkage releasing drug from a lysosomotropic conjugate. Biochem. Biophys. Res. Commun. 102 (3), 1048–1054.

Shimada, N., Ino, H., Maie, K., Nakayama, M., Kano, A., Maruyama, A., 2011. Ureido-derivatized polymers based on both poly (allylurea) and poly (L-citrulline) exhibit UCST-type phase transition behavior under physiologically relevant conditions. Biomacromolecules 12 (10), 3418–3422.

Solomatin, S.V., Bronich, T.K., Bargar, T.W., Eisenberg, A., Kabanov, V.A., Kabanov, A.V., 2003. Environmentally responsive nanoparticles from block ionomer complexes: effects of pH and ionic strength. Langmuir 19 (19), 8069–8076.

Son, S., Singha, K., Kim, W.J., 2010. Bioreducible BPEI-SS-PEG-cNGR polymer as a tumor targeted non-viral gene carrier. Biomaterials 31 (24), 6344–6354.

Štěpánek, M., Podhájecká, K., Tesarová, E., Procházka, K., Tuzar, Z., Brown, W., 2001. Hybrid polymeric micelles with hydrophobic cores and mixed polyelectrolyte/nonelectrolyte shells in aqueous media. 1. Preparation and basic characterization. Langmuir 17 (14), 4240–4244.

Stiriba, S.E., Frey, H., Haag, R., 2002. Dendritic polymers in biomedical applications: from potential to clinical use in diagnostics and therapy. Angew. Chem. Int. Ed. 41 (8), 1329–1334.

Stubbs, M., McSheehy, P.M., Griffiths, J.R., Bashford, C.L., 2000. Causes and consequences of tumour acidity and implications for treatment. Mol. Med. Today 6 (1), 15–19.

Sun, T.-M., Du, J.-Z., Yao, Y.-D., Mao, C.-Q., Dou, S., Huang, S.-Y., et al., 2011. Simultaneous delivery of siRNA and paclitaxel via a "two-in-one" micelleplex promotes synergistic tumor suppression. ACS Nano 5 (2), 1483–1494.

Sun, Y., Zou, W., Bian, S., Huang, Y., Tan, Y., Liang, J., et al., 2013. Bioreducible PAA-g-PEG graft micelles with high doxorubicin loading for targeted antitumor effect against mouse breast carcinoma. Biomaterials 34 (28), 6818–6828.

Sutton, D., Nasongkla, N., Blanco, E., Gao, J., 2007. Functionalized micellar systems for cancer targeted drug delivery. Pharm. Res. 24 (6), 1029–1046.

Taillefer, J., Jones, M.C., Brasseur, N., Van Lier, J., Leroux, J.C., 2000. Preparation and characterization of pH-responsive polymeric micelles for the delivery of photosensitizing anticancer drugs. J. Pharm. Sci. 89 (1), 52–62.

Taillefer, J., Brasseur, N., van Lier, J.E., Lenaerts, V., Garrec, D.L., Leroux, J.C., 2001. In-vitro and in-vivo evaluation of pH-responsive polymeric micelles in a photodynamic cancer therapy model. J. Pharm. Pharmacol. 53 (2), 155–166.

Takahashi, A., Yamamoto, Y., Yasunaga, M., Koga, Y., Kuroda, Ji, Takigahira, M., et al., 2013. NC-6300, an epirubicin-incorporating micelle, extends the antitumor effect and reduces the cardiotoxicity of epirubicin. Cancer Sci. 104 (7), 920–925.

Takei, Y.G., Aoki, T., Sanui, K., Ogata, N., Okano, T., Sakurai, Y., 1993. Temperature-responsive bioconjugates. 2. Molecular design for temperature-modulated bioseparations. Bioconjug. Chem. 4 (5), 341–346.

Tannock, I.F., Rotin, D., 1989. Acid pH in tumors and its potential for therapeutic exploitation. Cancer Res. 49 (16), 4373–4384.

Torchilin, V.P., 2005. Fluorescence microscopy to follow the targeting of liposomes and micelles to cells and their intracellular fate. Adv. Drug Deliv. Rev. 57 (1), 95–109.

Torchilin,V.P.,Trubetskoy,V.S.,Whiteman, K.R., Caliceti, P., Ferruti, P.,Veronese, F.M., 1995. New synthetic amphiphilic polymers for steric protection of liposomes in vivo. J. Pharm. Sci. 84 (9), 1049–1053.

Torchilin,V.P., 2001a. Structure and design of polymeric surfactant-based drug delivery systems. J. Control. Release 73 (2), 137–172.

Torchilin, V.P., Lukyanov, A.N., Gao, Z., Papahadjopoulos-Sternberg, B., 2001b. Immunomicelles: targeted pharmaceutical carriers for poorly soluble drugs. Proc. Natl. Acad. Sci. 100 (10), 6039–6044.

Torchilin,V.P., Levchenko, T.S., Lukyanov, A.N., Khaw, B.A., Klibanov, A.L., Rammohan, R., et al., 2001c. p-Nitrophenylcarbonyl-PEG-PE-liposomes: fast and simple attachment of specific ligands, including monoclonal antibodies, to distal ends of PEG chains via p-nitrophenylcarbonyl groups. Biochim. Biophys. Acta 1511 (2), 397–411.

Torchilin, V.P., Iakoubov, L.Z., Estrov, Z., 2003a. Therapeutic potential of antinuclear autoantibodies in cancer. Cancer Ther. 1 (1), 179.

Torchilin,V.P., Lukyanov, A.N., Gao, Z., Papahadjopoulos-Sternberg, B., 2003b. Immunomicelles: targeted pharmaceutical carriers for poorly soluble drugs. Proc. Natl. Acad. Sci. 100 (10), 6039–6044.

Tuzar, Z., 1996. Overview of polymer micelles. In: Webber, S., Munk, P., Tuzar, Z. (Eds.), Solvents and Self-Organization of Polymers. Springer, Netherlands, pp. 1–17.

Urry, D.W., 1992. Free energy transduction in polypeptides and proteins based on inverse temperature transitions. Prog. Biophys. Mol. Biol. 57 (1), 23–57.

van Sluis, R., Bhujwalla, Z.M., Raghunand, N., Ballesteros, P., Alvarez, J., Cerdan, S., et al., 1999. In vivo imaging of extracellular pH using 1 H MRSI. Magn. Reson. Med. 41 (4), 743–750.

Wang, G.,Tong, X., Zhao,Y., 2004. Preparation of azobenzene-containing amphiphilic diblock copolymers for light-responsive micellar aggregates. Macromolecules 37 (24), 8911–8917.

Wang, H.-X., Xiong, M.-H., Wang, Y.-C., Zhu, J., Wang, J., 2013. N-acetylgalactosamine functionalized mixed micellar nanoparticles for targeted delivery of siRNA to liver. J Control. Release 166 (2), 106–114.

Wang, L., Hao,Y., Li, H., Zhao,Y., Meng, D., Li, D., et al., 2015. Co-delivery of doxorubicin and siRNA for glioma therapy by a brain targeting system: angiopep-2-modified poly (lactic-co-glycolic acid) nanoparticles. J. Drug Target. 1–15. (ahead-of-print).

Wang, X.,Wang,Y., Chen, X.,Wang, J., Zhang, X., Zhang, Q., 2009. NGR-modified micelles enhance their interaction with CD13-overexpressing tumor and endothelial cells. J. Control. Release 139 (1), 56–62.

Wang, Y.-C., Liu, X.-Q., Sun, T.-M., Xiong, M.-H., Wang, J., 2008. Functionalized micelles from block copolymer of polyphosphoester and poly(ε-caprolactone) for receptor-mediated drug delivery. J. Control. Release 128 (1), 32–40.

Wei, H., Zhang, X.-Z., Zhou, Y., Cheng, S.-X., Zhuo, R.-X., 2006. Self-assembled thermoresponsive micelles of poly(N-isopropylacrylamide-b-methyl methacrylate). Biomaterials 27 (9), 2028–2034.

Wei, H., Chen, W.-Q., Chang, C., Cheng, C., Cheng, S.-X., Zhang, X.-Z., et al., 2008. Synthesis of star block, thermosensitive poly (L-lactide)-star block-poly (N-isopropylacrylamide-co-N-hydroxymethylacrylamide) copolymers and their self-assembled micelles for controlled release. J. Phys. Chem. C 112 (8), 2888–2894.

Werner, M.E., Karve, S., Sukumar, R., Cummings, N.D., Copp, J.A., Chen, R.C., et al., 2011. Folate-targeted nanoparticle delivery of chemo- and radiotherapeutics for the treatment of ovarian cancer peritoneal metastasis. Biomaterials 32 (33), 8548–8554.

Winkler, J., Martin-Killias, P., Pluckthun, A., Zangemeister-Wittke, U., 2009. EpCAM-targeted delivery of nanocomplexed siRNA to tumor cells with designed ankyrin repeat proteins. Mol. Cancer Ther. 8 (9), 2674–2683.

Xiao, Z., Ji, C., Shi, J., Pridgen, E.M., Frieder, J., Wu, J., et al., 2012. DNA self-assembly of targeted near-infrared-responsive gold nanoparticles for cancer thermo-chemotherapy. Angew. Chem. Int. Ed. 51 (47), 11853–11857.

Xing, L., Mattice, W.L., 1998. Large internal structures of micelles of triblock copolymers with small insoluble molecules in their cores. Langmuir 14 (15), 4074–4080.

Xu, M., Qian, J., Suo,A., Cui, N.,Yao,Y., Xu,W., et al., 2015. Co-delivery of doxorubicin and P-glycoprotein siRNA by multifunctional triblock copolymers for enhanced anticancer efficacy in breast cancer cells. J. Mat. Chem. B 3 (10), 2215–2228.

Yan, B., Boyer, J.-C., Branda, N.R., Zhao, Y., 2011. Near-infrared light-triggered dissociation of block copolymer micelles using upconverting nanoparticles. J. Am. Chem. Soc. 133 (49), 19714–19717.

Yang, S.R., Lee, H.J., Kim, J.-D., 2006. Histidine-conjugated poly (amino acid) derivatives for the novel endosomolytic delivery carrier of doxorubicin. J. Control. Release 114 (1), 60–68.

Yang, X., Iyer, A.K., Singh, A., Milane, L., Choy, E., Hornicek, F.J., et al., 2014. Cluster of differentiation 44 targeted hyaluronic acid based nanoparticles for MDR1 siRNA delivery to overcome drug resistance in ovarian cancer. Pharm. Res. 32 (6), 2097–2109.

Yang, X., Singh, A., Choy, E., Hornicek, F.J., Amiji, M.M., Duan, Z., 2015. MDR1 siRNA loaded hyaluronic acid-based CD44 targeted nanoparticle systems circumvent paclitaxel resistance in ovarian cancer. Sci. Rep. 5, 8509.

Yatvin, M., Kreutz, W., Horwitz, B., Shinitzky, M., 1980. pH-sensitive liposomes: possible clinical implications. Science 210 (4475), 1253–1255.

Yessine, M.-A., Dufresne, M.-H., Meier, C., Petereit, H.-U., Leroux, J.-C., 2007. Proton-actuated membrane-destabilizing polyion complex micelles. Bioconjug. Chem. 18 (3), 1010–1014.

Yin, T., Wang, L., Yin, L., Zhou, J., Huo, M., 2015. Co-delivery of hydrophobic paclitaxel and hydrophilic AURKA specific siRNA by redox-sensitive micelles for effective treatment of breast cancer. Biomaterials 61, 10–25.

Yoo, H.S., Park, T.G., 2004. Folate receptor targeted biodegradable polymeric doxorubicin micelles. J. Control. Release 96 (2), 273–283.

Yoo, H.S., Lee, E.A., Park, T.G., 2002. Doxorubicin-conjugated biodegradable polymeric micelles having acid-cleavable linkages. J. Control. Release 82 (1), 17–27.

Zeng, F., Lee, H., Allen, C., 2006. Epidermal growth factor-conjugated poly (ethylene glycol)-block-poly (δ-valerolactone) copolymer micelles for targeted delivery of chemotherapeutics. Bioconjug. Chem. 17 (2), 399–409.

Zhang, Z., Grijpma, D.W., Feijen, J., 2006. Thermo-sensitive transition of monomethoxy poly (ethylene glycol)-block-poly (trimethylene carbonate) films to micellar-like nanoparticles. J. Control. Release 112 (1), 57–63.

Zhao, B.-J., Ke, X.-Y., Huang, Y., Chen, X.-M., Zhao, X., Zhao, B.-X., et al., 2011. The antiangiogenic efficacy of NGR-modified PEG–DSPE micelles containing paclitaxel (NGR-M-PTX) for the treatment of glioma in rats. J. Drug Target. 19 (5), 382–390.

Zhao, X., Li, F., Li, Y., Wang, H., Ren, H., Chen, J., et al., 2015. Co-delivery of HIF1α siRNA and gemcitabine via biocompatible lipid-polymer hybrid nanoparticles for effective treatment of pancreatic cancer. Biomaterials 46, 13–25.

Zhu, L., Mahato, R.I., 2010. Targeted delivery of siRNA to hepatocytes and hepatic stellate cells by bioconjugation. Bioconjug. Chem. 21 (11), 2119–2127.

CHAPTER 6

Nanoparticle-Homing Polymers as Platforms for Theranostic Applications

Rimesh Augustine, Johnson V. John and Il Kim

Contents

6.1 INTRODUCTION

Nanotechnology is a rapidly developing branch of technology that deals with the study of materials in the nanosize range. Thus, the handling of numerous devices in nearly all fields has increased, specifically the use of electronic devices. Nanotechnology enables scientists to conduct research in nanodimensions and thus use cells, molecules, peptides, and proteins to develop new nanostructures by fabricating nanostructures on these biomolecules. These structures show enhanced properties in various medical applications (Peer et al., 2007). Over the past few years, many studies have focused on bionanotechnology to develop advanced platforms for biomedical applications. Many treatments are not effective because of the failure in identifying diseased body parts, particularly in cancer cell determination. Nanotechnology offers a unique prospect for drug delivery to cancer cells. Conventional chemotherapy shows several disadvantages, such as the nonspecificity of drug delivery, resulting in effects on noncancerous cells and decreased treatment efficiency (Li et al., 2014). Additionally, cancer drugs show low solubility in an aqueous medium. The short time for which these drugs are sustained in the body and low concentration of drug release in targeted cells led to the introduction of carrier molecules for effective drug delivery (Kwon, 2003; Luo and Prestwich, 2002). These nanocarriers loaded with drugs show enhanced pharmacokinetics, prolonged circulation time, increased solubility of drug, and stimuli–responsive

Nanotechnology-Based Approaches for Targeting and Delivery of Drugs and Genes.
DOI: http://dx.doi.org/10.1016/B978-0-12-809717-5.00007-5

203

characteristics of the polymer material, ensuring drug release in cancerous cells (Nehoff et al., 2014).

Gene therapy is similar to or more important than drug delivery in medical treatment because its successful delivery can prevent the need for surgery and drugs. In gene delivery, the required gene is delivered to the targeted location, where it replaces the mutated gene (Conde et al., 2014; Aoyama et al., 2015; Ashcroft, 2004). However, the main limitation of gene delivery is uncertain safety during delivery. Thus a biocompatible, safe, specific, and competent nanocarrier must be developed. The development of nanomaterials for the successful delivery of genes and replacement of mutated genes has gained attention worldwide. The successful delivery of genes depends on the size, shape, flexibility, architecture, elemental composition, and surface chemistry of the polymers designed as carriers (Kannan et al., 2014).

Nanoparticles (NPs) are expected to be good carriers for both gene and drug delivery because of their structural compatibility with the cellular structure and environment (Conde et al., 2012). Polymer nanocarriers are flexible and can be fabricated with required functional properties. Additionally, they can enable intracellular drug delivery by reducing the drug resistance of cancer stem-like cells (Kabanov et al., 2002). NPs 10–200 nm in size are suitable for passive targeting, accumulate in tumor tissues through an enhanced permeability and retention effect (EPR), and are capable of facilitating the cumulative release of drugs in cancer cells (Maeda et al., 2000). In humans, the biological processes leading to cancer occur on the nanoscale, and NPs 20 nm or lower in size can circulate throughout the body in the blood and can interact with biomolecules on cell surfaces and inside of cells. This makes it possible to detect cancer more precisely, even in the small volume inside cells. Cancer becomes stable in adverse environments and acquires resistance to drugs used for treatment by undergoing a series of mutations. NP–drug composites composed of various organic and inorganic materials are well known for their use in effective treatment, diagnosis, monitoring, and control of biological systems. Monitoring the response to treatment using new NPs has shown positive results, and these carriers are capable of loading multiple types of drugs.

Polymeric NPs used as drug and gene carriers are considered advantageous because of their biocompatibility and nontoxic nature. Various polymer NPs with different architectures, sizes, and surface properties have been designed and reported for their stimuli-responsive characteristics in targeted and cumulative drug release. Polypeptides have diverse chemical structures and biocompatibility and show high loading capacity for both hydrophobic and hydrophilic drugs, target molecular recognition sites, self-assembling properties, and stimuli-responsive characteristics. In addition to natural polymers, synthetic polymers such as poly(ethylene glycol) (PEG), poly(L-glutamic acid), cyclodextrin, poly(N-isopropylacrylamide) (PNIPAAm), poly(lactic acid) (PLA), poly(lactide-co-glycolide) (PLG), and poly(caprolactone) have been successfully exploited for drug and gene delivery. Additionally, combinations of synthetic and natural polymers have been evaluated to develop NP carriers. These combinations led to the development of diblock and triblock polymers with subsequent self-assembly derived from versatile morphologies on the nanoscale.

Inorganic NPs also have emerged as promising scaffolds for drug and gene delivery. Although noble metal NPs have been used for medicinal purposes since ancient times, the distinctive properties of inorganic NPs such as low inherent stability, high surface area to volume ratio, biocompatibility, efficient biodistribution, and controllable stability enable their use as conventional delivery vehicles. Gold, silver, platinum, iron, and carbon are widely used as delivery NPs in medical treatment. The synthesis of these NPs requires very expensive experimental equipment and the toxic reducing agents and stabilizing agents used ultimately cause environmental pollution (Sahu et al., 2013; Faramarzi and Forootanfar, 2011; Fayaz et al., 2011). Therefore, recent studies of NP synthesis have focused on biological pathways, which are cost-effective and use nontoxic and environmentally friendly materials. Various biological systems such as plants, bacteria, and yeast have been investigated for the efficient biosynthesis of metallic NPs (Zhao et al., 2013; Mittal et al., 2013). Although these syntheses are simple and inexpensive, researchers in the field of nanoengineering have experienced difficulty in controlling the size and shape of NPs. Additionally, inorganic NPs show some limitations as drug and gene carriers because chemical modification of the drugs is required for anchoring on the surface of nanomaterials, which typically requires a noncovalent encapsulation technique.

Polymer-supported metal NPs show several advantages over traditional polymers because of their unique surface properties (Abraham et al., 2007; Li et al., 2010). Currently, metal-containing polymers as novel biomedical materials are important in the field of medical treatment, particularly as therapeutic carriers (Table 6.1). This chapter concentrates on recent research efforts in the area of NPs supported on polymeric scaffolds for drug and gene delivery. The surface stability of NPs allows easy modification of their surface with polymers and other biological molecules. Biopolymers and metal NPs exhibit unique theranostic activities and, in combination, may exhibit a synergistic effect of the characteristic properties of both the polymer and metal NP.

6.2 NOBLE METAL NANOPARTICLES SUPPORTED ON DIFFERENT POLYMERS

Noble metals have been used since ancient times and are used in current medical treatment. The unique properties of metal NPs such as their electrochemical, optical, magnetic, and antimicrobial activities have resulted in their expanded use in different areas of research. However, only a few metals have been investigated for medical treatment

Table 6.1 Types of nanoparticles and their main biological applications

Sl. No.	Polymeric nanomaterial loaded with nanoparticle	Applications	Reference(s)
I. Gold			
1	Polyethylene glycol (PEG)–coated gold nanoparticle	Tumor-growth delay	Visaria et al. (2006)
2	PEG shell	Anticancer drug carrier	Park et al. (2009)
3	Thiolated PEG and methyl thioglycolate mixture	Acid-responsive drug release	Aryal et al. (2009)
4	Au nanocage with NIPAAm	IR-controlled drug release	Yavuz et al. (2009)
5	Lipid-coated Au	NIR–induced drug release	Braun et al. (2009)
6	Au-*p*(LA-DOX)-*b*-PEG-OH/FA	Anticancer drug delivery	Prabaharan et al. (2009)
7	DNA-capped Au, attached with FA and PEG	Drug delivery	Alexander et al. (2014)
8	PEG hybridized DNA–GNP	Intracellular drug carrier	Song et al. (2015)
II. Silver			
1	Poly(AM-*co*-AMPS) hydrogel	Drug delivery and antimicrobial	Ravindra et al. (2012)
2	Polyfilm of poly(allylaminehydrochloride) and dextran sulfate	Drug delivery	Sripriya et al. (2013)
3	PEG conjugated with curcumin	Drug delivery	Soumya and Hela (2013)
4	Chitosan	Rheumatoid arthritis inflammation treating	Prasad et al. (2013)
5	Lysine-conjugated Ag	Drug delivery	Bonor et al. (2014)
6	Ag coated on chitosan	Breast cancer treatment	Nayak et al. (2016)
III. Iron			
1	FA-PEG$_{114}$-PLA$_x$-PEG$_{46}$ acrylate	Drug delivery and MR imaging	Kim et al. (2008)
2	*p*(HFMA-*co*-VBK)-*g*-PEG	Magnetic resonance and optical imaging	Yan et al. (2014)
3	FA-conjugated PEG	Lung cancer imaging	Yoo et al. (2012)
4	PLA-magnetic nanoparticle	Drug release	Yang et al. (2015)
5	Polystyrene-grafted γ-Fe$_2$O$_3$	Drug delivery	Robbes et al. (2012)
6	Poly(NIPAM)	Drug delivery	Kaamyabi et al. (2016), Deka et al. (2011)
7	CS-PEG-PUP	Drug delivery	Prabha and Raj (2016)

and diagnosis because some metals are harmful to the human body. Recently, metals supported on polymers have become an active area of research, and studies typically aim to reduce the limitations of targeted delivery. Gold and silver are the major noble metals used in theranostic applications.

6.2.1 Gold nanoparticles supported on polymers

Among noble metal NPs, gold is the most widely studied and significant because of its low cytotoxicity, easily controlled shape, size, robust stability, and unique surface properties (Murphy et al., 2008; Fadeel and Garcia-Bennett, 2010; Dreaden et al., 2011; Rana et al., 2012; Giljohnn et al., 2010; Mieszawaka et al., 2013; Liu and Ye, 2013). Additionally, gold NPs can selectively target tumors via the EPR effect. The stimuli-responsive properties of biopolymers exhibit their function when the gold is supported on a polymer.

A new NP delivery system, developed by Visaria et al., consisted of 33-nm poly-ethylene glycol-coated colloidal gold NPs (PT–cAu–TNF-α). Each gold particle was loaded with several hundred tumor necrosis factor (TNF)-α molecules. PT–cAu–TNF-α in combination with hyperthermia induced significant delays in tumor growth and reduced tumor cell survival. TNF itself caused systemic toxicity, while studies of the injection of gold NP suggested that this material was safe (Visaria et al., 2006). TNF-α on gold surface acts as a therapeutic ligand that facilitates receptor targeting. The polyethylene glycol moiety plays a role in hydration and acts as a shield that protects the particle from detection and clearance by the reticuloendothelial system, thus ensuring prolonged circulation (Paciotti et al., 2004). Additionally, PEGylated AuNPs show excellent in vivo biodistribution and pharmacokinetic properties (Paciotti et al., 2006; James et al., 2007). Hyukjin et al. introduced multifunctional gold nano-probes for in vivo imaging of tumor and rheumatoid arthritis (RA) using NP surface energy transfer (NSET). These nanoprobes were synthesized by end-immobilizing a near-infrared fluorescence dye labeled with hyaluronic acid (HA) on the gold surface. NSET showed a lower noise to sound ratio and better covering distance of nearly 20 nm, which is much better than that of fluorescence resonance energy transfer. HA on the gold surface is nonimmunogenic and can bind many functional groups for multiple conjugation (Lee et al., 2008; Falcone et al., 2006; Kattumuri et al., 2007).

Park et al. reported a new type of gold nanocarrier prepared by covering the carrier with per-6-thio-β-cyclodextrin (SH-CD) as a drug pocket, PEG shell, and anti-epidermal growth factor receptor antibody as a targeting ligand (Park et al., 2009). A schematic representation of AuNP functionalization and drug release is shown in Fig. 6.1. The hydrophobic pocket of CD is set to encapsulate a variety of hydrophobic drugs, making gold nanocarriers a versatile platform for other anticancer drugs as well. AuNPs show prolonged circulation times and cumulative drug release in tumor cells.

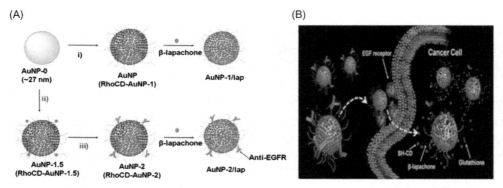

Figure 6.1 Schematic representation of the AuNP functionalization with β-lapachone (A) and drug release in cancer cells (B). *Adapted with permission from Park, C., Youn, H., Kim, H., Noh, T., Kook, Y.H., Oh, E.T., et al., 2009. Cyclodextrin-covered gold nanoparticles for targeted delivery of an anti-cancer drug. J. Mater. Chem. 19, 2310–2315. Copyright © 2009, Royal Society of Chemistry.*

In the same year, a drug-conjugated gold nanocarrier system was developed by Aryal et al., which showed powerful pH-responsive drug release and water-soluble characteristics. In this system, AuNPs were stabilized using an equimolar mixture of thiolated methoxy polyethylene glycol (MPEG-SH) and methylthioglycolate (MTG). The drug was conjugated to the MTG segment through a hydrazine linker; this bond was easily cleaved under acidic conditions, contributing to dramatic acid-responsive drug release properties. Moreover, the MPEG segment showed excellent water solubility and prolonged circulation time in the blood (Aryal et al., 2009). Yavuz et al. reported a gold nanocage platform for near-infrared light-controlled drug release. Ultraviolet light was typically used to remove the caging effect by photolysis and release the drug. However, the main disadvantage of this technique is that ultraviolet radiation causes cell damage and specific designs are required for different bioactive species. Gold nanocages have hollow interiors and porous walls that exhibit strong absorption in the near-infrared region, and heat generated by the photothermal effect dissipates into the surroundings. Smart polymers such as PNIPAAm are covalently anchored to the gold nanocage, which then undergo chain collapse; the cage interior is exposed and the bioactive species is released. The controlled release and transmission electron microscopy images of Au nanocages are shown in Fig. 6.2. The polymer shows reversible conformational changes upon near-infrared irradiation produced by the laser. Thus, the polymer can return to the original extended conformation after the laser is switched off to control drug release by regulating the irradiation time and intensity. This method is also suitable for in vivo studies unlike previous techniques, which are suited only for in vitro studies (Yavuz et al., 2009).

(A)

(B)

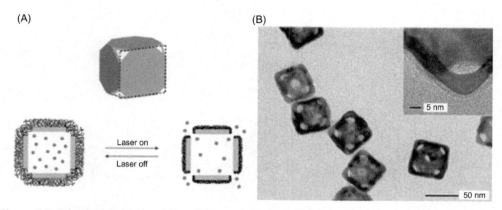

Laser on

Laser off

Figure 6.2 Schematic representation of the controlled release of the preloaded drug (A) and the TEM image of the Au nanocages (B). *Adapted with permission from Yavuz, M.S., Cheng, Y., Chen, J., Cobley, C.M., Zhang, Q., Rycenga, M., et al., 2009. Gold nanocages covered by smart polymers for controlled release with near-infrared light. Nat. Mater. 8, 935–939. Copyright © 2009, Rights managed by Nature Publishing Group.*

The pulsed near-infrared laser-induced time-dependent release of genes from a coated 40-nm gold shell was demonstrated by Braun et al. in the same year. The Tat-lipid formed an electrostatic bond with the siRNA and the lipid coating protected the RNA from enzymatic degradation. The Tat-lipid coating on gold mediated the cellular uptake of NPs, and release of siRNA from the nanoshell was attributed to surface-linker bond cleavage (Braun et al., 2009).

FA-conjugated Au-p(LA-DOX)-b-PEG-OH/FA NPs were synthesized in the same era by Prabaharan et al. for the targeted delivery of anticancer drugs to cancer cells. This system was formed with an Au core, hydrophobic inner shell formed by the poly(L-aspartate-doxorubicin), and hydrophilic outer shell formed by PEG and folate-conjugated PEG to prepare stable micelles in aqueous solution. The connection between the drug and hydrophobic inner shell was an acid-cleavable hydrazone linkage and thus under acidic pH conditions, the cancerous cellular environment facilitated drug release from the NP (Fig. 6.3). Additionally, cellular uptake of the micelle was assisted by folate receptor-mediated cellular endocytosis. Moreover, the FA-conjugated NP micelles exhibited higher cytotoxicity than FA-free NPs (Prabaharan et al., 2009).

The main advantage of AuNPs is their tunable surface charge, which enables control of cellular uptake and drug release. A study combining in vitro and mathematical modeling by Kim et al. suggested that positively charged particles showed the best delivery properties to the greatest number of tumor cells. Additionally, negatively charged particles showed the highest activity by delivering drugs in deep tissues. Positive particles were taken up by proliferating cells in the tumor and the uptake

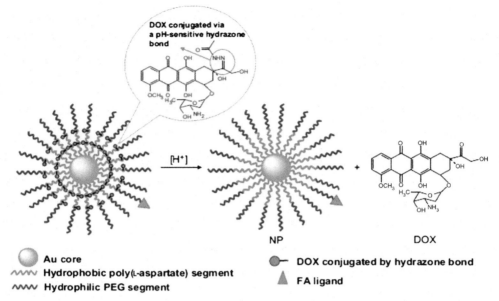

Figure 6.3 Schematic illustration of pH-triggered drug release from Au-*P*(LA-DOX)-*b*-PEG-OH/FA NP. *Adapted with permission from Prabaharan, M., Grailer, J.J., Pilla, S., Steeber, D.A., Gong, S., 2009. Gold nanoparticles with a monolayer of doxorubicin-conjugated amphiphilic block copolymer for tumor-targeted drug delivery. Biomaterials 30, 6065–6075. Copyright © 2009, Elsevier Ltd. All rights reserved.*

kinetics showed an irreversible trend, whereas negative particles entered cancer cells via diffusion and followed reversible kinetics. Thus, the tunable surface charge of AuNPs enabled controllable tissue penetration and drug release (Kim et al., 2010).

DNA-capped AuNPs as drug delivery vehicles were reported by Alexander et al. The AuNP was attached with folic acid (FA) and a thermo-responsive polymer or PEG. The drug-loaded vehicle showed greater cytotoxicity when 50% of DNA strands were attached to FA than when 100% of DNA strands were attached, indicating the significance of FA concentration in cytotoxicity. Confocal studies confirmed high drug accumulation inside the cell when using vehicle with an attached FA and thermo-responsive polymer (Alexander et al., 2014). Although DNA–gold NPs (DNA–GNPs) have been used for the past 10 years, their use has been limited by nonspecific interactions, low uptake, and low resistance to nuclease degradation. This drawback is overcome by introducing terminal PEGylation of the complementary DNA strand hybridized to a polyvalent DNA–GNP conjugate (Fig. 6.4). This NP shows approximately 10-fold higher stability against DNase I enzymatic digestion. This conjugate also shows excellent cellular uptake, making it a good intracellular drug carrier (Song et al., 2015).

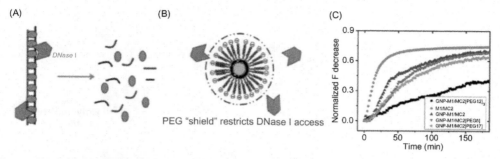

Figure 6.4 Schematic representation of e YO-PRO-1 loaded with dsDNA (A), PEGylated dsDNA (B), and normalized time-dependent fluorescence changes for the YO-PRO-1 loaded M1/MC2 and GNP — M1/MC2 (with or without PEGylation) conjugates after treatment with DNase I (C). *Adapted with permission from Song, L., Guo, Y., Roebuck, D., Chen, C., Yang, M., Yang, Z., et al., 2015. Terminal PEGylated DNA — gold nanoparticle conjugates offering high resistance to nuclease degradation and efficient intracellular delivery of DNA binding agents. ACS Appl. Mater. Interfaces 7, 18707–18716. Copyright © 2009, American Chemical Society.*

6.2.2 Silver nanoparticles supported on polymers

Silver has been used for a long period of time in the medical field to treat infectious diseases because this metal shows excellent antimicrobial properties. The significance of silver in medical treatment has resulted in the use of silver NPs for various medical applications. Moreover, the silver surface can be easily modified to confer targetability and stability to the NP because various organic molecules can be easily attached to the silver surface. Similar to gold, silver shows efficient, nontoxic, and tunable optical absorption properties. Recent studies have incorporated the unique physical, chemical, and photochemical properties of the silver NPs into the polymer materials to enhance drug and gene uptake into desired sites.

Recently, Ravindra et al. reported silver NPs embedded in poly(AM-*co*-AMPS) for use in drug delivery and antimicrobial applications. A curcumin loading and release study revealed that a silver NP-embedded polymer hydrogel showed higher holding capacity because of curcumin adsorption onto silver NPs as well as entrapping in the hydrogel to ensure prolonged circulation. Moreover, the silver NP-loaded hydrogel exhibited higher thermal stability and loading efficiency compared to other hydrogels (Ravindra et al., 2012). A simple method for incorporating silver NPs into multifunctional polyfilms was reported by Sripriya et al. (2013). Electrostatic adsorptions of poly(allylamine hydrochloride) and dextran sulfate (DS) were used to prepare multifunctional polyfilms and then silver NPs were incorporated for effective drug loading and delivery. The application of silver NPs as reducing agents was developed through a biological route from the *Hybanthus enneaspermus* leaf extract (Fig. 6.5). Silver NPs

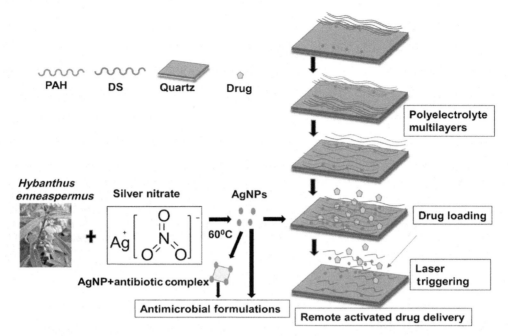

Figure 6.5 Schematic representation of laser-triggered drug release. *Adapted with permission from Sripriya, J., Anandhakumar, S., Achiraman, S., Antony, J.J., Siva, D., Raichur, A.M., 2013. Laser receptive polyelectrolyte thin films doped with biosynthesized silver nanoparticles for antibacterial coatings and drug delivery applications. Int. J. Pharm. 457, 206–213. Copyright © 2013 Elsevier B.V. All rights reserved.*

show significant energy–absorbing properties; thus, upon laser treatment, they absorb energy from the laser beam and transfer this energy to the film as heat. This ruptures the film, enabling remotely triggered drug delivery (Sripriya et al., 2013).

Ultrasound–sensitive nanocapsules were developed by Anandhakumar et al. for remotely activated drug release. Layer-by-layer assembly of polyelectrolytes was used to form the nanocapsules and then the silver NPs were embedded. The silver NPs were evenly distributed in the nanocapsule. Drug release occurred when the shell was ruptured into small fragments because of the presence of silver NPs. This feature of the nanocapsule embedded with silver NPs enables its use as a drug carrier vehicle (Anandhakumar et al., 2012). Novel polymer composite spheres composed of silver NPs have been developed at low cost using green synthesis strategies by Chen et al. in an antibacterial study. The synthesis was carried out without the use of reducing agents except a polymer colloidal support. The composite spheres showed very good stability and excellent antimicrobial activity. These properties depend on the size of the silver NPs and concentration of composite spheres (Chen et al., 2014).

Among metals, silver NPs show the best antibacterial, antiviral, and antifungal properties, and thus are widely used for drug and gene delivery to increase the efficacy of the drug by controlling microbial infections. Kumar et al. reported a low-cost and eco-friendly method for synthesizing silver NPs by using a cultural supernatant of *Delftia* sp. strain KCM-006, which acted as both a reducing and capping agent. Because no toxic reagents or other external reducing agents and surfactants were used in this synthesis strategy, the NP obtained was completely nontoxic and biodegradable, making this a completely green synthesis method. The results suggested that the silver NP miconazole drug conjugate had excellent stability under physiological conditions and good antifungal activity (Kumar and Poornachandra, 2015). A photoactivated nanocarrier was prepared by Brown et al. by functionalization of silver NPs with a thiol-capped nitrophenylethyl-linker containing oligonucleotides and was applied in antisense therapeutics. Particle size, shape, and distance of the photoactive molecule from the particle surface play a significant role in photoactivity and fluorescence of the NP. Moreover, the linker distance plays an important role in photoactivation. An ~2–3-nm linker distance was found to be suitable for UV-light exposure-mediated release of silver NP-tethered oligonucleotide therapeutics. A colored silver NP was prepared by Soumya et al. for targeted drug delivery. The nanoformulation consisted of a nanocarrier, cancer-specific receptor or general receptor, sensing material to determine whether the cell was normal, malignant, or benign, and stabilizing or binding agent. The silver NP was supported on PEG and conjugated with curcumin, folate receptor, and pemetrexed disodium to form a positively charged PEGylated NP with the folate receptor and curcuminoid. Upon treatment with negatively charged Prussian blue, the final nanoformulation for drug delivery was formed. In the presence of an external electrical potential, the nanoformulation disintegrated because of the loss of the negative charge from the binding agent, leading to drug release. In addition, by turning the voltage on and off, the amount of drug delivered and timing of the dose could be accurately controlled (Soumya and Hela, 2013).

Another green approach for the synthesis of chitosan-stabilized silver NPs was reported by Prasad et al. (2013). No external toxic materials were used in the synthesis stage, and only polysaccharide was used as a reducing and stabilizing agent. The NP formulations exhibited a synergistic effect in treating inflammation in RA and controlled drug delivery. The azathioprine-conjugated silver NPs showed a spherical morphology with sizes of 180–220 nm and an in vitro study reported a drug release rate of 67.34% (Prasad et al., 2013). Bonor et al. reported a quick and convenient method for synthesizing lysine-conjugated silver NPs in therapeutic applications. L-Lysine acts as a good pH-responsive carrier and the antibacterial, antiplatelet, and antiinflammatory characteristics of silver NPs together make the lysine-capped silver NP a promising material for drug delivery. Moreover, the authors suggested this as an outstanding method for synthesizing nanomaterials with exact size ranges and the materials produced were found to be biocompatible and nontoxic (Bonor et al., 2014).

A combinatorial approach was proposed by Nayak et al. for effective breast cancer treatment. The nanoformulation was prepared using a silver NP coated onto a biocompatible and biodegradable polymer such as chitosan, which was effective in targeted delivery and encapsulated vitamin E, vitamin C, and glucose. This combinatorial approach showed numerous advantages including improved efficacy against cancer cells and lower toxicity toward normal cells (Nayak et al., 2016).

6.3 OTHER METAL NANOPARTICLES SUPPORTED ON DIFFERENT POLYMERS

6.3.1 Iron (Fe) nanoparticles supported on polymers

Magnetite (Fe_3O_4) and maghemite (γ-Fe_2O_3) are two significant iron oxide NPs. Their strong paramagnetic effects and biocompatible nature make them important materials for medical applications. Although Cu, Co, and Ni show strong paramagnetic properties, they are not used in medical applications because of their toxicity and easily oxidized nature. Iron NPs are important in drug delivery and imaging because of their strong magnetic properties, appropriate surface properties, ease of attachment to the targeting ligand, good stability under physiological conditions, long circulation half-lives, and good biocompatibility. In drug delivery, the magnetic properties of iron NPs can be utilized to control the direction of drug carriers using an external magnet. Particularly, a current paradigm is developing a nanocarrier system with multifunctional features such as cancerous cell targetability, stimuli-responsive drug release properties, and imaging capabilities. Super paramagnetic iron oxide (SPIO) NPs are mainly used for magnetic resonance imaging (MRI).

A stable and multifunctional cancerous cell-targeting system such as NP vesicles was reported by Yang et al. for controlled drug release application while providing MR images. The polymer NPs were loaded with anticancer drugs doxorubicin and SPIO NPs for MRI. A double emulsion method was used to obtain worm-like polymeric vesicles (Fig. 6.6) of an amphiphilic heterobifunctional triblock attached to folate or methoxy (FA-PEG_{114}-PLA_x-PEG_{46}) acrylate. The outer PEG segments bearing a folate or methoxy group facilitated targetability and folate-mediated endocytosis for effective cellular uptake. Additionally, inner PEG-bearing acrylate contributed to the stability of the polymer vesicle in vivo. The SPIO NP was encapsulated into the aqueous core layer of the polymer vesicle with crosslinked inner PEG layer. The polymer worm-like vesicle showed excellent SPIO NP loading and a 9 wt% loading rate (Yang et al., 2010).

The fabrication of a multifunctional polymeric nanomedical platform for simultaneous optical imaging and magnetically guided drug delivery was proposed by Kim et al. (2008). The NP platform is made up with poly(D,L-lactic-co-glycolic acid) (PLGA) and the therapeutic agent was loaded into the polymer matrix. In the next step, superparamagnetic magnetic nanocrystals and semiconductor NPs (quantum

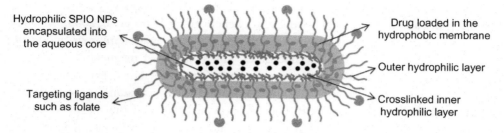

Hydrophilic SPIO NPs encapsulated into the aqueous core

Drug loaded in the hydrophobic membrane

Outer hydrophilic layer

Targeting ligands such as folate

Crosslinked inner hydrophilic layer

Figure 6.6 Illustration of the stable and tumor-targeting multifunctional worm-like polymer vesicle. *Adapted with permission from Yang, X., Grailer, J.J., Rowland, I.J., Javadi, A., Hurley, S.A., Steeber, D.A., et al., 2010. Multifunctional SPIO/DOX-loaded wormlike polymer vesicles for cancer therapy and MR imaging. Biomaterials 31, 9065–9073. Copyright © 2010 Elsevier Ltd. All rights reserved.*

dots) were loaded into the PLGA matrix. The magnetic nanocrystals facilitated both magnetic-directed drug delivery and MRI, whereas the quantum dots enabled optical imaging. Furthermore, the folate conjugation to the PLGA NP by PLL-PEG-FOL enabled targeting of the cancer cells, which had folate receptors on their surface. The ionic interaction between positively charged PLL and negatively charged PEG was suggested to be the binding force for the linkage; the presence of PEG in the NP enhanced biocompatibility and water dispersity (Kim et al., 2008).

Recently, self-assembled magnetic fluorescent polymeric micelles for MRI and optical imaging were reported by Yan et al. The polymeric nanocomposite has been reported to be stable and cytocompatible. The synthesis involved the self-assembly of amphiphilic poly(HFMA-*co*-VBK)-*g*-PEG copolymers and oleic acid-stabilized Fe_3O_4 NPs. The multifunctional polymer micelles exhibited interesting paramagnetic behaviors because of the presence of Fe_3O_4 at a saturation magnetization of approximately $9.61\,emu/g$ and transverse relaxivity rate of $157.44\,mM^{-1}\,s^{-1}$, and the small amount of fluorescent carbozole in the polymer shell presented the best fluorescent properties. Polymeric micelles thus had dual imaging probes with blue fluorescence (Yan et al., 2014). Yoo et al. demonstrated that folate-PEG-superparamagnetic iron oxide NPs could be used in lung cancer imaging. A stable superparamagnetic iron oxide NP (SPION) was designed by reacting FA-conjugated PEG with aminosilane immobilized SPIONs. The size of the FA-PEG-SPIONs was found to be stable for up to 8 weeks. Additionally, higher intercellular uptake via folate-mediated endocytosis was confirmed for this NP loaded with drug than that of the PEG-SPIONs (Yoo et al., 2012).

An efficient method was suggested by Guo et al. for the preparation of a monodisperse superparamagnetic single crystal magnetic NP with a mesoporous structure. The main advantages of this method are its simplicity, high quality, ease of scale-up, and good reproducibility. Dox uptake was found to be $40\,mg/g$ and was stored in the MSSN for release at specific sites. Most drugs were found to be released in $12\,h$, which is a favorable outcome in effective drug delivery. The morphology of Fe_3O_4 NPs was easily controlled by varying the amount of NaOH and 1,2-ethylenediamine (Guo et al., 2009). Magnetic liposomes for targeted gene delivery were reported by Hirao et al. to improve transfection using previously described cationic liposomes. The synthesized magnetic cationic liposomes have the smallest diameters of approximately 40 nm. The gene delivery efficiency of magnetic cationic liposomes was studied using plasmid DNA containing a luciferase gene in human osteosarcoma Saos-2 cells. The 1:5 ratio of DNA and the magnetic cationic liposome (MCL) mixture exhibited maximum luciferase activity in the absence of a magnetic field within 6 h of incubation and in the presence of a magnetic field within 30 min. Transfection was reported to be 3.5-fold greater than in the previous study and the apoptosis rate was increased from 2.4% to 18.9% by magnetic induction (Hirao et al., 2003).

An investigation of magnetic nanoparticle–polymer composites was reported by Urbina et al. (2008). Drug carrier efficiency was investigated using two types of magnetic NP composites with respective multiple controls such as magnetic and thermally induced drug release. The release of fluorescein isothiocyanate (FITC) from the magnetite or cobalt NP containing poly(methylmethacrylate) (PMMA) was influenced by changing the sample temperature or by exposure to an oscillating magnetic field. The investigation revealed that the release of FITC from magnetite–PMMA was thermally induced but not magnetically induced. Additionally, the release of FITC from cobalt–PMMA was not induced thermally or magnetically (Urbina et al., 2008). A detailed study of the impact of NPs on polymer nanocomposites was conducted using a simple PLA-magnetic NP. With variations in the composition of NP, the outer surface appearance as well as drug release behavior could be controlled. It was also found that the PLA-NP material at a concentration of 20% of magnetic nanoparticles in the composite showed an enhanced drug release rate that was approximately 200-fold greater in the first 30 min. Variation in drug release behavior is affected mainly by the different surface wettability and surface crystallization related to different interactions between PLA and magnetite NPs in various samples. Moreover, a 50% magnetic NP concentration resulted in aggregation of some magnetic NPs and resulted in greater crystal formation and lower hydrolysis (Yang et al., 2015).

A one-step template-free synthesis of reversibly swellable polymeric microcapsules with iron oxide magnetic NPs embedded in the polymeric system was reported by Koo et al. A double emulsion with a chloroform core and magnetic nanoparticle polymer shell was formed by emulsification of a mixture of liquid monomer and a magnetic NP in chloroform. This double emulsion was converted to microcapsules by exposure to UV light and the formed microcapsules showed reversible swelling under repetitive drying and addition of water. The intrinsic superparamagnetic feature of the magnetic NP in the microcapsule facilitated the magnetic field-controlled delivery of drugs loaded in the microcapsules (Koo et al., 2006). The possibility to modulate the charge surface of maghemite NPs from positive to negative was demonstrated by

Robbes et al. A multistep efficient "grafting from" method based on nitroxide-mediated polymerization resulted in well-defined polystyrene-grafted γ-Fe$_2$O$_3$. By maintaining colloidal stability, controlled molecular masses were successfully synthesized using either hydrogenated or deuterated monomers with a high conversion rate and low polydispersity index. Furthermore, the versatile properties of polystyrene-grafted magnetic NPs such as the tunability of core/shell size with synthesis parameters, scalability, and redispersity in different solvents made it a support for active species for drug delivery (Robbes et al., 2012).

Recently, through radical polymerization, a pH- and temperature-responsive magnetic NP polymer composite was synthesized for drug delivery application. The novel polymer composite [poly(NIPAAM@Fe$_3$O$_4$ MNPs/TMSPMC/DOX)] was synthesized via radical polymerization of NIPAAM and methacrylate-functionalized Fe$_3$O$_4$ nanoparticles/Dox complex using AIBN and EGDMA. The results showed that this composite had good drug loading and release profiles. Additionally, the lower critical solution temperature of the composite was 40°C, which favored targeted drug release. This composite also showed a better circulation time in the blood, making it a promising drug carrier (Kaamyabi et al., 2016). A magnetic molecularly imprinted polymer (MIP) was synthesized recently by Moghaddam et al. using polydopamine. The synthesized MIP was used to control the delivery of 5-fluorouracil (5-FU) in the presence of an external magnetic field. Tumor-growth delay, tumor doubling time, inhibition ratio, and histopathology were studied to estimate the antitumor efficiency of the 5-FU-imprinted polymer (IP). Based on the result, the 5-FU release efficiency of the 5FU-IP composite was found to be higher in the presence of an external magnetic field (Hashemi-Moghaddama et al., 2016).

The superparamagnetic iron oxide (Fe$_3$O$_4$) NP coated on a polymer composite of chitosan (CS), PEG, and polyvinylpyrrolidine (PVP) (CS-PEG-PVP) was also reported recently for drug delivery application. The anticancer drug curcumin was loaded into the polymer magnetic NP for the therapeutic study and drug release was carried out without any side effects and with greater efficiency. The particle size of the curcumin-loaded Fe$_3$O$_4$-CS-PEG-PVP was 183–390 nm and the zeta potential value was 26–41 mV (Prabha and Raj, 2016). A series of pH- and temperature-responsive polymer grafted iron oxide NPs were reported by Dutta et al. for theranostic application. Three water-soluble block polymers were prepared by the reversible addition fragmentation chain transfer polymerization technique in the presence of a macro chain transfer agent, poly(tert-butyl) acrylate. The product was coupled with an aminated iron oxide NP to obtain the NP with the desired qualities. Characterization and in vitro/in vivo studies suggested that by varying the monomer ratio of the block copolymer, it is possible to control the cloud point from 32°C to 43°C. The particles were in the nanosize with a mean diameter of 23–27 nm and polymer shell thickness of 3.5–4.5 nm. Additionally, the polymer showed good drug loading and drug release at the desired pH and temperature (Dutta et al., 2016).

A core–shell composed of biodegradable polymer and super paramagnetic iron oxide NPs using a Pickering emulsion was reported by Oka et al. (2015) for drug delivery purposes. The composite particle was composed of a pyrene-loaded core–shell composite particle composed of PHA particles and shell of assembled magnetic NPs. The dispersibility was found to be crucial in controlling the core–shell structure and the small amount of dispersant that could be used with the NP suspension. The drug was loaded into the core of the nanocomposite and the release was found to be the same as that under alkaline conditions, leading to hydrolysis of the core and release of the drug at a specific location (Oka et al., 2015). Superparamagnetic NPs decorated with PNIPAM were reported by Deka et al. (2011) for therapeutic applications. The NP had a diameter of 200 nm and the thermo-responsive characteristics of PNIPAM enabled controlled release of the cancer drug, Dox, in tumor cells. The NP showed enhanced efficacy in the presence of an external magnetic field (Deka et al., 2011).

A new type of biodegradable nanohydrogel was recently reported by Zhang et al. NP preparation was carried out via a facile method of reflux precipitation using an MAA monomer and the biodegradable crosslinker zinc dimethacrylate. The formed nanohydrogel showed pH/GSH dual-responsive characteristics and, after modification with FA, showed both outstanding colloidal stability and folate-targeted ability. The drug-loaded nanohydrogel exhibited good physiological stability and lower drug leakage. Moreover, drug release was found to be 90% within 48 h (Zhang et al., 2016). Recently, a selenium-containing thermogel for controlled drug delivery was reported. Cisplatin drug delivery was studied using a coordination-responsive thermogel containing selenium with a copolymer containing polyether and polyester Bi(mPEG-PLGA). The selenium was covalently conjugated to the hydrophobic end of the two mPEG-PLGA chains. The Bi(mPEG-PLGA)–Se conjugate was effectively coordinated with cisplatin in a time-dependent manner (Luan et al., 2015).

Another report showed that selenium, which does not belong to the main polymer chain, was effective for theranostic application. Han et al. suggested that polymeric superamphiphiles formed by the combination of selenium-containing surfactant and a double hydrophilic block copolymer could serve as an oxidation-responsive micelle. In the presence of a mild oxidation agent such as H_2O_2, selenide was oxidized to selenoxide, resulting in micelle disassembly. Thus, this controlled self-assembly and disassembly could be used to release a preloaded guest molecule in a mild oxidative environment (Han et al., 2010). As compared to selenium, the tellurium polymer conjugate could also be used as a nanocarrier because of similar or comparatively better oxidation properties and nontoxicity in humans. Tellurium is also oxidized under mild oxidation conditions and causes morphology changes in the micelle; further oxidation results in decomposition of the micelle, which is suitable for controlled drug and gene delivery. D-Glucose end-capped polylactide ruthenium cyclopentadienyl complex (RuPMC) was synthesized by Valente et al. as an anticancer drug delivery system. RuPMC was found to have good stability at physiological pH. UV-visible spectroscopy monitoring results revealed that RuPMC was stable under physiological conditions for nearly 72 h (Valente et al., 2013).

6.4 CONCLUSIONS

Noble metal NPs such as gold and silver have been used in biomedical applications for many years because of their well-known surface characteristics and antimicrobial properties. Nanotechnology development was an important turning point in the field of medicine and led to the concept of nanomedicine. The application of modern nanotechnology opened many avenues of research to use NPs for effective drug design, and drug and gene carrier development. Polymer NPs used as drug and gene carriers must show prolonged circulation, and studies have also focused on targeted cell binding and stimuli-responsive release under various internal and external conditions. Very recently, the combined use of polymer materials with metal NPs was reported from worldwide research groups. This combination of a polymer with a stimuli-responsive nature and metal NP with antimicrobial and magnetic properties enhances the colloidal stability of nanomaterials under biological conditions and shows many advantages over conventional polymers. A synergistic effect of both metal and polymers is achieved by using NP-homing polymers for theranostic applications. Recent reports have described the successful use of these nanomaterials for targeted drug and gene release, drug design, MRI, and optical imaging purposes.

While gold and silver are nontoxic to normal cells, the toxicity of other metals is a primary challenge limiting the use of NP-homing polymers, particularly in the human body. Thus, additional studies are required in these fields to decrease the toxicity of metals and prepare a common platform for loading different guest particles using simple modifications.

REFERENCES

Abraham, S., Kim, I., Batt, C.A., 2007. A facile preparative method for aggregation-free gold nanoparticles using poly(styrene-block-cysteine). Angew. Chem. Int. Ed. 46, 5720–5723.

Alexander, C.M., Hamner, K.L., Maye, M.M., Dabrowiak, J.C., 2014. Multifunctional DNA-gold nanoparticles for targeted doxorubicin delivery. Bioconjug. Chem. 25, 1261–1271.

Anandhakumar, S., Mahalakshmi, V., Raichur, A.M., 2012. Silver nanoparticles modified nanocapsules for ultrasonically activated drug delivery. Mater. Sci. Eng. C 32, 2349–2355.

Aoyama, Y., Kobayashi, K., Morishita, Y., Maeda, K., Murohara, T., 2015. Wnt11 gene therapy with adeno-associated virus 9 improves the survival of mice with myocarditis induced by coxsackievirus B3 through the suppression of the inflammatory reaction. J. Mol. Cell. Cardiol. 84, 45–51.

Aryal, S., Grailer, J.J., Pilla, S., Steeber, D.A., Gong, S., 2009. Doxorubicin conjugated gold nanoparticles as water-soluble and pH-responsive anticancer drug nanocarriers. J. Mater. Chem. 19, 7879–7884.

Ashcroft, R.E., 2004. Gene therapy in the clinic: whose risks? Trends Biotechnol. 22, 560–563.

Bonor, J., Reddy, V., Akkiraju, H., Dhurjati, P., Nohe, A., 2014. Synthesis and characterization of L-lysine conjugated silver nanoparticles smaller than 10 nM. Adv. Sci. Eng. Med. 6, 942–947.

Braun, G.B., Pallaoro, A., Wu, G., Missirlis, D., Zasadzinski, J.A., Tirrell, M., et al., 2009. Reich laser-activated gene silencing via gold nanoshell-siRNA conjugates. ACS Nano 3, 2007–2015.

Chen, G., Lu, J., Lamb, C., Yu, Y., 2014. A novel green synthesis approach for polymer nanocomposites decorated with silver nanoparticles and their antibacterial activity. Analyst 139, 5793–5799.

Conde, J., Doria, G., Baptista, P., 2012. Noble metal nanoparticles applications in cancer. J. Drug Deliv. Article ID751075, 12 pages.

Conde, J., Larguinho, M., Cordeiro, A., Raposo, L.R., Costa, P.M., Santos, S., et al., 2014. Gold-nanobeacons for gene therapy: evaluation of genotoxicity, cell toxicity and proteome profiling analysis. Nanotoxicology 8, 521–532.

Deka, S.R., Quarta, A., Corato, R.D., Riedinger, A., Cingolani, R., Pellegrino, T., 2011. Magnetic nano-beads decorated by thermo-responsive PNIPAM shell as medical platforms for the efficient delivery of doxorubicin to tumour cells. Nanoscale 3, 619–629.

Dreaden, E.C., Mackey, M.A., Huang, X., Kangy, B., El-Sayed, M.A., 2011. Beating cancer in multiple ways using nanogold. Chem. Soc. Rev. 40, 3391–3404.

Dutta, S., Parida, S., Maiti, C., Banerjee, R., Mandal, M., Dhara, D., 2016. Polymer grafted magnetic nanoparticles for delivery of anticancer drug at lower pH and elevated temperature. J. Colloid Interface Sci. 467, 70–80.

Fadeel, B., Garcia-Bennett, A.E., 2010. Better safe than sorry: understanding the toxicological properties of inorganic nanoparticles manufactured for biomedical applications. Adv. Drug Deliv. Rev. 62, 362–374.

Falcone, S.J., Palmeri, D., Berg, R., 2006. Biomedical applications of hyaluronic acid. ACS Symp. Ser. 934, 155–174.

Faramarzi, M.A., Forootanfar, H., 2011. Biosynthesis and characterization of gold nanoparticles produced by laccase from *Paraconiothyrium variabile*. Colloids Surf. B Biointerfaces 87, 23–27.

Fayaz, A.M., Girilal, M., Rahman, M., Venkatesan, R., Kalaichelvan, P.T., 2011. Biosynthesis of silver and gold nanoparticles using thermophilic bacterium *Geobacillus stearothermophilus*. Process Biochem. 46, 1958–1962.

Giljohnn, D.A., Seferos, D.S., Danial, W.L., Massich, M.D., Patel, P.C., Mirkin, C.A., 2010. Gold nanopar-ticles for biology and medicine. Angew. Chem. Int. Ed. 49, 3280–3294.

Guo, S., Li, D., Zhang, L., Li, J., Wang, E., 2009. Monodisperse mesoporous superparamagnetic single-crystal magnetite nanoparticles for drug delivery. Biomaterials 30, 1881–1889.

Han, P., Ma, N., Ren, H., Xu, H., Li, Z., Wang, Z., et al., 2010. Oxidation-responsive micelles based on a selenium-containing polymeric superamphiphile. Langmuir 26, 14414–14418.

Hashemi-Moghaddama, H., Kazemi-Bagsangania, S., Jamilib, M., Zavareh, S., 2016. Evaluation of magnetic nanoparticles coated by 5-fluorouracil imprinted polymer for controlled drug delivery in mouse breast cancer model. Int. J. Pharm. 497, 228–238.

Hirao, K., Sugita, T., Kubo, T., Igarashi, K., Tanimoto, K., Murakami, T., et al., 2003. Targeted gene delivery to human osteosarcoma cells with magnetic cationic liposomes under a magnetic field. Int. J. Oncol. 22, 1065–1071.

James, W.D., Hirsch, L.R., West, J.L., O'Neal, P.D., Payne, J.D., 2007. Application of INAA to the build-up and clearance of gold nanoshells in clinical studies in mice. J. Radioanal. Nucl. Chem. 271, 455–459.

Kaamyabi, S., Habibi, D., Amini, M.M., 2016. Preparation and characterization of the pH and thermosensi-tive magnetic molecular imprinted nanoparticle polymer for the cancer drug delivery. Bioorg. Med. Chem. Lett. 26, 2349–2354.

Kabanov, A.V., Batrakova, E.V., Alakhov, V.Y., 2002. Pluronic block copolymers for overcoming drug resis-tance in cancer. Adv. Drug Deliv. Rev. 54, 759–779.

Kannan, R.M., Nance, E., Kannan, S., Tomalia, D.A., 2014. Emerging concepts in dendrimer-based nano-medicine: from design principles to clinical applications. J. Intern. Med. 276, 579–617.

Kattumuri, V., Katti, K., Bhaskaran, S., Boote, E.J., Casteel, S.W., Fent, G.M., et al., 2007. Gum arabic as a phytochemical construct for the stabilization of gold nanoparticles: in vivo pharmacokinetics and X-ray-contrast-imaging studies. Small 3, 333–341.

Kim, J., Lee, J.E., Lee, S.H., Yu, J.H., Lee, J.H., Park, T.G., et al., 2008. Designed fabrication of a multi-functional polymer nanomedical platform for simultaneous cancer-targeted imaging and magnetically guided drug delivery. Adv. Mater. 20, 478–483.

Kim, B., Han, G., Toley, B., Kim, C., Rotello, V.M., Forbes, N.S., 2010. Tuning payload delivery in tumour cylindroids using gold nanoparticles. Nat. Nanotechnol. 5, 465–472.

Koo, H.Y., Chang, S.T., Choi, W.S., Park, J., Kim, D., Velev, O.D., 2006. Emulsion-based synthesis of reversibly swellable, magnetic nanoparticle-embedded polymer microcapsules. Chem. Mater. 18, 3308–3313.

Kumar, C.G., Poornachandra, Y., 2015. Biodirected synthesis of miconazole-conjugated bacterial silver nanoparticles and their application as antifungal agents and drug delivery vehicles. Colloids Surf. B Biointerfaces 125, 110–119.

Kwon, G.S., 2003. Polymeric micelles for delivery of poorly water-soluble compounds. Crit. Rev. Ther. Drug Carr. Syst. 20, 357–403.

Lee, H., Lee, K., Kim, I.K., Park, T.G., 2008. Synthesis, characterization, and *in vivo* diagnostic applications of hyaluronic acid immobilized gold nanoprobes. Biomaterials 29, 4709–4718.

Li, H., Jo, J.K., Zhang, L.D., Ha, C.S., Suh, H., Kim, I., 2010. Hyperbranched polyglycidol assisted green synthetic protocols for the preparation of multifunctional metal nanoparticles. Langmuir 26 (23), 18442–18453.

Li, H., John, J.V., Byeon, S.J., Heo, M.S., Sung, J.H., Kim, K.H., et al., 2014. Controlled accommodation of metal nanostructures within the matrices of polymer architectures through solution-based synthetic strategies. Prog. Polym. Sci. 39, 1878–1907.

Liu, A., Ye, B., 2013. Application of gold nanoparticles in biomedical researches and diagnosis. Clin. Lab. 59, 23–36.

Luan, J., Shen, W., Chen, C., Lei, K., Yu, L., Ding, J., 2015. Selenium-containing thermogel for controlled drug delivery by coordination competition. RSC Adv. 5, 97975–97981.

Luo, Y., Prestwich, G., 2002. Cancer-targeted polymeric drugs. Curr. Cancer Drug Targets 2, 209–226.

Maeda, H., Wu, J., Sawa, T., Matsumura, Y., Hori, K.J., 2000. Tumor vascular permeability and the EPR effect in macromolecular therapeutics: a review. J. Control. Release 65, 271–284.

Mieszawaka, A.J., Mulder, W.J., Fayad, Z.A., Cormode, D.P., 2013. Multifunctional gold nanoparticles for diagnosis and therapy of disease. Mol. Pharmacol. 10, 831–847.

Mittal, A.K., Chisti, Y., Banerjee, U.C., 2013. Synthesis of metallic nanoparticles using plant extracts. Biotechnol. Adv. 31, 346–356.

Murphy, C.J., Gole, A.M., Stone, J.W., Sisco, P.N., Alkilany, A.M., Goldsmith, E.C., et al., 2008. Gold nanoparticles in biology: beyond toxicity to cellular imaging. Acc. Chem. Res. 41, 1721–1730.

Nayak, D., Minz, A.P., Ashe, S., Rauta, P.R., Kumari, M., Chopra, P., et al., 2016. Synergistic combination of antioxidants, silver nanoparticles and chitosan in a nanoparticle based formulation: characterization and cytotoxic effect on MCF-7 breast cancer cell lines. J. Colloid Interface Sci. 470, 142–152.

Nehoff, H., Parayath, N.N., Domanovitch, L., Taurin, S., Greish, K., 2014. Nanomedicine for drug targeting: strategies beyond the enhanced permeability and retention effect. Int. J. Nanomedicine 9, 2539–2555.

Oka, C., Ushima, K., Horiishi, N., Tsuge, T., Kitamoto, Y., 2015. Core–shell composite particles composed of biodegradable polymer particles and magnetic iron oxide nanoparticles for targeted drug delivery. J. Magn. Magn. Mater. 381, 278–284.

Paciotti, G.F., Myer, L., Weinreich, D., Goia, D., Pavel, N., Mclaughlin, R.E., et al., 2004. Colloidal gold: a novel nanoparticle vector for tumor directed drug delivery. Drug Deliv. 11, 169–183.

Paciotti, G.F., Kingston, D.G.I., Tamarkin, L., 2006. Colloidal gold nanoparticles: a novel nanoparticle platform for developing multifunctional tumor-targeted drug delivery vectors. Drug Dev. Res. 67, 47–54.

Park, C., Youn, H., Kim, H., Noh, T., Kook, Y.H., Oh, E.T., et al., 2009. Cyclodextrin-covered gold nanoparticles for targeted delivery of an anti-cancer drug. J. Mater. Chem. 19, 2310–2315.

Peer, D., Karp, J.M., Hong, S., Farokhzad, O.C., Margalit, R., Langer, R., 2007. Nanocarriers as an emerging platform for cancer therapy. Nat. Nanotechnol. 2, 751–760.

Prabaharan, M., Grailer, J.J., Pilla, S., Steeber, D.A., Gong, S., 2009. Gold nanoparticles with a monolayer of doxorubicin-conjugated amphiphilic block copolymer for tumor-targeted drug delivery. Biomaterials 30, 6065–6075.

Prabha, G., Raj, V., 2016. Preparation and characterization of polymer nanocomposites coated magnetic nanoparticles for drug delivery applications. J. Magn. Magn. Mater. 408, 26–34.

Prasad, S.R., Elango, K., Damayanthi, D., Saranya, J.S., 2013. Formulation and evaluation of azathioprine loaded silver nanoparticles for the treatment of rheumatoid arthritis. Asian J. Biomed. Pharm. Sci. 3, 28–32.

Rana, S., Bajaj, A., Mout, R., Rotello, V.M., 2012. Monolayer coated gold nanoparticles for delivery applications. Adv. Drug Deliv. Rev. 64, 200–216.

Ravindra, S., Mulaba-Bafubiandi, A.F., Rajinikanth, V., Varaprasad, K., Reddy, N.N., Raju, C., 2012. Development and characterization of curcumin loaded silver nanoparticle hydrogels for antibacterial and drug delivery applications. J. Inorg. Organomet. Polym. Mater. 22, 1254–1262.

Robbes, A., Cousin, F., Meneau, F., Chevigny, C., Gigmes, D., Fresnais, J., et al., 2012. Controlled grafted brushes of polystyrene on magnetic γ-Fe$_2$O$_3$ nanoparticles via nitroxide-mediated polymerization. Soft Matter 8, 3407–3418.

Sahu, N., Soni, D., Chandrashekhar, B., Sarangi, B.K., Satpute, D., Pandey, R.A., 2013. Synthesis and characterization of silver nanoparticles using *Cynodon dactylon* leaves and assessment of their antibacterial activity. Bioprocess Biosyst. Eng. 36, 999–1004.

Song, L., Guo, Y., Roebuck, D., Chen, C., Yang, M., Yang, Z., et al., 2015. Terminal PEGylated DNA – gold nanoparticle conjugates offering high resistance to nuclease degradation and efficient intracellular delivery of DNA binding agents. ACS Appl. Mater. Interfaces 7, 18707–18716.

Soumya, R.S., Hela, P.G., 2013. Nano silver based targeted drug delivery for treatment of cancer. Pharm. Lett. J. 5, 189–197.

Sripriya, J., Anandhakumar, S., Achiraman, S., Antony, J.J., Siva, D., Raichur, A.M., 2013. Laser receptive polyelectrolyte thin films doped with biosynthesized silver nanoparticles for antibacterial coatings and drug delivery applications. Int. J. Pharm. 457, 206–213.

Urbina, M.C., Zinoveva, S., Miller, T., Sabliov, C.M., Monroe, W.T., Kumar, C.S.S.R., 2008. Investigation of magnetic nanoparticle-polymer composites for multiple-controlled drug delivery. J. Phys. Chem. C 112, 11102–11108.

Valente, A., Garcia, M.H., Marques, F., Rousseau, C., Zinck, P., 2013. First polymer "ruthenium–cyclopentadienyl" complex as potential anticancer agent. J. Inorg. Biochem. 127, 79–81.

Visaria, R.K., Griffin, R.J., Williams, B.W., Ebbini, E.S., Paciotti, G.F., Song, C.W., et al., 2006. Enhancement of tumor thermal therapy using gold nanoparticle–assisted tumor necrosis factor-A delivery. Mol. Cancer Ther. 5, 1014–1020.

Yan, K., Li, H., Li, P., Zhu, H., Shen, J., Yi, C., et al., 2014. Self-assembled magnetic fluorescent polymeric micelles for magnetic resonance and optical imaging. Biomaterials 35, 344–355.

Yang, F., Zhang, X., Song, L., Cui, H., Myers, J.N., Bai, T., et al., 2015. Controlled drug release and hydrolysis mechanism of polymer – magnetic nanoparticle composite. ACS Appl. Mater. Interfaces 7, 9410–9419.

Yang, X., Grailer, J.J., Rowland, I.J., Javadi, A., Hurley, S.A., Steeber, D.A., et al., 2010. Multifunctional SPIO/DOX-loaded wormlike polymer vesicles for cancer therapy and MR imaging. Biomaterials 31, 9065–9073.

Yavuz, M.S., Cheng, Y., Chen, J., Cobley, C.M., Zhang, Q., Rycenga, M., et al., 2009. Gold nanocages covered by smart polymers for controlled release with near-infrared light. Nat. Mater. 8, 935–939.

Yoo, M., Park, I., Lim, H., Lee, S., Jiang, H., Kim, Y., et al., 2012. Folate–PEG–superparamagnetic iron oxide nanoparticles for lung cancer imaging. Acta Biomater. 8, 3005–3013.

Zhang, Z., Wan, J., Sun, L., Li, Y., Guo, J., Wang, C., 2016. Zinc finger-inspired nanohydrogels with glutathione/pH triggered degradation based on coordination substitution for highly efficient delivery of anti-cancer drugs. J. Control. Release 225, 96–108.

Zhao, P., Li, N., Astruc, D., 2013. State of the art in gold nanoparticle synthesis. Coord. Chem. Rev. 257, 638–665.

CHAPTER 7

Polymeric Nanoparticles in Targeting and Delivery of Drugs

Sarita Rani, Ashok K. Sharma, Iliyas Khan, Avinash Gothwal,
Sonam Chaudhary and Umesh Gupta

Contents

Nanotechnology-Based Approaches for Targeting and Delivery of Drugs and Genes.
DOI: http://dx.doi.org/10.1016/B978-0-12-809717-5.00008-7

7.1 INTRODUCTION

7.1.1 History

In recent years, significant efforts has been dedicated to develop nanotechnology as a novel platform for drug delivery and targeting, since it offers a suitable means of delivering small molecular weight drugs, as well as macromolecules such as proteins, peptides or genes by either localized or targeted delivery to the tissue of interest (Moghimi et al., 2001). Polymers have been considered as an effectual drug delivery device over the past few decades and have played a pronounced role in targeted drug delivery, thus, increasing the therapeutic benefit, while minimizing side effects (Kreuter, 1994). These are widely used as biomaterials due to their constructive properties such as good biocompatibility, biodegradability and remarkable biomimetic character. Different kinds of polymers are used to prepare the polymeric nanoparticles (PNPs), among these, all polymer biodegradable polymers and their copolymers such as diblock, triblock, multiblock, or radial-block copolymer structures have been generally used to prepare PNPs and to encapsulate the active ingredients. The first PNPs were developed between 1960 and 1970 for therapeutic application; the first was polymeric micelle (Birrenbach and Speiser, 1976). Fig. 7.1 depicts the structure of PNPs.

7.1.2 Definition and classification

Nanoparticles generally vary in size from 10 to 1000 nm and largely depend upon on the tissue, target site and circulation for crossing different biological barriers (Couvreur, 1988). Free drug release rate is slower than the polymeric NPs bounded drug, which is depicted in Fig. 7.2. PNPs proved their effectiveness in stabilizing and protecting the drug molecules such as proteins, peptides, or DNA molecules from various environmental hazardous degradation (Soppimath et al., 2001). Hence, these polymeric NPs afford the potential for various protein and gene delivery also. The drug is entrapped, encapsulated, dissolved, or attached to a NPs matrix, depending upon the method of preparation; NPs, nanospheres or nanocapsules can be obtained as shown in Fig. 7.3. The term PNPs is a collective term given for any type of PNPs, but specifically for nanospheres and nanocapsules. Nanospheres are matrix particles, i.e., particles whose entire mass is solid and molecules may be adsorbed at the surface of the NPs. Nanocapsules are a type of vesicular systems, that act as a kind of reservoir, in which the entrapped substances are confined to a cavity consisting of a liquid core surrounded by a solid material shell. Nanomedicines of the dreadful diseases like cancer (Mu and Feng, 2003), AIDS (Coester et al., 2000), diabetes (Damge et al., 2007), malaria (Date et al., 2007), prion disease (Calvo et al., 2001), and tuberculosis (Ahmad et al., 2006) are in different trial phases for the testing and some of them are commercialized. For targeted delivery, persistence of NPs is required in systemic circulation of the body (Zambaux et al., 1999). Natural and synthetic, two types of polymers are used

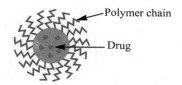

Figure 7.1 Polymeric nanoparticle.

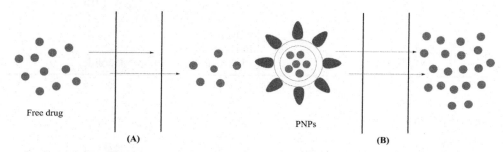

Figure 7.2 The Systematic representation of the drug from biological membrane. (A) Free drug release and (B) polymeric nanoparticles (*PNPs*) bounded drug release.

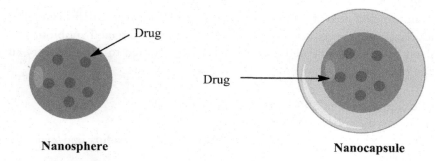

Figure 7.3 Nanosphere and nanocapsule.

in the nanodelivery system. Natural polymers are also called biopolymers. Different examples of natural and synthetic polymers are summarized in Table 7.1.

The PLGA (Poly-lactide-*co*-glycolic acid), PLA (Polylactic acid), and PLG (Poly-glycolic acid) polymers, being tissue-compatible, have been used earlier as sustained release formulations in various implantation and parenteral drug delivery applications. In addition, poly(e-caprolactone), PCL, was first reported by Pitt et al. (1979) for the controlled release (CR) of steroids as well as to deliver opthalmics (Calvo et al., 1996). Different merits and demerits of PNPs are described here:

Table 7.1 Natural and synthetic polymers

Natural polymers	Synthetic polymers
Cellulose	Poly(lactic acid) (PLA)
Starch	Poly(cyanoacrylates) (PACA)
Chitosan	Poly(ethylene glycol)
Carrageenan	Poly(vinyl alcohol) (PVA)
Aliginates	Poly(ethylene oxide) (PEO)
Xanthan gum	Poly(acaprolac-tone) (PCL)
Pectins	Poly(isobutylcynoacrylate) (PIBCA)
Gellan gum	Poly(amides) (HPMA)
–	Poly(acrylic acid)
–	Poly(anhydrides)
–	PLGA (Poly-lactide-*co*-glycolide)

7.1.3 Merits and demerits

Merits:

The unique properties mentioned below of polymeric NPs create better choices compared to conventional drug delivery systems (Brigger et al., 2002; Cui and Mumper, 2002).

- Increased stability and flexibility of its fabrication.
- Ability in targeted delivery makes them perfect candidates for delivery of vaccines, chemotherapeutic agents, contraceptives and delivery of targeted antibiotics.
- PNPs can be easily assimilated into other activities related to drug delivery, such as tissue engineering.

Demerits:

Apart from these merits NPs have some limitations also:

- Particle aggregation, owing to smaller size.
- Additionally, size and surface area readily affect the drug loading and make the system burst release.

7.2 METHODS OF PREPARATION

NPs can be from varieties of material like polymers, proteins, and polysaccharides. Numerous methods have been available to engineered NPs, depending on the physical and chemical properties of polymer and active ingredients (Cohen et al., 2000).

Various PNPs formulations have been available in the market for cancer diseases treatment listed in Table 7.2.

The selection of matrix material is a crucial step and depends upon so many factors like size, stability, solubility, surface properties such as charge and permeability, biocompatibility, degree of biodegradability, desired drug release profile, and antigenicity of

Table 7.2 PNPs mediated anticancer drug delivery

Category	Formulation	Drugs	Indication	Status	Reference
Anticancer drug	Combretastatin–doxorubicin nanocell	Combretastatin and doxorubicin	Lung carcinoma, melanoma and various cancer types	In vivo	Sengupta et al. (2005)
	Cationic core–shell nanoparticles	Paclitaxel and Bcl-2-targeted siRNA	Breast cancer	In vitro	Wang et al. (1996)
	Nanoparticle–aptamer bioconjugates	Doxorubicin and docetaxel	Prostate cancer and various cancer types	In vitro	Zhang et al. (2007)
	PLGA nanoparticle coencapsulating vincristine and verapamil	Vincristine and verapamil	Breast cancer	In vitro	Song et al. (2009)

DOX, doxorubicin; *Gem*, gemcitabine; *HPMA*: *N*-2 hydroxypropylmethacrylamide; *PCL*, polycaprolactone; *PLGA*, poly(lactic-*co*-glycolic acid); *WOR*, wortmannin,.

Table 7.3 Methods of preparation
Methods of preparation of PNP's

General methods	Advanced methods	Polymerization
1. Solvent evaporation	1. Sonication based system	1. Emulsion polymerization
2. Nanoprecipitation/solvent diffusion/interfacial deposition method	2. Core shell particulate system	1.1. Surfactant-free emulsion
3. Salting out	3. Microfluid system	1.2. Miniemulsion
4. Dialysis	4. Electrodropping system	1.3. Microemulsion
5. Supercritical fluid technology	–	1.4. Conventional
6. Coacervation/ionic gelation	–	1.5. Interfacial

the final product. The different methods of preparation of polymeric NPs are given in Table 7.3.

7.2.1 General methods

7.2.1.1 Solvent evaporation

7.2.1.1.1 Single emulsion

In this method of preparation as shown in Fig. 7.4, the first polymer is dissolved in any organic solvent (e.g., Ethyl Acetate, Acetone, DCM) for preparation of a single-phase solution followed by the addition of the drug to produce dispersion. This polymer and drug dispersion is emulsified into aqueous phase containing surfactants (Pluronic F-68,

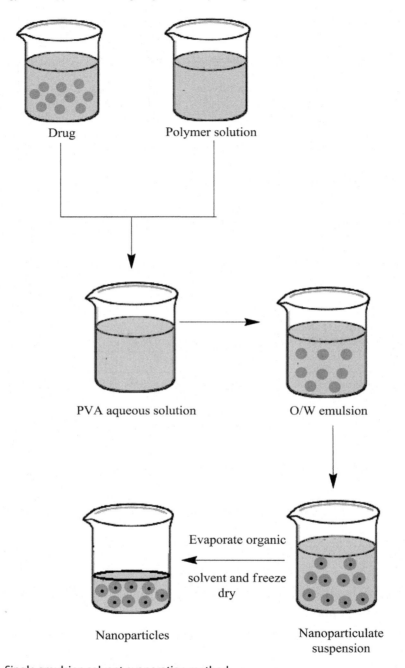

Figure 7.4 Single emulsion solvent evaporation method.

Polyvinyl alcohol) at suitable temperature, and then stirred. After that, the organic solvent is allowed to evaporate. In the first case, the emulsion is maintained at reduced atmospheric pressure with continuous stirring speed for evaporation of the organic solvent. In the second case, the emulsion is poured into aqueous phase to diffuse out the solvent associated with the oil droplet. The resulting PNPs are washed with water to remove excess amount of surfactants (Rao and Geckeler, 2011; Makadia and Siegel, 2011).

7.2.1.1.2 Double emulsion

In the double emulsion method, the drug is dissolved in aqueous phase followed by its addition into organic solvent containing polymer solution with continuous stirring to form W/O emulsion. The resultant emulsion is added into aqueous surfactants solution with continuous mixing at a suitable stirring speed. Then finally the organic solvent is allowed to evaporate to obtain the desired PNPs (Makadia and Siegel, 2011; Mirakabad et al., 2014).

7.2.1.2 Spray drying

Apart from the emulsification technique, spray-drying method is quick, appropriate and has few processing parameters. The drug-loaded microspheres are prepared by spraying a solid-in-oil dispersion or water-in-oil emulsion in a stream of heated airflow. After this, the polymer and drug are dissolved in organic solvent and mixed properly for homogeneous distribution of all particles. The resultant suspension system is spray-dried until no more production can be sprayed out and dried product can be collected (Makadia and Siegel, 2011).

7.2.1.3 Phase separation

The phase separation method involves the mixing into the polymer-drug solvent phase through continuous stirring, then extracting out the preformed mixture, resulting into phase separation of polymer by emerging drug-containing droplet. In this sequence, the droplet is dipped into a medium in which it is not soluble (both aqueous and organic) to quench these micro-droplets. The soaking time in the quenching bath controls the coarsening and hardness parameters of the droplet. Finally, it is collected by washing, sieving, filtration, centrifugation, and freeze drying for further use (Makadia and Siegel, 2011).

7.2.1.4 Nanoprecipitation (solvent displacement/solvent diffusion)

In this method, Fig. 7.5 represents that polymer and drug are dissolved in organic solvent and transferred into surfactants containing aqueous solution with constant speed. After addition, the resulting colloidal suspension is continuously stirred (800–900 rpm) for 4–5 h under fume hood, for removing the organic solvent completely (Ali et al., 2013).

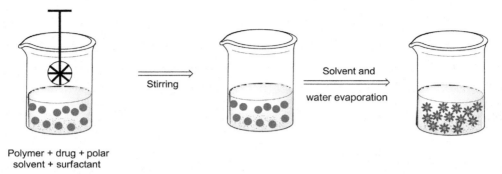

Polymer + drug + polar
solvent + surfactant

Figure 7.5 Solvent displacement method.

7.2.1.5 Dialysis

Dialysis is a simple method for the preparation of PNPs. This method involves the dissolution of polymer in an organic solvent. Dialysis is performed through the dialysis membrane against a nonsolvent miscible with the previous miscible solvent, and then dislocation of the solvent inside the membrane following the loss of solubility and then development of homogenous suspension of PNPs (Rao and Geckeler, 2011).

7.2.1.6 Salting out

Salting out method is an amended version of emulsion process that avoids addition of surfactant and chlorinated solvents, which are dangerous to the environment and physiological system (Bindschaedler et al., 1990). In this method, the emulsion is prepared with a polymeric solvent that is miscible with water like an ouzo effect, without using any high-shear forces (Ganachaud and Katz, 2005). Emulsification of the polymer solution in the aqueous phase is achieved by dissolving a high concentration of salt, which is selected for a strong salting-out effect in the aqueous phase. Magnesium acetate and calcium chloride are most commonly used suitable electrolytes (Allemann et al., 1992). A reverse salting out effect is obtained by dilution of the emulsion with an excess amount of water to the precipitation of the polymer dissolved in the emulsion system.

7.2.1.7 Supercritical fluid technology

Conventional methods are dangerous due to the requirement of organic solvents, which are dangerous to the environment as well as to physiological systems. So supercritical fluid (SCF) technology has been explored as an alternative and safe approach to prepare biodegradable micro and NPs because an SCF solvent remains at a temperature above its critical temperature, where fluid exits as a single phase regardless of pressure. Supercritical CO_2 is the most commonly used SCF due to its nontoxicity, nonflammability, and low price. The process of supercritical antisolvent employs a liquid solvent such as methanol, which is entirely miscible with the SCF, to dissolve the solute to be micronized. Insolubility of the solute in SCF carries the instant precipitation of the solute, resulting in the formation of NPs (Desai et al., 1997; Chernyak et al., 2001).

7.2.1.8 Coacervation/ionic gelation

For the past few decades, much more research has been focused on the preparation of NPs using biodegradable polymers such as gelatin, chitosan, and sodium alginate. Calvo et al. (1997a,b) developed a method for preparing hydrophilic chitosan NPs by ionic gelation. Generally mixing takes place between two aqueous phases, of which one is the polymer chitosan, a diblock copolymer ethylene oxide, or propylene oxide (PEO-PPO), and the other is a polyanion sodium tripolyphosphate. In this method coacervates are formed as a result of electrostatic interaction between two aqueous phases, whereas ionic gelation means liquid to gel happens due to ionic interaction conditions at room temperature.

7.2.2 Advanced methods

Apart from general methods, some advanced methods for NPs preparation are described here, e.g., sonication, core shell particulate, microfluidic, and electrodropping. Generally the drug-loaded NPs were prepared by dissolving the drug and polymer into the water-immiscible organic solvent and producing a nanoemulsion, as an example by the probe-sonication method (Hans and Lowman, 2002; Arshady, 1991; Bodmeier et al., 1991). The sonication process is a crucial step in the preparation of the sensitive drug-loaded nanoemulsion (Dendukuri and Doyle, 2009). In the core shell particulate method, a high electric field generates polymer liquids electrically charged in a capillary nozzle; this would lead to a cone-jet phenomenon at the tip of the nozzle. In a study, core–shell microencapsulation of photopolymer and ethylene glycol was carried out using electrified coaxial liquid jet (Choi et al., 2013). A microfluidic method has been developed to synthesize NPs inside microfluidic devices (Khan et al., 2004). The field of microfluidics plays a vital role in analytical tests for chemical and biological applications. Controlled environmental conditions, laminar flow, and continuous flow systems at the microfluidic length scale have all contributed to such applications (Furst et al., 1998, Singh et al., 2005).

7.2.3 Polymerization of monomers

The techniques discussed above involve the production of PNPs from preformed polymers and did not involve any polymerization processes. Polymerization of monomer technique has been reported for making poly (alkylcyanoacrylate) or polybutylcyanoacrylate NPs (Thickett and Gilbert, 2007). This technique involves like emulsion polymerization, conventional, surfactant free, miniemulsion, microemulsion, and interfacial polymerization (IP). Emulsion polymerization is the frequently used method that uses water as a dispersion medium. This method may be conventional and surfactant-free emulsion polymerization (Asua, 2004; Kreuter, 1982). In a conventional system, the ingredients comprise water, a monomer of low water solubility, water-soluble initiator and a surfactant. At the end of the reaction, PNPs are typically 102 nm in size, each containing many polymer chains. Phase separation and formation of solid particles can take place before or after the termination of the polymerization reaction (Munoz-Bonilla et al., 2010). Polystyrene (PS) (Zhang et al., 2004; Yih-Her et al., 2001), poly(vinylcarbazole), poly (methylmethacrylate) (PMMA), poly(ethylcyanoacrylate) (PECA), and poly(butylcyanoacrylate) NPs were produced by dispersion *via* surfactants into solvents, such as toluene, cyclohexane, and *n*-pentane. Surfactant-free polymerization systems utilize varying quantities of surfactants that need to be eliminated from the final product, but are hard to completely remove (Hearn et al., 1985; Song and Poehlein, 1990). Several mechanisms of nucleation and particle growth have been proposed for emulsion polymerization without an emulsifier, e.g., micellar-like nucleation (Goodall et al., 1977) and homogeneous nucleation (Landfester, 2001). Even with this success, several challenges still exist that greatly hinder the expanded utility of traditional surfactant-free emulsion polymerization, including the preparation of monodisperse and precisely controlled particle size (Zhang et al., 2001). A typical formulation used in miniemulsion polymerization consists of water, monomer mixture, costabilizer, surfactant, and initiator. The principal difference between these two techniques is the consumption of a low molecular mass compound as the costabilizer and also the use of ultrasound, a high-shear device (Chern and Sheu, 2001). The primary feature of this approach was that the SMA or DMA acted as a cosurfactant in the preparation of the miniemulsion and it became chemically incorporated into the emulsion polymer during polymerization (Ham et al., 2006). Ziegler et al. (2009) reported the synthesis of polystyrene and PMMA NPs using Lutensol AT 50.

Microemulsion polymerization is an innovative and effective approach for preparing nanosized polymer particles and has gained significant attention. Emulsion polymerization exhibits three reaction rate intervals, whereas only two are detected in microemulsion polymerization (Puig, 1996). Various types of surfactants and their combinations have been tested in the microemulsion polymerization of different monomers (Lapresta-Fernández et al., 2009; Dan et al., 2002; Babac et al., 2004; Jang et al., 2005; Reddy et al., 2009). The IP technique involves the polymerization of two reactive monomers, i.e., it takes place at the interface of the two liquids (Karode et al., 1998). The relative ease of obtaining IP has made it a preferred technique in many fields, ranging from encapsulation of pharmaceutical products (Hirech et al., 2003) to preparation of conducting polymers (Sree et al., 2002). Nanometer-sized hollow polymer particles were synthesized by engaging interfacial cross-linking reactions such as polyaddition and polycondensation (Crespy et al., 2007; Danicher et al., 2000; Torini et al., 2005).

7.3 PHYSIOCHEMICAL AND SURFACE PROPERTIES

PNPs have been characterized by their morphology as well as polymer composition. The distinctive sizes of NPs are agreeable to surface functionalization or alteration to achieve desired characteristics.

7.3.1 Morphology and shape

Morphology and shape factors, particularly particle size, are extremely important characteristics of NPs systems. Generally they determine the biological fate, toxicity, in vivo distribution, and the targeting aptitude of NPs systems. Morphology can also affect the stability, drug loading, and release of NPs. The release of the drug is affected by particle size. Formulation of NPs with absolutely smallest size possible is challenging work. Particle size also affects the polymer degradation (Tao and Desai, 2005). For instance, like the biodegradation rate of PLGA, the polymer was found to increase with increasing particle size in vitro. Therefore, it was assumed that larger size particles will contribute to faster polymer degradation as well as the drug release (Li et al., 2005; Maloney et al., 2005). Shape also plays a role, where spherical NPs experience faster uptake than rod-shaped NPs (Panyam and Labhasetwar, 2003). Generally NPs have comparatively higher intracellular uptake than microparticles with targeting efficacy. Desai et al. (1996) found that 100 nm NPs showed 2.5-fold more uptake than 1-μm microparticles, and 6-fold greater uptake than 10-μm microparticles in a Caco-2 cell line. In a subsequent study of Kroll et al. (1998), they concluded that the NPs penetrated through the submucosal layers in a rat in situ intestinal loop model, while microparticles were mostly localized in the epithelial lining. Owing to nanosize, NPs easily crosses the blood–brain barrier (BBB) through the opening of tight junctions by hyper-osmotic mannitol, which may be an effective target to treat diseases like brain tumors (Kreuter et al., 2003). Tween 80 coated NPs have been shown to cross the BBB (Zauner et al., 2001). Some cell line studies revealed that the status of the cell uptake depends on size (Redhead et al., 2001). Panyam et al. (2003) formulated varying sizes of PLGA particles and found that the polymer degradation rates in vitro were not significantly different for different size particles. To maintain the drug targeting by NPs, it is necessary to diminish the opsonization and to prolong the circulation of NPs in vivo. This can be achieved by surface engineering of NPs (Olivier, 2005).

7.3.2 Surface modification

The NPs surface modification is important to introduce drugs in the blood stream for "stealth" invisibility of the body's natural defense system (Storma et al., 1995). NPs can be easily identified by the host immune system when intravenously administered and cleared by phagocytes from the circulation (Muller et al., 1996). The mononuclear phagocytic system (MPS) eliminates them from the blood stream efficiently unless the particles are modeled to escape recognition. Longer circulation time increases the probability for the NPs to reach their target. Small particles (<100 nm) with a hydrophilic surface have the greatest ability to evade the MPS, because the foreign particles are rapidly cleared by MPS, one of the body's innate modes of defense. The process of opsonization is one of the most important biological barriers to NPs based controlled drug delivery. However, opsonin proteins present in the blood serum quickly bind to conventional nonstealth NPs, allowing macrophages of the MPS to easily recognize and remove these drug delivery devices before they can perform their designed therapeutic function (Owens and Peppas, 2006). Addition of PEG and PEG-containing copolymers to the surface of NPs results in an increase in the blood circulation half-life of the particles by several orders of magnitude. Small NPs (<100 nm) could not be recognized by cellular defense system after surface modification, and therefore, persist for a longer time in the body (Tobio et al., 2000). Surface modification by TPGS (D-α-Tocopheryl polyethylene glycol 1000 succinate), increases the adhesion of NPs to tumor cells surface (Lemarchand et al., 2005). PEG coating on the surface of PLA reduces the interaction between the NPs and the enzymes of the digestive fluids and increases uptake of encapsulated drug in the blood stream and lymphatic tissue (Ringe et al., 2004). Dextran was also used to coat the surface of PCL NPs. It was found that dextran coating on the surface of PCL-NPs inhibited protein adsorption (Shenoy and Amiji, 2005). It appears that the hydrophobic nature of most biodegradable particles would limit the applications. However, one may overcome concerns of clearance by the MPS through surface modification techniques. To further increase circulation time, the particles can be coated with molecules that provide them with a hydrophilic protective layer, such as PEG, or polyvinyl pyrrolidone (PVP). The accumulation of drug-loaded NPs at the target site is important, instead of their circulation and retention in the circulatory system of the body. This can be ensured by passive or active targeting (Soppimath et al., 2001).

7.4 DRUG LOADING AND DRUG RELEASE OF NANOPARTICLES

Drug loading depends on the method of preparation of NPs. Drug loading can be achieved by two methods:
- Incorporation method, in which incorporating occurs at the time of NPs production
- Adsorption/absorption technique

Many factors can affect the drug loading and entrapment efficiency, like molecular weight, polymer composition, the drug polymer interaction, and the presence of end-functional groups (Soppimath et al., 2001; Govender et al., 1999; Panyam et al., 2004). Surface modification by PEG has no or little effect on drug loading (Govender et al., 2000). Ionic interactions between the drug and matrix materials are suitable for small compounds, which can be a very effective way to increase the drug loading (Peracchia et al., 1997; Chen et al., 1994). In general, drug release rate depends on: (1) solubility of drug, (2) desorption of the surface bound drug, (3) NPs matrix erosion/degradation, (4) combination of diffusion/erosion process, and (5) drug diffusion through the NPs matrix. Thus solubility, diffusion and biodegradation of the matrix materials govern the release process surface of NPs (Chen et al., 2003; Magenheim et al., 1993). Diffusion is an important release mechanism in coated NPs (Peracchia et al., 1997). Addition of block copolymer like ethylene oxide-propylene oxide (PEO-PPO) to chitosan generally reduces the interaction of bovine serum albumin (BSA), a model drug due to competitive electrostatic interaction of PEO-PPO with chitosan (Calvo et al., 1997a,b).

7.5 RELEASE KINETICS

The drug release mechanisms are equally important as the drug polymer formulation because of the proposed application in sustained drug delivery. For manipulation of the rate and the timing of the drug release from NPs, a good understanding of the mechanisms of drug release is desired. There are five possible methods of drug release:
- Diffusion through the NPs matrix,
- Desorption of drug bound to the surface,
- NPs matrix erosion,
- A combined erosion–diffusion process,
- Diffusion through the polymer wall of nanocapsules.

7.5.1 Zero-order release

In zero-order release, drug released from formulations doesn't disaggregate and finally the drug release slowly which can be represented through following equations:

$$Q_0 - Q_t = K_0 t \tag{7.1}$$

By rearranging the Eq. (7.1) gives;

$$Q_t = Q_0 + K_0 t \tag{7.2}$$

Here, Q_t represents the amount of drug dissolved in time t, Q represents the initial concentration of drug and K_0 is the zero-order release constant expressed in units of concentration/time (Dash et al., 2010).

7.5.2 First-order release

First order release kinetics is generally used for absorption and elimination of some drugs. The release of drugs through first or der kinetics can be represented by the following equations:

$$\frac{dC}{dt} = -Kc \tag{7.3}$$

Here, K represents the first-order rate constant expressed (time^{-1}).

Eq. (7.3) can be written as;

$$\log C = \log C_0 - \frac{Kt}{2.303} \tag{7.4}$$

C_0 is the initial concentration of drugs, K is the first-order rate constant, and t is the time (Dash et al., 2010).

7.5.3 Higuchi release

The first model of drug release from matrix system was proposed by Higuchi, 1961a,b. This model is based on hypothesis of (1) drug concentration in the matrix system is more than drug solubility; (2) diffusion of drug takes place only in single or uni–direction; (3) diffusion of drug is constant; (4) swelling and dissolution of matrix system is negligible; (5) perfect sink conditions are always maintained in the release environment; (6) size of drug particles are smaller than matrix system thickness (Dash et al., 2010). Higuchi proposed the following equation (Higuchi, 1961a,b):

$$\frac{M_t}{A} = \sqrt{D(2c_0 - c_s)c_s t} \tag{7.5}$$

Here, M_t represents the amount of drug released at time t, A is the surface area of the matrix film, D is the diffusion of drug molecules in the matrix carrier, c_0 represents the initial drug concentration, and c_s represents the drug solubility in the matrix carrier (Siepmann and Siepmann, 2008).

7.5.4 Weibull release model

The Weibull equation is best described as a triphasic or sigmoidal release curve. The Weibull equation is applied to model release study with the drug delivery systems that follow:
- Erosion–dominated process coupled with least diffusive release
- Zero to least initial burst release
- Zero to least diffusion–mediated release

A Weibull equation was used to estimate the drug release from polymeric NPs at long and short duration of conditions.

$$\frac{X}{X_{\text{inf}}} = 1 - \exp\left[-\alpha(t^{\beta})\right] \tag{7.6}$$

Here, X is the drug release percentage at time t, X_{inf} is the 100% drug release, α is the scale factor equivalent to apparent release rate constant, and β is the shape factor.

In this equation, α explains the rate process and β explains the shape of the curve as exponential when $\beta = 1$, sigmoidal or S-shaped with upward curvature continued with a turning point when $\beta > 1$, and parabolic, with a higher initial slope and after that stable with exponential when $\beta < 1$ (D'Souza et al., 2005).

7.5.5 Hopfenberg model

Hopfenberg proposed a model to compare the drug release from the surface of polymer to how long a surface area remains constant and stable through the degradation. According to Hopfenberg the cumulative drug release at time t was explained as:

$$\frac{M_t}{M_\infty} = 1 - \left[1 - \frac{k_0 t}{C_L} a\right]^n \tag{7.7}$$

Here, k_0 is the zero-order release rate constant which explain the drug release from polymer surface, C_L is the initial amount of drug loading in the system, a is the radius of the system, and n is an exponent which varies with geometry, $n = 1$ for flat geometry, $n = 2$ for cylindrical geometry, and $n = 3$ for spherical geometry (Dash et al., 2010).

7.5.6 Hixson–Crowell model

Hixson and Crowell (1931) suggested that the particles area is proportional to the cube root of its volume. According to Hixson–Crowell:

$$W_0^{1/3} - W_t^{1/3} = \kappa t \tag{7.8}$$

Here, W_0 represent the initial concentration of drug in NPs, W_t denotes the remaining concentration of drug in NPs at time t, and κ (kappa) is a constant for surface–volume relationship (Dash et al., 2010).

7.5.7 Korsmeyer–Peppas model

Korsmeyer et al. (1983) derived a relationship that expressed the drug release from polymeric NPs system.

$$\frac{M_t}{M_\infty} = kt^n \tag{7.9}$$

Here, M_t is the concentration of the drug in the release medium at time t, M_∞ is the equilibrium concentration of drug in the release medium, k is the drug release rate constant, and n is the release exponent (Korsmeyer et al., 1983; Dash et al., 2010).

7.5.8 Baker–Lonsdale model

This release model was expressed by Baker and Lonsdale (1974) from the Higuchi model (Baker and Lonsdale, 1974) and expressed the drug release from spherical matrices according to the following equation:

$$f_1 = \frac{3}{2}\left[1 - \left(1 - \frac{M_t}{M_\infty}\right)^{\frac{2}{3}}\right]\frac{M_t}{M_\infty} = k_t \tag{7.10}$$

Here, M_t is the drug release at time t and M_∞ is the drug release at infinite time. The release rate constant k, corresponds to the slope. This equation has been used for linearization of release data from microcapsules or microspheres (Costa and Lobo, 2001; Dash et al., 2010).

7.5.9 Gompertz model

The in vitro dissolution rate is mostly expressed by an exponential model, which is generally known as the Gompertz model.

$$X(t) = X_{\text{Max}} \exp[-\alpha e^{\beta \log t}] \tag{7.11}$$

Here, $X(t)$ is the percent dissolved at time t divided by 100 and X_{Max} is the maximum dissolution, α determines the undissolved proportion at time interval t, β is the dissolution rate/unit of time expressed as shape parameter. The Gompertz model is more beneficial for comparing the release rate of drugs having good solubility and intermediate release rate profile (Dash et al., 2010).

7.6 POLYMERIC NANOPARTICLES IN TARGETED DRUG DELIVERY

Polymer in targeted drug delivery is a rapid emerging new technological discipline in which various therapeutic applications of nanoproducts are expected to overcome the patient complaints in healthcare. Drug targeting is defined as selective drug delivery to specific physiological sites, tissues, organs, or cells where a pharmacological response of drug is required as depicted in Fig. 7.6. Targeting may be active (ligand based) and passive (enhanced permeation and retention, EPR) (Masayuki, 2005).

7.6.1 Active and passive targeting

Active targeting denotes the effective bioactive concentration to the targeted site. These interactions e.g., include antigen–antibody and ligand–receptor interactions (Maeda, 2001; Danhier et al., 2010; Peer et al., 2007). Carriers may be classified as transferrin, antibodies, ferrite containing liposomes, and thermo-responsive carriers. In active targeting as shown in Fig. 7.7, the surface of NPs is coated with molecules (e.g., native ligands or antibodies) that can, in theory, increase the affinity of the NPs for specific cells or tissues, and potentially facilitate receptor remediated uptake. Often, active targeting is used in combination with NP formulations that are designed to enhance passive targeting. Two examples of such polymeric NPs systems currently in clinical trials are BIND-014 and CALAA-01 (Hrkach et al., 2012; Davis et al., 2010). These two formulations are surface-modified with both PEG and targeting molecules that bind to receptors that are enriched on some cancer cells. In the case of passive targeting, NPs accumulate at tumor sites because of the leaky vasculature that often characterizes

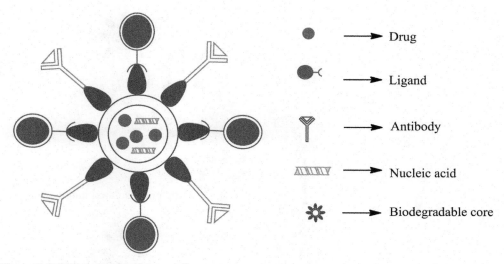

Figure 7.6 PNPs in targeted drug delivery.

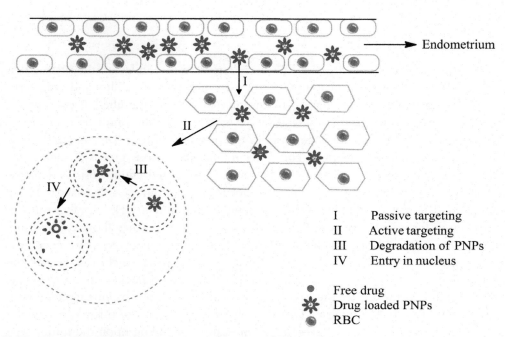

Figure 7.7 Active and passive targeting mechanism of PNPs.

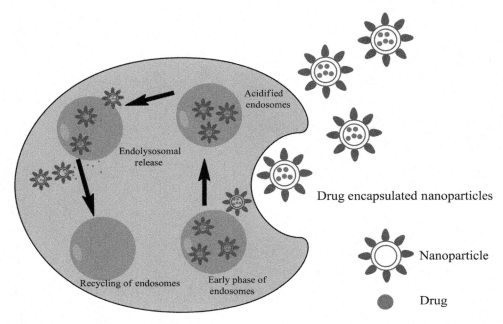

Figure 7.8 Intracellular uptake of PNPs by endocytosis.

tumor tissue. In active targeting, the NPs carry targeting molecules on the surface that are able to interact with the surrounding tissue. To increase the stability and long-term systemic circulation of NPs, surfaces of conventional NPs are modified. Passive targeting is also defined as a method in which the physical and chemical properties of carrier systems increase the target/nontarget ratio of the quantity of drug (Bhadra et al., 2002; Olivier, 2005). Uptake of particles could occur through various processes such as by phagocytosis, fluid phase pinocytosis or by receptor-mediated endocytosis as depicted in Fig. 7.8. The EPR effect was first described by (Matsumura and Maeda, 1986). Nanoparticulate systems take advantage of unique pathophysiologic characteristics of tumor vessels for passive targeting. When the tumor volume reaches above 2 mm–3 µm, diffusion limitation sets in, which eventually impairs nutrition intake, waste excretion, and oxygen delivery (Byrne et al., 2008). Such rapidly growing cancer cells recruit the generation of new blood vessels, a phenomenon called angiogenesis (or neovascularization). Aberrant tortuosity (twisted curves), abnormalities in the basement membrane, and the lack of pericytes lining endothelial cells are the features of this process, which results in leaky vessels with gap sizes of 100 nm–2 µm, depending upon the tumor type (Matsumura and Maeda, 1986). Moreover, such tumors exhibit poor lymphatic drainage due to the higher interstitial pressure at the core of the tumor than at the periphery.

This combination of leaky vasculature and poor lymphatic flow results in the (EPR) effect. NPs can preferentially localize in cancerous tissues owing to their size, smaller than blood vessel fenestration, and entrapped in the tumor due to higher retention ability than the normal tissues (Byrne et al., 2008).

7.7 CLINICAL APPLICATION OF POLYMERIC NANOPARTICLES

Nanotechnology is having a significant impact in medicine. NPs have been used in the clinical for anticancer drug and there are clear advantages of NPs with regard to reducing the side effects of drug cargo, enhanced tumor targeting, and, in some cases, therapeutic efficacy (Valadi et al., 2007). Researchers have begun to engineer exosomes for the targeted delivery of therapy, including siRNA, to cancer (Alvarez-Erviti et al., 2011). Chen et al. (2015) prepared ifosfamide-loaded poly (lactic-co-glycolic acid) (PLGA)-dextran NPs and evaluate its anticancer activity against multiple osteosarcoma cancer cells. The particle size of PD/IFS was observed to be 124 ± 3.45 nm with an excellent poly dispersity index of 0.124. Polymeric NPs are mostly reported to deliver the hydrophobic drugs; many researchers reported about the encapsulating hydrophilic drugs through surface attachment or polymer–drug conjugation methods (Zhang et al., 2007; Sengupta et al., 2005). Presently, many NPs based chemotherapeutic formulations have emerged on the market, and many are in the stages of clinical or preclinical development (Brannon-Peppas and Blanchette, 2004). Prominent examples of chemotherapeutic NPs include Doxil (a ~100-nm liposomal formulation of doxorubicin) (Northfelt et al., 1998) and Abraxane (a ~130-nm paclitaxel-bound protein particle) (Harries et al., 2005); both are drugs of choice as first-line treatments in various cancer types. Table 7.2 includes examples of different anticancer and PLGA formulation. Different therapeutic agents delivered by PNPs are discussed in Table 7.4.

7.7.1 Targeting of therapeutic agents through polymeric nanoparticles

Biodegradable PNPs are highly desired because they show potential ability in a drug delivery system. These nanomedicines are stable in blood, noninflammatory, nontoxic, nonimmunogenic, biodegradable and applicable to various biomolecules such as proteins, peptides, drugs, or nucleic acids (Rieux et al., 2006). Among various nanocarriers, polymer-based NPs and macromolecular approaches have resulted in improved drug delivery for diseases like cancers, diabetes, autoimmune disorders, and many more (Gothwal et al., 2015). For the past two decades, significant work has been conducted to develop the most effective nanomedicines from biocompatible and biodegradable nanopolymers (Kreuter, 1994). The therapeutic activity and stability of PCL nanomedicines are reasonably higher than PLGA nanomedicine (Kim and Lee, 2001). PLGA NPs have been mostly prepared by emulsification–diffusion (Sahana et al., 2008),

Table 7.4 Polymeric nanoparticles and encapsulation of drugs in target drug delivery

Polymer	Encapsulant	Method of preparation	Therapeutic improvement	Surface modification	In vivo	Reference
PLGA	9-Nitrocamptothecin	Nanoprecipitation	Controlled release			Derakhshandeh et al. (2007)
	Cisplatin	Double emulsion	Sustained release	PEG		Avgoustakis et al. (2002)
PLA	Haloperidol	Solvent evaporation	Slow release of drug up to 4 days			Budhian et al. (2005)
	Savoxepine	Salting out	Controlled drug release up to 1 week	PEG	Sustained plasma levels after intramuscular and intravenous injection	Jean-Christophe et al. (1996)
PCL	Tamoxifen	Solvent displacement	Preferential tumor targeting and circulating drug reservoir	Pluronics	Increased level of accumulation of the drug within tumor with time and extended their presence in circulation	Gao et al. (2005)
	Amphotericin B		Reduced side effects			Espuelas et al. (2007)
Chitosan	Insulin	Ionotropic gelation method	Oral absorption and oral bioactivity was increased		Nanoparticles adhere to intestinal epithelium and internalized by intestinal mucosa	Sarmento et al. (2007)
Gelatin	Didanosine	Double desolvation technique	Slow drug release up to 24 h	Manman	Higher accumulation of didanosine in brain	Soppimath et al. (2001)
PBCA	Ftorafur	Anionic polymerization	Controlled release of the drug up to 10 h			Arias et al. (2007)
PACA	Ampicillin		Inhibition of platelet aggregation			Gulyaev et al. (1999)

PACA, poly(cyanoacrylates); *PBCA*, poly-butyl-cyano-acrylate; *PCL*, polycaprolactone; *PLA*, polylactic acid; *PLGA*, poly-lactide-*o*-glycolide.

solvent emulsion–evaporation (Zambaux et al., 1999), interfacial deposition (Reis et al., 2006), and nanoprecipitation method (Barichello et al., 1999).

7.7.1.1 Poly-lactide-co-glycolic acid nanoparticles and therapeutic agents

The FDA approved the PLGA NPs for therapeutic purposes in human beings. Numerous cancer-related drugs have been incorporated through this NPs.

7.7.1.1.1 Anticancer agents

9-Nitrocamptothecin (9-NC) (derivative of camptothecin) and related analogs are a promising family of anticancer agents with a unique mechanism of action, targeting the nuclear enzyme topoisomerase-I. Unfortunately, all camptothecin derivatives undergo a pH-dependent rapid and reversible hydrolysis from closed lactone ring to the inactive hydroxyl carboxylated form with loss of antitumor activity. The delivery of lipophilic derivatives of 9-NC is quite challenging because of instability at biological pH and its low water solubility. PLGA has been used to encapsulate 9-NC successfully by nanoprecipitation methods having more than 30% encapsulation efficiency with its complete biological activity and without disturbing lactone ring (Derakhshandeh et al., 2007). The in vitro drug release profile showed a sustained 9-NC release up to 160 h indicating the suitability of PLGA NPs in controlled 9-NC release. Paclitaxel (commercially available as Taxol) binds with tubulin and interferes with the normal function of microtubule breakdown. This drug promotes the polymerization of tubulin causing cell death by disrupting the dynamics necessary for cell division. This drug has neoplastic activity against primary ovarian carcinoma breast and colon cancers. It is one of the potent anticancer agents but less useful for clinical administration due to its poor solubility. PLGA intermingled with vitamin E and tocopheryl polyethylene glycol succinate (TPGS) is widely used as encapsulating agents by solvent evaporation/extraction methods and in vitro CR of this drug (Wang et al., 1996). This formulation has shown reasonably good activity and much faster administration in comparison to traditional formulation. Using some additive to the PLGA NPs, 100% drug encapsulation efficiency was achieved with its full antitumor activity. The release kinetics include 50% burst release within 24 h and slow release for 1 month in vitro. It has also been demonstrated that incorporation of paclitaxel in the PLGA NPs strongly enhances its antitumoral efficacy as compared to free drug. This effect is more relevant for prolonged incubation times with cells (Fonseca et al., 2002). Another important anticancer drug is Cisplatin, known to interfere with cell division by mitosis and inhibit DNA synthesis. The damaged DNA elicits DNA repair mechanism. Cisplatin is a very potent anticancer drug, but the full therapeutic exploitation of this drug is limited due to its toxicity in healthy tissues, including renal and auditory toxicity, nausea and vomiting (Rosenberg, 1985). The targeted delivery of cisplatin to tumor cells would significantly reduce drug toxicity, and improve its therapeutic index.

PLGA–mPEG NPs encapsulating cisplatin is prepared by double emulsion methods. PLGA–methoxy(polyethylene glycol) (mPEG) NPs revealed prolonged drug residence in blood upon intravenous administration (Avgoustakis et al., 2002). However, at the targeted site it exhibited rapid degradation and sustained release profile. These characteristics help in preventing the tumor growth.

7.7.1.1.2 Antidiabetes agents

Four subcutaneous insulin injections per day are required to maintain serum glucose levels according to current dosage regimens. However, to mitigate the long-term complications of diabetes mellitus zinc insulin (1.6%) in PLGA with fumaric anhydride and iron oxide has been found effective for oral administration. This formulation is shown to have 11.4% of the efficacy of intraperitoneally delivered zinc insulin and is able to control plasma glucose levels when faced with a simultaneously administered glucose challenge (Kumar et al., 2006). The recent formulation of insulin-loaded NPs of PLGA is used to maintain the integrity of insulin during formulation and delivery.

7.7.1.1.3 Antipsychotic agents

Haloperidol is a primitive antipsychotic drug used for the treatment of schizophrenia and more acutely in the treatment of acute psychotic states. The delirium decanoate ester is used as a long-acting injection given every week to patients with schizophrenia. Haloperidol possesses a strong activity against delusions and hallucinations, most likely due to an effective dopaminergic receptor blockage in the mesocortex and the limbic system of the brain. PLGA has been used for encapsulation and for extended CR of haloperidol (Budhian et al., 2005). PLGA end-groups have a strong influence on haloperidol incorporation efficiency and its release from PLGA NPs. The hydroxyl terminated PLGA (uncapped) NPs have a drug incorporation efficiency of more than 30% as compared to only 10% with methyl terminated PLGA (capped) NPs. Haloperidol incorporation into PLGA with −COOH end-groups (uncapped) is three times more than in capped PLGA. Uncapped PLGA shows a lower initial burst release and a longer period of haloperidol release as compared to capped PLGA (Budhian et al., 2005).

7.7.1.1.4 Antibiotics

Gentamicin-loaded PLGA NPs were developed for the treatment of *Pseudomonas aeruginosa* infections. The formulation and the free drug were tested against biofilms in vitro as well as in a peritoneal murine infection model. The PLGA particles provided a sustained release of gentamycin, translating into a significantly enhanced antibiofilm effect compared to the free drug. In the in vivo model, the free and PLGA encapsulated gentamicin were equally potent in clearing the infection but 96 h after administering the formulations, the antibiofilm effect of the free gentamicin was significantly reduced while the biofilm formation was still largely inhibited by the nanoformulation. Control experiments showed that empty PLGA NPs exhibited no antimicrobial effect against planktonic *P. aeruginosa* cells (Abdelghany et al., 2012).

7.7.1.2 Polylactic acid nanoparticles and therapeutic agents

PLA polymer is biodegradable as well as biocompatible material that undergoes scission in the body to monomeric units of lactic acid as a natural intermediate in carbohydrate metabolism. Solvent evaporation, solvent displacement, salting out, and solvent diffusion are the extreme methods for preparation of PLA NPs (Fessi et al., 1989).

7.7.1.2.1 Antipsychotic drugs

Savoxepine acts via selective limbic dopamine D2 receptor blockade (Volz et al., 2002). Savoxepine-loaded nanospheres have been prepared by the salting out procedure. Drug loading reached up to 16.7% with encapsulation efficiency as high as 95%. In vitro release studies demonstrated that this type of drug carriers allows extended delivery of the drug over more than 1 week. In vivo, NPs loaded with neuroleptic compound savoxepine were able to provide sustained plasma levels after intramuscular and intravenous injection. Intramuscularly injected NPs remained at the site of injection, whereas intravenously injected NPs were located mostly in the macrophages (MPS) (Jean-Christophe et al., 1996). The PEG-6000 and PEG-20000 have been used to make an additional coating during the preparation and encapsulation of PLA NPs. In vitro, these coatings provided a protective barrier against extensive uptake by human monocytes, at least in plasma. Analysis of plasma proteins adsorbed on NPs and in vitro experiments on isolated cells revealed some differences between the opsonization process of plain and coated NPs (Jean-Christophe et al., 1996).

7.7.1.2.2 Anticancer agents

Oridonin (natural diterpenoid) induced growth arrest and apoptosis of cells from lymphoid malignancies in association with inhibition of NF-κB and down regulation of Bcl-2 family proteins (Ikezoe et al., 2005). The success of its clinical application is greatly limited by its poor water solubility and low therapeutic index. Oridonin-loaded poly(lactic acid) NPs were prepared by modified spontaneous emulsion solvent diffusion method. The entrapment efficiency and actual drug loading of the NPs were higher by as much as $91.88 \pm 1.83\%$ and $2.32 \pm 0.05\%$, respectively. The results of pharmacokinetics demonstrated that oridonin encapsulated on PLA NPs was remarkably effective for oridonin to prolong blood circulation time. The stable and high concentration of oridonin in liver, lung and spleen is reported after the intravenous administration of oridonin-PLA-NPs, while its distribution in heart and kidney is significantly decreased (Xing et al., 2007).

7.7.1.2.3 Proteins

A tadpole-shaped polymer mono (6-(2-aminoethyl) amino-6-deoxy)-cyclodextrin-PLA (CDen-PLA) was used to encapsulate the BSA successfully by the double emulsion method and the nanoprecipitation method. The encapsulation efficiency was up to 71.6%. The results showed that this polymeric system could load BSA effectively and BSA remained stable after it was released from the NPs. This report opens the door for the successful encapsulation and delivery of various disease-related proteins for protein therapy (Gao et al., 2005).

7.7.1.3 Polycaprolactone nanoparticles and therapeutic agents

Degradation of PCL by hydrolysis of its ester linkages makes attractive for delivery purposes. PCL-NPs have been prepared mostly by nanoprecipitation, solvent displacement and solvent evaporation.

7.7.1.3.1 Anticancer agents

Tamoxifen competitively binds to tumor-targeted tissue, resulting in a formation of nuclear complex that decreases DNA synthesis and inhibits estrogen effects (Shenoy and Amiji, 2005). It is a competitive nonsteroidal agent that generally competes with estrogen for binding sites in breast and other tissues. The drug tamoxifen causes cells to remain in the G0 and G1 phases of the cell cycle. Thus it prevents proliferation of pre-cancerous cells and distinguished from dividing cells. Tamoxifen-loaded polyethylene oxide (PEO) modified PCL were prepared by solvent displacement method. About 90% drug encapsulation efficiency has been achieved. The liver is a primary site of accumulation for the drug-loaded NPs after intravenous administration. Nearly 26% of the total activity could be recovered in a tumor at 6 h of postinjection for PEO-modified NPs. PEO-PCL NPs exhibited a significantly increased level of accumulation of the drug within a tumor with time, as well as extended their presence in the systemic circulation rather than the controls (unmodified NPs or the solution form) (Gao et al., 2005).

7.7.1.3.2 Antidiabetes agents

NPs prepared with a blend of biodegradable polyester PCL and a poly-cationic non-biodegradable acrylic polymer has been used as a drug carrier for oral administration of insulin in which the encapsulation efficiency was 90% (Damge et al., 2007). As per administration to diabetic rats, insulin NPs decreased fasted glycemia in a dose-dependent manner with a maximal effect observed with 100 IU/kg. These insulin–PCL NPs also increase serum insulin levels and improved the glycemic response to an oral glucose challenge for a prolonged period of time. FITC–insulin-loaded NPs strongly adhered to the intestinal mucosa and labeled insulin, either released and/or still inside NPs. This was mainly taken up by the cells of Peyer's patches. It is concluded that polymeric NPs allow the preservation of insulin's biological activity (Damge et al., 2007).

7.7.1.3.3 Antifungal agents

Amphotericin B (AmB) associates with ergosterol, a membrane chemical constituent of fungi, which creates a pore that leads to potassium leakage and ultimately death of fungi. Hydrogen bonding interactions among different groups like hydroxyl, carboxyl, and amino groups stabilize the channel in its open form, thus abolishing activity and allowing the cytoplasmic contents to leak out. PCL-NPs have been developed to improve antileishmanial action of AmB with concomitant reduction in the toxicity associated with it (Espuelas et al., 2007). Nanoencapsulated AmB was found to be 2–3 times effective than free AmB in terms of reducing parasite burden from leishmania-infected mice and also the side effects associated with AmB.

7.7.1.4 Chitosan and gelatin nanoparticles

Chitosan is a carbohydrate class of natural polymer that is prepared by the partial N-deacetylation of natural biopolymer chitin. Ionotropic gelation, emulsification solvent microemulsion, diffusion, and polyelectrolyte complex are methods widely used for NPs preparation (Tiyaboonchai, 2003).

7.7.1.4.1 Antidiabetic agents

Insulin was observed to be directly internalized by enterocytes in contact with intestine; retention of drugs at their absorptive sites by mucoadhesive carriers is a synergic factor. Insulin-loaded chitosan NPs markedly enhanced intestinal absorption of insulin following oral administration. Insulin internalization through enterocytes and uptake of insulin-loaded NPs by cells of Peyer's patches both are well known mechanisms of insulin absorption (Sarmento et al., 2007). Gelatin is widely used for its biodegradable, nontoxic, bioactive and inexpensive properties. Chemically gelatins have both cationic and anionic groups along with hydrophilic group. Mechanical properties, thermal properties, and swelling behavior depend significantly on the degree of crosslinking in gelatin. Gelatin NPs can be prepared by coacervation or emulsion method (Oppenhiem, 1981).

7.7.1.4.2 Anti-HIV agents

Due to the hydrophilic nature of didanosine, it crosses the BBB very slowly. This molecule was encapsulated on mannan-coated gelatin NPs by desolvation method (Soppimath et al., 2001). The extent of the presence of didanosine was greater in the spleen, lymph nodes, and brain after administration of mannan-coated gelatin NPs compared to that after injection in PBS (phosphate buffer) (Kaur et al., 2008).

7.7.1.5 Poly-alkyl-cyano-acrylates nanoparticles and therapeutic agents

Poly-alkyl-cyano-acrylates (PACA) is biodegradable as well as biocompatible, degraded by esterase enzyme in biological fluids that create toxic products and stimulate or damage the central nervous system (CNS). Due to this aforementioned reason this polymer is not authorized for human application. PACA NPs are prepared mostly by IP, emulsion polymerization, and nanoprecipitation for drug delivery.

7.7.1.5.1 Antibacterial agents

Ampicillin semisynthetic penicillin acts as a competitive inhibitor of the cell wall maker enzyme transpeptidase. There was no modification in molecular weight distribution in gel chromatography of poly-iso-butylcyanoacrylate (PIBC) NPs mechanically loaded with ampicillin. The release of antibiotic ampicillin was homogeneous.

7.7.1.5.2 Anticancer agents

The poly-butyl-cyano-acrylate (PBCA) encapsulated doxorubicin NPs have been reported to increase 60-fold in the brain after coated with polysorbate 80 (Gulyaev et al., 1999). Ftorafur is a type of substance being used in the treatment of cancer. It is a combination of tegafur and uracil. The tegafur is taken up by the cancer cells and breaks down into 5-FU, a substance that kills tumor cells. The uracil causes higher amounts of 5-FU to stay inside the cells and kill them (Arias et al., 2007). Poly-ethyl-2-cyanoacrylate (PE-2-CA) and PBC nanospheres were used to encapsulate ftorafur [Tegafur, 5-fluoro-1-(tetrahydro-2-furyl) uracil], a broad-spectrum antitumor drugs. With respect to the release profiles, ftorafur surface adsorption onto nanospheres led to a very rapid drug release in sink conditions. However, the drug incorporation into the NPs permitted a larger loading and a slower ftorafur release (Arias et al., 2007).

7.8 CONCLUSION AND FUTURE PROSPECTS

Nanotechnology covers a range of techniques and approaches that plays a crucial role in delivery of bioactives. Different nanocarriers in nanotechnological advancement offer a controlled or sustained release fashion of drugs and also protects the active ingredients from sudden clearance, premature degradation and toxic manifestations. This platform itself offers surface engineering of nanocarriers to eliminate or reduce the toxic effect of the drugs. Targeted delivery of active substances leads to increased drug concentrations at local site and significantly lowers the unwanted toxicity. In clinical scenario nanoparticle also showed its potential application in complex diseases like cancer, diabetes, CNS disorders, etc. Some of the nanotechnology-based products are on the market for commercial purposes, and some in different clinical trial stages for therapeutic applications. In the future, when this technology will be accepted in almost all respects, we can even have personalized medicines based on nanoparticulate platform. The uniqueness exists in these carriers due to the possibility of customized development and the selection of the polymers or nanoparticles of our choice. Finally, the successful attainment of this technology will possibly depend on some toxicological issues associated with the nanocarrier via its fate in the body and its possible polymeric constituents.

ACKNOWLEDGMENT

The authors are grateful and would like to acknowledge the University Grants Commission (UGC) New Delhi, India, and Science and Engineering Research Board (SERB), Department of Science and Technology (DST), New Delhi, India, for providing research funding to the corresponding author.

REFERENCES

Abdelghany, S.M., Quinn, D.J., Ingram, R.J., Gilmore, B.F., Donnelly, R.F., Taggart, C.C., et al., 2012. Gentamicin-loaded nanoparticles show improved antimicrobial effects towards *Pseudomonas aeruginosa* infection. Int. J. Nanomed. 7, 4053–4063.

Ahmad, Z., Pandey, R., Sharma, S., Khuller, G.K., 2006. Alginate nanoparticles as antituberculosis drug carriers: formulation development, pharmacokinetics and therapeutic potential. Indian J. Chest Dis. Allied Sci. 48 (3), 171–176.

Ali, H., Kalashnikova, I., White, M.A., Sherman, M., Rytting, E., 2013. Preparation, characterization, and transport of dexamethasone-loaded polymeric nanoparticles across a human placental in vitro model. Int. J. Pharm. 454 (1), 149–157.

Allemann, E., Gurny, R., Doelker, E., 1992. Preparation of aqueous polymeric nanodispersions by a reversible salting-out process: influence of process parameters on particle size. Int. J. Pharm. 87 (1–3), 247–253.

Alvarez-Erviti, L., Seow, Y., Yin, H., Betts, C., Lakhal, S., Wood, M.J.A., 2011. Delivery of siRNA to the mouse brain by systemic injection of targeted exosomes. Nat. Biotechnol. 29 (4), 341–345. [PubMed: 21423189].

Arias, J.L., Gallardo, V., Ruiz, M.A., Delgado, A.V., 2007. Ftorafur loading and controlled release from poly(ethyl-2cyanoacrylate) and poly(butylcyanoacrylate) nanospheres. Int. J. Pharm. 337 (1–2), 282–290.

Arshady, R., 1991. Preparation of biodegradable microspheres and microcapsules: polylactides and related polyesters. J. Control. Release 17 (1), 1–22.

Asua, J.M., 2004. Emulsion polymerization: from fundamental mechanisms to process developments. J. Polym. Sci. Part A Polym. Chem. 42, 1025–1041.

Avgoustakis, K., Belesti, A., Pangi, Z., Klepetsaines, P., Karyadas, A.J., Ithalkissos, D.J., 2002. PLGA–mPEG nanoparticles of cisplatin: in vitro nanoparticle degradation, in vitro drug release and in vivo drug residence in blood properties. J. Control. Release 79 (1–3), 123–135.

Babac, C., Guven, G., David, G., Simionescu, B.C., Piskin, E., 2004. Production of nanoparticles of methyl methacrylate and butyl methacrylate copolymers by microemulsion polymerization in the presence of maleic acid terminated poly(N-acetylethylenimine) macromonomers as cosurfactant. Eur. Polym. J. 40 (8), 1947–1952.

Baker, R.W., Lonsdale, H.S., 1974.. In: Tanquary, A.C., Lacey, R.E. (Eds.), Controlled Release of Biologically Active Agents. Plenum Press, New York.

Barichello, J.M., Morishta, M., Takaya, K., Nagai, T., 1999. Encapsulation of hydrophilic and lipophilic drugs in PLGA nanoparticles by the nanoprecipitation method. Drug Dev. Ind. Pharm. 25 (4), 471–476.

Bhadra, D., Bhadra, S., Jain, P., Jain, N.K., 2002. Pegnology: a review of PEGylated systems. Pharmazie 57 (1), 5–29.

Bindschaedler, C., Gurny, R., Doelker, E., 1990. Process for Preparing a Powder of Water-Insoluble Polymer Which Can Be Redispersed in a Liquid Phase, the Resulting Powder and Utilization Thereof. US Patent. 4968350, pp. 1–6.

Birrenbach, G., Speiser, P.P., 1976. Polymerized micelles and their use as adjuvants in immunology. J. Pharm. Sci. 65 (12), 1763–1766.

Bodmeier, R., Chen, H., Tyle, P., Jarosz, P., 1991. Spontaneous formation of drug-containing acrylic nanoparticles. J. Microencap. 8 (2), 161–170.

Brannon-Peppas, L., Blanchette, J.O., 2004. Nanoparticle and targeted systems for cancer therapy. Adv. Drug Deliv. Rev. 56 (11), 1649–1659.

Brigger, I., Dubernet, C., Couvreur, P., 2002. Nanoparticles in cancer therapy and diagnosis. Adv. Drug Deliv. Rev. 54 (5), 631–651.

Budhian, A., Siegel, S.J., Winey, K.I., 2005. Production of haloperidol-loaded PLGA nanoparticles for extended controlled drug release of haloperidol. J. Microencapsul. 22 (7), 773–785.

Byrne, J.D., Betancourt, T., Brannon-Peppas, L., 2008. Active targeting schemes for nanoparticle systems in cancer therapeutics. Adv. Drug Deliv. Rev. 60 (15), 1615–1626.

Calvo, P., Vila-Jato, J.L., Alonso, M.J., 1996. Comparative in vitro evaluation of several colloidal systems, nanoparticles, nanocapsules and nanoemulsions as ocular drug carriers. J. Pharm. Sci. 85 (5), 530–536.

Calvo, P., Remunan-Lopez, C., Vila-Jato, J.L., Alonso, M.J., 1997a. Novel hydrophilic chitosan-polyethylene oxide nanoparticles as protein carriers. J. Appl. Polym. Sci. 63 (1), 125–132.

Calvo, P., Remunan-Lopez, C., Vila-Jato, J.L., Alonso, M.J., 1997b. Chitosan and chitosan/ethylene oxide-propylene oxide block copolymer nanoparticles as novel carriers for proteins and vaccines. Pharm. Res. 14 (10), 1431–1436.

Calvo, P., Gouritin, B., Brigger, I., Lasmezas, C., Deslys, J.P., Williams, A., et al., 2001. PEGylated polycyanoacrylate nanoparticles as vector for drug delivery in prion diseases. J. Neurosci. Methods 111 (2), 151–155.

Chen, B., Yang, J.Z., Wang, L.F., Zhang, Y.V., Lin, X.J., 2015. Ifosfamide-loaded poly (lactic-co-glycolic acid) PLGA-dextran polymeric nanoparticles to improve the antitumor efficacy in Osteosarcoma. BMC Cancer 15, 752–761.

Chen, Y., McCulloch, R.K., Gray, B.N., 1994. Synthesis of albumin-dextran sulfate microspheres possessing favorable loading and release characteristics for the anti-cancer drug doxorubicin. J. Control. Release 31 (1), 49–54.

Chen, Y., Mohanraj, V.J., Parkin, J.E., 2003. Chitosan-dextran sulfate nanoparticles for delivery of an antiangiogenesis peptide. Lett. Pept. Sci. 10 (5), 621–629.

Chern, C.S., Sheu, J.C., 2001. Effects of carboxylic monomers on the styrene miniemulsion polymerizations stabilized by SDS/alkyl methacrylates. Polymer 42 (6), 2349–2357.

Chernyak, Y., Henon, F., Harris, R.B., Gould, R.D., Franklin, R.K., Edwards, J.R., et al., 2001. Formation of perfluoropolyether coatings by the rapid expansion of supercritical solutions (RESS) process Part 1: experimental results. Ind. Eng. Chem. Res. 40 (26), 6118–6126.

Choi, D.H., Subbiah, R., Kim, I.H., Han, D.K., Park, K., 2013. Dual growth factor delivery using biocompatible core-shell microcapsules for angiogenesis. Small 9 (20), 3468–3476.

Coester, C., Kreuter, J., Briesen, H.V., Langer, K., 2000. Preparation of avidin-labelled gelatin nanoparticles as carriers for biotinylated peptide nucleic acid (PNA). Int. J. Pharm. 196 (2), 147–149.

Cohen, H., Levy, R.J., Gao, J., Fishbein, I., Kousaev, V., Sosnowski, S., et al., 2000. Sustained delivery and expression of DNA encapsulated in polymeric nanoparticles. Gene Therapy 7 (22), 1896–1905.

Costa, P., Lobo, J.M.S., 2001. Modeling and comparison of dissolution profiles. Eur. J. Pharm. Sci. 13 (2), 123–133.

Couvreur, P., 1988. Polyalkylcyanoacrylates as colloidal drug carriers. Crit. Rev. Ther. Drug Carr. Syst. 5 (1), 1–20.

Crespy, D., Stark, M., Hoffmann-Richter, C., Ziener, U., Landfester, K., 2007. Polymeric nanoreactors for hydrophilic reagents synthesized by interfacial polycondensation on miniemulsion droplets. Macromolecules 40 (9), 3122–3135.

Cui, Z., Mumper, R.J., 2002. Plasmid DNA-entrapped nanoparticles engineered from microemulsion precursors: in vitro and in vivo evaluation. Bioconjug. Chem. 13 (6), 1319–1327.

D'Souza, S.S., Faraj, J.A., DeLuca, P.P., 2005. A model-dependent approach to correlate accelerated with real-time release from biodegradable microspheres. AAPS PharmSciTech 6 (4), E553–E564.

Damge, C., Maincent, P., Ubrich, N., 2007. Oral delivery of insulin associated to polymeric nanoparticles in diabetic rats. J. Control. Release 117 (2), 163–170.

Dan, Y., Yang, Y., Chen, S., 2002. Synthesis and structure of the poly (methyl methacrylate) microlatex. J. Appl. Polym. Sci. 85 (14), 2839–2844.

Danhier, F., Feron, O., Preat, V., 2010. To exploit the tumor microenvironment: passive and active tumor targeting of nanocarriers for anti-cancer drug delivery. J. Control. Release 148 (2), 135–146.

Danicher, L., Frere, Y., Calve, A.L., 2000. Synthesis by interfacial polycondensation of polyamide capsules with various sizes. Characteristics and properties. Macromol. Symp. 151, 387–392.

Dash, S., Murthy, P.N., Nath, L., Chowdhury, P., 2010. Kinetic modelling on drug release from controlled drug delivery systems. Acta Pol. Pharm. 67 (3), 217–223.

Date, A.A., Joshi, M.D., Patravale, V.B., 2007. Parasitic diseases: liposomes and polymeric nanoparticles versus lipid nanoparticles. Adv. Drug Deliv. Rev. 59 (6), 505–521.

Davis, M.E., Zuckerman, J.E., Choi, C.H., Seligson, D., Tolcher, A., Alabi, C.A., et al., 2010. Evidence of RNAi in humans from systemically administered siRNA via targeted nanoparticles. Nature 464, 1067–1070.

Dendukuri, D., Doyle, P.S., 2009. The synthesis and assembly of polymeric microparticles using microfluidics. Adv. Mater. 21 (41), 4071–4086.

Derakhshandeh, K., Erfan, M., Dadashzadeh, S., 2007. Encapsulation of 9-nitrocamptothecin, a novel anticancer drug, in biodegradable nanoparticles: factorial design, characterization and release kinetics. Eur. J. Pharm. Biopharm. 66 (1), 34–41.

Desai, M.P., Labhasetwar, V., Amidon, G.L., Levy, R.J., 1996. Gastrointestinal uptake of biodegradable microparticles: effect of particle size. Pharm. Res. 13 (12), 1838–1845.

Desai, M.P., Labhasetwar, V., Walter, E., Levy, R.J., Amidon, G.L., 1997. The mechanism of uptake of biodegradable microparticles in Caco-2 cells is size dependent. Pharm. Res. 14 (11), 1568–1573.

Espuelas, M.S., Legrand, P., Loiseau, P.M., Bories, C., Barate, G., Irache, J.M., 2007. In vitro antileishmanial activity of amphotericin B loaded in poly(epsilon-caprolactone) nanospheres. J. Drug Target. 10 (8), 593–599.

Fessi, H., Puisieux, F., Devissaguet, F., Ammoury, J., Benita, S., 1989. Nanocapsule formation by interfacial polymer deposition following solvent displacement. Int. J. Pharm. 55 (1), R1–R4.

Fonseca, C., Simoes, S., Gaspar, R., 2002. Paclitaxel loaded PLGA nanoparticles: preparation, physicochemical characterization and in vitro anti-tumoral activity. J. Control. Release 83 (2), 273–286.

Furst, E.M., Suzuki, C., Fermigier, M., Gast, A.P., 1998. Permanently linked monodisperse paramagnetic chains. Langmuir 14 (26), 7334–7336.

Ganachaud, F., Katz, J.L., 2005. Nanoparticles and nanocapsules created using the ouzo effect: spontaneous emulsification as an alternative to ultrasonic and high-shear devices. Chem. Phys. Chem. 6 (2), 209–216.

Gao, H., Yang, Y.N., Feng, G., Ma, J.B., 2005. Synthesis of a biodegradable tadpole-shaped polymer via the coupling reaction of polylactide onto mono (6-(2-aminoethyl)amino-6- deoxy)-beta-cyclodextrin and its properties as the new carrier of protein delivery system. J. Control. Release 107 (1), 158–173.

Goodall, A.R., Wilkinson, M.C., Hearn, J., 1977. On mechanism of emulsion polymerization of styrene in soap-free systems. J. Polym. Sci. Polym. Chem. Ed. 15 (9), 2193–2218.

Gothwal, A., Khan, I., Gupta, U., 2015. Polymeric micelles: recent advancements in the delivery of anticancer drugs. Pharm. Res. 33 (1), 18–39.

Govender, T., Stolnik, S., Garnett, M.C., Illum, L., Davis, S.S., 1999. PLGA nanoparticles prepared by nanoprecipitation: drug loading and release studies of a water soluble drug. J. Control. Release 57 (2), 171–185.

Govender, T., Riley, T., Ehtezazi, T., Garnett, M.C., Stolnik, S., Illum, L., et al., 2000. Defining the drug incorporation properties of PLA-PEG nanoparticles. Int. J. Pharm. 199 (1), 95–110.

Gulyaev, A.E., Gelperina, S.E., Skidan, I.N., Antropov, A.S., Kivman, G.Y., Kreuter, J., 1999. Significant transport of doxorubicin into the brain with polysorbate 80-coated nanoparticles. Pharm. Res. 16 (10), 1564–1569.

Ham, H.T., Choi, Y.S., Chee, M.G., Chung, I.J., 2006. Single wall carbon nanotubes covered with polystyrene nanoparticles by in-situ miniemulsion polymerization. J. Polym. Sci. Part. Polym. Chem. 44, 573–584.

Hans, M.L., Lowman, A.M., 2002. Biodegradable nanoparticles for drug delivery and targeting. Curr. Opin. Solid State Mater. Sci. 6 (4), 319–327.

Harries, M., Ellis, P., Harper, P., 2005. Nanoparticle albumin-bound paclitaxel for metastatic breast cancer. J. Clin. Oncol. 23 (31), 7768–7771.

Hearn, J., Wilkinson, M.C., Goodall, A.R., Chainey, M., 1985. Kinetics of the surfactant-free emulsion polymerization of styrene: the post nucleation stage. J. Polym. Sci. Polym. Chem. Ed. 23, 1869–1883.

Higuchi, T., 1961a. Physical chemical analysis of percutaneous absorption process from creams and ointments. J. Soc. Cosmet. Chem. 11, 85–97.

Higuchi, T., 1961b. Rate of release of medicaments from ointment bases containing drugs in suspensions. J. Pharm. Sci. 50, 874–875.

Hirech, K., Payan, S., Carnelle, G., Brujes, L., Legrand, J., 2003. Microencapsulation of an insecticide by interfacial polymerization. Powder. Technol. 130 (1–3), 324–330.

Hixson, A.W., Crowell, J.H., 1931. Dependence of reaction velocity upon surface and agitation. Ind. Eng. Chem. 23 (10), 923–931.

Hrkach, J., Hoff, V.D., Ali, M.M., 2012. Preclinical development and clinical translation of a PSMA-targeted docetaxel nanoparticle with a differentiated pharmacological profile. Sci. Transl. Med. 4 (128), 128–139.

Ikezoe, T., Yang, Y., Bandobashi, K., Saitto, T., Takemoto, S., Machida, H., et al., 2005. Oridonin, a diterpenoid purified from *Rabdosia rubescens*, inhibits the proliferation of cells from lymphoid malignancies in association with blockade of the NF-kappa B signal pathways. Mol. Cancer Ther. 4 (4), 578–586.

Jang, J., Bae, J., Ko, S., 2005. Synthesis and curing of poly(glycidyl methacrylate) nanoparticles. J. Polym. Sci. Part. Polym. Chem. 43 (11), 2258–2265.

Jean-Christophe, L., Allémann, E., Jaeghere, F.D., Doelker, E., Gurny, R., 1996. Biodegradable nanoparticles—from sustained release formulations to improved site specific drug delivery. J. Control. Release 39 (2–3), 339–350.

Karode, S.K., Kulkarni, S.S., Suresh, A.K., Mashelkar, R.A., 1998. New insights into kinetics and thermodynamics of interfacial polymerization. Chem. Eng. Sci. 53 (15), 2649–2663.

Kaur, A., Jain, S., Tiwary, A.K., 2008. Mannan-coated gelatin nanoparticles for sustained and targeted delivery of didanosine: in vitro and in vivo evaluation. Acta Pharm. 58 (1), 61–74.

Khan, S.A., Gunther, A., Schmidt, M.A., Jensen, K.F., 2004. Microfluidic synthesis of colloidal silica. Langmuir 20 (20), 8604.

Kim, S.Y., Lee, Y.M., 2001. Taxol-loaded block copolymer nanospheres composed of methoxy poly(ethylene glycol) and poly(epsilon-caprolactone) as novel anticancer drug carriers. Biomaterials 22 (13), 1697–1704.

Korsmeyer, R.W., Gurny, R., Doelker, E., Buri, P., Peppas, N.A., 1983. Mechanisms of solute release from porous hydrophilic polymers. Int. J. Pharm. 15 (1), 25–35.

Kreuter, J., 1982. The mechanism of termination in heterogeneous polymerization. J. Polym. Sci. Polym. Lett. Ed. 20, 543–545.

Kreuter, J., 1994. Nanoparticles. In: Kreuter, J. (Ed.), Colloidal Drug Delivery Systems. Marcel Dekker, New York, NY, pp. 219–342.

Kreuter, J., Ramge, P., Petrov, V., Hamm, S., Gelperina, S.E., Engelhardt, B., et al., 2003. Direct evidence that polysorbate-80-coated poly(butylcyanoacrylate) nanoparticles deliver drugs to the CNS via specific mechanisms requiring prior binding of drug to the nanoparticles. Pharm. Res. 20 (3), 409–416.

Kroll, R.A., Pagel, M.A., Muldoon, L.L., Roman-Goldstein, S., Fiamengo, S.A., Neuwelt, E.A., 1998. Improving drug delivery to intracerebral tumor and surrounding brain in a rodent model: a comparison of osmotic versus bradykinin modification of the blood-brain and/or blood-tumor barriers. Neurosurgery 43 (4), 879–886.

Kumar, P.S., Ramakrishna, S., Saini, R., Diwan, P.V.T., 2006. Influence of microencapsulation method and peptide loading on formulation of poly(lactide-co-glycolide) insulin nanoparticles. Pharmazie 61 (7), 613–617.

Landfester, K., 2001. Polyreactions in miniemulsions. Macromol. Rapid Commun. 22 (12), 896–936.

Lapresta-Fernández, A., Cywinski, P.J., Moro, A.J., Mohr, G.J., 2009. Fluorescent polyacrylamide nanoparticles for naproxen recognition. Anal. Bioanal. Chem. 395 (6), 1821–1830.

Lemarchand, C., Gref, R., Lesieur, S., Hommel, H., Vacher, B., Besheer, A., et al., 2005. Physico-chemical characterization of polysaccharide coated nanoparticles. J. Control. Release 108 (1), 97–111.

Li, Y., Duc, H.L.H., Tyler, B., Williams, T., Tupper, M., Langer, R., et al., 2005. In vivo delivery of BCNU from a MEMS device to a tumor model. J. Control. Release 106 (1–2), 138–145.

Maeda, H., 2001. The enhanced permeability and retention (EPR) effect in tumor vasculature: the key role of tumor-selective macromolecular drug targeting. Adv. Enzyme Regul. 41, 189–207.

Magenheim, B., Levy, M.Y., Benita, S., 1993. A new in vitro technique for the evaluation of drug release profile from colloidal carriers—ultrafiltration technique at low pressure. Int. J. Pharm. 94 (1–3), 115–123.

Makadia, H.K., Siegel, S.J., 2011. Poly Lactic-co-Glycolic acid (PLGA) as biodegradable controlled drug delivery carrier. Polymers 3 (3), 1377–1397.

Maloney, J.M., Uhland, S.A., Polito, B.F., Sheppard, N.F., Pelta, C.M., Santini, J.T.J., 2005. Electrothermally activated microchips for implantable drug delivery and biosensing. J. Control. Release 109 (1–3), 244–255.

Masayuki, Y., 2005. Drug targeting with nano-sized carrier systems. J. Artif. Organs 8 (2), 77–84.

Matsumura, Y., Maeda, H., 1986. A new concept for macromolecular therapeutics in cancer chemotherapy: mechanism of tumoritropic accumulation of proteins and the antitumor agent smancs. Cancer Res. 46, 6387–6392.

Mirakabad, F.S.T., Nejati-Koshki, K., Akbarzadeh, A., Yamchi, M.R., Milani, M., Zarghami, N., et al., 2014. PLGA-based nanoparticles as cancer drug delivery systems. Asian Pac. J. Can. Prev. 15 (2), 517–535.

Moghimi, S.M., Hunter, A.C., Murray, J.C., 2001. Long circulating and target specific nanoparticles: theory to practice. Pharmacol. Rev. 53 (2), 283–318.

Mu, L., Feng, S.S., 2003. A novel controlled release formulation for the anticancer drug paclitaxel (Taxol): PLGA nanoparticles containing vitamin E TPGS. J. Control. Release 86 (1), 33–48.

Muller, R.H., Maassen, S., Weyhers, H., Mehnert, W., 1996. Phagocytic uptake and cytotoxicity of solid lipid nanoparticles (SLN) sterically stabilized with poloxamine 908 and poloxamer 407. J. Drug Target. 4 (3), 161–170. [PubMed: 8959488].

Munoz-Bonilla, A., Herk, A.M.V., Heuts, J.P.A., 2010. Preparation of hairy particles and antifouling films using brush-type amphiphilic block copolymer surfactants in emulsion polymerization. Macromolecules 43, 2721–2731.

Northfelt, D.W., Dezube, B.J., Thommes, J.A., Miler, B.J., Fiscer, M., Kien, A., et al., 1998. PEGylated-liposomal doxorubicin versus doxorubicin, bleomycin and vincristine in the treatment of AIDS-related Kaposi's sarcoma: results of a randomized phase III clinical trial. J. Clin. Oncol. 16 (7), 2445–2451.

Olivier, J.C., 2005. Drug transport to brain with targeted nanoparticles. NeuroRx 2 (1), 108–119.

Oppenhiem, R.C., 1981. Paclitaxel loaded gelatin nanoparticles for intravesical bladder cancer therapy. Int. J. Pharm. 8, 217.

Owens, D.E., Peppas, N.A., 2006. Opsonization, biodistribution, and pharmacokinetics of polymeric nanoparticles. Int. J. Pharm. 307 (1), 93–102.

Panyam, J., Labhasetwar, V., 2003. Biodegradable nanoparticles for drug and gene delivery to cells and tissue. Adv. Drug. Deliv. Rev 55 (3), 329–347.

Panyam, J., Dali, M.M., Sahoo, S.K., Ma, W., Chakravarthi, S.S., Amidon, G.L., et al., 2003. Polymer degradation and in vitro release of a model protein from poly(D,L-lactide-co-glycolide) nano- and micropar-ticles. J. Control. Release 92 (1–2), 173–187.

Panyam, J., Williams, D., Dash, A., Leslie-Pelecky, D., Labhasetwar, V., 2004. Solid-state solubility influences encapsulation and release of hydrophobic drugs from PLGA/PLA nanoparticles. J. Pharm. Sci. 93 (7), 1804–18014.

Peer, D., Karp, J.M., Hong, S., Farokhzad, O.C., Margalit, R., Langer, R., 2007. Nanocarriers as an emerging platform for cancer therapy. Nat. Nanotechnol. 2 (12), 751–760.

Peracchia, M., Gref, R., Minamitake, Y., Domb, A., Lotan, N., Langer, R., 1997. PEG-coated nanospheres from amphiphilic diblock and multiblock copolymers: investigation of their drug encapsulation and release characteristics. J. Control. Release 46 (3), 223–231.

Pitt, C.G., Gratzi, M.M., Jeffcot, A.R., Zweidinger, R., Schindler, A., 1979. Sustained release drug delivery systems II: factors affecting release rate for poly(e-caprolactone) and related biodegradable polyesters. J. Pharm. Sci. 68 (12), 1534–1538.

Puig, J.E., 1996. Microemulsion polymerization (oil-in water) In: Salamone, J.C. (Ed.), Polymeric Materials Encyclopedia, vol. 6. CRC Press, Boca Raton, FL, pp. 4333–4341.

Rao, J.P., Geckeler, K.E., 2011. Polymer nanoparticles: preparation techniques and size-control Parameters. Prog. Polym. Sci. 36 (7), 887–913.

Reddy, K.R., Sin, B.C., Yoo, C.H., Sohn, D., Lee, Y., 2009. Coating of multiwalled carbon nanotubes with polymer nanospheres through microemulsion polymerization. J. Colloid Interface Sci. 340 (2), 160–165.

Redhead, H.M., Davis, S.S., Illum, L., 2001. Drug delivery in poly(lactide-co-glycolide) nanoparticles surface modified with poloxamer 407 and poloxamine 908: in vitro characterisation and in vivo evaluation. J. Control. Release 70 (3), 353–363.

Reis, C.P., Neufeld, R.J., Reberio, A.J., Veiga, F., 2006. Nanoencapsulation I. Methods for preparation of drug loaded polymeric nanoparticles. Nanomedicine 2 (1), 8–21.

Rieux, A.D., Garinot, M., Fievez, M., Schneider, Y.J., Preit, V., et al., 2006. Nanoparticles as potential oral delivery systems of proteins and vaccines: a mechanistic approach. J. Control. Release 116 (1), 1–27.

Ringe, K., Walz, C., Sabel, B., 2004. Nanoparticle drug delivery to the brain In: Nalwa, H.S. (Ed.), Encyclopedia of Nanoscience and Nanotechnology, 7. American Scientific Publishers, New York, NY.

Rosenberg, B., 1985. Fundamental studies with cisplatin. Cancer 55 (10), 2303–2306.

Sahana, D.K., Mitaal, G., Bhardwaj, V., Kumar, R.M.N.V., 2008. PLGA nanoparticles for oral delivery of hydrophobic drugs: influence of organic solvent on nanoparticle formation and release behavior in vitro and in vivo using estradiol as a model drug. J. Pharm. Sci. 97 (4), 1530–1554.

Sarmento, B., Ribeiro, A., Veiga, F., Sampaio, P., Neufeld, R., Ferreira, D., 2007. Alginate/chitosan nanoparticles are effective for oral insulin delivery. Pharm. Res. 24 (12), 2198–2206.

Sengupta, S., Eavarone, D., Capila, I., Zhao, G., Watson, N., Kiziltepe, T., et al., 2005. Temporal targeting of tumour cells and neovasculature with a nanoscale delivery system. Nature 436 (7050), 568–572.

Shenoy, D.B., Amiji, M.M., 2005. Poly(ethylene oxide)-modified poly(epsiloncaprolectone) nanoparticles for targeted delivery of tamoxifen in breast cancer. Int. J. Pharm. 293 (1–2), 261–270.

Siepmann, J., Siepmann, F., 2008. Mathematical modeling of drug delivery. Int. J. Pharm. 364 (2), 328–343.

Singh, H., Laibinis, P.E., Hatton, T.A., 2005. Rigid, super-paramagnetic chains of permanently linked beads coated with magnetic nanoparticles. Synthesis and rotational dynamics under applied magnetic fields. Langmuir 21 (24), 11500–11509.

Song, X.R., Cai, Z., Zheng, Y., He, G., Cui, F.Y., Gong, D.Q., et al., 2009. Reversion of multidrug resistance by co-encapsulation of vincristine and verapamil in PLGA nanoparticles. Eur. J. Pharm. Sci. 37 (3-4), 300–305.

Song, Z., Poehlein, G.W., 1990. Kinetics of emulsifier-free emulsion polymerization of styrene. J. Polym. Sci. Part. Polym. Chem. 28, 2359–2392.

Soppimath, K.S., Aminabhavi, T.M., Kulkarni, A.R., Rudzinski, W.E., 2001. Biodegradable polymeric nanoparticles as drug delivery devices. J. Control. Release 70 (1–2), 1–20.

Sree, U., Yamamoto, Y., Deore, B., Shiigi, H., Nagaoka, T., 2002. Characterisation of polypyrrole nano-films for membrane-based sensors. Synth. Met. 131 (1–3), 161–165.

Storma, G., Belliota, S.O., Lasicc, D.D., Daeman, T., 1995. Surface modification of nanoparticles to oppose uptake by the mononuclear phagocyte system. Adv. Drug Deliv. Rev. 17 (1), 31–48.

Tao, S.L., Desai, T.A., 2005. Micromachined devices: the impact of controlled geometry from cell-targeting to bioavailability. J. Control. Release 109 (1–3), 127–138.

Thickett, S.C., Gilbert, R.G., 2007. Emulsion polymerization: state of the art in kinetics and mechanisms. Polymer 48 (24), 6965–6991.

Tiyaboonchai, W., 2003. Chitosan nanoparticles: a promising system for drug delivery. Naresuan Univ. J. 11 (3), 51–66.

Tobio, M., Sánchez, A., Vila, A., Sorano, I., Evora, C., Vila-Jato, J.L., et al., 2000. The role of PEG on the stability in digestive fluids and in vivo fate of PEG–PLA nanoparticles following oral administration. Colloids Surf. B Biointerfaces 18 (3–4), 15–323.

Torini, L., Argillier, J.F., Zydowicz, N., 2005. Interfacial polycondensation encapsulation in miniemulsion. Macromolecules 38 (8), 3225–3236.

Valadi, H., Ekstrom, K., Bossios, A., Sjostrand, M., Lee, J.J., Lotval, J.O., 2007. Exosome-mediated transfer of mRNAs and microRNAs is a novel mechanism of genetic exchange between cells. Nat. Cell Biol. 9 (6), 654–U672. [PubMed: 17486113].

Volz, H.P., Moller, H.J., Gerebtzoff, A., Bischoff, S., 2002. Savoxepine versus haloperidol. Reasons for a failed controlled clinical trial in patients with an acute episode of schizophrenia. Eur. Arch. Psychiatry Clin. Neurosci. 252 (2), 76–80.

Wang, Y.M., Sato, H., Adachi, I., Horikoshi, I., 1996. Preparation and characterization of poly(lactic-co-glycolic acid) microspheres for targeted delivery of a novel anticancer agent, taxol. Chem. Pharm. Bull. (Tokyo) 44 (10), 1935–1940.

Xing, J., Zhang, D., Tan, T., 2007. Studies on the oridonin-loaded poly(D,L-lactic acid) nanoparticles in vitro and in vivo. Int. J. Biol. Macromol. 40 (2), 153–158.

Yih-Her, C., Yu-Der, L., Karlsson, O.J., Sundberg, D.C., 2001. Particle nucleation mechanism for the emulsion polymerization of styrene with a novel polyester emulsifier. J. Appl. Polym. Sci. 82 (5), 1061–1070.

Zambaux, M.F., Bonneaux, F., Gref, R., Dellacherie, E., Vigneron, C., 1999. Preparation and characterization of protein C-loaded PLA nanoparticles. J. Control. Release 60 (2–3), 179–188.

Zauner, W., Farrow, N.A., Haines, A.M., 2001. In vitro uptake of polystyrene microspheres: effect of particle size, cell line and cell density. J. Control. Release 71 (1), 39–51.

Zhang, G., Niu, A., Peng, S., Jiang, M., Tu, Y., Li, M., et al., 2001. Formation of novel polymeric nanoparticles. Acc. Chem. Res. 34 (3), 249–256.

Zhang, J., Cao, Y., He, Y., 2004. Ultrasonically irradiated emulsion polymerization of styrene in the presence of a polymeric surfactant. J. Appl. Polym. Sci. 94, 763–768.

Zhang, L., Radovic-Moreno, A.F., Alexis, F., Gu, F.X., Basto, P.A., Bagalkot, V., et al., 2007. Co-delivery of hydrophobic and hydrophilic drugs from nanoparticle-aptamer bioconjugates. Chem. Med. Chem. 2 (9), 1268–1271.

Ziegler, A., Landfester, K., Musyanovych, A., 2009. Synthesis of phosphonate functionalized polystyrene and poly(methyl methacrylate) particles and their kinetic behavior in miniemulsion polymerization. Colloid Polym. Sci. 287 (11), 1261–1271.

CHAPTER 8

Solid Lipid Nanoparticles for Targeting and Delivery of Drugs and Genes

Rakesh K. Tekade, Rahul Maheshwari, Muktika Tekade and Mahavir B. Chougule

Contents

Nanotechnology-Based Approaches for Targeting and Delivery of Drugs and Genes.
DOI: http://dx.doi.org/10.1016/B978-0-12-809717-5.00010-5

8.1 INTRODUCTION

Over the last two decades in particular, several colloidal carriers have been investigated, resulting in more than 32,000 articles in ScienceDirect (Gorain et al., 2016; Dua et al., 2016). They mainly involved liposomes (Maheshwari et al., 2012; Maheshwari et al., 2015a), niosomes (Maheshwari et al., 2015b), dendrimers (Tekade and Chougule, 2013), nanoparticles (Soni et al., 2016; Sharma et al., 2015), and many other vectors (Mody et al., 2014; Choudhury et al., 2016). These formulations suffer from low physical stability, opsonization, and low entrapment efficiency (EE), which further limit clinical availability and commercialization of these formulations (Kesharwani et al., 2015b). Various methods of determination for these carrier systems were also developed (Tekade et al., 2013). In 1990, an alternate to these delivery systems, the solid lipid (solid at physiological temperature) based delivery system was developed. Solid lipid nanoparticles (SLNs) (as depicted in Fig. 8.1) were introduced, which in addition possess potential for drug targeting and controlled release with promising long-term stability (Gajbhiye et al., 2007, 2009).

SLNs are a safe and flexible carrier for the delivery of drug, gene, and nucleic acid and safe for particular administration routes (Muller et al., 2011). New technologies have been developed for SLNs production, and are currently under investigation to obtain the optimum encapsulation of different drug categories and to deliver the bioactive within the desired site (Beloqui et al., 2014).

This chapter presents state-of-the-art information related to various aspects of SLNs viz., SLNs morphology, its structural characteristics, methods of production, ingredients used in the preparation, characterization, incurred limitations and approaches to overcome these limitations. Understanding of biopharmaceutics and pharmacokinetics of SLNs-based formulations is imperative, hence the formulations are included. The chapter also elaborates on the advanced SLNs formulations like semisolid formulations along with their prospective administration routes. We also discuss the medical applications and market scenario including the patents to exploit these SLNs in the pharmaceutical and biomedical market due to nontoxic and long-term stability characteristics. A conclusive discussion on prospective scope and future applications of SLNs in the pharmaceutical and biomedical field is also included.

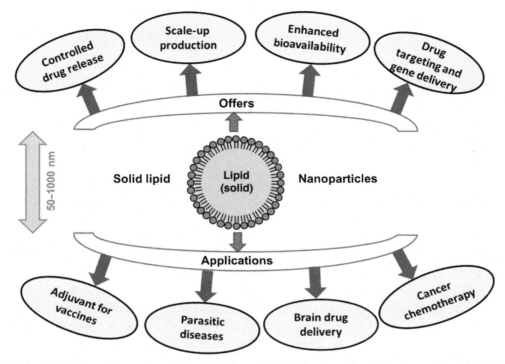

Figure 8.1 Schematic presentation of different advantages and applications of solid lipid nanoparticles.

Before exploring the technical components on SLNs, we anticipate a refresher on historical aspects for better understanding of the subject matter. Therefore, the following section of the chapter presents a brief overview on the historical outlook of various lipid–based nanocarriers.

8.1.1 Historical background of lipid carriers

Being the major component of cellular membrane and having an important role in formation of lipid bilayers, phospholipid as a class of lipid molecules has been extensively studied in the last few years (Muller et al., 2011). When we talk about lipid and lipid-based drug delivery systems, phospholipid is a class of choice due to the variety of properties they possess like multifunctionality, amphiphilic nature, and biocompatibility (Maheshwari et al., 2012).

Lipid-based delivery systems include microemulsion, microspheres, vesicle-based and SLNs. Microemulsion was first described by Hoar and Schulman (1943) as transparent water in oil dispersion and renamed as microemulsion by Schulman and coworkers (1959). Many authors have reported microemulsion as a carrier for enhancement of drug solubility/dissolution and absorption via oral route (Sahu et al., 2015). Domb and Maninar (1990) filed a patent on (phospholipid based drug delivery system in which drug is entrapped inside phospholipid core and the surface of which is covered with surfactant layer having diameter between 1 and 250 μm. The system enables delivery of poorly water soluble drugs due to high dispersibility in hydrophilic systems and controlled particle diameter. Later on lipid implants (1998) were introduced for controlled drug delivery system (Allababidi and Shah, 1998).

To overcome the problems (established production method, percent entrapment efficiency (%EE), scale-up manufacturing) associated with liposomes, lipospheres, and microemulsions, the SLNs delivery system has emerged (Lim et al., 2012; Dhakad et al., 2013). SLNs are not just the frozen emulsion or destabilized vesicle system, rather the result of high-pressure homogenization. They combined many advantages of liposomes, lipospheres and at the same time avoid many drawbacks of these carrier systems. Surface characteristics of SLNs provide the facility of surface modification to enable delivery of gene and targeted chemotherapeutics. Various aspects of SLNs surface characteristics and structures are discussed in Section 8.2.

8.2 MORPHOLOGY AND STRUCTURE OF SOLID LIPID NANOPARTICLES

SLPs in general, found spherical in shape with smooth surface and mean diameter between 50 and 1000 nm. Physicochemical characteristics of SLNs surface also affect their behavior both in vivo and in vitro. The major components of typical SLNs formulation includes lipid that is solid at room temperature, emulsifiers and sometimes a combination of these two; active pharmaceutical ingredient(s) (API) and suitable solvent system for solubilizing the lipid and nonlipid phase (Fig. 8.2).

Commonly, the lipids generally regarded as safe are considered while formulating SLNs. These include fatty acids/esters/alcohols, triglycerides (e.g., campritol 888 ATO, dynasan 112, caprylic capric), waxes (carnauba, bees), and cholesterol and cholesterol butyrate. Generally used emulsifiers are pluronic F-68, F127, Sodium Cholate and poloxamer-188. Sometimes organic salts, ionic polymers, and surface modifiers are used to solve the specific purpose (like surface functionality, targeting, PEGylation). Based on the position of drug molecule inside the structure, SLNs may be classified into three different models, viz.: (1) API loaded shell model; (2) API loaded core model; and (3) homogeneous matrix model.

The API loaded shell model firmed when the hot liquid droplets cooled promptly as a result of phase separation. Liquid precipitation mechanism is the leading phenomenon that occurs during rapid cooling and SLNs firms upon complete cooling. This model is suitable for topical skin treatment. In the API loaded core model, the

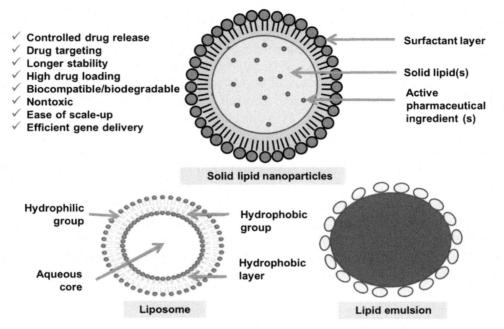

Figure 8.2 Schematic presentation of complete structure of solid lipid nanoparticles with advantages over liposomes and lipid emulsion.

reverse of the recrystalline mechanism as we saw above is applicable here and the API should crystallize prior to the liquid, which is mandatory. Opposite to the shell model, this model is meant for the sustained release of drug and applies Fick's law of diffusion. On the other hand, the homogeneous matrix model is for highly lipophilic drugs that dispersed homogeneously inside the lipid matrix and uses cold homogenization as preparation methodology. Jenning et al. (2000) developed this type of model using prednisolone as the drug. Apart from structural benefits, SLNs also provides ease in production and can be prepared using a simple homogenization technique, which is also the general method for the production of emulsion and microspheres. A detailed insight upon different production techniques along with generally used ingredients are presented in this section.

8.3 INGREDIENTS AND PREPARATION OF SOLID LIPID NANOPARTICLES

8.3.1 General ingredients

SLNs mainly consist of solid lipid(s), surfactant(s), cosurfactant (if required) along with APIs such as drugs, gene, DNA, nucleotides, plasmid, vaccines, proteins. The lipids employed in the preparation of SLNs are solid at physiological temperature. Depending upon structural multiplicity, lipids employed in the preparation are majorly divided into fatty acids, fatty esters, fatty alcohols, triglycerides, or partial glycerides. Some investigators also used waxes in the production of lipidic nanoparticles (Madureira et al., 2015). An exhaustive list of ingredients commonly used in the production of SLNs is given in Table 8.1.

Table 8.1 Lipids, surfactants, and excipients used in the production of solid lipid nanoparticles

Ingredients	Examples	Reference
Lipids/lipid matrices	Behenic acid; Caprylic/capric triglyceride (Miglyol 812); Cetlypalmitate; Cholesterol; Dynsan 112 (Glyceryltrilaurate); Dynsan 114 (Glyceryltrimyristate); Dynsan 116 (Glyceryltripalmitate); Dynsan 118 (Glyceryltristearate); Witeposol bases; Imwitot 900 (Glycerylmonostearate); Compritol 888 ATO (Glycerylbehenate); Precirol ATO 5 (Glycerylpalmitostearate); Hardened fat (Witepsol E 85) Witepsol E 85/cetyl alcohol (75:25); Witepsol H5; Witepsol W 35; Monostearatemonocitrate glycerol (Acidan N12); Monosyeol (Propylene glycol palmitic stearate); Precirol ATO 5 (mono, di, triglycerides of C16–C18 fatty acids); Softisan 142/cetyl alcohol (75:25); Softisan 142; Solid paraffin; Stearic acid; Superpolystate; Synrowax; HRSC (mixture of glycerol tribehenate and calcium behenate)	Kathe et al. (2015)
Emulsifiers and coemulsifiers	Phosphatidyl choline 95% (Epikuron 200); Soy lecithin (Lipoid S 75, Lipoid S 100); Egg lecithin (Lipoid E 80); Poloxamer-188 (Pluronic F-68); Poloxamer 407; Poloxamine 908; Cremophor EL; Solutol HS 15; Lecithin; Tyloxapol; Polysorbate 20; Polysorbate 60; Polysorbate 80; Sodium cholate; Sodium glycocholate; Taurodeoxycholic acid sodium; Butanol and Butyric acid; Cetylpyridinium chloride; Sodium dodecyl sulfate; Sodium oleate; Polyvinyl alcohol	Muller et al. (2011)
Charge modifiers	Stearylamine; Dicetylphosphate; Dipalmitoylphosphatidyl choline (DPPC); Dimyristoylphophatidyl glycerol (DMPG); Disetostyrylphosphatidyl ethanolamine (DSPE)	Tabatt et al. (2004)
Long circulating agents	Polyethyleneglycol; Poloxamer	Kathe et al. (2015)
Preservative	Thiomersal	Kathe et al. (2015)

8.3.2 Preparation of solid lipid nanoparticles

8.3.2.1 High-pressure homogenization

In high-pressure homogenization, liquid or dispersion is forced at high pressure (100–2000 bar) through a gap of few micrometers. A very high shear stress and cavitations forces are responsible for reduced particle size. This process can be operated at hot or cold temperature (Fig. 8.3A), and has emerged as a consistent strategy for the production of SLNs with prospective scale-up capability (Kathe et al., 2015).

8.3.2.1.1 Hot homogenization

In this case, the lipid(s) and drug(s) are melted usually 5–10°C above the transition temperature of the lipid. Typical lipid concentrations are in the range of 5–20%w/v. The aqueous phase containing surface active agent is then incorporated to the lipid phase (at the same temperature) and a hot preemulsion is obtained by high-speed stirring (Mohamed et al., 2013).

8.3.2.1.2 Cold homogenization

In this case, first a suspension is formed by melting lipid(s) and drug(s) together, followed by rapid grinding under liquid nitrogen forming SLNs. The homogenization is carried out with the solid lipid including the drug, i.e., a high-pressure milling of a suspension. The homogenization conditions are generally five cycles at 500 bars. This technique is exclusively employed for temperature-sensitive or hydrophilic drugs (Shah et al., 2011).

8.3.2.2 Solvent emulsification/evaporation technique

This technique is based on precipitation of lipids in o/w emulsions. The lipid phase is dissolved in a water-immiscible organic solvent and is then emulsified in an aqueous phase before evaporation of the solvent under reduced pressure (Fig. 8.3B). The main advantage of this technique is the avoidance of high temperature; therefore, it can be applied with thermosensitive drugs, although the complete removal of residual solvent from final dispersion is difficult due to solubilization of a minute quantity of lipid in the solvent (Negi et al., 2013).

8.3.2.3 Microemulsion technique

This method is based on the dilution of microemulsions containing a molten lipid, surfactant, cosurfactant and water. The hot microemulsion is then dispersed in cold water (2–3°C) under stirring and the excess water removed by ultrafiltration or lyophilization to increase the particles concentration (Fig. 8.3C). Due to the dilution step, the achievable lipid contents are considerably lower when compared to high-pressure homogenization-based formulations. Furthermore, high concentrations of surfactants and co-surfactants are required for the production process, which is undesirable with respect to regulatory requirements (Kathe et al., 2015).

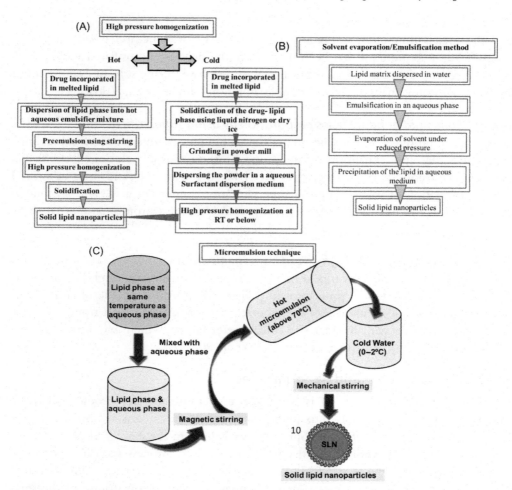

Figure 8.3 (A) Schematic presentation of complete flow diagram of high-pressure homogenization at elevated (high) and low (cold) temperature, (B) solvent evaporation method, and (C) microemulsion technique.

8.3.2.4 Solvent injection technique

This technique is based on lipid precipitation from a dissolved lipid in solution. In this case, a solution of the lipid in a water-miscible solvent is quickly injected into a water phase with/without surfactant. This technique is more advantageous over the other methods since it avoids high-pressure mechanism and uses safe organic solvents. It is also easy to scale-up and a fast tool to produce SLNs avoiding technically sophisticated instruments (Shah et al., 2011).

8.3.2.5 Double emulsion method

For the preparation of hydrophilic loaded SLNs, a novel method based on solvent emulsification–evaporation has been used. Here the drug is encapsulated with a stabilizer to prevent drug partitioning to external water phase during solvent evaporation in the external water phase of w/o/w double emulsion (Muller et al., 2011).

8.3.2.6 Solid lipid nanoparticles preparation by using supercritical fluid

SLNs can be produced employing one of the techniques of supercritical fluid technology, that is the rapid expansion of supercritical carbon dioxide solutions (RESS). The RESS process consists of two steps: (1) dissolving the solid lipid and drug in an SCF, and (2) formation of particles due to supersaturation. Supercritical CO_2 is an innovative solvent as it is safe, inexpensive, readily available, and a promising alternative for many unsafe and toxic solvents (Ciftci and Temelli, 2016).

8.3.2.7 Ultrasonication or high-speed homogenization

In this method lipid phase is mixed in aqueous phase containing a high percentage of surfactant, and subjected to sonication to prepare SLNs. Although the ultrasonication facilities are available in almost all the laboratories, this method is in trouble because it produces particles with broader size (micrometer range), which may further lead to physical instabilities like particle growth (Das et al., 2011).

8.3.3 Drying of solid lipid nanoparticles

Drying of the liquid dispersion is advantageous because the dry product (powders, granules) can be filled suitably into appropriate medium. Drying is also an alternate to lyophilization, to be cost effective. Spray-drying is used in the case of SLNs because it translates a liquid dispersion into a dry product in a single process and can form a fine, dust-free product (Kathe et al., 2015). Generally, the technique uses four different steps: (1) atomization of the feed into a spray, (2) spray-air contact, (3) drying of the spray, and (4) separation of the dried product from the drying gas.

Drying is a suitable method for the formulations oral and topical SLNs, but the formulation should be sterilized when the parenteral administration is required. Therefore, Section 8.3.4 is important, about knowing and understanding the sterilization.

8.3.4 Sterilization and storage of solid lipid nanoparticles

In the case of pulmonary and parenteral administration of SLNs, sterilization is an important step to consider because autoclaving is not possible with SLNs due to its production parameters. For instance, autoclaving (121°C) may affect the release profile of SLNs system due to a high temperature effect. Another way to sterilize SLNs is by filtration. It is highly recommended to filter them in the liquid phase; this permits even particles with a size larger than the pores in the filter to be filtered (El-Salamouni et al., 2015).

8.4 PRODUCTION OF SOLID LIPID NANOPARTICLES-BASED HYBRID PREPARATIONS

8.4.1 Production of semisolid preparations containing solid lipid nanoparticles

The more efficient and rapid one-step process was developed to produce SLNs, especially semisolid formulations. The process involves melting of the lipid followed by dispersing it in hot surfactant dispersion usually above the 10°C of the melting point of lipid using homogenizer at 9500 rpm for 1 min. This process requires 30–50%w/v lipid concentration. This dispersion is then subjected to a high-pressure homogenizer. Three cycles at 500 bar and 85°C are performed. After the completion of one cycle, the former liquid o/w dispersion becomes quite viscous. This viscous dispersion is used for the remaining two cycles. Final viscous hot o/w nanoemulsion is left to cool at room temperature. The lipid droplets recrystallize to solid nanoparticles forming a gel network with a semisolid consistency (Abdel-Salam et al., 2016).

8.5 CHARACTERIZATION OF SOLID LIPID NANOPARTICLES

8.5.1 Particle size and zeta potential (ξ)

Dynamic light scattering (DLS) and laser diffraction are the methods mainly used for particle size and charge analysis. DLS operates at a few nanometers to about 3 μm. A laser diffraction technique is used to analyze quite bigger particles (more than 3 μm) and based on the dependence of the diffraction angle on the particle radius (Chopdey et al., 2015; Chougule et al., 2014). Both the methods detect light scattering effects which are used to calculate particle size.

8.5.2 Electron microscopy

Evaluation of morphology and surface topography is possible with scanning electron microscopy (SEM) and transmission electron microscopy (TEM). These are the advanced technologies that can confirm the formation of SLNs with surface structural properties and size (Ghanghoria et al., 2016; Huang et al., 2015; Gandhi et al., 2014). One can measure the stability of particles using TEM and SEM and analyze the structural changes over time.

8.5.3 Atomic force microscopy

More advanced than TEM and SEM, the atomic force microscopy (AFM) technique utilizes a probe tip with atomic scale sharpness across a sample to produce a topological map based on the forces at play between the tip and the surface. The probe can be dragged across the sample (contact mode), or allowed to hover just above (noncontact mode), with the exact nature of the particular force employed serving to distinguish among the sub techniques. That atomic level resolution is accessible with this method, which along with the ability to map a sample according to properties in addition to size, e.g., colloidal attraction or resistance to deformation, makes AFM an important tool.

8.5.4 Entrapment efficiency

Generally, the entrapment of SLNs was measured using centrifugation or micro-centrifugation techniques. In this technique samples were centrifuged at 10,000–25,000 rpm for 5–30 min. The amount of free compound was determined in the clear supernatant by ultraviolet spectrophotometer or high performance liquid chromatography using supernatant of nonloaded nanoparticles as basic correction (Jain and Tekade 2013; Kayat et al., 2011; Tekade et al., 2009).

8.5.4.1 Factors affecting drug release from the solid lipid nanoparticles

A major restriction factor in the entrapment of hydrophilic drugs into SLNs is their poor loading and retention within the lipid matrix, at the time of preparation. Many investigators point the influence of lipid type (velocity of crystallization, hydrophilicity, and self-emulsifying properties of the lipid) on the particle size of the SLNs.

8.5.5 Polymorphic form and degree of crystallinity

The geometric scattering of radiation from crystal planes within a solid allows the presence or absence of the former to be determined, thus permitting the degree of crystallinity to be assessed. Another method that is a little different from its implementation with bulk materials, DSC can be used to determine the nature and speciation of crystallinity within nanoparticles through the measurement of glass and melting point temperatures and their associated enthalpies.

An array of reports is available that discusses the advantages of SLNs over other nanocarrier systems such as lipid emulsion and liposomes. Only a few reports are available that describe the limitations of this carrier system. The following section is dedicated to the discussion of the limitations of SLNs.

8.6 LIMITATIONS OF SOLID LIPID NANOPARTICLES AND APPROACHES TO OVERCOME

Since SLNs are mostly composed of solid lipid, the possibility of degradation and instability may be an issue and points to consider include high pressure–induced drug degradation, the coexistence of different lipid modifications and different colloidal species, the low drug-loading capacity and the kinetics of distribution processes.

8.6.1 High pressure–induced drug degradation

Molecular weight and structure are the leading cause for the drug degradation, and high-pressure homogenization has been demonstrated to minimize the molecular weight of polymers. High molecular weight composites and long chain atoms are more susceptible than low molecular weight substances with a spherical shape, although many reports reveal that high-pressure homogenization–induced drug degradation is not a problem with the majority of the bioactive (Kathe et al., 2015). However, the high molecular weight compounds like DNA, albumin, dextrose are very susceptible to lysis; therefore, a different method should be adopted to utilize these molecules to incorporate into SLNs.

8.6.2 Lipid crystallization and drug incorporation

Another important point to be considered is lipid crystallization. The relation between lipid modification and drug incorporation has been investigated for decades. The characterization of lipid modifications is well established. Methods are mainly based on X-ray and differential scanning calorimetric measurements. However, most of the data have been extracted from investigations on bulk lipids. The performance of SLNs might differ considerably because of the nanosize of the carrier and the large amount of surface active agents that are required to stabilize the colloidal lipid dispersion. Therefore, lipid crystallization and drug incorporation influence lipid particle characteristics. The major aspects that should be kept in mind in the discussion of drug entrapment inside SLNs are: (1) existence of super-cooled melts; (2) presence of several lipid modifications; (3) shape of lipid nanodispersions; and (4) gelation phenomena.

8.6.3 Coexistence of several colloidal species

Coexistence of many colloidal particles inside SLNs is not explored much in the literatures, although it is an important point to consider. Surfactants are embedded on the lipid surface and core as well. Mixed micelles have to be considered in glycocholate/lecithin stabilized and related systems. Micelles, mixed micelles and liposomes are known to solubilize drugs and are, therefore, alternative drug incorporation sites. Only the detection of the presence of several colloidal species is not sufficient to describe the structure of colloidal lipid dispersions, because dynamic phenomena are very important for drug stability and drug release. Therefore, the kinetics of distribution processes has to be considered. For example, the degradation of hydrolysable drugs will be faster for water solubilized and surface localized molecules compared to molecules in the lipid matrix.

The kinetics of the degradation will be determined by: (1) the chemical reactivity of the drug, and (2) the concentration of the drug in the aqueous medium or at the lipid/ water interface. Unstable drugs will hydrolyze rapidly in contact with water and, therefore, the distribution equilibrium of the drug between the different environments will be distorted. Carrier systems will be protective only if they prevent the redistribution of the drug. Of course, increasing the matrix viscosity will decrease the diffusion coefficient of the drug inside the carrier and, therefore, SLNs are expected to be superior to lipid nanoemulsions.

To develop a successful delivery system, complete clarity in regards to bits in vitro as well as in vivo fate must be described. Pharmacokinetic studies are important to set up clinical investigations on developed formulations. Hence, Section 8.7 discusses various routes of administration and implications of its fate under in vivo environment.

8.7 ROUTE OF ADMINISTRATION, BIOPHARMACEUTIC AND PHARMACOKINETIC OF SOLID LIPID NANOPARTICLES

8.7.1 Oral administration

Oral administration of SLNs is helpful especially in the case of insulin delivery (Dwivedi et al., 2016a). Insulin as a noninvasive therapy for diabetes mellitus is still a challenge. Different formulation approaches have been investigated to overcome the gastrointestinal tract barriers for the delivery of insulin via the oral route, among them SLNs represent a promising approach. Zhang et al. (2012) formulated SLNs tailored with stearic acid–octaarginine employing emulsion solvent diffusion method and evaluated as a vehicle for oral administration of insulin (stearic acid–octaarginine insulin SLNs). Their in vitro results demonstrated that SLNs and stearic acid–octaarginine could partially protect insulin from proteolysis when subjected to a degradation experiment using enzymes. Their in vivo investigations suggested a potential hypoglycemic effect in diabetic rats in comparison to controls. These results display that stearic acid–octaarginine decorated SLNs increase the oral absorption of insulin.

8.7.2 Parenteral administration

The parenteral route is essential in different fatal complications to get prompt relief. SLNs are suitable vehicles for intravenous administration of some drugs that cause pain and inflammation at the site of administration like diazepam, and also for drugs which exhibit toxicity with additives employed to dissolve it such as etoposide. Yang and coworkers designed camptothecin-entrapped SLNs using high-pressure homogenization as a fabrication method and administered them intravenously into mice (Yang et al., 1999). The outcomes revealed that in the organs investigated, the area under curve (AUC)/dose and the mean residence time of camptothecin-entrapped SLNs were greater than those of camptothecin dispersion, especially in brain, heart and in organs containing reticuloendothelial cells. Among the tested organs, AUC ratio of camptothecin-entrapped SLNs to camptothecin solution was highest in the brain.

These results indicate that SLNs are promising drug targeting systems for lipophilic antitumor drugs and may also allow reduction in dose and a decrease in systemic toxicity. Intravenous injection of paclitaxel (PTX)-bearing SLNs led to greater and sustained plasma levels of PTX. Both PTX non-PEGylated and PEGylated SLNs showed a decreased uptake by the liver and spleen macrophages and higher uptake in the brain. In another study, doxorubicin (DOX) PEGylated and non-PEGylated SLNs give greater concentrations of (DOX) in the brain of rats and rabbits investigated, following intravenous injection. The amount of DOX in the brain increased as the concentration of the

PEGylating agent increased. Cardiotoxicity of SLNs was lower when compared with that of DOX solution.

8.7.3 Topical and transdermal applications

Due to the nanodimensional structure, SLNs are able to cross stratum corneum and thereby uplift the penetration of entrapped drug into the viable skin. Sustained release of the drug from SLNs supplies the drug to the skin over a prolonged period and thereby reduces systemic absorption. SLNs showed occlusive properties as a result of film formation on the skin, which reduces transdermal water loss. Increase of water content in the skin reduces the symptoms of atopic eczema and also improves the appearance of healthy human skin. Occlusion also favors the drug penetration into the skin. After applying the topical dosage form to the skin, water evaporates from the preparation and SLNs are exposed to electrolytes present in the surface of skin. These factors induce polymorphic transitions and as a result, drug expulsion takes place. Controlled release of drug from SLNs can be obtained by controlling polymorphic transitions, using surfactant mixtures.

Upon application, SLNs transform slowly to the stable polymorph and sustain release is expected. Application of a conventional o/w cream did not change skin structure, whereas application of SLNs containing cream increased the thickness of the stratum corneum, which in turn improves the penetration of the drug. Therefore, penetration characteristics of the drug can be altered by SLNs.

8.7.4 Ocular delivery

Due to the complex anatomy of the eye, ocular administration remains challenging. Urgent attention to develop a capable delivery tool to enhance ocular bioavailability and decrease both local and systemic cytotoxicity is required. Several nano-based formulation approaches have successfully been applied to overcome ocular barriers (Maheshwari et al., 2015b). Due to their nontoxic character, SLNs are thought to be a potential delivery system since the discovery. They can be employed in ocular delivery of drugs to increase the corneal absorption and enhance the bioavailability of the drug inside the eye.

8.7.5 Rectal delivery

Rectal delivery of the drug is the preferred route of administration in pediatric patients due to ease of application. Conventional rectal formulations may contain organic solvents such as ethanol, acetone, benzyl alcohol and propylene glycol, which may be toxic. On other hand, SLNs are made of physiological lipids and therefore safe as potential delivery system for the rectal administration of drugs. Also, lipid matrix that is solid at body temperature is not an advantageous system for rectal delivery of drugs, even if delivered as submicron dispersions. Thus, low melting point lipids must be selected for formulation of rectal delivery dispersions to achieve prolonged release of the drug as well as higher absorption and bioavailability. Recently, Mohamed et al. developed metoclopramide SLNs for rectal administration. Their finding suggests 80% drug release in vitro, when Poloxamer 407 is used as surfactant. The formulation is also shown to be the best as the same % gastric emptying was found when compared to marketed metoclopramide suppository with sustained-release effect (Mohamed et al., 2013).

SLNs are rapidly developing an era of nanomaterials with huge applications in drug delivery, clinical medicine and research as well as in other varied arenas of biological and biomedical sciences. There are several potential applications of SLNs, and some key points are discussed in Section 8.8.

8.8 MEDICAL APPLICATIONS OF SOLID LIPID NANOPARTICLES

Due to the versatile properties and flexible surface morphology, SLNs have been explored widely for vivid medical applications (Table 8.2). Targeting is an important aspect and specially required for the delivery of anticancer molecules; due to surface modification flexibility, SLNs can be targeted using suitable targeting legend. SLNs can be easily prepared using inexpensive, safe, stable, and biodegradable materials, using the cheaper techniques like high-pressure homogenization, and can be loaded internally or externally with APIs for controlled delivery. The controlled-release mechanism of SLNs, in turn, offers numerous clinical benefits for the treatment and management of a variety of clinical symptoms with administration of a fewer number of doses.

8.8.1 Cancer chemotherapy

The fight against cancer has always been a priority for the scientific community involved in medical research (Tekade et al., 2015c; Youngren et al., 2013). Anticancer molecules have a profound effect on cancerous cells as well as normal cells (Mansuri et al., 2016; Kurmi et al., 2010; Tekade et al., 2008a,b). Many times, combined chemotherapy is desired to treat cancer. Therefore, an urgent delivery/carrier option is required to deliver the drug to the desired location to avoid toxic effects on normal cells. Since the early 1990s, a number of SLNs or SLNs-based systems for the delivery of anticancer drugs have been successfully prepared and evaluated.

Recently, Zhuang et al. developed some promising anticancer drug SLNs for clinical application. They formulated SLNs carrying mitoxantrone, PTX, MTX and investigated their cytotoxic effects on the human MCF-7 breast cancer cell line (Zhuang et al., 2012). Their results indicated that the mean tumor size of mice treated with SLNs was significantly smaller than that with free drug ($P < .05$). Additionally, the percent inhibition of mice treated with SLNs was obviously lower than that with free drug ($P < .05$). Therefore, the conclusion can be drawn that anticancer drugs carried by SLNs, including mitoxantrone, MTX and PTX, may be more effective than free anticancer drugs for breast cancer treatment.

Table 8.2 Solid lipid nanoparticles (SLNs) explored as delivery system for the treatment of various diseases/infections

Drug delivered	Targeted disease	SLNs composition	Reference
Alendronate sodium	Bone diseases	Compritol and precirol	Ezzati Nazhad Dolatabadi et al. (2014)
Lopinavir	HIV infections	Stearic acid	Negi et al. (2013)
Risperidone	Schizophrenia	Compritol 888 ATO	Silva et al. (2012)
Rifampicin	Bacterial infection	Compritol 888 ATO	Singh et al. (2013)
Paclitaxel, doxorubicin	Solid tumors	Glyceryl monostearate	Miao et al. (2013)
Doxorubicin	Cancer	Docosahexaenoic acid	Mussi et al. (2013)
Terbinafine hydrochloride	Skin infections	Compritol and precirol	Vaghasiya et al. (2013)
Aceclofenac	Inflammation in rheumatoid arthritis	Compritol and precirol	Chawla and Saraf (2012)
Insulin	Diabetes mellitus	Phosphatidyl choline	Liu et al. (2008)
Stavudine, delavirdine, and saquinavir	Human immunodeficiency virus (HIV)	Compritol 888 ATO, tripalmitin, and cacao butter	Kuo and Chung (2011)
Norfloxacin	Bacterial infection	Stearic acid	Wang et al. (2012)
Gemcitabine	Solid tumors	4-(N)-stearoyl	Wonganan et al. (2013)
Rifampicin, isoniazid, pyrazinamide	Tuberculosis	Stearic acid	Pandey and Khuller (2006)
Clotrimazole	Skin infections	Glyceryl tripalmitate and tyloxapol	Souto and Muller (2005)
Ketoconazole	Skin infections	Glyceryl behenate and sodium deoxycholate	Souto and Muller (2005)
Econazole nitrate	Skin infections	Glycerol palmitostearate	Sanna et al. (2007)
Ciprofloxacin hydrochloride	Microbial infection	Stearic acid, soy phosphatidylcholine, and sodium taurocholate	Jain (2007)
Tobramycin	Microbial infection	Stearic acid, soy phosphatidylcholine, and sodium taurocholate	Cavalli et al. (2002)

In another study, the cytotoxicity of SLNs formulations carrying cholesteryl butyrate, DOX or PTX was evaluated on the human HT-28 colorectal cancer cell line. The results showed that SLNs of cholesteryl butyrate and DOX exhibited significantly higher cytotoxicity than the equivalent amount of free drug. The 50% inhibitory concentration (i.e., IC_{50}) values for HT-28 cell growth of SLNs drug formulations were both lower than the corresponding conventional drug solutions.

Taveira and coworkers (2012), recently developed cationic SLNs composed of stearic acid and glyceryl behenate 1:2 using Poloxamer and cetylpyridinium chloride as surfactant and cosurfactant, respectively. The optimized formulation elicited greater DOX loading and significant increase in cytotoxicity in B16F10 murine melanoma culture cell line. Methotrexate loaded SLNs, prepared by coacervation, showed an increased cytotoxicity toward MCF-7 and Mat B-III cell lines compared with free drug (Battaglia et al., 2011). The in vivo animal study showed that after intravenous administration, higher blood levels were achieved and major drug accumulation within breast cancer tumor tissue was shown compared with drug solution alone.

8.8.2 Brain targeting

The blood–brain barrier is a distinctive barrier formed by the endothelial cells and presents challenges to brain delivery of drugs, mainly the transport into the brain through both physical barriers, i.e., tight junctions and metabolic barrier, various enzymes (Dwivedi et al., 2016b). SLNs are one of the most recent brain targeting strategies for the delivery of drugs and to overcome the hurdles of blood–brain barrier.

Yusuf et al. recently formulated surface-modified SLNs, prepared using piperine by emulsification–solvent diffusion technique, and Tween 80 as a coating agent (Yusuf et al., 2013). Piperine is a natural alkaloid having a potent antioxidant effect, with potential applications in Alzheimer's disease, since it readily crosses the blood–brain barrier. Due to intense first-pass metabolism, the administration of piperine for brain delivery is not straightforward. Piperine was successfully targeted to the brain, and was found to be effective at a low dose (2 mg/kg body weight) in a Tween 80 solid lipid formulation; this proved to be successful in providing effective delivery across the blood–brain barrier, with a generous payload and good delivery capabilities.

In another study tristearin–tricaprin-based nanostructured lipid carriers were developed as carriers for bromocriptine, used in treating Parkinson's disease (Esposito et al., 2008). Bromocriptine, a dopamine agonist, marketed more than 30 years ago, possesses a slow onset of action (1–2 h) and prolonged half-life (3–5 h), which probably explains the lower dyskinesiogenic potential compared to L-3,4-dihydroxyphenylalanine using longer acting dopamine agonists further reduces dyskinesia compared to L-3,4-dihydroxyphenylalanine. In this context, SLNs encapsulation represents a novel strategy to obtain stable plasma levels and increase the drug's half-life. The nanostructured lipid carrier was formulated and characterized and its potential in the symptomatic control of Parkinson's disease was evaluated in a rat model, after intraperitoneal administration. The results indicated attenuated akinesia in hemi–Parkinsonian rats, which was more evident for the bromocriptine-loaded nanostructured lipid carrier than for the free bromocriptine; this might be due to the different pharmacokinetic profiles of the two formulations.

8.8.3 Pulmonary disorders

Pulmonary delivery is of increasing interest nowadays, not only because of the treatment for airborne diseases, but also for the systemic administration of drugs. To take full advantage of lung delivery, a suitable delivery system with appropriate physico-chemical properties is necessary and SLNs is most fitted on this ground. In a study, Liu et al. (2008), prepared insulin-loaded SLNs by micelle-double emulsion method and investigated the stability of SLNs during nebulization, the deposition properties and the hypoglycemic effect after intrapulmonary application. In comparison with a pulmonary applied insulin phosphate buffer solution, insulin-SLNs showed higher bio-availability of insulin in lungs. Bi et al. (2009) spray freeze dried insulin-loaded SLNs in order to get a dry powder formulation. In vivo studies on diabetic rats showed pro-longed hypoglycemic effect after intra-tracheal instillation, which is in accordance with the study carried out by Liu and coworkers (2008).

8.8.4 Protein and peptide delivery

The first attempts to encapsulate peptide drugs in SLNs were those of Morel et al., who used the warm w/o/w microemulsion-based technique to incorporate [D-Trp-6] LHRH and thymopentin (Morel et al., 1996). Encapsulation efficiencies were gener-ally low even when peptide lipophilicity was increased by forming an ionic pair with a counter ion. Similar results were obtained with cyclosporine A (CyA), a hydropho-bic peptide that was also incorporated using a single (w/o) warm microemulsion. In another attempt to produce insulin-loaded SLNs, Trotta et al. (2005) used an o/w emulsion–diffusion method, obtaining an encapsulation efficiency of about 80%.

In a comprehensive study Saraf et al. (2006) produced positively and negatively charged lipid microparticles by the solvent evaporation technique (w/o/w) containing protein antigens (HBsAg) for intranasal immunization against hepatitis B. The mean par-ticle size obtained ($1.6 \pm 0.08\,\mu m$) was appropriate for mucosal uptake, which was dem-onstrated by ex vivo with alveolar macrophages. When given by intranasal to rats both anionic and cationic microparticles elicited higher specific mucosal antibody responses, when compared with the free and the alum-adsorbed antigen. This response was also higher than that obtained upon i.m. vaccination with the same formulations. The spe-cific anti-HBsAg levels induced via intranasal route by the cationic microparticulate for-mulation were the highest in the nasal, pulmonary, and salivary glands, as determined in nasal washes, in bronchoalveolar lavages and saliva, respectively. Only in serum, the intra-muscularly administered alum-adsorbed antigen resulted in a superior IgG level.

8.8.5 Nucleic acid delivery

The delivery of nucleic acid such as plasmids, antisense oligonucleotides, and small inter-fering RNA (siRNA), for the therapeutic benefits is limited due to the presence of nucleases in blood, cytosol, and lysosomes. Apart from that, various biological barriers also present problems in nucleic acid delivery. The application of SLNs into nucleic acid delivery has not been much explored, although SLNs have shown promising prospects in drug delivery. Tabatt and coworkers (2004) showed that the nature of the cationic lipid determined DNA binding by SLNs. It will be relevant to investigate the effects of the chain length and saturation of the cationic lipid on DNA binding and delivery with SLNs, which will also be influenced by the relative fluidity as well as the biodegradability of the matrix lipid.

Another group of researchers further improved the loading efficiency of siRNA into SLNs by preparing siRNA cationic polymer complexes in oil prior to the mixing with solid lipids, generating siRNA loaded nanostructured lipid carrier (Xue and Wong, 2011). The nanostructured lipid carrier formulation showed high encapsulation efficiency of the siRNA, while the release kinetics could be manipulated by varying the oil content and, to a minor extent, the solid lipid composition. This points to the importance of carrier internal structures of nanostructured lipid carriers in determining its degradation. While the nanostructured lipid carrier appears to be promising for generating sustained siRNA release, its suitability for the prolonged delivery of other types of nucleic acids, including plasmid DNA, remains to be investigated.

8.8.6 Cosmeceuticals

Lipid nanoparticles proved to have a synergistic effect of UV scattering when used as vehicles for molecular sunscreens. Advantages taken from these observations are the possibility to reduce the concentration of the molecular sunscreen, consequently its potential side effects, as well as the costs of formulation of expensive sunscreens. In addition, lipid nanoparticles can be explored to formulate sunscreen products with lower and medium sun protection factors. The loading capacity of lipid nanoparticles depends mainly on the miscibility of the active in the lipid selected for their production.

The first two cosmetic products based on nanostructured lipid carrier technology were introduced to the market by the company Dr. Rimpler GmbH in Wedemark/ Hannover, Germany. The products NanoRepair Q10 cream and NanoRepair Q10 Serum (Dr. Kurt Richter Laboratorien GmbH, Berlin, Germany) were introduced to the cosmetic market in October 2005 revealing the success of lipid nanoparticles in the antiageing field (Muller et al., 2007). The Chemisches Laboratorium Dr. Kurt Richter GmbH (Berlin, Germany) has reached the cosmetic market with nanostructured lipid carrier concentrate formulations (NanoLipid Q10 CLR and NanoLipid Restore CLR (Dr. Kurt Richter Laboratorien GmbH, Berlin, Germany)) during April 2006 in Barcelona. SLNs are a source of promising delivery for the local economy and for good cosmetic products. It is important that the therapeutic ingredient should penetrate the skin deeply, but not so deeply to avoid entry into the systemic circulation.

8.8.7 Drug targeting

Drug targeting is an approach to selectively deliver the therapeutics to a desired site of action, while sparing the normal cells and tissues. This strategy employs nanomedicine to combat the limitations of conventional drug delivery (Tekade et al., 2014a,b;

Prajapati et al., 2009; Kesharwani et al., 2015a,b). These nanomedicines contain therapeutic agents and are targeted to diseased tissue, where higher concentration of the drug is needed. The goal of a targeted delivery approach is to localize, target and sustain drug action with the least interaction of drugs with the nondiseased tissues (Tekade et al., 2015a,b, 2016; Tekade, 2014).

Among the nanoparticulate carriers, SLNs hold great promise for reaching the goal of controlled and site-specific drug delivery, and hence attract wide attention of researchers (Abdel-Salam et al., 2016; Beloqui et al., 2014). All these platforms must be compatible with the physiological environment and prevent undesirable interactions with the immune system. When developing new strategies in drug and gene delivery, avoiding immune stimulation or suppression is an important consideration, whereas in adjutants for vaccine therapies, immune activation is desired (Peer, 2012).

8.8.7.1 Antibody mediated targeted drug delivery using lipidic carriers

Precise targeted drug delivery to specific sites or organs in the body is so important. To achieve this goal, the host immune response to tumor-specific antigens in chemotherapy and other disorders could be a very specific and low toxic option in medicine applications. Thus, antibodies that target tumor-specific and tumor-associated antigens can be applied in targeted drug delivery (Casi and Neri, 2012). Lim and coworkers have conjugated antihuman fibrinogen and intercellular adhesion molecule antibodies onto acoustically reflective liposomes composed of phospholipids and cholesterol. This group investigated that incorporation of drugs into these antibodies conjugated carriers may be a potential future application to actively target drugs to the atherosclerotic plaques (Lim et al., 2012).

Peer et al. described general concepts of the immune system and the interaction of subsets of leukocytes with lipid-based nanoparticles. Furthermore, they also studied the immunological toxicities and suggested the methods to manipulate leukocytes functions using lipid-based nanoparticles (Peer, 2012). In another study, Kuo and Ko (2013) demonstrated the capability of 83-14 monoclonal antibody-grafted SLNs to improve the brain-targeting delivery of saquinavir. This group investigated the endocytosis of 83-14 monoclonal antibody-grafted saquinavir-loaded SLNs into human brain-microvascular endothelial cells by staining cell nuclei, insulin receptors, and drug carriers. This group demonstrated that the increase in the concentration of surface 83-14 monoclonal antibody enhanced the percentage of surface nitrogen permeability across the blood–brain barrier and up-taking by human brain-microvascular endothelial cells as a result of targeting delivery for designed carrier (Kuo and Ko, 2013).

8.8.7.2 Magnetic targeting drug delivery

Using magnetic nanoparticles, either for diagnostic or treatment purposes, was the center of interest during the last two decades. Pang and coworkers (2009) prepared magnetic SLNs containing ibuprofen by an emulsification dispersion ultrasonic method. They demonstrated that changing the SLNs formulation offers a tool to control the magnetic properties and percentage of magnetite-loading for composite lipid particles. Drug % EE was determined to be almost 80% by magnetite-loaded SLNs. The ibuprofen magnetite-loaded SLNs, together with the magnetite nanoparticles, were about 7 nm in diameter, shown to be spherical and uniform by TEM, superconducting quantum interference device results reveals that at blocking temperature of 86K, ibuprofen-magnetite-loaded SLNs behaved like super paramagnetic. Hsu and Su (2008) developed magnetic lipid nanoparticles that could serve as controlled delivery carriers to release encapsulated drugs in a desired manner. The nanoparticles are composed of multiple drugs in lipid matrices, which are solid at body temperature and melt around 45–55°C.

8.8.7.3 pH-sensitive solid lipid nanoparticles

pH-sensitive carriers are another option for targeted drug delivery which were used by Siddiqui and coworker to show that, the combination of docosahexaenoic acid, a long-chain (C_{22}) polyunsaturated fatty acid, and anticancer drugs increases the sensitivity of tumors to chemotherapy. In an investigation, a group of investigators formulated aqueous dispersions of lipid nanoparticles employing a pH-sensitive derivative of phosphatidyl ethanolamine for pH-sensitive nanoparticles preparation. SLNs were prepared using polysorbate 80 as the surfactant and tripalmitin glyceride and N-glutarylphosphatidyl ethanolamine as the lipid phase. Their results revealed that SLNs was capable of controlling the release of triamcinolone acetonide under acidic condition (Kashanian et al., 2011).

8.8.7.4 Cationic solid lipid nanoparticles in drug delivery

Cationic SLNs may make possible the delivery of genetic material in cancer treatment. The same in vivo transfection efficacy were observed in cationic SLNs and liposome formulated with the same cationic lipids. The intrinsic toxicity of the delivery vehicle may also be minimized using a good combination of two-tailed cationic lipids. More recently Carrillo and coworkers used cationic SLNs for brain-targeted DNA delivery (Carrillo et al., 2013). Cationic SLNs have been proven to be effective in transfecting African green monkey kidney fibroblast-like cells (Cos-1) and human broncho-epithelial cells. Heydenreich and coworkers (2003) prepared the SLNs which consisted mainly of stearylamine and different triglycerides. Purification methods such as ultrafiltration, dialysis, and ultracentrifugation were evaluated and compared with the physical stability of the nanoparticles and cellular toxicity. The cellular toxicity was affected by both the purification method and the SLNs composition.

8.8.8 Gene delivery

Successful gene therapy is defined as expression of the therapeutic gene in the target organ or tissue. Viral gene delivery (or viral vector) and nonviral gene delivery (or nonviral vector) are two types of gene therapy techniques. In the first type, the genes are transferred by a virus into the cells because of virus ability in penetrating cells (Singh et al., 2013). Even though high levels of gene expression are reported with viral gene delivery, it can have oncogenic and immunogenic effects, and induce inflammation that render transgene expression transient. Nonviral vectors can overcome some of these concerns; compared to viral vectors they have considerable manufacturing and safety advantages.

A variety of nonviral delivery systems have been demonstrated for efficient delivery of siRNAs and DNA including cationic polymers, liposomes, cationic lipids, cell-penetrating peptide, carbon nanotubes, and so on. SLNs as gene delivery systems have attracted increasing attention in recent years. Cationic SLNs usually have been used for gene delivery due to the possible electrostatic interaction between the negative charges of the DNA and the positive charges of the lipid, which allow the formation of a complex called lipoplex. Carrillo et al. (2013) demonstrated cationic SLNs capable of forming a complex with DNA plasmids.

In another study, Montana et al. (2007) utilized cationically modified SLNs as RNA carriers and evaluated their potential as a nonviral vector for gene delivery. Future gene expression study definitely will open a new horizon for the achievement of rational design approaches in the development of cationic SLNs as effective gene delivery systems due to remarkable capabilities such as penetrating into cells, capability to achieve spatially and temporally controlled release for targeted gene silencing.

Tremendous biomedical success of nanotechnology-based products in the market can be conferred by market availability and upcoming patent scenarios on innovative nanomedicines/nanocosmeceuticals. It may be noted that the commercial feasibility is the most desired characteristic of a novel formulation. A system that has great efficacy but poor commercial feasibility is merely futile. Section 8.9 presents the product commercialization aspects of SLNs as well as patents filed on SLNs-based nanoformulations.

8.9 MARKETED PRODUCTS AND PATENTS

The commercial feasibility of any delivery system is governed by the availability of a large scale production method, yielding a product of a quality that is acceptable by the regulatory authorities (e.g., Food & drug administration) and cost of the material. Considering these facts, the scenario is promising in the case of SLNs. As SLNs require easily available and reasonably priced triglyceride lipids, the material cost is much less than carriers like Poly Lactic Glycolic Acid (PLGA), Poly Lactic Acid (PLA), Polycaprolactone (PCL), or phospholipids.

Moreover, SLNs are mainly manufactured by using high-pressure homogenizers which have been in use for the manufacturing of parenteral nutrition products for many years. They may be used even without any modification (e.g., for production of SLNs by hot homogenization technique). Scale-up and manufacturing are also possible when microemulsions are used as templates for SLNs production. Various patents are available on SLNs and some of them are elaborated in Table 8.3.

Table 8.3 List of patents on SLNs

Patent no.	Title of patent	Inventor/applicant	Filing year	Reference
EP0167825	Lipid nanopellets for oral administration	Speiser Peter	1985	E.P. Patent No. 167825 (1990)
US5250236	Method for producing solid lipid microspheres having a narrow size distribution	Maria R. Gasco	1991	U.S. Patent No. (1993)
US5667800	Topical preparation containing a suspension of solid lipid particles	Tom De Vringer	1991	U.S. Patent No. (1997)
WO9305768	Medication vehicles made of solid lipid particles	Stefan Lucks, Rainer Müller	1992	W.O. Patent No. (1996)
US 5785976	Solid lipid particles, particles of bioactive agents and methods for the manufacture and use	Kirsten Westesen, Britta Siekmann	1994	U.S. Patent No. (1998)
DE19825856	New topical formulation which includes active agent as liquid lipid nanoparticles in an oil–in–water emulsion	Labtec Gmbh	1998	D.E. Patent No. (1999)
US6551619	Pharmaceutical cyclosporine formulation with improved biopharmaceutical properties, improved physical quality and greater stability, and method for producing said formulation	Lawrence John Penkler, Rainer Helmut Müller, Stephan Anton Runge, Vittorino Ravelli	1999	U.S. Patent No. (2003)
WO0006120	Lipid emulsion and solid lipid nanoparticle as a gene or drug carrier	Seo Young Jeong, Ick Chan Kwon, Hesson Chung	1999	W.O. Patent No. (2000)
DE19952410 B4	Sunscreen preparations comprising SLNs	Hansen Peter, Heppner Andrea, Schumann Christof	1999	D.E. Patent No. (2001)
US6770299	Lipid matrix–drug conjugates particle for controlled release of active ingredient	Rainer H. Muller, Carsten Olbrich	2000	U.S. Patent No. (2004)

(Continued)

Table 8.3 List of patents on SLNs   (Continued)

Patent no.	Title of patent	Inventor/applicant	Filing year	Reference
US6814959	UV radiation reflecting or absorbing agents, protecting against harmful UV radiation and reinforcing the natural skin barrier	Rainer H. Muller, Wissing Sylvia, Mader Karsten	2000	U.S. Patent No. (2004)
US7153525	Microemulsions as precursors to solid nanoparticles	Russell John Mumper, Michael Jay	2001	U.S. Patent No. (2006)
CA2524589	Compositions for the targeted release of fragrances and aromas	Gerd Dahms, Andreas Jung, Holger Seidel	2003	C.A. Patent No. (2004)
US2006024374	Pharmaceutical compositions suitable for the treatment of ophthalmic diseases	Maria Gasco, Marco Saettone, Gian Zara	2003	U.S. Patent No. (2006)
US7147841	Formulation of UV absorbers by incorporation in SLNs	Bernd Herzog	2003	U.S. Patent No. (2006)
US2006222716	Colloidal solid lipid vehicle for pharmaceutical use	Joseph Schwarz, Michael Weisspapir	2005	U.S. Patent No. (2006)
US2006008531	Method for producing solid lipid composite drug particles	Boris Shekunov, Pratibhash Chattopadhyay, Robert Huff	2005	U.S. Patent No. (2006)
US20080311214A1	Polymerized SLNs for oral or mucosal delivery of therapeutic proteins and peptides	Kollipara Koteswara Rao	2006	U.S. Patent No. (2006)
US11921634	Use of SLNs comprising cholesteryl propionate and cholesteryl butyrate	Maria Rosa Gasco	2006	U.S. Patent No. (2009)
EP2413918 A1	SLNs encapsulating minoxidil, and aqueous suspension containing same	Karine Padois, Fabrice Pirot, Françoise Falson	2010	E.P. Patent No. (2010)
EP2549977 A2	Lipid nanoparticle capsules	Petit Josep LLuis Viladot, González Raquel Delgado, Botello Alfonso Fernández	2011	E.P. Patent No. (2011)

8.10 SUMMARY AND CONCLUSION

SLNs symbolize a solid-state lipid-based matrix system which can be prepared with a well-established method such as high-pressure homogenization permitting production at bulk scale. Based on the formulation composition of the SLNs, too many homogenization cycles or too high pressure may affect the stability and integrity of SLNs. Sterilization of SLNs containing sterically stabilizing poloxamer was not possible by autoclaving, possibly due to the reduced steric stabilization at increased temperatures. Substitutes are sterilization by filtration or gamma irradiation. Lecithin stabilized SLNs can be autoclaved similarly to parenteral fat emulsion which is advantageous for production on industrial scale. To explore the use of SLNs as an alternative delivery system further investigations are currently being performed with regard to lipid matrixes, stabilizing surfactants, drug-loading capacity and related release profiles, crystallization state of the particles and effect of drugs on this state, optimization of lyophilization and spray-drying and determination of in vitro toxicity.

Special emphasis is given to surface modification of SLNs particles for site-specific drug delivery. SLNs targeting can be accomplished by antibody, cationic lipid, pH sensitive lipid mediated attachment. The expected low toxicity is in agreement with the lack of in vitro cytotoxicity in cell cultures. These potential characteristics qualify SLNs as competent nanocarriers in the expansion of targeted delivery systems for the clinical trial investigations. The sustained release potential of SLNs also provides effective delivery of drug to the target tissue. Although to become a futuristic delivery system, extensive research is required for the enhancement of quality, efficacy, and safety profile of drugs/gene/nucleic acids using SLNs.

ACKNOWLEDGMENT

The author RKT acknowledges the support by Fundamental Research Grant (FRGS) scheme of Ministry of Higher Education, Malaysia to support research on gene delivery. The authors would like to acknowledge International Medical University, Malaysia for providing research support on cancer and arthritis research. We also acknowledge internal grants to Dr. Tekade from the IMU-JC for providing start-up financial support to our research group.

The authors acknowledge the support of the National Institute of General Medical Science of the National Institutes of Health under award number SC3GM109873 to Dr. Chougule. The authors acknowledge Hawai'i Community Foundation, Honolulu, HI 96813, USA, for research support on lung cancer, mesothelioma, and asthma projects (Leahi Fund 15ADVC-74296) in 2015, 2013, and 2011, respectively. The authors would like to acknowledge the 2013 George F. Straub Trust and Robert C. Perry Fund of the Hawai'i Community Foundation, Honolulu, HI 96813, USA for research support on lung cancer to Dr. Chougule's lab. The authors also acknowledge a seed grant from the Research Corporation of the University of Hawai'i at Hilo, Hilo, HI 96720, USA, and The Daniel K. Inouye College of Pharmacy, University of Hawaii at Hilo, Hilo, HI 96720, USA, for providing start-up financial support to their research group. The authors acknowledge the donation from Dr. Robert S. Shapiro, MD, Dermatologist, Hilo, HI 96720, USA in support of development of nanotechnology-based medicines.

REFERENCES

Abdel-Salam, F.S., Elkheshen, S.A., Mahmoud, A.A., Ammar, H.O., 2016. Diflucortolone valerate loaded solid lipid nanoparticles as a semisolid topical delivery system. Bull. Faculty Pharm. Cairo Univ. 54 (1), 1–7.

Allababidi, S., Shah, J.C., 1998. Efficacy and pharmacokinetics of site-specific cefazolin delivery using biodegradable implants in the prevention of post-operative wound infections. Pharm. Res. 15 (2), 325–333.

Battaglia, L., Serpe, L., Muntoni, E.E., Zara, G., Trotta, M., Gallarate, M., 2011. Methotrexate-loaded SLNs prepared by coacervation technique: in vitro cytotoxicity and in vivo pharmacokinetics and biodistribution. Nanomedicine 6, 1561–1573.

Beloqui, A., Solinis, M.A., Delgado, A., Evora, C., Del Pozo-Rodriguez, A., Rodriguez-Gascon, A., 2014. Biodistribution of Nanostructured Lipid Carriers (NLCs) after intravenous administration to rats: influence of technological factors. Eur. J. Pharm. Biopharm. 84, 309–314.

Bi, R., Shao, W., Wang, Q., Zhang, N., 2009. Solid lipid nanoparticles as insulin inhalation carriers for enhanced pulmonary delivery. J. Biomed. Nanotechnol. 5, 84–92.

Carrillo, C., Hernandez, N., Garcia-Montoya, E., Pérez-Lozano, P., SuneNegre, J.M., Tico, J.R., et al., 2013. DNA delivery via cationic solid lipid nanoparticles (SLNs). Eur. J. Pharm. Sci. 49 (2), 157–165.

Casi, G., Neri, D., 2012. Antibody–drug conjugates: basic concepts, examples and future perspectives. J. Control. Release 161 (2), 422–428.

Cavalli, R., Gasco, M.R., Chetoni, P., Burgalassi, S., Saettone, M.F., 2002. Solid lipid nanoparticles (SLNs) as ocular delivery systems for tobramycin. Int. J. Pharm. 238, 241–245.

Chawla, V., Saraf, S.A., 2012. Rheological studies on solid lipid nanoparticle based carbopol gels of aceclofenac. Colloids Surf. B Biointerfaces 92, 293–298.

Chopdey, P.K., Tekade, R.K., Mehra, N.K., Mody, N., Jain, N.K., 2015. Glycyrrhizin conjugated dendrimer and multi-walled carbon nanotubes for liver specific delivery of doxorubicin. J. Nanosci. Nanotechnol. 15 (2), 1088–1100.

Choudhury, H., Gorain, B., Chatterjee, B., Mandal, U.K., Sengupta, P., Tekade, R.K., 2016. Pharmacokinetic and pharmacodynamic features of nanoemulsion following oral, intravenous, topical and nasal route. Curr Pharm. Des.

Chougule, M.B., Tekade, R.K., Hoffmann, P.R., Bhatia, D., Sutariya, V.B., Pathak, Y., 2014. Nanomaterial-based gene and drug delivery: pulmonary toxicity considerations. Biointeract. Nanomater. 225–248.

Ciftci, O.N., Temelli, F., 2016. Formation of solid lipid microparticles from fully hydrogenated canola oil using supercritical carbon dioxide. J. Food Eng. 178, 137–144.

Dahms, G., Seidel, H., Jung, A., 2004. Canadian patent No. 2524589. Canadian Intellectual Property Office, Gatineau, Canada.

Das, S., Ng, W.K., Kanaujia, P., Kim, S., Tan, R.B., 2011. Formulation design, preparation and physicochemical characterizations of solid lipid nanoparticles containing a hydrophobic drug: effects of process variables. Colloids Surf B Biointerfaces 88 (1), 483–489.

Dhakad, R.S., Tekade, R.K., Jain, N.K., 2013. Cancer targeting potential of folate targeted nanocarrier under comparative influence of tretinoin and dexamethasone. Curr. Drug Deliv. 10 (4), 477–491.

Domb, A.J., Maniar, M. 1991. WIPO patent No. 1991007171. World Intellectual Property Organization, Geneva, Switzerland.

Dua, K., Shukla, S.D., Tekade, R.K., Hansbro, P.M., 2016. Whether a novel drug delivery system can overcome the problem of biofilms in respiratory diseases? Drug Deliv. Transl. Res. 7, 179–187.

Dwivedi, P., Tekade, R.K., Jain, N.K., 2016a. Nanoparticulate carrier mediated intranasal delivery of insulin for the restoration of memory signaling in Alzheimer's disease. Curr. Nanosci. 9, 46–55.

Dwivedi, N., Shah, J., Mishra, V., Mohd Amin, M.C., Iyer, A.K., Tekade, R.K., et al., 2016b. Dendrimer-mediated approaches for the treatment of brain tumor. J. Biomater. Sci. 27 (7), 557–580.

El-Salamouni, N.S., Farid, R.M., El-Kamel, A.H., El-Gamal, S.S., 2015. Effect of sterilization on the physical stability of brimonidine-loaded solid lipid nanoparticles and nanostructured lipid carriers. Int. J. Pharm. 496 (2), 976–983.

Esposito, E., Fantin, M., Marti, M., Drechsler, M., Paccamiccio, L., Mariani, P., et al., 2008. Solid lipid nanoparticles as delivery systems for bromocriptine. Pharm. Res. 25, 1521–1530.

Ezzati Nazhad Dolatabadi, J., Hamishehkar, H., Eskandani, M., Valizadeh, H., 2014. Formulation, characterization and cytotoxicity studies of alendronate sodium-loaded solid lipid nanoparticles. Colloids Surf. B Biointerfaces 117, 21–28.

Falson, F., Padois, K., Fabrice, F., 2010. European Patent No. 2413918. European Patent Office, Munich, Germany.

Gajbhiye, V., Vijayaraj Kumar, P., Tekade, R.K., Jain, N.K., 2007. Pharmaceutical and biomedical potential of PEGylated dendrimers. Curr. Pharm. Design 13 (4), 415–429.

Gajbhiye, V., Palanirajan, V.K., Tekade, R.K., Jain, N.K., 2009. Dendrimers as therapeutic agents: a systematic review. J. Pharm. Pharmacol. 61 (8), 989–1003.

Gandhi, N.S., Tekade, R.K., Chougule, M.B., 2014. Nanocarrier mediated delivery of siRNA/miRNA in combination with chemotherapeutic agents for cancer therapy: current progress and advances. J. Control. Release 194, 238–256.

Gasco, M.R., 1993. US patent No. 5250236. US Patent and Trademark Office, Washington DC.

Gasco, M.R., 2009. US patent No. 11921634. US Patent and Trademark Office, Washington DC.

Gasco, M.R., Saettone, M.F., Zara, G., 2006. US patent No. 2006024374. US Patent and Trademark Office, Washington DC.

Ghanghoria, R., Tekade, R.K., Mishra, A.K., Chuttani, K., Jain, N.K., 2016. Luteinizing hormone-releasing hormone peptide tethered nanoparticulate system for enhanced antitumoral efficacy of paclitaxel. Nanomedicine (Lond.) 11 (7), 797–816.

Gorain, B., Choudhury, H., Tekade, R.K., Karan, S., Jaisankar, P., Pal, T.K., 2016. Comparative biodistribution and safety profiling of olmesartan medoxomil oil-in-water oral nanoemulsion. Regul. Toxicol. Pharmacol. 82, 20–31.

Heppner, A., Hansen, P., Schumann, C., 2001. Deutsches Patent No. 19952410. German Patent and Trade Mark Office, Munich, Germany.

Herzog, B., 2006. US Patent No. 20067147841. US Patent and Trademark Office, Washington DC.

Heydenreich, A.V., Westmeier, R., Pedersen, N., Poulsen, H.S., Kristensen, H.G., 2003. Preparation and purification of cationic solid lipid nanospheres—effects on particle size, physical stability and cell toxicity. Int. J. Pharm. 254 (3), 83–87.

Hoar, T.P., Schulman, J.H., 1943. Transparent water-in-oil dispersions, the oleopathic hydro-micelle. Nature 152, 102–103.

Hsu, M.H., Su, Y.C., 2008. Iron-oxide embedded solid lipid nanoparticles for magnetically controlled heating and drug delivery. Biomed. Microdevices 10 (6), 785–793.

Huang, C.W., Kearney, V., Moeendarbari, S., Jiang, R.Q., Christensen, P., Tekade, R.K., et al., 2015. Hollow gold nanoparticals as biocompatible radiosensitizer: an in vitro proof of concept study. J. Nano Res. 32, 106–112.

Jain, K.K., 2007. Applications of nanobiotechnology in clinical diagnostics. Clin. Chem. 53 (11), 2002–2009.

Jain, N.K., Tekade, R.K., 2013. Dendrimers for enhanced drug solubilization. Drug Deliv. Strategies Poorly Water-Soluble Drugs, 373–409.

Jenning, V., Schafer-Korting, M., Gohla, S., 2000. Vitamin A-loaded solid lipid nanoparticles for topical use: drug release properties. J. Control. Release 66, 115–126.

Jeong, S.Y., Kwon, I.C., Chung, H., 2000. WIPO Patent No. 00006120. World Intellectual Property Organization, Geneva, Switzerland.

Josep, P., Raquel, D., Alfonso, F., 2011. European Patent No. 2549977. European Patent Office, Munich, Germany.

Kashanian, S., Azandaryani, A.H., Derakhshandeh, K., 2011. New surface-modified solid lipid nanoparticles using N-glutaryl phosphatidylethanolamine as the outer shell. Int. J. Nanomedicine 6, 2393–2401.

Kathe, N., Henriksen, B., Chauhan, H., 2015. Physicochemical characterization techniques for solid lipid nanoparticles: principles and limitations. ChemInform 46 (24), 1565–1575.

Kayat, J., Gajbhiye, V., Tekade, R.K., Jain, N.K., 2011. Pulmonary toxicity of carbon nanotubes: a systematic report. Nanomedicine 7 (1), 40–49.

Kesharwani, P., Tekade, R.K., Jain, N.K., 2015a. Generation dependent safety and efficacy of folic acid con-jugated dendrimer based anticancer drug formulations. Pharm. Res. 32 (4), 1438–1450.

Kesharwani, P., Tekade, R.K., Jain, N.K., 2015b. Dendrimer generational nomenclature: the need to harmo-nize. Drug Discov. Today 20 (5), 497–499.

Koteswara Rao, K., 2006. US Patent No. 20080311214. US Patent and Trademark Office, Washington DC.

Kuo, Y.C., Chung, C.Y., 2011. Solid lipid nanoparticles comprising internal Compritol 888 ATO, tripalmi-tin and cacao butter for encapsulating and releasing stavudine, delavirdine and saquinavir. Colloids Surf. B Biointerfaces 88 (2), 682–690.

Kuo, Y.C., Ko, H.F., 2013. Targeting delivery of saquinavir to the brain using 83-14 monoclonal antibody-grafted solid lipid nanoparticles. Biomaterials 34 (20), 4818–4830.

Kurmi, B.D., Kayat, J., Gajbhiye, V., Tekade, R.K., Jain, N.K., 2010. Micro-and nanocarrier-mediated lung targeting. Expert Opin. Drug Deliv. 7 (7), 781–794.

Labtec, G.F.T., 1999. Deutsches Patent No. 825856. German Patent and Trade Mark Office, Munich, Germany.

Lim, S.B., Banerjee, A., Önyüksel, H., 2012. Improvement of drug safety by the use of lipid-based nanocar-riers. J. Control. Release 163 (1), 34–45.

Liu, J., Gong, T., Fu, H., Wang, C., Wang, X., Chen, Q., et al., 2008. Solid lipid nanoparticles for pulmonary delivery of insulin. Int. J. Pharm. 356 (1–2), 333–344.

Madureira, A.R., Campos, D.A., Fonte, P., Nunes, S., Reis, F., Gomes, A.M., et al., 2015. Characterization of solid lipid nanoparticles produced with carnauba wax for rosmarinic acid oral delivery. RSC Advances 5, 22665–22673.

Maheshwari, R.G.S., Tekade, R.K., Sharma, P.A., Gajanan, D., Tyagi, A., Patel, R.P., et al., 2012. Ethosomes and ultradeformable liposomes for transdermal delivery of clotrimazole: a comparative assessment. Saudi Pharm. J. 20, 161–170.

Maheshwari, R., Tekade, M., Sharma, P.A., Tekade, R.K., 2015a. Nanocarriers assisted siRNA gene therapy for the management of cardiovascular disorders. Curr. Pharm. Design 21 (30), 4427–4440.

Maheshwari, R.G., Thakur, S., Singhal, S., Patel, R.P., Tekade, M., Tekade, R.K., 2015b. Chitosan encrusted nonionic surfactant based vesicular formulation for topical administration of ofloxacin. Sci. Adv. Mater. 7, 1163–1176.

Mansuri, S., Kesharwani, P., Tekade, R.K., Jain, N.K., 2016. Lyophilized mucoadhesive-dendrimer enclosed matrix tablet for extended oral delivery of albendazole. Eur. J. Pharm. Biopharm. 102, 202–213.

Miao, J., Du, Y.Z., Yuan, H., Zhang, X.G., Hu, F.Q., 2013. Drug resistance reversal activity of anticancer drug loaded solid lipid nanoparticles in multi-drug resistant cancer cells. Colloids Surf. B Biointerfaces 110, 74–80.

Mody, N., Tekade, R.K., Mehra, N.K., Chopdey, P., Jain, N.K., 2014. Dendrimer, liposomes, carbon nanotubes and PLGA nanoparticles: one platform assessment of drug delivery potential. AAPS PharmSciTech 15 (2), 388–399.

Mohamed, R.A., Abass, H.A., Attia, M.A., Heikal, O.A., 2013. Formulation and evaluation of metoclo-pramide solid lipid nanoparticles for rectal suppository. J. Pharm. Pharmacol. 65 (11), 1607–1621.

Montana, G., Bondi, M.L., Carrotta, R., Picone, P., Craparo, E.F., San Biagio, P.L., et al., 2007. Employment of cationic solid–lipid nanoparticles as RNA carriers. Bioconjug. Chem. 18 (2), 302–308.

Morel, S., Ugazio, E., Cavalli, R., Gasco, M.R., 1996. Thymopentin in solid lipid nanoparticles. Int. J. Pharm. 132, 259–261.

Muller, R.H., Carsten, O., 2004. US Patent No. 20046770299. US Patent and Trademark Office, Washington DC.

Muller, R.H., Lucks, J.S., 1996. WIPO Patent No. 93/05768. World Intellectual Property Organization, Geneva, Switzerland.

Muller, R.H., Wissing, S., Mader, K., 2004. US Patent No. 20046814959. US Patent and Trademark Office, Washington DC.

Muller, R.H., Petersen, R.D., Hommoss, A., Pardeike, J., 2007. Nanostructured lipid carriers (NLC) in cosmetic dermal products. Adv. Drug Deliv. Rev. 59, 522–530.

Muller, R.H., Shegokar, R., Keck, C.M., 2011. 20 years of lipid nanoparticles (SLNs and NLC): present state of development and industrial applications. Curr. Drug Discov. Technol. 8, 207–227.

Mumper, R.J., Jay, M., 2006. US Patent No. 20067153525. US Patent and Trademark Office, Washington DC.

Mussi, S.V., Silva, R.C., Oliveira, M.C., Lucci, C.M., Azevedo, R.B., Ferreira, L.A., 2013. New approach to improve encapsulation and antitumor activity of DOX loaded in solid lipid nanoparticles. Eur. J. Pharm. Sci. 48 (1–2), 282–290.

Negi, J.S., Chattopadhyay, P., Sharma, A.K., Ram, V., 2013. Development of solid lipid nanoparticles (SLNs) of lopinavir using hot self nano-emulsification (SNE) technique. Eur. J. Pharm. Sci. 48 (1–2), 231–239.

Pandey, R., Khuller, G.K., 2006. Oral nanoparticle-based antituberculosis drug delivery to the brain in an experimental model. J. Antimicrob. Chemother. 57, 1146–1152.

Pang, X., Cui, F., Tian, J., Chen, J., Zhou, J., Zhou, W., 2009. Preparation and characterization of magnetic solid lipid nanoparticles loaded with ibuprofen. Asian J. Pharm. Sci. 4 (2), 132–137.

Peer, D., 2012. Immunotoxicity derived from manipulating leukocytes with lipid-based nanoparticles. Adv. Drug Deliv. Rev. 64 (15), 1738–1748.

Penkler, L.J., Muller, R.H., Runge, S.A., Ravelli, V., 2003. US Patent No. 20036551619. US Patent and Trademark Office, Washington DC.

Prajapati, R.N., Tekade, R.K., Gupta, U., Gajbhiye, V., Jain, N.K., 2009. Dendimer-mediated solubilization, formulation development and in vitro-in vivo assessment of piroxicam. Mol. Pharm. 6 (3), 940–950.

Sahu, G.K., Sharma, H., Gupta, A., Kaur, C.D., 2015. Advancements in microemulsion based drug delivery systems for better therapeutic effects. Int. J. Pharm. Sci. Dev. Res. 1 (1), 8–15.

Sanna, V., Gavini, E., Cossu, M., Rassu, G., Giunchedi, P., 2007. Solid lipid nanoparticles (SLNs) as carriers for the topical delivery of econazole nitrate: in vitro characterization, ex-vivo and in-vivo studies. J. Pharm. Pharmacol. 59, 1057–1064.

Saraf, S., Mishra, D., Asthana, A., Jain, R., Singh, S., Jain, N.K., 2006. Lipid microparticles for mucosal immunization against hepatitis B. Vaccine 24, 45–56.

Schulman, J.H., Stoeckenius, W., Prince, L.M., 1959. Mechanism of formation and structure of micro emulsions by electron microscopy. J. Phys. Chem. 63, 1677–1680.

Schwarz, J., Weissapir, M., 2006. US Patent No. 20,06,222,716. US Patent and Trademark Office, Washington, DC.

Shah, C., Shah, V., Upadhyay, U., 2011. Solid lipid nanoparticles: a review current. Pharm. Res. 1 (4), 351–368.

Sharma, P.A., Maheshwari, R.G.S., Tekade, M., Tekade, R.K., 2015. Nanomaterial based approaches for the diagnosis and therapy of cardiovascular diseases. Curr. Pharm. Design 21 (30), 4465–4478.

Shekunov, B.Y., Chatopadhyay, P., Huff, R.W., 2006. US Patent No. 2006008531. U.S. Patent and Trademark Office, Washington, DC.

Silva, A.C., Kumar, A., Wild, W., Ferreira, D., Santos, D., Forbes, B., 2012. Long-term stability, biocompatibility and oral delivery potential of risperidone-loaded solid lipid nanoparticles. Int. J. Pharm. 436 (1–2), 798–805.

Singh, H., Bhandari, R., Kaur, I.P., 2013. Encapsulation of Rifampicin in a solid lipid nanoparticulate system to limit its degradation and interaction with Isoniazid at acidic pH. Int. J. Pharm. 446 (1–2), 106–111.

Soni, N., Soni, N., Pandey, H., Maheshwari, R., Kesharwani, P., Tekade, R.K., 2016. Augmented delivery of gemcitabine in lung cancer cells exploring mannose anchored solid lipid nanoparticles. J. Colloid Interface Sci. 481, 107–116.

Souto, E.B., Muller, R.H., 2005. SLNs and NLC for topical delivery of ketoconazole. J. Microencapsul. 22, 501–510.

Speiser, P., 1990. European Patent No. 0167825. European Patent Office, Munich, Germany.

Tabatt, K., Sameti, M., Olbrich, O., Müller, R.H., Lehr, C.M., 2004. Effect of cationic lipid and matrix lipid composition on solid lipid nanoparticle-mediated gene transfer. Eur. J. Pharm. Biopharm. 57, 155–162.

Taveira, S., Araújo, L., de Santana, D., Nomizo, A., De Freitas, L.A., Lopez, R.F., 2012. Development of cationic solid lipid nanoparticles with factorial design-based studies for topical administration of doxorubicin. J. Biomed. Nanotechnol. 8, 219–228.

Tekade, R.K., 2014. Editorial: Contemporary siRNA therapeutics and the current state-of-art. Curr. Pharm. Design 21 (31), 4527–4528.

Tekade, R.K., Chougule, M.B., 2013. Formulation development and evaluation of hybrid nanocarrier for cancer therapy: Taguchi orthogonal array based design. BioMed Res. Int., 2013–2031.

Tekade, R.K., Kumar, P.V., Jain, N.K., 2008a. Dendrimers in oncology: an expanding horizon. Chem. Rev. 109 (1), 49–87.

Tekade, R.K., Dutta, T., Tyagi, A., Bharti, A.C., Das, B.C., Jain, N.K., 2008b. Surface-engineered dendrimers for dual drug delivery: a receptor up-regulation and enhanced cancer targeting strategy. J. Drug Target. 16 (10), 758–772.

Tekade, R.K., Dutta, T., Gajbhiye, V., Jain, N.K., 2009. Exploring dendrimer towards dual drug delivery: pH responsive simultaneous drug-release kinetics. J. Microencapsul. 26 (4), 287–296.

Tekade, R.K., Emanuele, A.D., Elhissi, A., Agrawal, A., Jain, A., Arafat, B.T., et al., 2013. Extraction and RP-HPLC determination of taxol in rat plasma, cell culture and quality control samples. J. Biomed. Res. 27, 394–405.

Tekade, R., Xu, L., Hao, G., Ramezani, S., Silvers, W., Christensen, P., et al., 2014a. A facile preparation of radioactive gold nanoplatforms for potential theranostic agents of cancer. J. Nucl. Med. 55 (1), 1047.

Tekade, R.K., Youngren-Ortiz, S.R., Yang, H., Haware, R., Chougule, M.B., 2014b. Designing hybrid onconase nanocarriers for mesothelioma therapy: a Taguchi orthogonal array and multivariate component driven analysis. Mol. Pharm. 11 (10), 3671–3683.

Tekade, R.K., Youngren-Ortiz, S.R., Yang, H., Haware, R., Chougule, M.B., 2015a. Albumin-chitosan hybrid onconase nanocarriers for mesothelioma therapy. Cancer Res. 75 (15 Suppl.), 3680.

Tekade, R.K., Maheshwari, Rahul, G.S., Sharma, P., Tekade, M., Singh Chauhan, A., 2015b. siRNA therapy, challenges and underlying perspectives of dendrimer as delivery vector. Curr. Pharm. Design 21 (31), 4614–4636.

Tekade, R.K., Tekade, M., Kumar, M., Chauhan, A.S., 2015c. Dendrimer-stabilized smart-nanoparticle (DSSN) platform for targeted delivery of hydrophobic antitumor therapeutics. Pharm. Res. 32 (3), 910–928.

Tekade, R.K., Tekade, M., Kesharwani, P., D'Emanuele, A., 2016. RNAi-combined nano-chemotherapeutics to tackle resistant tumors. Drug Discov. Today 21 (11), 1761–1774.

Trotta, M., Cavalli, R., Carlotti, M.E., Battaglia, L., Debernardi, F., 2005. Solid lipid micro-particles carrying insulin formed by solvent-in-water emulsion–diffusion technique. Int. J. Pharm. 288, 281–288.

Vaghasiya, H., Kumar, A., Sawant, K., 2013. Development of solid lipid nanoparticles based controlled release system for topical delivery of terbinafine hydrochloride. Eur. J. Pharm. Sci. 49 (2), 311–322.

Vringer, T.D., 1997. US Patent No. 5667800. U.S. Patent and Trademark Office, Washington, DC.

Wang, Y., Zhu, L., Dong, Z., Xie, S., Chen, X., Lu, M., et al., 2012. Preparation and stability study of norfloxacin-loaded solid lipid nanoparticle suspensions. Colloids Surf. B Biointerfaces 98, 105–111.

Westesen, K., Siekmann, B., 1998. US Patent No5785976. U.S. Patent and Trademark Office, Washington, DC.

Wonganan, P., Lansakara, P.D., Zhu, S., Holzer, M., Sandoval, M.A., Warthaka, M., et al., 2013. Just getting into cells is not enough: mechanisms underlying 4-(N)-stearoyl gemcitabine solid lipid nanoparticle's ability to overcome gemcitabine resistance caused by RRM1 overexpression. J. Control. Release 169 (1–2), 17–27.

Xue, H.Y., Wong, H.L., 2011. Tailoring nanostructured solid–lipid carriers for time-controlled intracellular siRNA kinetics to sustain RNAi-mediated chemosensitization. Biomaterials 32 (10), 2662–2672.

Yang, S., Zhu, J., Lu, Y., Liang, B., Yang, C., 1999. Body distribution of camptothecin solid lipid nanoparticles after oral administration. Pharm. Res. 16 (5), 751–757.

Youngren, S.R., Tekade, R.K., Gustilo, B., Hoffmann, P.R., Chougule, M.B., 2013. STAT6 siRNA matrix-loaded gelatine nanocarriers: formulation, characterization, and ex vivo proof of concept using adenocarcinoma cells. BioMed Res. Int. 2013, 858956.

Yusuf, M., Khan, M., Khan, R.A., Ahmed, B., 2013. Preparation, characterization, in vivo and biochemical evaluation of brain targeted Piperine solid lipid nanoparticles in an experimentally induced Alzheimer's disease model. J. Drug Target. 21, 300–311.

Zhang, Z.H., Zhang, Y.L., Zhou, J.P., Lv, H.X., 2012. Solid lipid nanoparticles modified with stearic acid-octaarginine for oral administration of insulin. Int. J. Nanomedicine 7, 3333–3339.

Zhuang, Y., Xu, B., Huang, F., Wu, J., Chen, S., 2012. Solid lipid nanoparticles of anticancer drugs against MCF-7 cell line and a murine breast cancer model. Pharmazie 67, 925–929.

CHAPTER 9

Lipid-Based Nanoparticles for Targeted Drug Delivery of Anticancer Drug

Ranjita Shegokar, Rajani Athawale, Nalini Kurup, Rongbing Yang and
Mahavir B. Chougule

Contents

9.1 INTRODUCTION

9.1.1 Cancer disease

Cancer is the second leading cause of worldwide mortality (Cara, 2015). About 1.66 million new cases were diagnosed in the United States in 2015, and the probable mortality due to cancer is about 0.59 million. In 2015, a total of 1.36 million cancer deaths were estimated in the EU in 2015 (Siegel, Miller, and Jemal, 2015). Cancer is a collective and related disease that is caused due to the interruption of a normal cell cycle. Cancer cells, which have escaped from a routine apoptosis process, begin to proliferate in an uncontrollable manner. Furthermore, these cells eventually spread in other organs (National Cancer Institution, 2016). Although it is difficult to identify the

Nanotechnology-Based Approaches for Targeting and Delivery of Drugs and Genes.
DOI: http://dx.doi.org/10.1016/B978-0-12-809717-5.00011-7

specific reason for the cancer, the understanding on the general risk factors is growing. Researchers reported risk factors that can be described in two aspects, external or internal. The external risk factors, also known as environmental factors, include the exposure to chemicals or other substances (Danaei et al., 2005). The internal risk factors include genetic heritage and certain behaviors. Most of the time, cancer is a result of multiple factors (Holschneider and Berek, 2000; McPherson, Steel, and Dixon, 2000).

Cancer can be classified based on the origin of organ or tissue, such as lung cancer, brain cancer, etc. (World Health Organization (WHO), 2013). Based on the cell type, cancer can be divided into carcinoma, sarcoma and so on. According to the morphology of the cancer cells, it can be further categorized into adenocarcinoma, large cell carcinoma, etc. (American Joint Committee on Cancer, 2002). Recently, the development of the genetic profile in cancer cells has added more methods to separate the various cancers by identifying the mutation or expression of genes, such as triple negative breast cancer, KRAS mutation lung cancer, etc. (Shigematsu et al., 2005; Sørlie et al., 2001). Lately, the location of metastasis was proposed to include into the classification of cancer. For example, the lung cancer has a subtype of brain metastasis which means this type of lung cancer tends to spread into the brain mostly (Goldberg, Contessa, Omay, and Chiang, 2015); breast cancer lung metastasis has its special gene profile (Leone and Leone, 2015). Although more and more methods were developed for a better classification, the current complex categorizing system has made accurate diagnosis difficult, which is based on the treatment plan. Therefore, the accurate diagnosis of cancer, especially early stages of cancer, is crucial for patient survival rate.

Current treatments of cancer include radiation therapy, chemotherapy, immunotherapy, hormone therapy, surgery, stem cell transplant, precision medicine, and targeted therapy (Sawant and Shegokar, 2014). Radiation therapy uses high doses of radiation to kill the cancer cells, causing the tumor to shrink locally or systematically (Delaney, Jacob, Featherstone, and Barton, 2005). Chemotherapy is a chemo-drug-based treatment that kills the cancer cells by specific mechanism of action (Carter and Slavik, 1974). The development of chemo-drug resistance has been reported in many cases and generally causes the recurrence of cancer. Chemotherapy extends the life span by few years but eventually developed drug resistant can cause death (Cancer multidrug resistance, 2000). The immunotherapy boosts the immune system of cancer patients to fight cancer cells while hormone therapy alters the amount or behavior of hormones in patients (Rosenberg, Yang, and Restifo, 2004). The side effects of radiation therapy, chemotherapy, immunotherapy, and hormone therapy were major obstacles for the efficacy of these therapies (Carelle et al., 2002; Trotti et al., 2000). Surgery is used to remove the cancer cells/tissue from patients via either scalpels or its alternatives. However, surgery also poses the danger of release of cancer cells into the blood stream during the procedure, resulting in metastasis (Ben-Eliyahu, 2003; Ben-Eliyahu, Page, Yirmiya, and Shakhar, 1999). Stem cell transplant is the procedure to restore the blood-forming stem cells in cancer patients. However, the chances of complications have diminished the efficacy of this therapy, such as mucositis and sinusoidal obstruction syndrome in the early stage, and chronic graft-versus-host disease in a delayed manner (Lee et al., 2010). In general, most of these therapies cause some level of side effects, mainly due to the off-target effect. The treatment involves not only removal or killing of the cancer cells but also sacrificing the normal cells. Therefore, the target therapy, which could separate the cancer cells from normal cells, was a better choice for the cancer treatment. Precision medicine has been proposed due to its specificity to individual patients with various markers. It is more like a diagnosis tool instead of a treatment. Targeted therapy was designed to target the change of the cancer cells specifically. This type of therapy has combined the diagnosis and treatment together. It can recognize the cancer cells and kill them while normal cells were spared. Although one of these particular therapies might work, most patients will get two or more treatments combined in order to maximum the therapeutic effect and to prolong the life span of cancer patients.

9.1.2 Cancer targeting strategies

The concept of the targeted therapy was introduced as "magic bullet" by Ehrlich in the early 1900s (Strebhardt and Ullrich, 2008; Winau, Westphal, and Winau, 2004). Since then, many "magic bullets" were approved by the FDA and applied clinically to save patients' lives (Table 9.1). The targeting strategies developed for the cancer treatment could be divided into three categories: pharmacological targeting, passive targeting and active targeting (Danhier, Feron, and Preat, 2010; Torchilin, 2010).

Based on the pharmaceutical targeting agents, a number of therapeutic agents were designed. The small molecule drug, such as everolimus, and mTOR inhibitor, has obtained approval from the FDA for the treatment of HER2 positive breast cancer (Motzer et al., 2008; Pouliot and Pantuck, 2009). The antibody of ErbB2/HER2 is the first antibody therapy that the FDA approved for cancer treatment (Hu et al., 2008; Lemmens, Segers, Demolder, and De Keulenaer, 2006). This antibody was specifically targeted to breast cancer patients who have HER2 biomarker. Although the antibody has very high specificity targeting to molecules, the aggregation, stability, and fast clearance has limited the therapeutic effects (Parakh, Parslow, Gan, and Scott, 2016; Scott, Wolchok, and Old, 2012). The peptides, as a fragment of the antibody or other proteins, have shown a high level of binding affinity to the targeted molecule while maintaining low aggregation and high stability (Xiao et al., 2015). Gene therapy is another targeting strategy (Nastiuk and Krolewski, 2016). US-FDA (United States Food and Drug Administration) has supplied the framework to assess the gene therapy products, meaning more gene therapy products could be expected in the following years.

Table 9.1 Marketed approved nanotechnology-based products

Product	Delivery system	Use	Manufacturer	Approved in
Adagen	Pegylated	i.v.	Enzon	1990
Megace ES	Nanocrystal	Oral	Elan Corp/Par Pharma	1993
Oncaspar	PEG conjugate	Subcutaneous	Sigma–Tau Pharma	1994
Abelcet	Phospholipid complex	i.v.	Sigma–Tau Pharma	1995
Copaxone	Polypeptides complex in suspension	Subcutaneous	Teva Pharma	1996
Amphotec	Disc-shaped NP	Subcutaneous	Sequus	1996
DaunoXome	Liposome	i.v.	Gilead Sciences	1996
AmBisome	Liposome	i.v.	Gilead Sciences	1997
Inflexal V	PEG-liposome	Intramuscular	Crucell	1997
Ontak	Nanoparticle	i.v.	Seragen	1999
DepoCyt	Liposome	Intrathecal	Skypharma	1999
Visudyne	Liposome	i.v.	Valeant Pharmaceuticals	2000
Rapamune	Nanocrystals	Oral	Elan nanosystems	2000
PEG-Intron	Pegylated complex	Subcutaneous	Nektar	2001
Pegasys	Pegylated complex	Subcutaneous	Nektar/Hoffmann-La Roche	2002
Neulasta	Pegylated complex	Subcutaneous	Amgen	2002
Estrasorb	Micellar NP	Topical	Espirit pharma	2002
Somavert	Polymer protein congugate	Subcutaneous	Nektar/Pfizer	2003
Emend	Nanocrystals	Oral	Elan nanosystems	2003
DepoDur	Liposome	i.v.	Pacira/EKR Therapeutics	2004
Macugen	PEGylated aptamer	Injection	OSI Pharmaceuticals/Pfizer	2004
TriCor	Nanocrystals	Oral	Abbott Élan/Alkermes	2004
Abraxane	Album particles	Intravenous	Celgene	2005
Triglide	Nanocrystals	Oral	Skyepharma/First horizon Pharmaceuticals	2005
Elestrin	Phosphate NP	Transdermal	BioSanté	2006
Mircera	PEGylated NP	Injection	Roche	2007
Cimzia	PEGylated NP	Injection	UCB Pharma S.A.	2008
Invega Sustenna	Extended release nanocrystals		Elan nanosystems	2009
Marqibo	Liposome	i.v.	Talon	2012

9.1.3 Passive and active targeting

The passive targeting strategy has been exploited for over 30 years (Matsumura and Maeda, 1986). In the passive targeting strategy, the physical properties of the therapeutic agents facilitates the accumulation at the cancer site (Iyer, Khaled, Fang, and Maeda, 2006). In tumor vasculature, a known fenestration causes the imperfection of tumor blood vessels and poor lymphatic drainage (Maeda, Wu, Sawa, Matsumura, and Hori, 2000). The combination of these two phenomena was coined as the enhanced permeation and retention (EPR) effect. Since then, researchers have been conducting a detailed study on the EPR effect. It was reported that the fenestrations in the capillaries can reach sizes ranging from 200 to 2000 nm, depending on the tumor type, its environment and its localization (Maeda, 2013, 2014). The nanocarriers of size >150 nm are rapidly cleared from the circulation, whereas nanocarriers of size <30 nm rapidly undergo hepatic and renal clearance (Ma et al., 2013; Walkey, Olsen, Guo, Emili, and Chan, 2012). The improved efficacy of drug-loaded nanocarriers associated with EPR effects (size <100 nm) is due to the leaky tumor vasculature and improved pharmacokinetic parameters (Patel, Spencer, Chougule, Safe, and Singh, 2012; Petanidis et al., 2016).

Considering passive targeting, the accumulation of the nanoparticles or other therapeutic agents facilitates their efficient localization in the tumor interstitium but cannot further promote their uptake by cancer cells (Iyer et al., 2006; Maeda et al., 2000). Therefore, the active targeting strategy was adopted by actively targeting therapeutic agents to receptors or other surface membrane proteins which are overexpressed on target neoplastic cells. The delivery of the drug to uniquely identifiable cells or even subcellular sites reduces the unwanted systemic exposure of cytotoxic drugs to minimize the side effects (Bazak, Houri, El Achy, Kamel, and Refaat, 2015). Currently, the active targeting strategy has included antibodies and their fragment, aptamers, and small molecules. These ligands specifically bind with the membrane bounded receptor or proteins which could trigger the internalization via binding. Although actively targeting the cell surface protein could enhance the uptake by the targeted cells, the low penetration and low accumulation in the tumor site is still an issue. Therefore, the active targeting strategy can include the environmental targeting and the tumor associated stromal cells targeting. By tuning the surface charge of the therapeutic agents, the interaction between cells and the agents could be enhanced (He, Hu, Yin, Tang, and Yin, 2010). The enzymatic or pH responsive excipients or drugs are being explored to modulate the release of the drug at the targeted site (Kulsh, 1997; Li et al., 2016; Tran et al., 2015).

9.2 NANOPARTICLES FOR CANCER THERAPY

Nanotechnology has been well developed since Dr. Richard Feynman introduced the concept of, "There's Plenty of Room at the Bottom," especially in its medical applications. The first US–FDA approved nano-based delivery system was in 1983. Since then, nanocarriers have been used in the clinic and have shown enhanced efficacy and reduced side effects of therapeutic agents (Agostinelli, Vianello, Magliulo, Thomas, and Thomas, 2015). The nanoparticles used for cancer therapy in the clinic and approved by the FDA are listed in Table 9.1.

There are two main aspects for applying nanoparticles in cancer treatment: one is diagnosis and secondly therapy. Diagnostic nanoparticles were designed to identify the neoplastic cells by visualizing the pathologies (Brigger, Dubernet, and Couvreur, 2012). The improvement of the understanding of physiological principles of cancer could improve the accuracy of the diagnosis, which could lead to more efficient treatment plan. Nanoparticles were engineered as a contrast agents for functional and molecular imaging with the current imaging techniques, such as magnetic resonance imaging (MRI), computed tomography (CT), ultrasound (US), optical imaging (OI), and photoacoustic imaging (PAI), as well as positron emission tomography (PET) and single photon emission computed tomography (SPECT) (Yiyao Liu, Miyoshi, and Nakamura, 2007). Various imaging techniques supply different focuses and strengths on the imaging due to utilization of various principles. Based on the building materials, these nanoparticles include synthetic or natural polymers, liposomes, magnetic nanoparticles, metal nanoparticles, carbon nanostructures, etc. (Cerqueira, Lasham, Shelling, and Al-Kassas, 2015; Steichen, Caldorera-Moore, and Peppas, 2013). Generally, the ideal diagnostic nanoparticles should have a rapid and highly site-specific contrast enhancement. However, the application of nanoparticles on diagnostics is limited clinically due to the complex requirements on their pharmacokinetic properties and elimination. Therefore, the majority of nanoparticle formulations currently used in the clinics are for therapeutic purposes. The ideal nanoparticles for therapy could improve the accumulation and release of active agents at the neoplastic site, increase therapeutic efficacy, and reduce the incidence and intensity of side effects by reducing their localization in healthy tissues. By adopting the various targeting strategies, versatile nanoparticles could realize these characteristics. The nanoparticles designed with passive targeting properties could improve the accumulation at pathological sites. The environmental responsive nanoparticles could boost the release rate of therapeutic agents. The active targeting strategy has improved the uptake by the neoplastic cells. With combinations of these various strategies, nanoparticles could be able to avoid the healthy tissue while eliminating the tumor cells (Fig. 9.1).

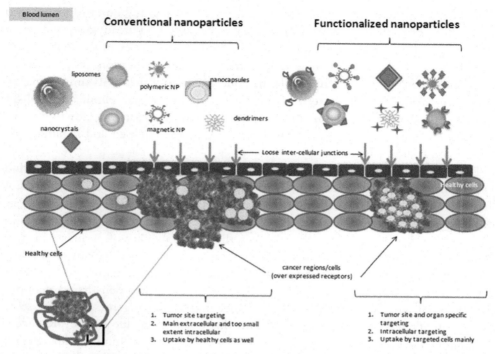

Figure 9.1 Conventional and targeted nanoparticle-based approaches for the cancer therapy. *With permission from Sawant, S., Shegokar, R., 2014. Cancer research and therapy: where are we today? Int. J. Cancer Ther. Oncol. 2 (4), 02048.*

9.3 CHALLENGES IN DELIVERY OF A PROMISING ANTICANCER AGENT—PACLITAXEL

Paclitaxel, also known as Taxol or Taxol A, is one of the broad spectrum toxoid anticancer drugs. It was used in first-line and subsequent chemotherapy for the treatment of advanced carcinoma of various cancers, such as ovary, breast, AIDS-related Kaposi's sarcoma, nonsmall cell lung cancer, head and neck cancers, and brain tumors. It kills the cancer cells by interfering with their growth. As an antimicrotubule agent, paclitaxel promotes the assembly of tubulin dimers into microtubules, prevents depolymerization of microtubules, and disturbs the normal dynamic reorganization of the microtubule network in vital interphase and mitotic cellular functions. As a result of the paclitaxel binding to tubulin, cell division is inhibited (Qian et al., 2006). Paclitaxel can bind with B-cell Leukemia 2 to induce the apoptosis. However, the normal cells would be affected by this drug at the same time and thus the severe side effects usually occur.

Paclitaxel is a hydrophobic anticancer drug which is highly bound to plasma protein (88% ~ 98%) (Kumar, Walle, Bhalla, and Walle, 1993). Its steady-state volume of distribution is large but it does not cross the blood–brain barrier in significant quantities (Fellner et al., 2002). Single-dose intravenous administration of paclitaxel at $135 \sim 350\,mg/m^2$ produces a mean plasma concentration of $0.23 \sim 10\,\mu m$. An average distribution half-life of paclitaxel is about 0.34 h and an average elimination half-life is about 5.8 h. The typical doses and dosing schedules are $175\,mg/m^2$ for three times weekly and $80 \sim 100\,mg/m^2$ for the weekly schedule (Sonnichsen and Relling, 1994; Spencer and Faulds, 1994).

The systematic administration of paclitaxel cause various adverse effects including the hypersensitivity reactions, myelosuppression, bradycardia and hypotension. These side effects vary from patient to patient but most of the cases are dose and dosing schedule related. Therefore, the efficacy of paclitaxel is limited due to limiting doses. Beside pharmacological challenges, the poor solubility of paclitaxel in water (<0.03 mg/L) is another factor to limit its efficacy against cancer (Alani et al., 2010; Hamada, Ishihara, Masuoka, Mikuni, and Nakajima, 2006). In commercial preparations, cremophor is used as a cosolvent system (cremophor with dehydrated alcohol at a 50:50 v/v ratio) to increase the paclitaxel concentration to 6 mg/mL. However, severe hypersensitivity reactions are associated with this preparation due to the presence of cremophor and therefore, it is not suitable for intravenous administration (Beijnen et al., 1994; Hitt, 1994). Inspired by the high binding to serum, paclitaxel albumin formulation was developed and approved by the FDA in 2005. Currently, paclitaxel available in the marketplace contains a mixture of Cremophor EL and dehydrated alcohol. Cremophor EL at higher concentration shows hypersensitivity reactions. The formulation also has a tendency to precipitate in aqueous infusion due to a decrease in solubility of the drug. Thus, various attempts have been made to develop lipid–based drug delivery systems and are published in literature.

Lipid–based nanoparticles offer a solution to the above challenges, and for that reason it is a widely studied for paclitaxel drug delivery system (MuÈller, MaÈder, and Gohla, 2000). Among the various nanoparticles used in targeted delivery approaches, lipid–based nanoparticles reserved their own benefits and successes in drug delivery to target sites. These lipid nanoparticles include solid lipid–based nanoparticles (SLNs), combination of solid and liquid lipid–nanostructured lipid carriers (NLCs), solid to semisolid lipid–based liposomes, and nanoemulsions (NEs), and self-nanoemulsifying drug delivery systems (SNEDDS), which are mainly based on liquid lipids (Mulder, Strijkers, van Tilborg, Griffioen, and Nicolay, 2006; Muller and Keck, 2004; Puri et al., 2009). Lipid nanocarriers offer a wide variety of advantages, starting from solubilization, encapsulation, surface functionalization (e.g., antibody, ligands), processing and tunability at both hot/cold conditions, biodegradability (Li, Wang, Sun, and Wang, 2016). A wide range of particle size (micron to nano size) is mainly based on the production method processing used and long-term stability. Lipid nanoparticles are easy to produce and most of the preparation methods avoid use of solvents.

To date, paclitaxel was successfully encapsulated in lipid nanoparticles and tested in clinics. However, as per our information no reference is available that reviews all the available research findings on paclitaxel delivery in the form of lipid-based nanoparticles. Therefore, the main focus of this chapter is to compile the findings on paclitaxel encapsulated lipid-based drug delivery systems, and discuss the overall success along with translational application to the clinic.

9.4 DRUG DELIVERY CARRIERS BASED ON LIPIDS (SOLID AND/OR LIQUID)

9.4.1 Nanoemulsions

The delivery of various actives using liquid lipids (or oils) in different emulsion forms was studied for many decades (Mulder et al., 2006; Puri et al., 2009). Nanoemulsions are thermodynamically stabilized disperse systems of oil in water. These novel formulations enhance drug delivery when given orally, parenterally and dermally. In contrast to microemulsions, nanoemulsions diluted with water remain stable without changing the droplet size distribution; this stability is influenced by changes in temperature and pH.

Paclitaxel nanoemulsions (22 nm) composed of labrasol, D-α-tocopheryl polyethylene glycol 1000 succinate and labrafil, when administered orally at a dose of 10 mg/kg paclitaxel, were rapidly absorbed reaching a steady-state value in half an hour, which were constant up to 18 h and amounted to an absolute bioavailability of 70.62%. This increase in bioavailability could be due to the inhibition of P-glycoprotein efflux by D-α-tocopheryl polyethylene glycol 1000 succinate and labrasol, which have contributed to the enhanced peroral bioavailability of paclitaxel (Khandavilli and Panchagnula, 2007). A similar improvement upon nanoemulsion administration was noted by Tiwari and Amiji (2006) and Dias, Carvalho, Rodrigues, Graziani, and Maranhao (2007) in respective studies on paclitaxel nanoemulsions.

In another study, paclitaxel nanoemulsions of mean particle size 25 nm showed significantly reduced drug resistance and improved cytotoxicity (IC50 of 5.39 μg/mL for nanoemulsion) from 101.45 μg/mL (pure paclitaxel) in drug resistant MCF-7/ADR cells. The reduced paclitaxel resistance could be due to nanoemulsion–induced inhibition of P-gp activity, thereby improved anticancer activity in a resistant tumor xenograft model (Bu et al., 2014).

Lipid nanoemulsion of paclitaxel oleate, a paclitaxel derivative, was tested intravenously in rabbits undergoing heterotopic heart transplantation upon intravenous administration; NE exhibited reduced degree of coronary stenosis and macrophage infiltration (Carvalho, Maranhao, and Stolf, 2014). In another separate study, a novel

PEG400-mediated lipid nanoemulsion of paclitaxel showed potential application when tested in vitro and in vivo. A significant increase in tumor uptake showed enhanced antitumor efficacy on bearing A2780 or Bcap-37 tumor nude mice compared to conventional paclitaxel loaded lipid nanoemulsion. A novel PEG400 lipid nanoemulsion showed no signs of haematolysis and intravenous irritation and exhibited similar cytotoxicity to marketed formulation Taxol against HeLa cells (Jing et al., 2014).

Other studies include high pressure homogenization (HPH) processed paclitaxel encapsulated vitamin E nanoemulsion which exhibited higher cytotoxicity in breast cancer cell line (MCF-7) compared to pure paclitaxel and marketed formulation (Taxol) (Pawar et al., 2014). Synergistic nanoemulsion-based delivery of paclitaxel and etoposide to reduce toxicity and increase the therapeutic action as the form of combined chemotherapy resulted in significant improvement in anticancer effects compared to marketed and pure drug forms (Kretzer, Maria, and Maranhao, 2012; Lee, Maturo, Rodriguez, Nguyen, and Shorr, 2011), coadministration of curcumin/ceramide and paclitaxel in nanoemulsion form to increase anticancer efficacy (Desai, Vyas, and Amiji, 2008; Ganta and Amiji, 2009; Ganta, Devalapally, and Amiji, 2010). Published studies on nanoemulsion confirms the potential of stabilized nanodroplets of oil in delivering paclitaxel to various target sites with reduced toxicity and improved cellular uptake.

9.4.2 Self-nano emulsifying drug delivery systems (SNEDDS)

SNEDDS are transparent thermodynamically stable nanoemulsions of size less than 100 nm. They exhibit an increased rate of absorption, enhanced solubilization and good stability; however, they are susceptible to extreme temperature and pH conditions. In general, high dose drugs are not suitable to be developed as SNEDDS due to drug expulsion, although dissolving the drug in suitable oil could be an option. SNEDDS consisting of drug, oil, surfactant, solubilize and are spontaneously formed from oil in water nanoemulsions on dilution with water with gentle stirring. Many studies have been reported on paclitaxel SNEDDS for enhancement of bioavailability (Gao et al., 2003; Oostendorp et al., 2011; Veltkamp et al., 2006; Zhang, Tang, Gong, Yan, and Zhang, 2006). Sun et al. developed paclitaxel SNEDDS using isopropyl myristate, cremophor and PEG 400. The developed formulations showed significantly higher drug release to that of drug suspension (Sun et al., 2011). In another study, a solid nanoemulsion preconcentrate of paclitaxel composed of oil, surfactant and cosurfactant was formulated using a solid carrier (polyoxyethylene) by fusion method (Li et al., 2009). The preconcentrate is a SNEDDS when it comes in contact with GI fluid. The excipients used possess property of modulating P-glycoprotein present in the lumen. In the ex vivo absorption studies the permeability coefficient of preconcentrate was found to be higher than the marketed formulation of paclitaxel. Furthermore, a biodistribution study indicated higher systemic exposure of paclitaxel from the preconcentrate and reduced cellular toxicity on MCF-7 cell line (Ahmad et al., 2014). Mixed nonionic surfactant systems were used to produce paclitaxel SNEDDS (Lo, Chen, Lee, Han, and Li, 2010) while Gursoy et al. studied the effects of various excipients on cytotoxicity of SNEDDS (Gursoy, Garrigue, Razafindratsita, Lambert, and Benita, 2003). To overcome side effects associated with Cremophor EL and ethanol in Taxol, medium chain triglyceride-based bioequivalent SNEDDS were developed (Patel, Patil, Mehta, Gota, and Vavia, 2013). Overall, SNEDDS are limitedly studied for delivery of paclitaxel.

9.4.3 Liposomes and lipid nanocapsules

Liposomes are spherical vesicles with several concentric layers composed of phospholipids and cholesterol containing both a polar end and nonpolar end. Depending on the solubility and partition coefficient of the drug, liposome can encapsulate hydrophilic and/or lipophilic drugs. In cancer, liposome enhances the therapeutic efficacy of the drug due to enhanced permeability and EPR by traveling via extravasation into interstitial space. Liposomes in a size range of 50–100 nm are the most suitable candidates for cytotoxic drug delivery due to improved bioavailability performance and reduced toxicity. Liposomes can facilitate cellular uptake and alter the pharmacokinetics of the drugs when surface-modified by various targeting ligands (Malam, Loizidou, and Seifalian, 2009). A liposome surface decorated with PEG (PEGylated) can significantly improve the half-life of liposomes (>200 nm) in systemic circulation. Furthermore, the PEGylation approach can help to facilitate liposomal drug delivery by reducing multidrug resistance due to the over-expression of drug efflux transporter pumps such as P-glycoprotein.

Ahmad et al. developed nanosomal lipid suspension of paclitaxel (100 nm) to eliminate the usage of cremophor and alcohol from the currently marketed formulation for treatment of breast cancer. The suspension was prepared by HPH using soy phosphatidylcholine and sodium cholesteryl sulphate as lipid phase in an aqueous medium. Nanosomal suspension exhibited a superior biological response rate and disease control rate in patients compared to Taxol (a marketed paclitaxel product). No hypersensitivity reaction in patients to the lipids used in the nanosuspensions confirmed its safety (Ahmad, Sheikh, Ali, Paithankar, and Mehta, 2015).

In another work, in vitro release and cytotoxic effect of HSPC (hydrogenated natural soybean phospholipids/cholesterol) based conventional liposomes and PEGylated liposomes (220–250 nm) were studied. Paclitaxel loaded conventional liposomes composed of HSPC and cholesterol (at various ratios 1:10, 1:20 and 1:30) were formed by rotary film hydration method and HPH. Increase in the cholesterol concentration showed increase in the stability but reduction in the drug entrapment capacity. Paclitaxel and cholesterol both are lipophilic molecules and there might be competition for the hydrophobic space in lipid, which may be the cause for the less drug entrapment beyond certain limit. Furthermore, PEGylated liposomes were formed using optimized ratios of DSPE-PEG2000. The stability of the formulation was challenged using sodium sulphate ranging from 0.5 to 2 m; DSPE-PEG2000 at 6 mol% showed maximum stability. Both conventional and PEGylated liposomes showed similar release profiles at pH 7.4 and showed Higuchi diffusion pattern of drug release. In vitro cellular cytotoxicity studies in B16F10 melanoma cells revealed dose-dependent toxicity for both liposomal formulations compared to toxicity of pure paclitaxel solution. At 48 h, liposomal formulations showed similar toxicity profile to that of paclitaxel solution (Shenoy, Gude, and Murthy, 2011). PEGylation of nanoliposomes using PEG 2000 was carried out and further size reduction was done using probe sonicator. The average particle size for nanoliposomes and PEGylated liposomes was around 421 nm and 370 nm, respectively. The cytotoxicity by MTT assay for PEGylated liposome showed low IC50 value compared to drug solution, thereby confirming the safety of developed liposomes (Esfahani et al., 2014).

The effects of pH sensitive paclitaxel liposomes (100–200 nm with zeta potential of −12 mV) on B16F1 were studied by Karanth et al. Liposomes were composed of dioliyl phosphatidyl ethanolamine (DOPE), 1,2-Dimyristyoyl-*sn*-glycero-3-phosphoglycerol, sodium salt of DMPG (1,2-Dimyristoyl-*sn*-glycero-3-phosphoglycerol), hydrogenated soya phosphatidylcholine, cholesterol and cholesteryl hemisuccinate. Lighten tube studies using B16F1 showed significant changes in cellular morphology. A marked change in cellular morphology at subtoxic concentrations was noted for pH sensitive liposomes. Wound assay and colony formation assay showed similar efficacy and confirmed superior performance of pH sensitive liposomes compared to pure paclitaxel solution. Conventional formulations showed less inhibition compared to pH sensitive liposomes. pH-sensitive liposomes arrested B16F1 cells in G2-M phase cell cycle similar to the drug solution; significant dose-dependent inhibition was observed. The pH sensitive liposomes showed destabilization at cytosolic acidic conditions, and were found to be more cytotoxic compared to conventional liposomes (Karanth and Murthy, 2009). In a separate study, three factors, three-level Box-Behnken Design was used to find the effect of variables on formation of pH sensitive liposomes (Rane and Prabhakar, 2013).

Kunstfeld et al. (Kunstfeld et al., 2003) investigated the effects of cationic paclitaxel liposomes (180–200 nm) on tumor angiogenesis and melanoma growth in a humanized SCID mouse model. Human cadaveric skin was grafted in Female CB17 SCID mice of 6-week of age to generate humanized SCID model. A separate pocket was created by surgical procedure in the mouse epidermis and dermis to graft the skin. After 2 weeks, A-375 melanoma cells in PBS were injected into the grafted portion to form nodules. The cationic liposomes showed superior tumor inhibition and improved survival rate compared with conventional liposomes or control group which received glucose solution. The tumor treated with cationic liposomes showed a growth retardant effect and a decrease in the size of necrotic capsule. Antiproliferative efficacy in endothelial cell culture of both cationic liposomes and paclitaxel solution was similar. Cell proliferative assay in A-375 cell culture showed decreased mitotic index of endothelial cells as these vessels are the preferential targets for cationic liposomes, although mitotic index of normal endothelium cells remains unchanged. Pharmacokinetic studies showed similar biological effects as obtained in vitro. In addition to cationic liposomes, various other paclitaxel-loaded surface-modified liposomes decorated with hyaluronic acid (Ravar et al., 2016), peptide (Cao et al., 2015; Du et al., 2014; Liu et al., 2015; Luo et al., 2013), siRNA(Yin et al., 2014), triphenyl phosphonium (Biswas, Dodwadkar, Deshpande, and Torchilin, 2012; Solomon, Shah, and D'Souza, 2013), Palmitoyl ascorbate (Sawant, Vaze, Rockwell, and Torchilin, 2010) and folic acid (Niu et al., 2011; Zhao et al., 2010) are reported.

Lipid nanocapsules (10–1000 nm) are lipid-based drug delivery systems composed of a shell and a core in which the drug is placed. The core acts as a reservoir for the drug and the shell is a protective membrane. The various formulation techniques adopted to formulate nanocapsules are nanoprecipitation, emulsion–diffusion, emulsion–coacervation, double emulsification, polymer coating and layer-by-layer. Nanocapsulated systems offer high drug encapsulation efficiency, low polymer content, biocompatibility, biodegradability, reduced toxicity, controlled release and ability to target specific tissues with reduced irritation. The stability of nanocapsules is dependent on temperature and pH. The major disadvantage of lipid nanocapsules is that it requires an exhaustive purification process due to the inherent generation of contaminants during the formulation of nanocapsules. There have been attempts to develop novel lipid nanocapsules capable of linking antibodies to give active immune nanocapsules as anticancer therapies (Sanchez-Moreno et al., 2012). Only a few reports are available on paclitaxel delivery using lipid nanocapsules (Groo et al., 2013; Hureaux et al., 2010; Lacoeuille et al., 2007; Peltier, Oger, Lagarce, Couet, and Benoit, 2006; Roger, Lagarce, Garcion, and Benoit, 2010). Lipid nanocapsules, because of their colloidal size, are able to interact favorably with the mucosal barriers and able to protect the drug from the harsh acidic environment of the gastrointestinal tract. In addition, paclitaxel was loaded in multifunctional lipid nanocapsules (Balzeau et al., 2013; Joshi, Kaviratna, and Banerjee, 2013; Lollo et al., 2015) and as ready-to-use nanovectors for the aerosol delivery have been studied (Hureaux et al., 2009).

9.4.4 Lipid nanoparticles

SLNs (solid lipid nanoparticles) and NLCs (nanostructred lipid nanocarriers) are the colloidal carriers that are widely studied and are emerging as multifunctional drug delivery vehicles for hydrophobic drugs. It is one of the most promising drug delivery systems with a proven record of improved drug delivery and biocompatibility. SLNs are composed of physiological lipids dispersed in water or in an aqueous surfactant solution and offer several advantages, such as improved feasibility of incorporation of lipophilic drugs, enhanced biocompatibility because of biodegradable or physiological lipids, possibility to control the release potential for site-specific drug delivery, tunable particle size by optimizing process conditions and composition, and the ability to lyophilize to further improve physical stability. SLNs offer easy scale-up and manufacturing at industry batch size (R.H. Muller, Shegokar, and Keck, 2011). The disadvantages of SLNs involve rapid temperature changes that can lead to drug degradation, rapid drug leakage, lipid crystallization, gelation phenomena and the coexistence of several other colloidal species.

A variety of lipids including lipid acids, mono- (glycerol monostearate), di- (glycerol bahenate) or triglycerides (tristearin), glyceride mixtures or waxes (e.g., cetylpalmitate) and several others are available in the market. Lipid nanoparticles are stabilized by the biocompatible surfactants(s) of choice (nonionic or ionic). Lipids most commonly used are triglyceride esters of hydrogenated fatty acids, including hydrogenated cottonseed oil (Lubritab or Sterotex), hydrogenated palm oil (Dynasan P60 or Softisan), hydrogenated castor oil (Cutina HR), and hydrogenated soybean oil (Sterotex HM, or Lipo), as typical examples. Various emulsifiers and their combination (Pluronic F 68, F 127) have also been added to stabilize the lipid dispersion by more efficiently preventing particle agglomeration. Since high lipid crystallinity is the major cause of burst release of the drug from SLNs, this undesirable phenomenon may be minimized by choosing lipids that do not form optimum matrix, including mono- or diglycerides, or triglycerides with chains of different lengths. For this reason, in formulation design use of more complex lipids is recommended for higher drug loading.

Paclitaxel-loaded SLN composed of glyceryl monostearate were prepared by the solvent emulsification technique. Powdered X-ray diffraction studies confirmed the amorphous nature of the drug and reduced intensity of lipid peak, thereby indicating the decreased crystallinity of lipid and better drug loading (Dinda, Biswal, Chowdhury, and Mohapatra, 2013) (Table 9.2).

Long-circulating SLNs were developed using Brij 78 (polyoxyethylene (20) stearyl ether) and using poloxamer F68 and PEG–DSPE (phospholipid substituted with PEG). Two surfactants had a distinct effect on the particle size of nanoparticles and drug loading. The diameter of SLN using Brij 78 as surfactant and F 68 as surfactant were 103 and 220 nm; the drug content was 58% and 75%, respectively. The in vitro release studies using multicompartmental rotating cell showed slower drug release from SLN composed of Brij 78 than F68 when following the Weibull equation. The pharmacokinetic studies at a dose level of 10 mg/kg showed biphasic drug release pattern with rapid terminal elimination phase from systemic circulation. The F68 composed SLN showed longer blood circulation time compared to Brij 78, which could be due to the presence of long PEG resulting in a thicker cloud of hydrophilic surface which repels the plasma proteins more compared to Brij 78 SLN (Chen, Yang, Lu, and Zhang, 2001).

Table 9.2 List of anticancer drugs incorporated in lipid nanoparticles and liposomes

DDS	Drug used	Target organ	Typical particle size
SLN/NLC	Capecitabine pro drug	Nonspecific	700 nm
	Carmustine	Brain	110 ~ 250 nm
	DH-I-180-3	Breast	200 nm
	Docetaxel	Colon, brain	76 ~ 250 nm
	Doxorubicin	Lung, ovarian, local solid tumor	60 − 370 nm
	Emodin	Breast	40 ~ 225 nm
	Etoposide	Brain, lung	40 ~ 235 nm
	Gemcitabine	Deoxycytidine kinase deficient cancer	150 ~ 175 nm
	Gemcitabine	RRM1-overexpressing tumor cells	186 nm
	Mitoxantrone	Breast	61 ~ 79 nm
	Tocotrienol	Breast	110 ~ 580 nm
	Verapamil	Multidrug resistance breast cancer	250 ~ 350 nm
	Vorinostat	CD44 overexpressing cancer cells	100 nm
Liposomes	Camptothecin	Nonspecific	150 nm
	Cantharidin	Breast	100 ~ 170 nm
	Celecoxib	Breast	106 ~ 110 nm
	Cetuximab	Colorectal	120 nm
	Chrysophsin-1	Multidrug resistant cancer	95 ~ 105 nm
	Cisplatin	Colon carcinoma ovarian	70 − 170 nm
	Daunorubicin and quinacrine	Breast	95 ~ 200 nm
	Docetaxel	Lung breast brain	100-220 nm
	Doxorubicin	CD44 expressing cancer cells, lung, breast, ovarian cancer, colon	50 ~ 500 nm
	Epirubicin	Breast, intestine, multidrug resistant cancer	95 ~ 2500 nm
	Irinotecan	Ovarian	150 ~ 180 nm
	Oxaliplatin	Colorectal	120 nm
	Sorafenib	Rectal cell carcinoma	101 ~ 229 nm
	Tamoxifen	Breast	155 ~ 225 nm
	Topotecan	Breast	65 ~ 110 nm
	Vinblastine	Lung	100 nm
	Vincristine and quercetin	Breast	130 nm

Lee et al. have developed sterically stabilized paclitaxel SLN for parenteral administration. The SLNs were composed of trimyristin as a solid core, egg phosphatidylcholine and PEGylated phospholipid distearoyl phosphatidyl ethanol amine—N-Poly (ethylene glycol) 2000 as a stabilizer. The research work is based on the hypothesis that triglyceride having a fatty acid chain can at least partially solubilize paclitaxel at elevated temperature and drug immobilization inside the solid core would increase stability. Also the phospholipid was used as a stabilizer. The reason for using the same is liposomes or even mixed micelles might coexit along with SLN dispersion developed during the studies. Also it is reported that more concentration of egg phosphatidylcholine was required along with PEGylated phospholipid to avoid gel-forming tendency. It was also found that the drug release of SLN in the plasma was very slow and difficult to measure the concentration of paclitaxel due to limitation of assay method (Lee, Lim, and Kim, 2007). In another study, a combination of stearic acid, phospholipid DPPC (Dipalmitoyl phosphatidylcholine) and DPPG (1,2-dipalmitoyl-sn-glucero-3-phospho-glycerol) sodium was successfully used to form SLN by microemulsification technique. Soya lecithin and sodium glycolate were used as surfactant. Further in vitro cytotoxicity studies and in vivo tissue targeting study confirmed the localization of SLN formulations inside the tumor (Patil and Joshi, 2013).

Cellular uptake and cytotoxicity of paclitaxel-loaded SLN was investigated by Yuan et al. SLNs were formulated by using different lipids such as glycerol tristearate, monostearin, stearic acid and Compritol 888 ATO. The functionality of SLN upon incorporation of ODA-FITC (octadecylamine-fluorescein isothiocyanate) and paclitaxel was evaluated. The incorporation of ODA-FITC into SLN enhanced the hydrophilicity. As a result, the size of fluorescent SLN was smaller than paclitaxel loaded SLN. SLN prepared with glycerol tristearate had the smallest particle size compared to SLN prepared with monostearin, stearic acid and Compritol 888 ATO. Based on optimized formulation for paclitaxel, drug-loaded monostearin SLNs were further evaluated and conjugated with PEG-Stearic acid and folic acid-stearic acid to enhance cellular uptake and cytotoxicity of the drug mediated by folate receptor. Modified SLNs were found to have smaller particle size may be due to enhanced hydrophilicity of the surface. The cellular uptake in A549 cells for both SLN was time dependent. The uptake was highly enhanced due to increased hydrophilicity of the surface. Florescent SLN were accumulated in the cytoplasm. Since the action of target of paclitaxel is in cytoplasmic microtubules, this could be an ideal carrier. Paclitaxel monostearin SLN conjugated with FA-SA showed higher cytotoxicity compared to paclitaxel solution which might have contributed to the faster cellular uptake due to enhanced endocytosis mediated by the folate receptor, whereas cytotoxicity of monostearin SLN modified with PEG-SA was not comparable with above formulations (Yuan et al., 2008).

Cell penetrating peptides (CPPs) are relatively short peptides consisting of 5–40 amino acid residuals and have the ability to cross the plasma membrane and enter into cell interiors. CPP promotes intracellular delivery of covalently or noncovalently conjugated bioactive drugs. Zhang et al. studied the Coumarin-6 and paclitaxel-loaded SLN surface modified with CPP (stearic acid octa arginine, SA-R8) and their effect on the cellular uptake of SLN in A549 cell lines. The amount and binding efficiency of SA-R8 attached to the SLN was related to initial SA-R8 concentration. Zeta potential values for Coumarin-6-loaded SLN and paclitaxel loaded SLN were less than 30 mV, confirming the absence of stearic stabilization. The zeta potential of Coumarin-6-loaded SLN and paclitaxel loaded SLN was increased with increase in amount of CPP. The initial particle size of paclitaxel loaded SLN (135 nm) was higher than Coumarin-6-loaded SLN (116 nm) having similar lipid matrix. Also no significant change in the particle size and polydispersity index was recorded after SA-R8 modification. CPP modified SLN showed slower release than coumarion-6-SLN and paclitaxel-SLN at pH 7.8 at 40 h. The cell viability of A-549 cells exposed to marketed paclitaxel (Taxol), paclitaxel-SLN and CPP coated paclitaxel-SLN showed dose-dependent cytotoxicity. The percent survival of A 549 cells after exposure to CPP-paclitaxel-SLN was decreased with increase in the amount of SA-R8. The cytotoxicity of Taxol was stronger than paclitaxel SLN (Zhang et al., 2013).

Similarly, Baek et al. attempted to surface-modify stearic acid-based SLN using 2-hydroxy propoyl–β-cyclodextrin (HPCD) and found that particle size had great impact on cellular uptake. Paclitaxel-loaded SLN showed prolonged cytotoxic effect in MCF-7 cell lines. Apoptosis studies were carried out by flow cytometry with annexin V–FITC/PI treatment with paclitaxel solution, HPCD modified—paclitaxel loaded SLN for 8 h. It showed an increase in the population of annexin-V-positive cells to 26% and 44.5% indicating the late occurrence and thus late apoptosis in MCF-7 cells. Also the paclitaxel solution and HPCD modified paclitaxel-loaded SLN showed increase in the population of PI positive cells to 17% and 20.9%. Comparative data of paclitaxel solution with modified SLN exhibited 44.5% apoptic cell death and 20.9% necrotic cell death. The pharmacokinetic studies were carried out at a dose level of 5 mg/kg body weight for paclitaxel solution; paclitaxel-SLN and HPCD modified paclitaxel-SLN. All the formulations showed biphasic decline. HPCD-modified SLN exhibited delayed blood clearance compared to paclitaxel solution owing to more stabilization of SLN by HPCD. Higher AUC in plasma studies were reported for HPCD-modified SLN followed by paclitaxel-SLN alone and lower AUC for paclitaxel solution. Both paclitaxel-SLN and HPCD-modified SLN showed insignificant changes in kidney and plasma creatinine levels suggested no kidney toxicity compared to conventional formulations. Antitumor efficacy in mice showed no significant changes in the size of tumor. However, tumor volume was remarkably smaller for mice treated with HPCD-modified SLN after 27 days. These findings indicated greater antitumor efficacy due to high cellular uptake and sustained action shown by HPCD-modified SLN (Baek, Kim, Park, and Cho, 2015).

Low-density lipoprotein (LDL) mimetic paclitaxel–loaded SLN were investigated as potential tumor targeted delivery by Kim et al. SLNs were composed of LDL mimicking composition-cholesteryl oleate and triolein as core lipid and DOPE while cholesterol and DC-cholesterol (carbamoyl cholesterol hydro-chloride) was used for surface modification. At later stage attempt was done to attach ligand antibody to paclitaxel-SLN by using heterofunctional cross-linking agent. LDL mimicking paclitaxel-loaded SLN in a hydrophobic lipid core possessed amphiphilic layer. PEG derivative was covalently attached to the surface to further enhance in vivo compatibility. Furthermore, antibody was conjugated to the distal ends of the PEG molecule. Three formulations were developed namely non-PEGylated SLN, PEGylated SLN, PEGylated SLN-antibody. Cellular uptake was drastically enhanced with ligand attached SLN in H1975 cancer cells. Also the pharmacokinetic and tissue distribution studies in mice indicated early phase accumulation of nanoparticles in liver for all three formulations, but more with non-PEGylated SLN. Radiolabeling studies further confirmed target specificity and significantly higher levels of ligand attached to SLN in tumor cells. The radio-activity values for paclitaxel SLN and PEGylated SLN were not significant compared to antibody attached SLN. Preclinical in vivo animal studies showed enhanced half-life of paclitaxel for more than 40 h due to stability, delayed clearance, longer circulation and more payload (Kim et al., 2015).

SLN coated with hyaluronic acid to target CD 44 overexpressed in B16F10 melanoma cells were investigated by Shen et al. As cancer stem cells represent abundant CD 44 receptor its binding with hyaluronic acid can arrest cell growth. Highly stable spherical-shaped cationic SLNs/paclitaxel was prepared by film–ultrasonic method and showed a particle size around 161 nm, and after surface modification 190 nm. B16F10-CD44 expressed cells showed faster uptake of surface-conjugated SLN compared to conventional nanoparticles and nanoparticles coated with the similar mucopolysaccharide heparin confirming HA-mediated mechanism though CD44 interaction resulting in rapid binding. Paclitaxel loaded in native and surface-modified SLNs in the dose range of 2–20 μg/mL showed dose-dependent cytotoxicity against B16F10-CD44+ and A549 cell lines. The use of hyaluronic acid SLNs facilitated rapid cell proliferation inhibition when tested in vitro. Immunostaining experiments further confirmed that drug-loaded surface-modified SLNs induced high chromatin condensation, nuclear fragmentation and rapid apoptotic body formation in larger number of cells than the mock, paclitaxel and paclitaxel SLN indicated greater proliferative effects. Furthermore, superior invasion and adhesion were noted for surface-modified SLNs in addition to significant attenuation of cell motility and migration. A biodistribution study upon i.v. injection in mice confirmed high accumulation of SLN compared to free drug, surface modification with hyaluronic acid further enhanced the drug accumulation in lung (Shen, Shi, Zhang, Gong, and Sun, 2015). Co-administration of paclitaxel with other drugs (Bae, Lee, Lee, Park, and Nam, 2013; Baek and Cho, 2015;

Table 9.3 Solid Lipids, and oils explored in preparation of lipid nanoparticle for delivery of various drugs

Class	Examples	Drug used	Particle size
Solid lipids	Cetylpalmitate	Plasmid DNA	198 ~ 400 nm
	Compritol 888 ATO	Celecoxib	217 nm
	Glycerol palmitostearate	Paclitaxel	92.6 ~ 114 nm
	Palmitic acid	Insulin	88.2 ~ 109 nm
	Precirol ATO 5	Itraconazole	120 ~ 350 nm
	Stearic acid	Insulin	88.2 ~ 109 nm
	Tripalmitin	Ketoprofen and indomethacin	30 nm
	Tristearin	Ketoprofen and indomethacin	30 nm
Liquid lipids (oils)	Oleic acida	Itraconazole	120 ~ 350 nm
	Miglyol 812	Celecoxib	217 nm
	Castor oil	Tilmicosin	103 ~ 211 nm
Surfactants	Poloxamer 188	Epirubicin	223 ~ 231 nm
	Polysorbate 20a	Itraconazole	120 ~ 350 nm
	Polysorbate 80a	Paclitaxel	92.6~114 nm
	Polyvinyl alcohol	Rifampicin, isoniazid and pyrazinamide	1.1 ~ 2.1 μm
	Sodium cholate	Thymopentin	4.1 ± 0.1 μm
	Sodium glycocholate	Ketoprofen and indomethacin	30 nm
	Sodium taurocholate	Celecoxib	217 nm
	Sorbitan trioleatea	Plasmid DNA	198 ~ 400 nm
	Soybean lecithina	Epirubicin	223 ~ 231 nm
	Soybean phosphatidylcholine	Thymopentin	4.1 ± 0.1 μm

Shi et al., 2014; Yu et al., 2012) and surface modification with folate (Yu et al., 2012), lactoferrin (Pandey, Gajbhiye, and Soni, 2015), monosyl (Pandey et al., 2015) are reported. Several other PEGylated lipid nanoparticles are successfully tested both in vitro and in vivo (Chen et al., 2001; Lee et al., 2007; Li, Eun, and Lee, 2011; Wu, Tang, and Yin, 2010) (Table 9.3).

9.4.5 Nanostructured lipid carriers

NLCs are designed to overcome challenges associated with SLNs like drug loading, lipid crystallization over time and drug expulsion. For NLCs, the highest drug load could be achieved by mixing solid lipids with small amounts of liquid state lipids (oils). They are sort of analogous to w/o/w emulsions, as it is a matrix composed of oil-in-solid lipid-in-water dispersion (Table 9.4).

Table 9.4 Commercial paclitaxel compositions

Name of product	Availability	Producer
Taxol (R)	Taxol is available in 30 mg (5 mL), 100 mg (16.7 mL), and 300 mg (50 mL) multidose vials. Each mL of sterile nonpyrogenic solution contains 6 mg paclitaxel, 527 mg of purified Cremophor EL (polyoxyethylated castor oil) and 49.7% (v/v) dehydrated alcohol	Bristol-Myers Squibb
Other generic brands: Paclitaxel medac 6 mg/mL concentrate for solution for infusion	It is available as injection concentrate and is composed of Macrogolglycerol ricinoleate 527 mg/mL and ethanol, anhydrous. Other ingredients include citric acid Available as 5, 16.7, 25, 50, and 100 mL vial	Medac GmbH, Actavis UK Ltd, Peckforton Pharmaceuticals, Hospira UK Ltd., Accord healthcare Ltd., and several others
Abraxane 5 mg/mL powder for suspension for infusion	Abraxane is albumin-bound paclitaxel nanoparticles (130 nm). Paclitaxel exists in the particles in a noncrystalline, amorphous state. It is available as a white to yellow, sterile, lyophilized powder for reconstitution with 20 mL of 0.9% Sodium Chloride Injection, USP prior to intravenous infusion Each vial contains 100 mg of paclitaxel formulated as albumin bound nanoparticles Each vial contains 250 mg of paclitaxel formulated as albumin bound nanoparticles After reconstitution, each mL of suspension contains 5 mg of paclitaxel formulated as albumin bound nanoparticles Each single-use vial contains 100 mg of paclitaxel (bound to human albumin) and approximately 900 mg of human albumin (containing sodium caprylate and sodium acetyltryptophanate). Each milliliter (mL) of reconstituted suspension contains 5 mg paclitaxel formulated as albumin-bound particles. The reconstituted suspension has a pH of 6–7.5 and an osmolality of 300–360 mOsm/kg	Celgene Ltd.

Acessed on 05.04.2016, 01.42am https://pubchem.ncbi.nlm.nih.gov/compound/paclitaxel#section=Top.

Target-specific cholesterol-based paclitaxel NLCs were developed by Rezazadon et al. to selectively bind LDL receptors overexpressed on the malignant cells. The efficacy of the NLC's (200 nm) was evaluated in HT 29 cell lines and cytotoxicity was found to be comparable with Cremophor EL-based paclitaxel formulations (Rezazadeh, Emami, and Varshosaz, 2012). Similarly, Jaber et al. developed NLC formulations composed of oleic acid and cholesterol. The hypothesis of the research group was based on the fact that LDL receptors easily engulf the cholesterol rich nanoemulsions. Therefore, oleic acid at 15% and 30% to lipid ratio of 5% and 10% was used to form nanostructured carriers which can be readily taken up by cells (Emami, Rezazadeh, Varshosaz, Tabbakhian and Aslani, 2012).

NLCs were prepared by using the emulsion solvent diffusion method and evaporation method. The oleic acid and surfactant concentration had a significant effect on the particle size distribution. As oleic acid concentration increased up to 30% particle size was found to decrease. The drug release from NLCs was dependent on the drug: lipid ratio used. Furthermore, NLCs were lyophilized and the effect of various cryoprotectants like PEG 4000, Avicel RC591, aerosil, sorbitol and dextrose on particle size distribution along with effect of processing variables was studied. Sorbitol and aerosil showed less change in particle size and zeta potential upon redispersion in distilled water (Emami, Rezazadeh, Varshosaz, Tabbakhian, and Aslani, 2012). Recently, a coadministration of paclitaxel and doxorubicin in the form of NLC was successfully tested and showed improved anticancer effects (Wang et al., 2015). Attempts were also made to deliver paclitaxel via surface-modified NLCs systems to selectively target the drug (Taratula, Kuzmov, Shah, Garbuzenko, and Minko, 2013; Yang et al., 2013; Yang, Xie, Zhang, and Mei, 2015) (Table 9.5).

9.5 MARKETED PACLITAXEL PRODUCTS

The drug paclitaxel is highly lipophilic in nature and insoluble in water. Its current marketed form as an intravenous infusion paclitaxel is solubilized by using Cremophor EL, which is a nonionic surfactant and helps in the micellar solubilization. Cremophor is polyoxyl-ethylated castor oil in which fatty acid esters of glycerol represent the hydrophobic portion, and the polyethylene glycol represent the hydrophilic portion of the surfactant having an HLB value of 12–14. In addition, dehydrated alcohol is used as the thinning agent and a cosolvent.

Paclitaxel formulated for parenteral administration is used as a concentrate for intravenous infusion and used upon dilution. It is marketed by Teva, Activis, Bristol-Myers Squibb and several others. As per the US-FDA, medicine compendium of United Kingdom, Rx list records, the other marketed formulations of paclitaxel injection (6 mg/mL) by companies such as Actavis Totowa, Fresenius Kabi Oncology, Hospira, Mylan Labs, Sandoz Inc., Teva Pharma, etc. All marketed preparations are composed of dehydrated alcohol and polyoxyl 35 castor oil (Cremophor EL). Most

Table 9.5 Various paclitaxel ANDA and NDA applications (referred from FDA database)

Proprietary name	Dosage form	Package description	Strength	Application	Labeler name	Start date
Paclitaxel	Injection	16.7 mL in 1 vial (0069-0076-01)	100 mg/16.7 mL	ANDA	Pfizer Laboratories Div Pfizer, Inc.	09/30/2011
Paclitaxel	Injection	50 mL in 1 vial (0069-0078-01)	300 mg/50 mL	ANDA	Pfizer Laboratories Div Pfizer, Inc.	09/30/2011
Paclitaxel	Injection	5 mL in 1 vial (0069-0079-01)	30 mg/5 mL	ANDA	Pfizer Laboratories Div Pfizer, Inc.	09/30/2011
Paclitaxel	Injection solution concentrate	1 vial, multidose in 1 carton (0703-4764-81) > 5 mL in 1 vial, multidose	6 mg/mL	ANDA	Teva Parenteral Medicines, Inc.	01/03/2016
Paclitaxel	Injection solution concentrate	1 vial, multidose in 1 carton (0703-4764-01) > 5 mL in 1 vial, multidose	6 mg/mL	ANDA	Teva Parenteral Medicines, Inc.	11/03/2009
Paclitaxel	Injection solution concentrate	1 vial, multidose in 1 carton (0703-4766-81) > 16.7 mL in 1 vial, multidose	6 mg/mL	ANDA	Teva Parenteral Medicines, Inc.	03/03/2016
Paclitaxel	Injection solution concentrate	1 vial, multidose in 1 carton (0703-4766-01) > 16.7 mL in 1 vial, multidose	6 mg/mL	ANDA	Teva Parenteral Medicines, Inc.	09/24/2008

Paclitaxel	Injection solution concentrate	1 vial, multidose in 1 carton (0703-4767-01) > 25 mL in 1 vial, multidose	6 mg/mL	ANDA	Teva Parenteral Medicines, Inc.	10/23/2008
Paclitaxel	Injection solution concentrate	1 vial, multidose in 1 carton (0703-4768-81) > 50 mL in 1 vial, multidose	6 mg/mL	ANDA	Teva Parenteral Medicines, Inc.	03/03/2016
Paclitaxel	Injection solution concentrate	1 vial, multidose in 1 carton (0703-4768-01) > 50 mL in 1 vial, multidose	6 mg/mL	ANDA	Teva Parenteral Medicines, Inc.	09/16/2009
Paclitaxel	Injection solution	1 vial in 1 carton (25021-213-05) > 5 mL in 1 vial	6 mg/mL	ANDA	Sagent Pharmaceuticals	09/27/2011
Paclitaxel	Injection solution	1 vial in 1 carton (25021-213-17) > 16.7 mL in 1 vial	6 mg/mL	ANDA	Sagent Pharmaceuticals	09/27/2011
Paclitaxel	Injection solution	1 vial in 1 carton (25021-213-50) > 50 mL in 1 vial	6 mg/mL	ANDA	Sagent Pharmaceuticals	09/27/2011
Paclitaxel	Injection	1 vial, multidose in 1 carton (44567-504-01) > 5 mL in 1 vial, multidose	6 mg/mL	NDA	WG Critical Care, LLC	07/11/2013

(Continued)

Table 9.5 Various paclitaxel ANDA and NDA applications (referred from FDA database) (Continued)

Proprietary name	Dosage form	Package description	Strength	Application	Labeler name	Start date
Paclitaxel	Injection	1 vial, multidose in 1 carton (44567–505–01) > 16.7 mL in 1 vial, multidose	6 mg/mL	NDA	WG Critical Care, LLC	07/11/2013
Paclitaxel	Injection	1 vial, multidose in 1 carton (44567–506–01) > 50 mL in 1 vial, multidose	6 mg/mL	NDA	WG Critical Care, LLC	07/11/2013
Paclitaxel	Injection solution	1 vial, multidose in 1 carton (45963–613–53) > 16.7 mL in 1 vial, multidose	6 mg/mL	ANDA	Actavis Pharma, Inc.	01/05/2015
Paclitaxel	Injection solution	1 vial, multidose in 1 carton (45963–613–56) > 5 mL in 1 vial, multidose	6 mg/mL	ANDA	Actavis Pharma, Inc.	01/05/2015
Paclitaxel	Injection solution	1 vial, multidose in 1 carton (45963–613–59) > 50 mL in 1 vial, multidose	6 mg/mL	ANDA	Actavis Pharma, Inc.	01/05/2015
Paclitaxel	Injection solution	1 vial, multidose in 1 carton (61703–342–09) > 5 mL in 1 vial, multidose	6 mg/mL	ANDA	Hospira Worldwide, Inc.	05/08/2002

Paclitaxel	Injection solution	1 vial, multidose in 1 carton (61703–342–22) > 16.7 mL in 1 vial, multidose	6 mg/mL	ANDA	Hospira Worldwide, Inc.	05/08/2002
Paclitaxel	Injection solution	1 vial, multidose in 1 carton (61703–342–50) > 50 mL in 1 vial, multidose	6 mg/mL	ANDA	Hospira Worldwide, Inc.	05/08/2002
Paclitaxel	Injection solution	1 vial in 1 box (63323–763–05) > 5 mL in 1 vial	6 mg/mL	ANDA	APP Pharmaceuticals, LLC	03/20/2009
Paclitaxel	Injection solution	1 vial in 1 box (63323–763–16) > 16.7 mL in 1 vial	6 mg/mL	ANDA	APP Pharmaceuticals, LLC	03/20/2009
Paclitaxel	Injection solution	1 vial in 1 box (63323–763–50) > 50 mL in 1 vial	6 mg/mL	ANDA	APP Pharmaceuticals, LLC	03/20/2009
Paclitaxel	Injection solution	1 vial, multidose in 1 carton (63323–763–06) > 5 mL in 1 vial, multidose	6 mg/mL	ANDA	Fresenius Kabi USA, LLC	08/15/2014
Paclitaxel	Injection solution	1 vial, multidose in 1 carton (63323–763–17) > 16.7 mL in 1 vial, multidose	6 mg/mL	ANDA	Fresenius Kabi USA, LLC	08/15/2014

(Continued)

Table 9.5 Various paclitaxel ANDA and NDA applications (referred from FDA database) (Continued)

Proprietary name	Dosage form	Package description	Strength	Application	Labeler name	Start date
Paclitaxel	Injection solution	1 vial, multidose in 1 carton (63323–763–52) > 50 mL in 1 vial, multidose	6 mg/mL	ANDA	Fresenius Kabi USA, LLC	08/15/2014
Paclitaxel	Injection solution	1 vial, multidose in 1 carton (66758–043–01) > 5 mL in 1 vial, multidose	6 mg/mL	ANDA	Sandoz, Inc.	01/02/2008
Paclitaxel	Injection solution	1 vial, multidose in 1 carton (66758–043–02) > 16.7 mL in 1 vial, multidose	6 mg/mL	ANDA	Sandoz, Inc.	01/02/2008
Paclitaxel	Injection solution	1 vial, multidose in 1 carton (66758–043–03) > 50 mL in 1 vial, multidose	6 mg/mL	ANDA	Sandoz, Inc.	01/02/2008
Paclitaxel	Injection	1 vial in 1 carton (67457–434–51) > 50 mL in 1 vial	300 mg/50 mL	ANDA	Mylan Institutional LLC	09/30/2011
Paclitaxel	Injection	1 vial in 1 carton (67457–449–17) > 16.7 mL in 1 vial	100 mg/16.7 mL	ANDA	Mylan Institutional LLC	09/30/2011
Paclitaxel	Injection	1 vial in 1 carton (67457–471–52) > 5 mL in 1 vial	30 mg/5 mL	ANDA	Mylan Institutional LLC	09/30/2011
ABRAXANE	Injection, powder, lyophilized, for suspension	1 vial, single–use in 1 carton (68817–134–50) > 20 mL in 1 vial, single-use	100 mg/20 mL	NDA	Abraxis BioScience, LLC	02/10/2005

National Drug Code Directory, accessed on 05.04.2016, 02.15 am, http://www.accessdata.fda.gov/scripts/cder/ndc/default.cfm.

of the paclitaxel injection 1 mL concentrate for solution for infusion contains 6 mg paclitaxel. Commercially available paclitaxel injectable concentrate is a viscous solution and must be used upon dilution with 5% dextrose and 0.9% sodium chloride injection or 5% dextrose and Ringer's injection. In general, solutions containing 0.6 or 1.2 mg of paclitaxel per mL maintain a pH of 4.4–5.6 for up to 27 h Abraxane (R) (Abraxis Bioscience, USA) currently represent the only albumin-based nanotechnology therapy approved for the treatment of metastatic breast cancer, lung cancer and pancreatic cancer in the United States, Europe, and other markets around the world. It is manufactured using the patented lab technology and contains albumin-bound paclitaxel nanoparticles. It is free of solvents. The current patent expiration date is February 21, 2026, and its exclusivity expiry is September 6, 2020. Abraxane avoids cremophor-related toxicities such as severe hypersensitivity including anaphylactic shocks and death. The Rx website (www.rxlist.com) The Internet Drug Index, provides ADME information of Abraxane. The pharmacokinetic studies revealed the AUC was dose-proportional and the pharmacokinetic data of administered Abraxane for 30 min was comparable to a paclitaxel injection for about 3-h infusion. The dose of Abraxane was 260 mg/m^2 and that of paclitaxel injection was 175 mg/m^2. Also Abraxane showed a larger clearance and higher volume of distribution than the paclitaxel injection. The drug paclitaxel (from Abraxane formulation) was uniformly distributed in blood cells and plasma. The unbound fraction of paclitaxel was higher in plasma with Abraxane as compared to paclitaxel injection. Paclitaxel from Abraxane is primarily metabolized in human liver microsomes during in vitro studies. The average total clearance of paclitaxel from Abraxane is from 13 to 30 L/h/m^2, half-life is from 13 to 27 h. Post 30 min infusion of Abraxane, almost 4% of unchanged drug and <1% of metabolites is excreted in urine. Almost 20% of the total dose given was excreted through the fecal route. The advantages of Abraxane over Taxol (R) are that it permits administration of a higher dose with comparative toxicity. The intratumor drug concentration is higher, decreases neutropenia, and produces transient neuropathy.

9.6 CONCLUSION

The lipidic nanoparticle delivery systems is a suitable drug delivery carrier for hydrophobic drugs such as paclitaxel. The biocompatibility, superior in vivo performance, use of safer excipients used in the lipidic nanoparticle systems makes them an attractive approach to target cancer with minimal toxic effects on normal cells or tissues. The performance of lipidic nanoparticles can be further optimized by surface modification using specific targeting ligands which can enhance the site-specific targeting to tumor cells over normal cells. Lipid nanoparticles like liposomes, SLNs and NLCs can be used for hydrophobic drug delivery to cancerous tissues and have minimal side effects and formulation issues like use of higher concentration of toxic solubilizer associated with available commercial paclitaxel solution. The commercial success of albumin–bound paclitaxel nanoparticles provides a rationale for further development of nanoparticle system for cancer therapy. Research till date on paclitaxel lipid-based nanoparticles shows that the further collaborative effort by scientific community is needed to create fundamental understanding, translational application, and commercial production of lipid nanoparticles.

ACKNOWLEDGMENT

The authors acknowledge the support of the National Institute of General Medical Science of the National Institutes of Health under award number SC3GM109873. The authors acknowledge Hawai'i Community Foundation, Honolulu, HI 96813, USA, for research support on lung cancer, mesothelioma, and asthma projects (Leahi Fund 15ADVC-74296) in 2015, 2013, and 2011, respectively. The authors would like to acknowledge the 2013 George F. Straub Trust and Robert C. Perry Fund of the Hawai'i Community Foundation, Honolulu, HI 96813, USA, for research support on lung cancer. The authors also acknowledge a seed grant from the Research Corporation of the University of Hawai'i at Hilo, Hilo, HI 96720, USA, and The Daniel K. Inouye College of Pharmacy, University of Hawaii at Hilo, Hilo, HI 96720, USA, for providing start-up financial support to their research group. The authors acknowledge the donation from Dr. Robert S. Shapiro, MD, Dermatologist, Hilo, HI 96720, USA in support of development of nanotechnology-based medicines.

REFERENCES

Agostinelli, E., Vianello, F., Magliulo, G., Thomas, T., Thomas, T.J., 2015. Nanoparticle strategies for cancer therapeutics: nucleic acids, polyamines, bovine serum amine oxidase and iron oxide nanoparticles (Review). Int. J. Oncol. 46 (1), 5–16. http://dx.doi.org/10.3892/ijo.2014.2706.

Ahmad, A., Sheikh, S., Ali, S., Paithankar, M., Mehta, A., et al., 2015. Nanosomal paclitaxel lipid suspension demonstrates higher response rates compared to paclitaxel in patients with metastatic breast cancer. J. Cancer Sci. Ther. 7, 116–120.

Ahmad, J., Mir, S.R., Kohli, K., Chuttani, K., Mishra, A.K., Panda, A.K., et al., 2014. Solid-nanoemulsion preconcentrate for oral delivery of paclitaxel: formulation design, biodistribution, and gamma scintigraphy imaging. BioMed. Res. Int. 2014, 984756. http://dx.doi.org/10.1155/2014/984756.

Alani, A.W., Rao, D.A., Seidel, R., Wang, J., Jiao, J., Kwon, G.S., 2010. The effect of novel surfactants and Solutol HS 15 on paclitaxel aqueous solubility and permeability across a Caco-2 monolayer. J. Pharm. Sci. 99 (8), 3473–3485. http://dx.doi.org/10.1002/jps.22111.

American Joint Committee on Cancer, 2002. AJCC Cancer Staging Handbook: TNM Classification of Malignant Tumors. 2002, sixth ed. Springer Science and Business Media, New York.

Bae, K.H., Lee, J.Y., Lee, S.H., Park, T.G., Nam, Y.S., 2013. Optically traceable solid lipid nanoparticles loaded with siRNA and paclitaxel for synergistic chemotherapy with in situ imaging. Adv. Healthcare Mater. 2 (4), 576–584. http://dx.doi.org/10.1002/adhm.201200338.

Baek, J.S., Cho, C.W., 2015. Controlled release and reversal of multidrug resistance by co-encapsulation of paclitaxel and verapamil in solid lipid nanoparticles. Int. J. Pharm. 478 (2), 617–624. http://dx.doi.org/10.1016/j.ijpharm.2014.12.018.

Baek, J.S., Kim, J.H., Park, J.S., Cho, C.W., 2015. Modification of paclitaxel-loaded solid lipid nanoparticles with 2-hydroxypropyl-beta-cyclodextrin enhances absorption and reduces nephrotoxicity associated with intravenous injection. Int. J. Nanomed. 10, 5397–5405. http://dx.doi.org/10.2147/IJN.S86474.

Balzeau, J., Pinier, M., Berges, R., Saulnier, P., Benoit, J.P., Eyer, J., 2013. The effect of functionalizing lipid nanocapsules with NFL-TBS.40-63 peptide on their uptake by glioblastoma cells. Biomaterials 34 (13), 3381–3389. http://dx.doi.org/10.1016/j.biomaterials.2013.01.068.

Bazak, R., Houri, M., El Achy, S., Kamel, S., Refaat, T., 2015. Cancer active targeting by nanoparticles: a comprehensive review of literature. J. Cancer Res. Clin. Oncol. 141 (5), 769–784. http://dx.doi. org/10.1007/s00432-014-1767-3.

Beijnen, J.H., Huizing, M.T., ten Bokkel Huinink, W.W., Veenhof, C.H., Vermorken, J.B., Giaccone, G., et al., 1994. Bioanalysis, pharmacokinetics, and pharmacodynamics of the novel anticancer drug paclitaxel (Taxol). Semin. Oncol. 21 (5 Suppl 8), 53–62.

Ben-Eliyahu, S., 2003. The promotion of tumor metastasis by surgery and stress: immunological basis and implications for psychoneuroimmunology. Brain, Behav. Immunity 17 (Suppl 1), S27–S36.

Ben-Eliyahu, S., Page, G.G., Yirmiya, R., Shakhar, G., 1999. Evidence that stress and surgical interventions promote tumor development by suppressing natural killer cell activity. Int. J. Cancer 80 (6), 880–888.

Biswas, S., Dodwadkar, N.S., Deshpande, P.P., Torchilin, V.P., 2012. Liposomes loaded with paclitaxel and modified with novel triphenylphosphonium-PEG-PE conjugate possess low toxicity, target mitochondria and demonstrate enhanced antitumor effects in vitro and in vivo. J. Control. Release 159 (3), 393–402. http://dx.doi.org/10.1016/j.jconrel.2012.01.009.

Brigger, I., Dubernet, C., Couvreur, P., 2012. Nanoparticles in cancer therapy and diagnosis. Adv. Drug Deliv. Rev. 64, 24–36.

Bu, H., He, X., Zhang, Z., Yin, Q., Yu, H., Li, Y., 2014. A TPGS-incorporating nanoemulsion of paclitaxel circumvents drug resistance in breast cancer. Int. J. Pharm. 471 (1–2), 206–213. http://dx.doi. org/10.1016/j.ijpharm.2014.05.039.

Cancer multidrug resistance, NATURE BIOTECHNOLOGY VOL 18 SUPPLEMENT 2000 IT 18- IT 20. http://www.nature.com/nbt/journal/v18/n10s/pdf/nbt1000_IT18.pdf.

Cao, J., Wang, R., Gao, N., Li, M., Tian, X., Yang, W., et al., 2015. A7RC peptide modified paclitaxel liposomes dually target breast cancer. Biomater. Sci. 3 (12), 1545–1554. http://dx.doi.org/10.1039/ c5bm00161g.

Cara, E., 2015. Cancer Now Second Leading Cause Of Death Worldwide: 14.9 Million Cases and 8.2 Million Deaths In 2013. http://www.medicaldaily.com/cancer-now-second-leading-cause-death-worldwide-149-million-cases-and-82-million-335846, acessed on 11 April 2016.

Carelle, N., Piotto, E., Bellanger, A., Germanaud, J., Thuillier, A., Khayat, D., 2002. Changing patient perceptions of the side effects of cancer chemotherapy. Cancer 95 (1), 155–163. http://dx.doi.org/10.1002/ cncr.10630.

Carter, S.K., Slavik, M., 1974. Chemotherapy of cancer. Ann. Rev. Pharmacol. 14 (1), 157–183.

Carvalho, P.O., Maranhao, R.C., Stolf, N.A., 2014. A lipid nanoemulsion carrying paclitaxel improves the gene expression of inflammatory factors of heart grafts in rabbits. J. Thorac. Cardiovas. Surg. 148 (4), 1765–1766. http://dx.doi.org/10.1016/j.jtcvs.2014.05.026.

Cerqueira, B.B., Lasham, A., Shelling, A.N., Al-Kassas, R., 2015. Nanoparticle therapeutics: technologies and methods for overcoming cancer. Eur. J. Pharm. Biopharm. 97 (Pt A), 140–151. http://dx.doi. org/10.1016/j.ejpb.2015.10.007.

Chen, D.B., Yang, T.Z., Lu, W.L., Zhang, Q., 2001. In vitro and in vivo study of two types of long-circulating solid lipid nanoparticles containing paclitaxel. Chem. Pharm. Bull. 49 (11), 1444–1447.

Danaei, G., Vander Hoorn, S., Lopez, A.D., Murray, C.J., Ezzati, M., group, C.R.A.C., 2005. Causes of cancer in the world: comparative risk assessment of nine behavioural and environmental risk factors. Lancet 366 (9499), 1784–1793.

Danhier, F., Feron, O., Preat, V., 2010. To exploit the tumor microenvironment: passive and active tumor targeting of nanocarriers for anti-cancer drug delivery. J. Control. Release 148 (2), 135–146. http:// dx.doi.org/10.1016/j.jconrel.2010.08.027.

Delaney, G., Jacob, S., Featherstone, C., Barton, M., 2005. The role of radiotherapy in cancer treatment. Cancer 104 (6), 1129–1137.

Desai, A., Vyas, T., Amiji, M., 2008. Cytotoxicity and apoptosis enhancement in brain tumor cells upon coadministration of paclitaxel and ceramide in nanoemulsion formulations. J. Pharm. Sci. 97 (7), 2745–2756. http://dx.doi.org/10.1002/jps.21182.

Dias, M.L., Carvalho, J.P., Rodrigues, D.G., Graziani, S.R., Maranhao, R.C., 2007. Pharmacokinetics and tumor uptake of a derivatized form of paclitaxel associated to a cholesterol-rich nanoemulsion (LDE) in patients with gynecologic cancers. Cancer Chemother. Pharmacol. 59 (1), 105–111. http://dx.doi. org/10.1007/s00280-006-0252-3.

Dinda, A., Biswal, I., Chowdhury, P., Mohapatra, R., 2013. Formulation development and evaluation of pacli-taxel loaded solid lipid nanoparticles using glyceryl monostearate. J. Appl. Pharm. Sci. 3 (8), 133–138.

Du, R., Zhong, T., Zhang, W.Q., Song, P., Song, W.D., Zhao, Y., et al., 2014. Antitumor effect of iRGD-mod-ified liposomes containing conjugated linoleic acid-paclitaxel (CLA-PTX) on B16-F10 melanoma. Int. J. Nanomed. 9, 3091–3105. http://dx.doi.org/10.2147/IJN.S65664.

Emami, J., Rezazadeh, M., Varshosaz, J., Tabbakhian, M., Aslani, A., 2012. Formulation of LDL targeted nanostructured lipid carriers loaded with paclitaxel: a detailed study of preparation, freeze drying con-dition, and in vitro cytotoxicity. J. Nanomater. Article ID 358782.

Esfahani, M., Alavi, E., Akbarzadeh, A., Ghassemi, S., Saffari, Z., Farahnak, M., et al., 2014. Pegylation of nanoliposomal paclitaxel enhances its efficacy in breast cancer. Trop. J. Pharm. Res. 13 (8), 1195–1198.

Fellner, S., Bauer, B., Miller, D.S., Schaffrik, M., Fankhanel, M., Spruss, T., et al., 2002. Transport of pacli-taxel (Taxol) across the blood-brain barrier in vitro and in vivo. J. Clin. Investigat. 110 (9), 1309–1318. http://dx.doi.org/10.1172/JCI15451.

Ganta, S., Amiji, M., 2009. Coadministration of Paclitaxel and curcumin in nanoemulsion formulations to overcome multidrug resistance in tumor cells. Mol. Pharm. 6 (3), 928–939. http://dx.doi.org/10.1021/mp800240j.

Ganta, S., Devalapally, H., Amiji, M., 2010. Curcumin enhances oral bioavailability and anti-tumor thera-peutic efficacy of paclitaxel upon administration in nanoemulsion formulation. J. Pharm. Sci. 99 (11), 4630–4641. http://dx.doi.org/10.1002/jps.22157.

Gao, P., Rush, B.D., Pfund, W.P., Huang, T., Bauer, J.M., Morozowich, W., et al., 2003. Development of a supersaturable SEDDS (S-SEDDS) formulation of paclitaxel with improved oral bioavailability. J. Pharm. Sci. 92 (12), 2386–2398. http://dx.doi.org/10.1002/jps.10511.

Goldberg, S.B., Contessa, J.N., Omay, S.B., Chiang, V., 2015. Lung cancer brain metastases. Cancer J. 21 (5), 398–403. http://dx.doi.org/10.1097/PPO.0000000000000146.

Groo, A.C., Saulnier, P., Gimel, J.C., Gravier, J., Ailhas, C., Benoit, J.P., et al., 2013. Fate of paclitaxel lipid nanocapsules in intestinal mucus in view of their oral delivery. Int. J. Nanomed. 8, 4291–4302. http://dx.doi.org/10.2147/IJN.S51837.

Gursoy, N., Garrigue, J.S., Razafindratsita, A., Lambert, G., Benita, S., 2003. Excipient effects on in vitro cytotoxicity of a novel paclitaxel self-emulsifying drug delivery system. J. Pharm. Sci. 92 (12), 2411–2418. http://dx.doi.org/10.1002/jps.10501.

Hamada, H., Ishihara, K., Masuoka, N., Mikuni, K., Nakajima, N., 2006. Enhancement of water-solubility and bioactivity of paclitaxel using modified cyclodextrins. J. Biosci. Bioeng. 102 (4), 369–371. http://dx.doi.org/10.1263/jbb.102.369.

He, C., Hu, Y., Yin, L., Tang, C., Yin, C., 2010. Effects of particle size and surface charge on cellular uptake and biodistribution of polymeric nanoparticles. Biomaterials 31 (13), 3657–3666. http://dx.doi.org/10.1016/j.biomaterials.2010.01.065.

Hitt, C., 1994. Paclitaxel: a new antineoplastic agent. Connecticut Med. 58 (3), 175–177.

Holschneider, C.H., Berek, J.S., 2000. Ovarian cancer: epidemiology, biology, and prognostic factors. Paper presented at the Seminars in surgical oncology.

Hu, S., Zhu, Z., Li, L., Chang, L., Li, W., Cheng, L., et al., 2008. Epitope mapping and structural analysis of an anti-ErbB2 antibody A21: molecular basis for tumor inhibitory mechanism. Proteins 70 (3), 938–949. http://dx.doi.org/10.1002/prot.21551.

Hureaux, J., Lagarce, F., Gagnadoux, F., Rousselet, M.C., Moal, V., Urban, T., et al., 2010. Toxicological study and efficacy of blank and paclitaxel-loaded lipid nanocapsules after i.v. administration in mice. Pharm. Res. 27 (3), 421–430. http://dx.doi.org/10.1007/s11095-009-0024-y.

Hureaux, J., Lagarce, F., Gagnadoux, F., Vecellio, L., Clavreul, A., Roger, E., et al., 2009. Lipid nanocap-sules: ready-to-use nanovectors for the aerosol delivery of paclitaxel. Eur. J. Pharm. Biopharm. 73 (2), 239–246. http://dx.doi.org/10.1016/j.ejpb.2009.06.013.

Iyer, A.K., Khaled, G., Fang, J., Maeda, H., 2006. Exploiting the enhanced permeability and retention effect for tumor targeting. Drug Discov. Today 11 (17), 812–818.

Jing, X., Deng, L., Gao, B., Xiao, L., Zhang, Y., Ke, X., et al., 2014. A novel polyethylene glycol mediated lipid nanoemulsion as drug delivery carrier for paclitaxel. Nanomed. Nanotechnol. Biol. Med. 10 (2), 371–380. http://dx.doi.org/10.1016/j.nano.2013.07.018.

Joshi, N., Kaviratna, A., Banerjee, R., 2013. Multi trigger responsive, surface active lipid nanovesicle aerosols for improved efficacy of paclitaxel in lung cancer. Integrat. Biol. Quant. Biosci. Nano Macro 5 (1), 239–248. http://dx.doi.org/10.1039/c2ib20122d.

Karanth, H., Murthy, R., 2009. Action of Paclitaxel pH-Sensitive Liposomes on B16F1 Melanoma Cells. NSTI-Nanotech 2, 104–107.

Khandavilli, S., Panchagnula, R., 2007. Nanoemulsions as versatile formulations for paclitaxel delivery: peroral and dermal delivery studies in rats. J. Invest. Dermatol. 127 (1), 154–162. http://dx.doi.org/10.1038/sj.jid.5700485.

Kim, J.H., Kim, Y., Bae, K.H., Park, T.G., Lee, J.H., Park, K., 2015. Tumor-targeted delivery of paclitaxel using low density lipoprotein-mimetic solid lipid nanoparticles. Mol. Pharm. 12 (4), 1230–1241. http://dx.doi.org/10.1021/mp500737y.

Kretzer, I.F., Maria, D.A., Maranhao, R.C., 2012. Drug-targeting in combined cancer chemotherapy: tumor growth inhibition in mice by association of paclitaxel and etoposide with a cholesterol-rich nanoemulsion. Cell. Oncol. 35 (6), 451–460. http://dx.doi.org/10.1007/s13402-012-0104-6.

Kulsh, J., 1997. Targeting a key enzyme in cell growth: a novel therapy for cancer. Med. Hypoth. 49 (4), 297–300.

Kumar, G.N., Walle, U.K., Bhalla, K.N., Walle, T., 1993. Binding of taxol to human plasma, albumin and alpha 1-acid glycoprotein. Res. Commun. Chem. Pathol. Pharmacol. 80 (3), 337–344.

Kunstfeld, R., Wickenhauser, G., Michaelis, U., Teifel, M., Umek, W., Naujoks, K., et al., 2003. Paclitaxel encapsulated in cationic liposomes diminishes tumor angiogenesis and melanoma growth in a "humanized" SCID mouse model. J. Invest. Dermatol. 120 (3), 476–482. http://dx.doi.org/10.1046/j.1523-1747.2003.12057.x.

Lacoeuille, F., Hindre, F., Moal, F., Roux, J., Passirani, C., Couturier, O., et al., 2007. In vivo evaluation of lipid nanocapsules as a promising colloidal carrier for paclitaxel. Int. J. Pharm. 344 (1–2), 143–149. http://dx.doi.org/10.1016/j.ijpharm.2007.06.014.

Lee, J.S., Hong, J.M., Moon, G.J., Lee, P.H., Ahn, Y.H., Bang, O.Y., 2010. A long-term follow-up study of intravenous autologous mesenchymal stem cell transplantation in patients with ischemic stroke. Stem Cells 28 (6), 1099–1106. http://dx.doi.org/10.1002/stem.430.

Lee, K.C., Maturo, C., Rodriguez, R., Nguyen, H.L., Shorr, R., 2011. Nanomedicine-nanoemulsion formulation improves safety and efficacy of the anti-cancer drug paclitaxel according to preclinical assessment. J. Nanosci. Nanotechnol. 11 (8), 6642–6656.

Lee, M.K., Lim, S.J., Kim, C.K., 2007. Preparation, characterization and in vitro cytotoxicity of paclitaxel-loaded sterically stabilized solid lipid nanoparticles. Biomaterials 28 (12), 2137–2146. http://dx.doi.org/10.1016/j.biomaterials.2007.01.014.

Lemmens, K., Segers, V.F., Demolder, M., De Keulenaer, G.W., 2006. Role of neuregulin-1/ErbB2 signaling in endothelium-cardiomyocyte cross-talk. J. Biol. Chem. 281 (28), 19469–19477. http://dx.doi.org/10.1074/jbc.M600399200.

Leone, J.P., Leone, B.A., 2015. Breast cancer brain metastases: the last frontier. Exp. Hematol. Oncol. 4, 33. http://dx.doi.org/10.1186/s40164-015-0028-8.

Li, J., Wang, F., Sun, D., Wang, R., 2016. A review of the ligands and related targeting strategies for active targeting of paclitaxel to tumours. J. Drug Target. 1–13. http://dx.doi.org/10.3109/1061186X.2016.1154561.

Li, N., Huang, C., Luan, Y., Song, A., Song, Y., Garg, S., 2016. Active targeting co-delivery system based on pH-sensitive methoxy-poly(ethylene glycol)-poly(epsilon-caprolactone)-poly(glutamic acid) for enhanced cancer therapy. J. Colloid Interface Sci. 472, 90–98. http://dx.doi.org/10.1016/j.jcis.2016.03.039.

Li, P., Hynes, S.R., Haefele, T.F., Pudipeddi, M., Royce, A.E., Serajuddin, A.T., 2009. Development of clinical dosage forms for a poorly water-soluble drug II: formulation and characterization of a novel solid microemulsion preconcentrate system for oral delivery of a poorly water-soluble drug. J. Pharm. Sci. 98 (5), 1750–1764. http://dx.doi.org/10.1002/jps.21547.

Li, R., Eun, J.S., Lee, M.K., 2011. Pharmacokinetics and biodistribution of paclitaxel loaded in pegylated solid lipid nanoparticles after intravenous administration. Arch. Pharm. Res. 34 (2), 331–337. http://dx.doi.org/10.1007/s12272-011-0220-2.

Liu, Y., Mei, L., Yu, Q., Xu, C., Qiu, Y., Yang, Y., et al., 2015. Multifunctional tandem peptide modified paclitaxel-loaded liposomes for the treatment of vasculogenic mimicry and cancer stem cells in malignant glioma. ACS Appl. Mater. Interfaces 7 (30), 16792–16801. http://dx.doi.org/10.1021/acsami.5b04596.

Liu, Y., Miyoshi, H., Nakamura, M., 2007. Nanomedicine for drug delivery and imaging: a promising avenue for cancer therapy and diagnosis using targeted functional nanoparticles. Int. J. Cancer 120 (12), 2527–2537.

Lo, J.T., Chen, B.H., Lee, T.M., Han, J., Li, J.L., 2010. Self-emulsifying O/W formulations of paclitaxel prepared from mixed nonionic surfactants. J. Pharm. Sci. 99 (5), 2320–2332. http://dx.doi.org/10.1002/jps.21993.

Lollo, G., Vincent, M., Ullio-Gamboa, G., Lemaire, L., Franconi, F., Couez, D., et al., 2015. Development of multifunctional lipid nanocapsules for the co-delivery of paclitaxel and CpG-ODN in the treatment of glioblastoma. Int. J. Pharm. 495 (2), 972–980. http://dx.doi.org/10.1016/j.ijpharm.2015.09.062.

Luo, L.M., Huang, Y., Zhao, B.X., Zhao, X., Duan, Y., Du, R., et al., 2013. Anti-tumor and anti-angiogenic effect of metronomic cyclic NGR-modified liposomes containing paclitaxel. Biomaterials 34 (4), 1102–1114. http://dx.doi.org/10.1016/j.biomaterials.2012.10.029.

Ma, N., Ma, C., Li, C., Wang, T., Tang, Y., Wang, H., et al., 2013. Influence of nanoparticle shape, size, and surface functionalization on cellular uptake. J. Nanosci. Nanotechnol. 13 (10), 6485–6498.

Macrophage, G., 1994. Chemotherapy and Drug Resistance.

Maeda, H., 2013. The link between infection and cancer: tumor vasculature, free radicals, and drug delivery to tumors via the EPR effect. Cancer Sci. 104 (7), 779–789. http://dx.doi.org/10.1111/cas.12152.

Maeda, H., 2014. Research spotlight: emergence of EPR effect theory and development of clinical applications for cancer therapy. Therap. Deliv. 5 (6), 627–630. http://dx.doi.org/10.4155/tde.14.36.

Maeda, H., Wu, J., Sawa, T., Matsumura, Y., Hori, K., 2000. Tumor vascular permeability and the EPR effect in macromolecular therapeutics: a review. J. Control. Release 65 (1), 271–284.

Malam, Y., Loizidou, M., Seifalian, A.M., 2009. Liposomes and nanoparticles: nanosized vehicles for drug delivery in cancer. Trends Pharm. Sci. 30 (11), 592–599. http://dx.doi.org/10.1016/j.tips.2009.08.004.

Matsumura, Y., Maeda, H., 1986. A new concept for macromolecular therapeutics in cancer chemotherapy: mechanism of tumoritropic accumulation of proteins and the antitumor agent smancs. Cancer Res. 46 (12 Part 1), 6387–6392.

McPherson, K., Steel, C., Dixon, J., 2000. Breast cancer—epidemiology, risk factors, and genetics. BMJ 321 (7261), 624–628.

Motzer, R.J., Escudier, B., Oudard, S., Hutson, T.E., Porta, C., Bracarda, S., et al., 2008. Efficacy of everolimus in advanced renal cell carcinoma: a double-blind, randomised, placebo-controlled phase III trial. Lancet 372 (9637), 449–456. http://dx.doi.org/10.1016/S0140-6736(08)61039-9.

MuÈller, R.H., MaÈder, K., Gohla, S., 2000. Solid lipid nanoparticles (SLN) for controlled drug delivery–a review of the state of the art. Eur. J. Pharm. Biopharm. 50 (1), 161–177.

Mulder, W.J., Strijkers, G.J., van Tilborg, G.A., Griffioen, A.W., Nicolay, K., 2006. Lipid-based nanoparticles for contrast-enhanced MRI and molecular imaging. NMR Biomed. 19 (1), 142–164.

Muller, R.H., Keck, C.M., 2004. Challenges and solutions for the delivery of biotech drugs–a review of drug nanocrystal technology and lipid nanoparticles. J. Biotechnol. 113 (1), 151–170.

Muller, R.H., Shegokar, R., Keck, C.M., 2011. 20 years of lipid nanoparticles (SLN and NLC): present state of development and industrial applications. Curr. Drug Discov. Technol. 8 (3), 207–227.

Nastiuk, K.L., Krolewski, J.J., 2016. Opportunities and challenges in combination gene cancer therapy. Adv. Drug Deliv. Rev. 98, 35–40. http://dx.doi.org/10.1016/j.addr.2015.12.005.

National Cancer Institution, 2016. What Is Cancer? Retrieved April 26th, 2016, from http://www.cancer.gov/about-cancer/what-is-cancer.

Niu, R., Zhao, P., Wang, H., Yu, M., Cao, S., Zhang, F., et al., 2011. Preparation, characterization, and antitumor activity of paclitaxel-loaded folic acid modified and TAT peptide conjugated PEGylated polymeric liposomes. J. Drug Target. 19 (5), 373–381. http://dx.doi.org/10.3109/1061186X.2010.504266.

Oostendorp, R.L., Buckle, T., Lambert, G., Garrigue, J.S., Beijnen, J.H., Schellens, J.H., et al., 2011. Paclitaxel in self-micro emulsifying formulations: oral bioavailability study in mice. Invest. New Drugs 29 (5), 768–776. http://dx.doi.org/10.1007/s10637-010-9421-7.

Pandey, V., Gajbhiye, K.R., Soni, V., 2015. Lactoferrin-appended solid lipid nanoparticles of paclitaxel for effective management of bronchogenic carcinoma. Drug Deliv. 22 (2), 199–205. http://dx.doi.org/10.3109/10717544.2013.877100.

Parakh, S., Parslow, A.C., Gan, H.K., Scott, A.M., 2016. Antibody-mediated delivery of therapeutics for cancer therapy. Exp. Opin. Drug Deliv. 13 (3), 401–419. http://dx.doi.org/10.1517/17425247.2016.1124854.

Patel, A.R., Spencer, S.D., Chougule, M.B., Safe, S., Singh, M., 2012. Pharmacokinetic evaluation and in vitro–in vivo correlation (IVIVC) of novel methylene-substituted 3,3' diindolylmethane (DIM). Eur. J. Pharm. Sci. 46 (1-2), 8–16. http://dx.doi.org/10.1016/j.ejps.2012.01.012.

Patel, K., Patil, A., Mehta, M., Gota, V., Vavia, P., 2013. Medium chain triglyceride (MCT) rich, paclitaxel loaded self nanoemulsifying preconcentrate (PSNP): a safe and efficacious alternative to Taxol. J. Biomed. Nanotechnol. 9 (12), 1996–2006.

Patil, S., Joshi, H., 2013. Lipid shell modified with combination of lipid and phospholipids in solid lipid nanoparticles for engineered specificity of paclitaxel in tumor bearing mice. Int. J. Drug Deliv. 5, 196–205.

Pawar, V.K., Panchal, S.B., Singh, Y., Meher, J.G., Sharma, K., Singh, P., et al., 2014. Immunotherapeutic vitamin E nanoemulsion synergies the antiproliferative activity of paclitaxel in breast cancer cells via modulating Th1 and Th2 immune response. J. Control. Release 196, 295–306. http://dx.doi.org/10.1016/j.jconrel.2014.10.010.

Peltier, S., Oger, J.M., Lagarce, F., Couet, W., Benoit, J.P., 2006. Enhanced oral paclitaxel bioavailability after administration of paclitaxel-loaded lipid nanocapsules. Pharm. Res. 23 (6), 1243–1250. http://dx.doi.org/10.1007/s11095-006-0022-2.

Petanidis, S., Kioseoglou, E., Domvri, K., Zarogoulidis, P., Carthy, J.M., Anestakis, D., et al., 2016. In vitro and ex vivo vanadium antitumor activity in (TGF-beta)-induced EMT. Synergistic activity with carboplatin and correlation with tumor metastasis in cancer patients. Int. J. Biochem. Cell Biol. 74, 121–134. http://dx.doi.org/10.1016/j.biocel.2016.02.015.

Pouliot, F., Pantuck, A.J., 2009. Words of wisdom. Re: efficacy of everolimus in advanced renal cell carcinoma: a double-blind, randomised, placebo- controlled phase III trial. Eur. Urol. 55 (6), 1482–1484.

Puri, A., Loomis, K., Smith, B., Lee, J.-H., Yavlovich, A., Heldman, E., et al., 2009. Lipid-based nanoparticles as pharmaceutical drug carriers: from concepts to clinic. Crit. Rev. Therap. Drug Carrier Systems 26 (6), 523–580.

Qian, X., LaRochelle, W.J., Ara, G., Wu, F., Petersen, K.D., Thougaard, A., et al., 2006. Activity of PXD101, a histone deacetylase inhibitor, in preclinical ovarian cancer studies. Mol. Cancer Therap. 5 (8), 2086–2095. http://dx.doi.org/10.1158/1535-7163.MCT-06-0111.

Rane, S., Prabhakar, B., 2013. Optimization of paclitaxel containing pH sensitive liposomes by 3 factor, 3 level box-behnken design. Indian J. Pharm. Sci. 75 (4), 420–426.

Ravar, F., Saadat, E., Gholami, M., Dehghankelishadi, P., Mahdavi, M., Azami, S., et al., 2016. Hyaluronic acid-coated liposomes for targeted delivery of paclitaxel, in-vitro characterization and in-vivo evaluation. J. Control. Release 229, 10–22. http://dx.doi.org/10.1016/j.jconrel.2016.03.012.

Rezazadeh, M., Emami, J., Varshosaz, J., 2012. Cellular uptake of targeted nanostructured lipid carrier (NLC) and cytotoxicity evaluation of encapsulated paclitaxel in HT29 cancer cells. Res. Pharm. Sci. 7 (5), S191.

Roger, E., Lagarce, F., Garcion, E., Benoit, J.P., 2010. Reciprocal competition between lipid nanocapsules and P-gp for paclitaxel transport across Caco-2 cells. Eur. J. Pharm. Sci. 40 (5), 422–429. http://dx.doi.org/10.1016/j.ejps.2010.04.015.

Rosenberg, S.A., Yang, J.C., Restifo, N.P., 2004. Cancer immunotherapy: moving beyond current vaccines. Nat. Med. 10 (9), 909–915.

Sanchez-Moreno, P., Ortega-Vinuesa, J.L., Martin-Rodriguez, A., Boulaiz, H., Marchal-Corrales, J.A., Peula-Garcia, J.M., 2012. Characterization of different functionalized lipidic nanocapsules as potential drug carriers. International J. Mol. Sci. 13 (2), 2405–2424. http://dx.doi.org/10.3390/ijms13022405.

Sawant, S., Shegokar, R., 2014. Cancer research and therapy: where are we today? Int. J. Cancer Ther. Oncol. 2 (4), 02048.

Sawant, R.R., Vaze, O.S., Rockwell, K., Torchilin, V.P., 2010. Palmitoyl ascorbate-modified liposomes as nanoparticle platform for ascorbate-mediated cytotoxicity and paclitaxel co-delivery. Eur. J. Pharm. Biopharm. 75 (3), 321–326. http://dx.doi.org/10.1016/j.ejpb.2010.04.010.

Scott, A.M., Wolchok, J.D., Old, L.J., 2012. Antibody therapy of cancer. Nat. Rev. Cancer 12 (4), 278–287. http://dx.doi.org/10.1038/nrc3236.

Shen, H., Shi, S., Zhang, Z., Gong, T., Sun, X., 2015. Coating solid lipid nanoparticles with hyaluronic acid enhances antitumor activity against melanoma stem-like cells. Theranostics 5 (7), 755–771. http://dx.doi.org/10.7150/thno.10804.

Shenoy, S., Gude, P., Murthy, R., 2011. Investigations on paclitaxel loaded HSPC based conventional and PEGylated liposomes: in vitro release and cytotoxic studies. Asian J. Pharm. Sci. 6 (1), 1–7.

Shi, S., Han, L., Deng, L., Zhang, Y., Shen, H., Gong, T., et al., 2014. Dual drugs (microRNA-34a and paclitaxel)-loaded functional solid lipid nanoparticles for synergistic cancer cell suppression. J. Control. Release 194, 228–237. http://dx.doi.org/10.1016/j.jconrel.2014.09.005.

Shigematsu, H., Lin, L., Takahashi, T., Nomura, M., Suzuki, M., Wistuba, I.I., et al., 2005. Clinical and biological features associated with epidermal growth factor receptor gene mutations in lung cancers. J. Natl. Cancer Inst. 97 (5), 339–346.

Siegel, R.L., Miller, K.D., Jemal, A., 2015. Cancer statistics 2015. CA Cancer J. Clin. 65 (1), 5–29.

Solomon, M.A., Shah, A.A., D'Souza, G.G., 2013. In Vitro assessment of the utility of stearyl triphenyl phosphonium modified liposomes in overcoming the resistance of ovarian carcinoma Ovcar-3 cells to paclitaxel. Mitochondrion 13 (5), 464–472. http://dx.doi.org/10.1016/j.mito.2012.10.013.

Sonnichsen, D.S., Relling, M.V., 1994. Clinical pharmacokinetics of paclitaxel. Clin. Pharm. 27 (4), 256–269. http://dx.doi.org/10.2165/00003088-199427040-00002.

Sørlie, T., Perou, C.M., Tibshirani, R., Aas, T., Geisler, S., Johnsen, H., et al., 2001. Gene expression patterns of breast carcinomas distinguish tumor subclasses with clinical implications. Proc. Natl. Acad. Sci. 98 (19), 10869–10874.

Spencer, C.M., Faulds, D., 1994. Paclitaxel. A review of its pharmacodynamic and pharmacokinetic properties and therapeutic potential in the treatment of cancer. Drugs 48 (5), 794–847.

Steichen, S.D., Caldorera-Moore, M., Peppas, N.A., 2013. A review of current nanoparticle and targeting moieties for the delivery of cancer therapeutics. Eur. J. Pharm. Sci. 48 (3), 416–427. http://dx.doi.org/10.1016/j.ejps.2012.12.006.

Strebhardt, K., Ullrich, A., 2008. Paul Ehrlich's magic bullet concept: 100 years of progress. Nat. Rev. Cancer 8 (6), 473–480. http://dx.doi.org/10.1038/nrc2394.

Sun, M., Han, J., Guo, X., Li, Z., Yang, J., Zhang, Y., et al., 2011. Design, preparation and in vitro evaluation of paclitaxel-loaded self-nanoemulsifying drug delivery system. Asian J. Pharm. Sci. 6 (1), 18–25.

Taratula, O., Kuzmov, A., Shah, M., Garbuzenko, O.B., Minko, T., 2013. Nanostructured lipid carriers as multifunctional nanomedicine platform for pulmonary co-delivery of anticancer drugs and siRNA. J. Control. Release 171 (3), 349–357. http://dx.doi.org/10.1016/j.jconrel.203.04.018.

Tiwari, S.B., Amiji, M.M., 2006. Improved oral delivery of paclitaxel following administration in nanoemulsion formulations. J. Nanosci. Nanotechnol. 6 (9-10), 3215–3221.

Torchilin, V.P., 2010. Passive and active drug targeting: drug delivery to tumors as an example. Handbook Exp. Pharmacol. 197, 3–53. http://dx.doi.org/10.1007/978-3-642-00477-3_1.

Tran, T.H., Ramasamy, T., Choi, J.Y., Nguyen, H.T., Pham, T.T., Jeong, J.H., et al., 2015. Tumor-targeting, pH-sensitive nanoparticles for docetaxel delivery to drug-resistant cancer cells. Int. J. Nanomed. 10, 5249–5262. http://dx.doi.org/10.2147/IJN.S89584.

Trotti, A., Byhardt, R., Stetz, J., Gwede, C., Corn, B., Fu, K., et al., 2000. Common toxicity criteria: version 2.0. an improved reference for grading the acute effects of cancer treatment: impact on radiotherapy. Inter. J. Radiat. Oncol. Biol. Phys. 47 (1), 13–47.

Veltkamp, S.A., Thijssen, B., Garrigue, J.S., Lambert, G., Lallemand, F., Binlich, F., et al., 2006. A novel self-microemulsifying formulation of paclitaxel for oral administration to patients with advanced cancer. Br. J. Cancer 95 (6), 729–734. http://dx.doi.org/10.1038/sj.bjc.6603312.

Walkey, C.D., Olsen, J.B., Guo, H., Emili, A., Chan, W.C., 2012. Nanoparticle size and surface chemistry determine serum protein adsorption and macrophage uptake. J. Am. Chem. Soc. 134 (4), 2139–2147. http://dx.doi.org/10.1021/ja2084338.

Wang, Y., Zhang, H., Hao, J., Li, B., Li, M., Xiuwen, W., 2015. Lung cancer combination therapy: co-delivery of paclitaxel and doxorubicin by nanostructured lipid carriers for synergistic effect. Drug Deliv. 1–6. http://dx.doi.org/10.3109/10717544.2015.1055619.

Winau, F., Westphal, O., Winau, R., 2004. Paul Ehrlich--in search of the magic bullet. Microbes Infection / Institut Pasteur 6 (8), 786–789. http://dx.doi.org/10.1016/j.micinf.2004.04.003.

World Health Organization (WHO), 2013. International Classification of Diseases for Oncology. http://apps.who.int/iris/bitstream/10665/96612/1/9789241548496_eng.pdf, III edition, First revision (accessed on 11.04.2016).

Wu, L., Tang, C., Yin, C., 2010. Folate-mediated solid-liquid lipid nanoparticles for paclitaxel-coated poly(ethylene glycol). Drug Dev. Industrial Pharm. 36 (4), 439–448. http://dx.doi.org/10.3109/03639040903244472.

Xiao, Y.F., Jie, M.M., Li, B.S., Hu, C.J., Xie, R., Tang, B., et al., 2015. Peptide-based treatment: a promising cancer therapy. J. Immunol. Res. 2015, 761820. http://dx.doi.org/10.1155/2015/761820.

Yang, X.Y., Li, Y.X., Li, M., Zhang, L., Feng, L.X., Zhang, N., 2013. Hyaluronic acid-coated nanostructured lipid carriers for targeting paclitaxel to cancer. Cancer Lett. 334 (2), 338–345. http://dx.doi.org/10.1016/j.canlet.2012.07.002.

Yang, Y., Xie, X., Zhang, H., Mei, X., 2015. Photo-responsive and NGR-mediated multifunctional nanostructured lipid carrier for tumor-specific therapy. J. Pharm. Sci. 104 (4), 1328–1339. http://dx.doi.org/10.1002/jps.24333.

Yin, T., Wang, P., Li, J., Wang, Y., Zheng, B., Zheng, R., et al., 2014. Tumor-penetrating codelivery of siRNA and paclitaxel with ultrasound-responsive nanobubbles hetero-assembled from polymeric micelles and liposomes. Biomaterials 35 (22), 5932–5943. http://dx.doi.org/10.1016/j.biomaterials.2014.03.072.

Yu, Y.H., Kim, E., Park, D.E., Shim, G., Lee, S., Kim, Y.B., et al., 2012. Cationic solid lipid nanoparticles for co-delivery of paclitaxel and siRNA. Eur. J. Pharm. Biopharm. 80 (2), 268–273. http://dx.doi.org/10.1016/j.ejpb.2011.11.002.

Yuan, H., Miao, J., Du, Y.Z., You, J., Hu, F.Q., Zeng, S., 2008. Cellular uptake of solid lipid nanoparticles and cytotoxicity of encapsulated paclitaxel in A549 cancer cells. Int. J. Pharm. 348 (1–2), 137–145. http://dx.doi.org/10.1016/j.ijpharm.2007.07.012.

Zhang, X.N., Tang, L.H., Gong, J.H., Yan, X.Y., Zhang, Q., 2006. An alternative paclitaxel self-emulsifying microemulsion formulation: preparation, pharmacokinetic profile, and hypersensitivity evaluation. PDA J. Pharm. Sci. Technol. / PDA 60 (2), 89–94.

Zhang, Y.L., Zhang, Z.H., Jiang, T.Y., Ayman, W., Jing, L., Lv, H.X., et al., 2013. Cell uptake of paclitaxel solid lipid nanoparticles modified by cell-penetrating peptides in A549 cells. Die Pharmazie 68 (1), 47–53.

Zhao, P., Wang, H., Yu, M., Cao, S., Zhang, F., Chang, J., et al., 2010. Paclitaxel-loaded, folic-acid-targeted and TAT-peptide-conjugated polymeric liposomes: in vitro and in vivo evaluation. Pharma. Res. 27 (9), 1914–1926. http://dx.doi.org/10.1007/s11095-010-0196-5.

CHAPTER 10

Bioadhesive Polymers for Targeted Drug Delivery

Lalit Kumar, Shivani Verma, Bhuvaneshwar Vaidya and Vivek Gupta

Contents

Nanotechnology-Based Approaches for Targeting and Delivery of Drugs and Genes.
DOI: http://dx.doi.org/10.1016/B978-0-12-809717-5.00012-9

10.1 INTRODUCTION

The process of fixing of two surfaces with one another is known as "adhesion." Adhesion can occur in biological settings, known as "bioadhesion"; and at the mucosal membrane, termed as "mucoadhesion." Furthermore, bioadhesion may also involve binding of polymer (natural or synthetic) to a biological substrate (Andrews et al., 2009). The most common biological substrate is the mucosal layer leading to the use of the term mucoadhesion. Internal organs of the body like the buccal cavity, gastrointestinal tract (GIT), eye, ear, vagina, rectum, and nose are usually covered with a gel-like structure known as mucin (Kingshott and Griesser, 1999). Given the widespread presence of mucin in the physicological system, all bioadhesives should have the capability to interact with the mucin layer for appropriate attachment. Due to the binding properties of mucus, it acts as a link between adhesive and membrane. Polymers mostly used as bioadhesives show binding with mucin and also a lack of deep penetration into underlying epithelial cells (Henriksen et al., 1996). The process of bioadhesion/mucoadhesion has been widely explored for site-specific delivery of various bioactive molecules through the use of bioadhesive/mucoadhesive polymers in various pharmaceutical formulations. A bioadhesive delivery system residing on a biological surface allows the localized therapeutic delivery by releasing bioactive molecule in the vicinity of the site of action, thus promoting bioavailability enhancement (Bernkop-Schnürch et al., 1998). Fig. 10.1 shows the mechanism of bioadhesion between a dosage form and mucosal surface. Various advantages of bioadhesive systems include:

- Reduction in dosing frequency of bioactive molecules due to increased residence time and controlled release at target site;
- Improved bioavailability of bioactive molecules at lower concentration due to prolonged contact time;
- Avoiding first-pass metabolism;
- Site-specific targeting of bioactive molecules e.g., buccal mucosa;
- Prevention of enzymatic degradation of protein and peptide drug due to high intimacy between delivery vehicle and absorbing surface (Woodley, 2001).

Orabase was the first marketed bioadhesive formulation prepared by using natural gums as bioadhesive materials. Currently, a large number of bioadhesive products are commercially available due to ease of manufacturing, minimal toxicity, and ease of administration (Kumar et al., 2014). They can be administered via (1) vaginal route in the form of suppositories, vaginal gel, and pessaries; (2) oral route using tablets and lozenges, and (3) nasal route through gels, pumps, and sprays. The soft tissues covered with mucus secreted by goblet cells present in their surrounding or secretory glands are the main target sites for bioadhesive drug delivery systems (Asane et al., 2008). This chapter provides a comprehensive description of the physicochemical aspects of bioadhesive polymers and their role in efficient drug delivery to the mucosal target site.

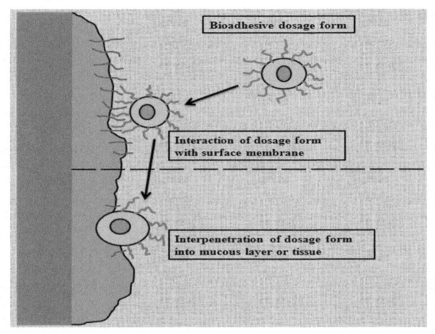

Figure 10.1 Mechanism of bioadhesion. Interaction occurs between surface membrane covering mucus layer followed by interpenetration of dosage form either into mucosal layer or in target tissue.

10.2 MUCUS: A BASIC NEED FOR PHYSIOLOGICAL FUNCTIONING AND BIOADHESION

Mucus is an adherent secretion synthesized by specialized goblet cells with a complex viscous nature. Goblet cells usually cover all the organs having exposure to the external environment and typically exist as glandular columnar epithelium cells. Mucus performs various functions, such as acting as a barrier to pathogen or toxic materials, lubricant for passage of various materials like nutrients, and hydration of epithelial cellular layer (Bansil and Turner, 2006). Mucus is composed of about 95% water and very high molecular weight (2–14 x 10^6 g/mol) glycoprotein known as mucin. Mucus may also exist as "viscoelastic substance" containing small proportions of proteins, lipids and mucopolysaccharides (nearly 1%) along with water and glycoproteins (Davies and Viney, 1998). Mucin glycoproteins show association with one another through noncovalent bonds, thus forming a complex entangled network in the mucus structure. This complex entangled network is responsible for the rheological properties of mucus. The presence of sialic acid and sulfate groups on the glycoprotein molecules enables the mucin to act as anionic polyelectrolyte at neutral pH (Capra et al., 2007). Mucus may also contain nonmucin components like lipids, polysaccharides, lysozymes, and secretory IgA responsible for bacteriostatic capability of mucus (Fiebrig et al., 1995). The storage sites for mucus are submucosal and goblet cells. Mucin glycoprotein provides a negative charge to mucus, the shielding of which is done by calcium ions, thus promoting close packing of mucus in goblet cells (Kocevar–Nared et al., 1997). Variations in the exact composition of mucus may happen due to change in secretion sites, existence of any disease, or change in its mechanical and physiological role (Willits and Saltzman, 2001).

10.3 THEORIES OF BIOADHESION

Over the years, the mechanism of bioadhesion has been explained by various theories like electrostatic, wettability, diffusion interpenetration, adsorption, and fracture theory (Fig. 10.2) (Andrews et al., 2009). Each theory represents a supplementary process involved in different phases of substrate mucus interaction (Madsen et al., 1998).

10.3.1 The electrostatic theory

The electrostatic theory of bioadhesion states that the transfer of electrons between mucoadhesive dosage form and mucus, occurring due to differences in their electronic structures, is responsible for the process of mucoadhesion. Formation of the double layer of electric charges occurs at the interface between mucus and bioadhesive dosage

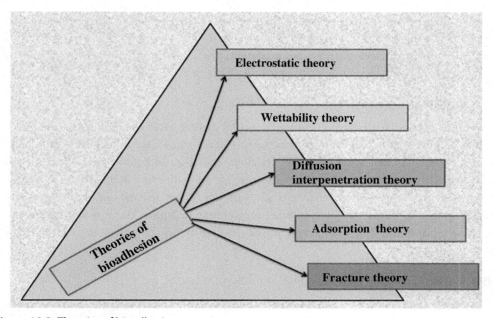

Figure 10.2 Theories of bioadhesion process.

form due to transfer of electrons between the two. Strong attractive forces are generated within this double-layered region due to this process (Dodou et al., 2005).

10.3.2 The wettability theory

According to this theory, "spreadability" of a bioadhesive drug delivery system across a biological substrate affects bioadhesion. This theory is applicable only to low viscosity or low mucoadhesive systems. The wettability theory states that bioadhesive system penetrates to the irregular substrate surface followed by hardening and anchoring itself towards the surface. Spreadability and wettability are the critical parameters governing adhesive performance of such delivery systems. Low interfacial tension between bioadhesive system and substrate surface increases the wettability and spreadability (Ugwoke et al., 2005). Bioadhesive polymers having the same functional group as that of the mucus layer show a very high degree of polymer spreadability over mucosal surface due to increased miscibility (Shojaei and Li, 1997).

10.3.3 The diffusion interpenetration theory

According to this theory, bioadhesion is due to time-dependent diffusion of bioadhesive polymeric chains into the mucus layer consisting of a complex glycoprotein network. Diffusion-based penetration of the polymeric chains into substrate depends upon the diffusion coefficient of both the interacting polymer and the substrate (Lee et al., 2000). Furthermore, cross-linking density, molecular weight, chain mobility/flexibility, temperature, expansion capacity of both networks govern effective penetration of polymer into the mucus network (Jabbari and Peppas, 1995). Interpenetration and polymer mobility is reduced due to excessive chain cross-linking. Maximum interpenetration and bioadhesive strength are achieved when solubility parameter is similar for interacting polymer and mucus glycoprotein (Vasir et al., 2003).

10.3.4 The adsorption theory

This theory states that bioadhesion is a result of surface interactions between the polymer and mucus substrate. Surface interaction may occur due to primary or secondary bonding. Primary bonds include covalent, metallic, and ionic bonds, which may provide permanent interaction between polymer and substrate (Kinloch, 1980). Bonds arising due to hydrophobic interactions, van der Waals forces, and hydrogen bond are secondary in nature; surface interaction generated due to them is semipermanent (Ahagon and Gent, 1975). In most cases, mucoadhesion involves secondary bonds (Jiménez-Castellanos et al., 1993).

10.3.5 The fracture theory

This theory demonstrates the relationship between adhesive bonds and forces required to separate both surfaces from one another, i.e., fracture theory relates the strength of mucus adhesive bonds with the force required for polymer detachment. It is found that longer polymeric network strands and reduced degree of cross-linking within polymer lead to greater work fracture (Gu et al., 1988).

10.4 EFFECT OF POLYMER PROPERTIES ON BIOADHESION PROCESS

The interaction of bioadhesive polymers with substrate mucus is affected by various polymer properties as explained in different theories of bioadhesion. Fig. 10.3 enlists various polymer properties affecting the bioadhesion process. Keeping polymeric properties in mind, the formulation of effective bioadhesive systems should be carefully studied. Various polymer properties affecting the bioadhesion process are discussed below.

10.4.1 Concentration of polymer

The process of bioadhesion is significantly affected by polymer concentration. Usually, the physical state of final formulation plays an important role in determining optimum polymer concentration. In a semisolid formulation, the polymer used should be in an optimum amount because at concentration beyond this causes reduction in bioadhesion due to availability of fewer polymeric chains for interpenetration with mucus. For a solid formulation like a tablet, the strength of bioadhesion increases as the concentration of mucoadhesive polymer rises (Ugwoke et al., 2005).

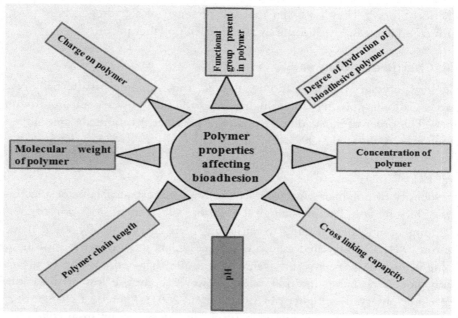

Figure 10.3 Properties of polymers affecting process of bioadhesion.

10.4.2 Degree of hydration of bioadhesive polymer

There are many bioadhesive polymers that show excellent adhesive properties in the dry state. Capillary attractive force and osmotic force existing between dry polymer and wet mucosal surface may be responsible for bioadhesion in this case (Sigurdsson et al., 2006). Some polymers in wet condition show better bioadhesion due to attachment of swollen polymers to mucosal substrate. The hydration of polymer promotes interpenetration and relaxation of polymeric chains. However, excess hydration of a polymer may cause reduction in the bioadhesion due to slippery mucilage formation (Mortazavi and Smart, 1993).

10.4.3 Functional group present in polymer

It has theoretically been demonstrated that secondary noncovalent bonding between polymer and substrate is responsible for bioadhesion. Polymers having hydrophilic functional groups like $-NH_2$, $-COOH$, $-SO_4H$, and $-OH$ show effective noncovalent bonding due to formation of hydrogen bonds. So, polymers having acceptable hydrogen bonding functional groups interact more strongly with mucus glycoproteins (Madsen et al., 1998). Bioadhesive polymers usually contain numerous polar functional groups leading to formation of hydrophilic network in their domain. Beside physical entanglements, bioadhesive polymers interact with mucus through secondary chemical bond formation, generating a weak cross-linked network (Capra et al., 2007). Hagesaether and Sande (2007) reported the significance of hydrogen bonding in the bioadhesion process. They observed that the presence of a hydrogen bond disruptor like urea could decrease the bioadhesive strength of polymer pectin.

10.4.4 Charge on polymer

Bioadhesive polymers may be cationic, anionic, and nonionic on the basis of overall charge. Cationic polymers show higher mucoadhesive power compared to anionic polymers. The order of mucoadhesive power of polymer is cationic > anionic > nonionic polymers (Lehr et al., 1992).

10.4.5 pH

Charge density of polymeric macromolecules is an important parameter affecting the process of bioadhesion. The physiological pH value affects the dissociation of the functional group of polymers leading to variation in its charge density value (Park and Robinson, 1984). Carboxylated mucoadhesive polymers show effective bioadhesion at pH value below the respective pK_a value. It is found that optimization of the mucoadhesion process occurs at low pH values; however, complete loss of mucoadhesion process is not observed at higher pH (Riley et al., 2001). There is a change in spatial confirmation of a polymer containing "COO^-" functional group from coiled form to rod-like form at higher pH values, thus promoting their interpenetration and interdiffusion (Sudhakar et al., 2006). A positively charged polymer like chitosan may show strong bioadhesion process at elevated pH values because of the formation of polyelectrolyte complex with negatively charged mucin (Peppas and Haung, 2004).

10.4.6 Molecular weight of polymer

Usually for effective entanglement, molecular weight of polymer should be high. However, use of bioadhesive polymer at optimum molecular weight may promote the process of bioadhesion (Huang et al., 2000). Optimum molecular weights of bioadhesive polymers like poly (acrylic) acid and polyethylene oxide are 750,000 and 4,000,000 respectively (Andrews et al., 2009). Dextran polymers may show similar bioadhesive strength at the molecular weights of 200,000 and 19,500,000 (Yang and Robinson, 1998).

10.4.7 Polymer chain length

Polymer chains must have critical length to produce effective bioadhesion. A polymer chain length is influenced by its shape and size; therefore, these critical points should also be taken into consideration (Mortazavi and Smart, 1994).

10.4.8 Cross-linking capacity of polymer

Cross-linking capacity of the mucoadhesive polymer should be very high for effective bioadhesion. Highly cross-linked polymer shows high swelling in the presence of water, and high swelling increases the surface area of polymer promoting its mucus interpenetration (Imam et al., 2003).

10.5 BIOADHESIVE POLYMERS USED FOR DELIVERY OF THERAPEUTIC MOLECULES

Polymers that are employed for production of the bioadhesive drug delivery platforms are broadly classified into (1) nonspecific bioadhesive polymer (old generation), and (2) specific bioadhesive polymers (new generation) (Peppas and Haung, 2004) (Fig. 10.4).

10.5.1 Nonspecific bioadhesive polymers (old generation)

Conventional mucoadhesive polymers may be anionic, cationic, and nonionic on the basis of the charge carried by them. Furthermore, cationic and anionic polymers are employed for mucoadhesive platforms because of their high mucoadhesive potential (Ludwig, 2005).

10.5.1.1 Cationic polymers

Chitosan is an example of cationic polymers widely used for mucoadhesive investigations. It is produced from deacetylation of chitin, which is the second-most abundant polysaccharide in the world (He et al., 1998). Chitosan has a film-forming capacity

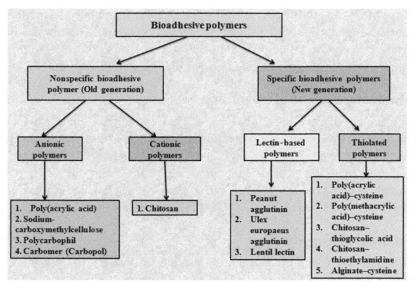

Figure 10.4 Overview of various bioadhesive polymers. Specific polymers are center of attraction for scientists due to their high bioadhesive strength and specificity to target.

and it is widely used as a strengthening agent in paper manufacturing, and is the dye binder in the textile industry (Bautista-Baños et al., 2006). Due to its hypolipidimic effect, it is an important constituent of dietary materials. Chitosan has good biodegradable, toxicological, and biocompatible characteristics, thus promoting its use as a bioadhesive polymer (Dodane and Vilivalam, 1998). The ionic interaction between an amino functional group of chitosan and sulfonic acid or sialic acid component of mucus is responsible for mucoadhesion (Bernkop-Schnürch, 2005). Sufficient chain flexibility provided by linearity of chitosan molecules promotes its interpenetration in the mucus layer (El-Kamel et al., 2002). Chitosan-based bioadhesive systems also promote absorption of various bioactive molecules through a paracellular route, due to the anionic charge neutralization existing between mucosal cells in the tight junction area (Bravo-Osuna et al., 2007). Chitosan is a very beneficial polymer as it can be modified easily by the addition of various chemical groups into its structure, especially at the C-2 position. Various pharmaceutical challenges can be overcome by using chemically modified chitosan as the mucoadhesive polymer of choice (Onishi and Machida, 1999).

10.5.1.2 Anionic polymers

Anionic bioadhesive polymers are extensively used because of their low toxicity profile and rich mucoadhesive functionality. Such polymers have sulfate and carboxyl functional groups in their molecular structure, providing them a negative charge. Examples of such polymers are poly (acrylic acid) (PAA), sodium carboxymethylcellulose (NaCMC), Polycarbophil, and carbomer (Carbopol) (Fefelova et al., 2007). Anionic polymers interact through strong hydrogen bonding with mucus covering. PAA derivatives, carbomers, and polycarbophil are especially used to develop mucoadhesive systems for effective delivery of drugs to the GI tract (Singla et al., 2006). The US FDA has categorized PAA polymers as a GRAS (*Generally Recognized As Safe*) class for delivery of dosage forms via the oral route. PAA polymers have properties like transparency, wide molecular weight range, nontoxicity, nonirritancy (Ugwoke et al., 1999). Polycarbophil polymer shows high swelling index under neutral pH condition and very low solubility in water. The high swelling index promotes efficient entanglement of polymer within the mucus (Robinson and Mlynek, 1995). Carbomer has an equivalent use compared to polycarbophil but its swelling index is slightly low compared to polycarbophil. The cross-linking capacity of carbomer is slightly low compared to polycarbophil and it is cross-linked with allylpentaerythritol or allyl sucrose, while cross-linking of polycarbophil is carried out with divinyl glycol. Polycarbophil and carbomer interact with mucus through a hydrogen bond between carboxylic group and mucosal surface (Khutoryanskiy, 2007; Andrews et al., 2009).

10.5.2 Specific bioadhesive polymers (new generation)

Nonspecific polymers may be less effective as they may sometimes attach to a site that is not the desired target (off-target binding). At the same time, nonspecific polymers are highly susceptible to mucus turnover rates. At a higher mucus turnover rate, efficiency of bioadhesion of nonspecific polymers decreases. So, new generation-specific bioadhesive polymers have been used recently by pharmaceutical scientists due to their independency from mucus turnover rates and effective targeting of mucus surface based upon the presence of carbohydrates and protein composition on it (Andrews et al., 2009).

10.5.2.1 Thiolated polymers

Thiolated polymers include derivatives of various hydrophilic polymers like chitosan or polyacrylates (Leitner et al., 2003). The main examples of thiolated polymers are poly(acrylic acid)–cysteine, poly(methacrylic acid)–cysteine, chitosan–thioglycolic acid (TGA), chitosan–thioethylamidine, and alginate–cysteine. Thiolated polymers show improvement in residence time and bioavailability due to the formation of covalent bonds between cysteine-rich parts of the mucus layer (Albrecht et al., 2005). Thiol group is responsible for this covalent bonding. Thiomers also have the capability to become covalently anchored in the mucus layer due to formation of disulfide bonds (Roldo et al., 2004). Disulfide bonding also influences the drug release mechanism of a mucoadhesive system because of the increase in cross-linking and rigidity. Thiolated new generation polymers usually show diffusion controlled drug release mechanism (Bernkop-Schnürch et al., 2004).

10.5.2.2 Lectin-based polymers

Lectins are chemically proteins which are involved in various phenomena like biological adhesion and adherence. More specifically, they are structurally diverse glycoproteins having the capability to bind with carbohydrate residues (Clark et al., 2000). Lectins can reside on a cell surface during initial mucosal cell adherence. However, in the case of receptor-mediated adherence, they show the internalization effect by following endocytosis. Examples of various lectin-based polymers used in bioadhesive platforms are peanut agglutinin, ulex europaeus agglutinin, and lentil lectin (Lehr, 2000). Lectin-based bioadhesive polymers show targeted specific bioadhesion phenomenon along with controlled delivery of bioactive molecules active cell mediated uptake process (Mansuri et al., 2016). Lectins do not show premature inactivation by the shed off mucus as observed in the case of first-generation polymers, thus promoting high distribution of lectin-based delivery systems through reversible adherence (Wirth et al., 2002). Despite all these advantages, lectins may also show various toxicological and immunological problems. It is also reported that lectin-induced antibiotics have the capability to block the interaction between bioadhesive surface and lectin-based drug delivery systems (Lis and Sheron, 1986).

10.6 EVALUATION PARAMETERS OF BIOADHESIVE POLYMER-BASED DRUG DELIVERY SYSTEMS

For development of an effective bioadhesive drug delivery system, evaluation of its bioadhesive property is imperative (Metia and Bandyopadhyay, 2008). Adhesion strength of a bioadhesive system can be evaluated by various in vitro and in vivo tests (Fig. 10.5).

10.6.1 In-vitro evaluation parameters

Various in-vitro evaluation parameters for bioadhesive polymers reported in the literature are discussed below.

10.6.1.1 Measurement of detachment force

This test involves measurement of the force required to separate two parallel glass slides covered with polymer and mucus layer respectively (Kumar et al., 2014). This method includes attachment of the glass plate covered with polymer through microforce balance and its immersion in the mucus sample under the controlled environmental condition. Detachment force is measured as force required to pull the plate out of the mucus sample (Chowdary and Rao, 2004).

10.6.1.2 Tensile strength measurement

Tensile strength measurement is carried out by using different equipment like M30K, JJ Lloyd Instruments Ltd., in which aqueous dispersion of a bioadhesive polymer is

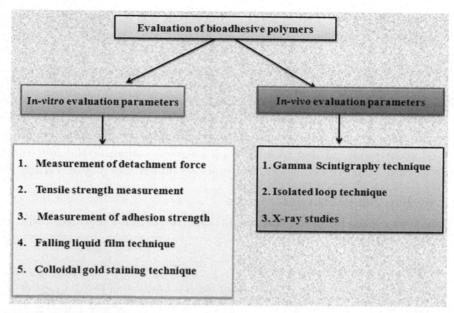

Figure 10.5 Evaluation of bioadhesive polymers. Various in vitro and in vivo parameters are utilized to evaluate bioadhesive potency of polymers.

placed between two discs made up of polyoxymerhylene (Mansuri et al., 2016). The upper disc shows movement while the lower disc is stationary, as it is fixed on a stationary frame of the machine. After application of the tensile force, maximum force required for detaching next to the fracture is calculated by using a force displacement curve. A force that is required to detach bioadhesive cup in a perpendicular fashion from bovine buccal mucosa is termed as tensile strength (Dyvik and Graffner, 1992).

10.6.1.3 Measurement of adhesion strength
Adhesion strength measurement can be carried out by using the fluorescent probe method. Probes are used to analyze the polymer and mucin interaction. This method involves labeling of pyrene as a probe over mucosal surface. Polymer adhesion to the mucosal surface causes a change in degree of fluorescence, which is proportional to the polymer binding (Asane et al., 2008).

10.6.1.4 Falling liquid film technique
This method involves in situ quantification of adherence of particles on a mucosal surface. Briefly, a particulate system like microspheres in suspension form are allowed to flow down through an inclined plastic slide covered with mucosal membrane. The

difference between applied microsphere amount and flowed microsphere amount gives a value of adhering microspheres (Shahi et al., 2011).

10.6.1.5 Colloidal gold staining technique

This technique is used for quantitative comparison of bioadhesive properties of various hydrogels. This method involves interaction of mucin-gold conjugates with hydrogel surface leading to red coloration. Mucin–gold conjugates are composed of red colloidal gold particles having adsorbed mucin molecules in the surface. Measurement of intensity of red color produced due to interaction between conjugate and mucoadhesive hydrogel gives quantitative evaluation of bioadhesive property of hydrogel (Park, 1989).

10.6.2 In-vivo evaluation parameters

Various in-vivo evaluation parameters for bioadhesive polymers reported in the literature are discussed below.

10.6.2.1 Gamma scintigraphy technique

An in-vivo distribution pattern of any dosage form is confirmed by using gamma scintigraphy technique. This method involves the addition of a radioactive tracer to any dosage form and detection of rays emitted through the gamma ray camera (Meseguer et al., 2004). Activity of formulation at the desired target site can be determined by using scintigrams. Target sites are also known as regions of interest (ROIs). ROIs related to bioadhesion are drawn through the gamma image quantification done by computer software programs (Pund et al., 2011).

10.6.2.2 Isolated loop technique

Lehr et al. (1990) investigated intestinal transit of bioadhesive microspheres by using the isolated ileal loop model. In this technique, the bioadhesive property of microspheres was calculated in terms of mean residence time, followed by injection into the in situ perfused gut segment.

10.6.2.3 X-ray studies

GI transit time of bioadhesive formulations can be confirmed through X-ray inspection by coating the formulations with radio-opaque markers like barium sulfate. X-ray photographs are taken from an animal body at different time intervals to give an idea about GI transit time (Senthil et al., 2010).

10.7 UTILITY OF BIOADHESIVE POLYMERS IN DRUG DELIVERY

Bioadhesive polymer-based formulations can be delivered through various routes like vaginal, buccal, rectal, nasal, ocular, and gastrointestinal. In the following paragraphs applications of bioadhesive polymer-based delivery systems are discussed.

10.7.1 Applications in vaginal drug delivery

The vagina is a fibromuscular and tubular organ having a length of about 9 centimeters extending from the cervix of uterus to the vaginal vestibules. The vagina is generally considered as mucosal tissue without gland and vaginal secretions are a mixture of various fluids from a number of sources. Vaginal mucus coating plays an important role in various physiological functions along with drug absorption. The pH of vaginal fluid ranges from 4.5 to 5.5 and menstrual cycle has a distinct impact on its rheology, composition, and volume (Hussain and Ahsan, 2005). Vaginal drug delivery offers various advantages like avoidance of enzymatic degradation, drug interactions, and first-pass effect. Beside this vaginal formulation generate patient incompliance due to high dosing frequency, night time dosing, and dripping creams. To avoid these issues, a bioadhesive drug delivery system would be an effective choice (Merabet et al., 2005). Kast et al. (2002) prepared and evaluated chitosan-TGA conjugate-based bioadhesive tablets for vaginal drug delivery. Two chitosan-TGA conjugates, i.e., conjugate A (having 160 µmol thiol group per gram polymer), and conjugate B (having 280 µmol thiol group per gram polymer), were used to prepare tablets. Disintegration time of the bioadhesive tablets prolonged 1.6-fold for conjugate A and 100-fold for conjugate B. Chitosan-TGA conjugate B-based tablet showed 26 times longer adhesion time compared to unmodified polymers (Fig. 10.6).

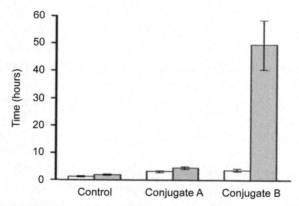

Figure 10.6 Comparison of the bioadhesive properties of chitosan-TGA conjugates and controls influenced by the presence of clotrimazole. Represented values are means (±S.D.; $n = 3-5$) of the total time determined in in vitro adhesion studies on the rotating cylinder at pH 6.0 (100 mM acetate buffer) at 37°C. White bars indicate the adhesion time of tablets (30 mg) containing the control polymer and the conjugates, respectively; gray bars indicate the adhesion time of tablets comprising 25 mg of polymer (control or conjugate) and 5 mg of clotrimazole. *Used with permission from Kast, C.E., Valenta, C., Leopold, M., Bernkop-Schnürch, A., 2002. Design and in vitro evaluation of a novel bioadhesive vaginal drug delivery system for clotrimazole. J. Control. Release. 81(3):347–354.*

Furthermore, the efficacy of the bioadhesive liposomal gel containing clotrimazole and metronizole was checked by Pavelic et al. (2005) for local treatment of vaginitis. In vitro release studies showed that 30% of the original entrapped clotrimazole (or 50% metronidazole) was still retained in the gel. Stability studies predicted the gel to preserve the original size distribution of incorporated liposomes. Baloglu et al. (2006) developed and evaluated bioadhesive vaginal tablets by using mixtures of different polymers like Carbopol 934 (Cp), pectin (Pc), hydroxypropylmethylcellulose (HPMC), sodium carboxymethylcellulose (Na CMC), and guar gum (GG) in different ratios. Tablets showed good bioadhesive strength in bovine vagina and non-Fickian mechanism of the drug release. All formulations were found histologically safe except one containing a high amount of guar gum. Table 10.1 gives an overview of various bioadhesive systems studied for vaginal drug delivery.

10.7.2 Applications in buccal drug delivery

The oral route is the most patient-compliant route for delivery of various therapeutic active molecules as it avoids discomfort, pain, and chances of infections caused by injections. However, the utility of the oral route has been hampered by some associated disadvantages like acidic degradation of bioactive molecules in the stomach, least macromolecular permeability through intestinal epithelium, and degradation through proteolytic enzymes (Wilding et al., 1994). Orally, drugs can be delivered through buccal or sublingual routes. Sublingual mucosa shows rapid absorption and rise in bioavailability of drugs as it is highly permeable in nature. On the other hand, the permeability of buccal mucosa is lower, leading to less absorption and reduction in bioavailability of drugs (Zhou, 1994). It has been observed that the sublingual mucosa is not suitable for delivery of bioadhesive systems as it is uneven in distribution and it shows mobility due to constant washing through secreted saliva. Therefore, buccal mucosa is preferred over sublingual mucosa to deliver bioadhesive formulations (Chen et al., 2015). Perioli et al. (2004) developed a metronidazole mucoadhesive tablet composed of a mixture of cellulose and polyacrylic derivatives to treat periodontal disease. Mucoadhesive tablets composed of hydroxyethylcellulose (HEC) and carbomer 940 in 2:2 ratio showed sustained release up to 12 h with buccal concentration always higher than its MIC (Fig. 10.7). Furthermore, evaluation of nicotine hydrogen tartrate (NHT) mucoadhesive delivery system composed of chitosan and carbomer in different ratio was carried out by Ikinci et al. (2006). A prepared tablet was compared with transdermal patch in vivo. Time to reach the C_{max} was 2.9 ± 0.2 h and 11.5 ± 1.3 h, and AUC_{0-24} values were 59.3 ± 5.1 ng $\times$ h $\times$ mL^{-1} and 204.1 ± 31.2 ng $\times$ h $\times$ mL^{-1} for buccal tablet and transdermal patch, respectively.

Vishnu et al. (2007) developed a mucoadhesive patch loaded with carvedilol for buccal delivery. A prepared buccal patch showed 38.69% + 6.61% drug permeation through porcine buccal membrane in 4 h and raised bioavailability of drug 2.3-fold

Table 10.1 Bioadhesive systems for vaginal drug delivery

Drug	Delivery system	Bioadhesive polymer	Sophisticated technique	Key findings	References
Povidone iodine	Tablets	Polyvinyl pyrrolidone	High-performance liquid chromatography (HPLC)	Prepared formulation sustained the release of polymer complex with iodine for 8h along with rapid disintegration and good bioadhesive strength	Garg et al. (2007)
Natamycin–γcyclodextrin	Tablets	Carbopol 934P	HPLC, nuclear magnetic resonance (NMR), differential scanning calorimetric (DSC), powder X-ray diffraction (PXRD)	Complexing of drug with γcyclodextrin increased antimycotic activity compared to free drug and also sustained the release up to 8h. Inclusion of carbopol 934P in formulation increased bioadhesive strength	Cevher et al. (2008)
Ketoconazole	Effervescent tablet	Carbomer (Carbopol 974P, Carbopol 934P), hydroxypropylmethyl cellulose (HPMC) and hydroxypropyl cellulose (HPC)	Ultraviolet (UV) spectroscopy	Formulation containing 100mg of effervescent and carbopol 934P: HPC in ratio of 1:9 showed maximum bioadhesion and sustained release of drug in rat vaginal tissue	Wang and Tang (2008)

(*Continued*)

Table 10.1 Bioadhesive systems for vaginal drug delivery (Continued)

Drug	Delivery system	Bioadhesive polymer	Sophisticated technique	Key findings	References
Itraconazole	Vaginal film	Hydroxypropyl methylcellulose E15	XRD, DSC	Bioadhesive polymer kept film up to 7 h on the vaginal mucosa of rat and showed improved therapeutic benefits of drug against *Candida albicans* vaginitis	Dobaria et al. (2009)
'ʒazido-3'-deoxythymidine (AZT) and polystyrene sulfonate (PSS)	Intravaginal bioadhesive polymeric device (IBPD)	Pentaerythritol polyacrylic acid (APE-PAA)	Ultraperformance liquid chromatography (UPLC), DSC, X-ray imaging	X-ray imaging showed adherence of polymer coated IBPD on vaginal tissue for 30 days period along with controlled drug release over 9 weeks	Ndesendo et al. (2010)
Hexyl-aminolevulinate	Pellets	Carbopol 934	HPLC	At carbopol content of 8% w/w, pellets showed optimum bioadhesive strength, drug release, and high stability	Hiorth et al. (2013)
Oxybutynin	Gel	Chitosan, hydroxypropyl methylcellulose (HPMC K100M) and Poloxamer 407 (Pluronic F 127)	HPLC, transmission electron microscope (TEM)	Vaginal gel prepared from HPMC K100M showed highest value of AUC, relative bioavailability, and bioadhesive strength	Tuğcu-Demiröz et al. (2013)

Drug	Formulation	Polymer	Characterization technique	Observation	Reference
Curcumin	Liposomes	Chitosan and carbopol	Liquid chromatography-mass spectrometry (LC–MS)	Both chitosan and carbopol coatings significantly increased curcumin permeability through vaginal mucus in vitro compared to curcumin solution and noncoated liposomes	Berginc et al. (2014)
Clindamycin hydrochloride	In-situ gel	Hydroxypropyl methycellulose and gellan gum	Fourier transform infrared (FT-IR) spectroscopy, UV-spectroscopy	Prepared formulation was found nonirritant along with good bioadhesive characteristics and showed sustained release of drugs up to 12h	Patel and Patel (2015)
Nystatin	Vaginal films	Carboxy methyl derivative of fenugreek gum (CMFG)	DSC, FT-IR and X-ray diffraction	Film having 5% w/v polymer and 2% v/v glycerol showed optimum bioadhesive strength and sustained drug release up to 5h	Bassi and Kaur (2015)

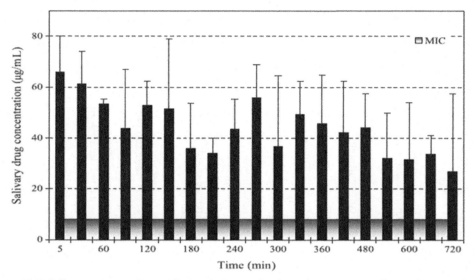

Figure 10.7 Salivary concentration with 20 mg metronidazole mucoadhesive tablet 11 ($n = 3$, $a = 0.05$). *Used with permission from Perioli, L., Ambrogi, V., Rubini, D., Giovagnoli, S., Ricci, M., Blasi, P., et al., 2004. Novel mucoadhesive buccal formulation containing metronidazole for the treatment of periodontal disease. J. Control. Rel. 95(3):521–533.*

compared to oral solution. Later on, Garg et al. (2007) prepared bioadhesive film composed of hydroxypropyl methylcellulose and polycarbophil polymers for buccal delivery of NHT. The formulation showed suitable adhesion and an initial burst release of 40% drug in the first 15 min followed by a total 80% drug release in a characteristic manner until 4 h, which is the desired time of application. Table 10.2 summarizes bioadhesive formulations used for buccal delivery of therapeutic agents.

10.7.3 Applications in rectal drug delivery

The rectal route serves as an alternative to oral or invasive administration. The rectal route of drug delivery is utilized when oral medication is not possible and the patient has difficulty swallowing due to nausea and vomiting produced in patients after oral administration. The rectal route offers various advantages like avoidance of first-pass effect, absorption of medicament into the lymphatic system, and absorption enhancement (Van Hoogdalem et al., 1991). The last portion of intestine is known as the rectum, having a length of 12–18 cm along with two or three curves within its lumen generated through submucosal folds. The rectal wall is composed of cylindrical epithelial cells and goblet cells which secrete mucus (Andrew, 2010). The surface area available for drug absorption from mucosal layer is 200–400 cm². Rectal mucus has a volume of 1–3 mL and its pH is about 7.5–8. Therefore, the rectal route has been

Table 10.2 Bioadhesive formulations used for buccal delivery of therapeutic agents

Drug	Delivery system	Bioadhesive polymer	Sophisticated technique	Key findings	References
5-Fluorouracil	Bioadhesive gel	Poloxamer 407, HPMC K 15M, Gantrez S-97 (polymethylvinylether-ω-maleic anhydride)	XRD, DSC, FT-IR	Bioadhesive strength of gel increased with increasing concentration of HPMC K 15M and Gantrez S-97 and gel showed sustained release of the drug up to 8h along with high buccal mucosal permeability	Dhiman et al. (2008)
Clotrimazole: hydroxypropyl-beta-cyclodextrin (CTZ-HPbetaCD)	Tablets	Xanthan gum, carbopol 974P	XRD, FT-IR	Buccal bioadhesive tablets prolonged the duration of antifungal activity of drug and found effective compared to local oral medication	Singh et al. (2008)
Ondansetron hydrochloride	Tablets	Carbopol (CP 934), sodium alginate, sodium carboxymethyl cellulose low viscosity (SCMC LV), hydroxypropyl methylcellulose (HPMC 15cps)	Ultraviolet (UV) spectroscopy	Bioadhesive tablets showed high drug permeation through the bovine buccal mucosa and sustained the release up to 500min in vitro. Tablets show high stability in human saliva	Hassan et al. (2009)
Diltiazem hydrochloride	Tablets	Hydroxypropyl methyl cellulose (HPMC) K4M and Carbopol 934	HPLC	Rise in polymer concentration retarded drug release from tablets and bioavailability of bioadhesive tablet was 1.56 folds higher compared to the oral tablets	Shayeda et al. (2009)

(Continued)

Table 10.2 Bioadhesive formulations used for buccal delivery of therapeutic agents (Continued)

Drug	Delivery system	Bioadhesive polymer	Sophisticated technique	Key findings	References
Carbenoxolone sodium	Buccal discs	Pectin	Ultraviolet (UV) spectroscopy	Reduction of pectin to lactose ratio in disc produced high drug dissolution rate and low bioadhesive power and disc with sweetener showed higher drug release compared to disc without sweetener	Wattana-korn et al. (2010)
Amitriptyline	Tablets	High-viscosity hydroxypropylmethyl cellulose (HPMC–K4M), sodium carboxymethyl cellulose (NaCMC), HPMC of lower-viscosity grade (HPMC–E5LV)	Ultraviolet (UV) spectroscopy	Tablets prepared from HPMC–E5LV showed high mucoadhesive strength and high release rate compared to tablets composed of HPMC–K4M	Movass-aghian et al. (2011)
Sumatriptan succinate	Tablets	Hydroxyl propyl methyl cellulose K4M, sodium carboxy methyl cellulose, Carbopol	FT-IR, ultraviolet (UV) spectroscopy	Tablets showed high bioadhesive strength, greater buccal permeation, and followed zero order kinetics by a diffusion mechanism type	Prasanna et al. (2011)

Drug	Dosage form	Polymers	Characterization	Results	Reference
Rizatriptan benzoate	Buccal film	Tamarind seed xyloglucan (TSX), carbopol 934P (CP), ethyl cellulose	DSC	Buccal film showing the presence of 4% (w/v) tamarind seed xyloglucan and 0.5% (w/v) carbopol 934P in the layer containing the drug and ethyl cellulose in 1% (w/v) concentration in backing layer enhanced drug penetration up to 93.45% in the porcine buccal mucosa	Avachat et al. (2013)
Timolol maleate	Tablets	Carbopol 974P (CP-974P) and sodium alginate (SA)	FT-IR, DSC, P-XRD	Bioadhesive tablets showed 98.67% drug release for 8 h, bioadhesive strength of 0.088 N, and absence of crystallinity changes of the drug	Gaikwad et al. (2014)
Buspirone hydrochloride	Buccal discs	Hydroxypropylmethyl cellullose (HPMC), mannitol (diluent and pore former)	DSC, FT-IR	Prepared bioadhesive buccal disc showed 10 folds increase in bioavailability of drug compared to oral bioavailability of drug and high content of HPMC in buccal disc increased bioadhesion strength	Jaipal et al. (2015)

preferred to deliver bioadhesive formulations (Lakshmi Prasanna et al., 2012) (Table 10.3). Koffi et al. (2008) prepared and evaluated thermosensitive and bioadhesive gel of quinine for rectal delivery. Polymers employed for formulation of system were poloxamer, HPMC, and propanediol-1,2. Gel composed of ternary system 16/0.5/30 ((poloxamer (16%)/HPMC (0.5%)/propanediol-1,2 (30%)) showed good bioadhesive strength in vivo and higher average values of MRT and T_{max} (9.1 $\pm$ 0.2 h and 30 min, respectively) compared to rectal solution (6.9 $\pm$ 0.9 h and 15 min, respectively). Fig. 10.8 shows plasma concentration profile after administration of intravenous solution, rectal solution, and rectal gels.

Furthermore, Yong et al. (2003) prepared poloxamer-based bioadhesive gel of diclofennac sodium for rectal delivery. Gel containing less than 1.0% sodium chloride showed good bioadhesive strength, higher initial plasma concentration, and faster T_{max} of drug compared to solid suppositories of diclofennac sodium in rats.

10.7.4 Applications in nasal drug delivery

Nasal mucosa may be a prominent choice for systemic drug delivery as this route is open to self-medication and is painless (Illum, 2003). The nose is usually considered a local drug delivery route. This route of delivery is preferred in management of conditions like severe nausea and vomiting (Behl et al., 1998). In comparison to the oral route, drugs like protein and peptide administered through nasal delivery systems show increased absorption, which may be due to high epithelial permeability/porosity along with a limited enzymatic activity. By utilizing various nasal mucoadhesive polymers, controlled-release formulation can be developed and there are a large number of polymers that come under the category of GRAS (Suzuki and Makino, 1999). Several studies have also categorized nasal mucosa as the preferred site for immunization purpose (Lemoine et al., 1998). Illum et al. (2001) prepared bioadhesive starch microspheres loaded with insulin and investigated their effect on the nasal absorption enhancement capacity of various enhancers with insulin in sheep as the animal model. Lysophosphatidylcholine, glycodeoxycholate and sodium taurodihydroxyfusidate were used as nasal absorption enhancers. Administration of freeze-dried starch microspheres along with glycodeoxycholate showed highest nasal bioavailability of insulin. Fig. 10.9 shows the plasma insulin levels in the sheep following nasal administration of insulin through various formulations.

The effect of multiple nasal administrations of bioadhesive powder on the insulin bioavailability was investigated by Callens et al. (2003). Insulin was administered in rabbits through two bioadhesive platforms consisting of a cospray dried mixture of amioca starch and carbopol 974P (1/3) and a physical mixture of drum dried waxy maize starch and carbopol 974P (9/1), respectively. Both formulations showed effective reduction in blood glucose level for 8 days. When the formulations were not administered from day 2 until day 7, the bioavailability on the eighth day compared with the first day of administration

Table 10.3 Summary of bioadhesive formulations for rectal drug delivery

Drug	Delivery system	Bioadhesive polymer	Sophisticated technique	Key findings	References
Etodolac	Thermogel	Poloxamer, hydroxypropylmethyl cellulose, polyvinyl pyrrolidone, methyl cellulose, hydroxyethylcellulose, carbopol	Ultraviolet (UV) spectroscopy	Prepared thermogel showed good bioadhesive strength, drug release through the Fickian diffusion, and no morphological damage to rectal tissues of rats	Barakat (2009)
5- Fluorouracil (5-FU)	Matrix tablet	Chitosan and polycarbophil	FT-IR, DSC, XRD	Formulations containing chitosan and polycarbophil interpolyelectrolyte complex showed controlled release of drug at all pH conditions related to vaginal, buccal, and rectal region with lack of initial burst release and pH dependency	Pendekal and Tegginamat (2012)
Epirubicin (Epi)	Liquid and solid suppository	Pluronic (Plu) and pH-sensitive polyacrylic acid (PAA)	FT-IR, HPLC	Prepared suppositories showed high in vitro bioadhesive strength, higher value of relative bioavailability and AUC along with efficient reduction in tumor cells in Balb/c mice	Lo et al. (2013)

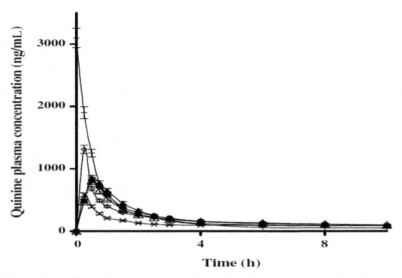

Figure 10.8 Variations of the plasma concentration of quinine versus time after intravenous bolus administration or rectal administration in albino rabbits. Intravenous solution (+), rectal solution (×), rectal gels 0/0.5/30 (◇), 16/0.5/30 (▲), 16/1/30 (●) and 17/0.5/30 (△). $n = 6$, mean±SD. *Used with permission from Koffi, A.A., Agnely, F., Besnard, M., Kablan Brou, J., Grossiord, J.L., Ponchel, G., 2008. In-vitro and in vivo characteristics of a thermogelling and bioadhesive delivery system intended for rectal administration of quinine in children. Eur. J. Pharm. Biopharm. 69(1):167–175.*

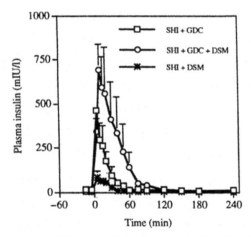

Figure 10.9 The plasma insulin levels in sheep following nasal administration of insulin in combination with glycodeoxycholate as a solution formulation, with starch microspheres as a freeze-dried powder or with glycodeoxycholate and starch microspheres as a freeze dried powder. SHI, Sodium insulin; DSM, Starch microspheres; GDC, Glycodeoxycholate. *Used with permission from Illum, L, Fisher, A.N., Jabbal-Gill, I., Davis, S.S., 2001. Bioadhesive starch microspheres and absorption enhancing agents act synergistically to enhance the nasal absorption of polypeptides. Int. J. Pharm. 222(1):109–119.*

was not changed. Later on, Varshosaz et al. (2006) prepared chitosan-based bioadhesive gel for nasal delivery. Nasal insulin absorption was significantly enhanced after administration of a gel containing 2% medium molecular weight chitosan along with EDTA, and it effectively reduced blood glucose level compared to an intravenous formulation in rats. The role of bioadhesive formulations for nasal drug delivery is summarized in Table 10.4.

10.7.5 Applications in ocular drug delivery

Topical administration of drugs to the eyes is well accepted and many ocular disorders are treated with topical administration of various drugs. Various physiological activities of eyes including reflex lachrymation, blinking, and rapid drainage are responsible for poor bioavailability of drugs to the eye (Topalkara et al., 2000). Frequent administration of the eye drops in the ocular cavity can cause cellular damage at the ocular surface along with induction of many side effects (Baudouin, 1996). For enhancement of ocular bioavailability, the residence time of drugs in ocular cavity should be increased. Various semisolid formulations like ointments or gels have the capability to interact with eyes in a sustained fashion, although they may cause various problems like blurred vision, sticky sensation, and ocular discomfort (Dudinski et al., 1983). The concept of bioadhesion is somewhat new in the field of the ocular dosage form. Weyenberg et al. (2006) evaluated different bioadhesive mini-tablet formulations based on amioca starch, carbopol 974P, and waxy maize starch (DDWM) for ocular drug delivery. Mini-tablets containing cospray (CS) dried amioca starch and carbopol in 15% w/w concentration showed the highest mucoadhesive strength to corneal mucosa and prolonged drug release up to 12 h, which was higher compared to other mini-tablet formulations (Fig. 10.10).

Further, Motwani et al. (2008) performed in vitro optimization studies of gatifloxacin-loaded chitosan-sodium alginate bioadhesive nanoparticles. Drug-loaded nanoparticles showed particles size variation in the range 205–572 nm along with a range of 0.325–0.489 and 17.6–47.8 mV for the polydispersity index and zeta potential respectively. Nanoparticles loaded with drug showed high mucoadhesive strength and quick release of drug for an initial 1 h and sustained drug release for next the 24 h through the mechanism of non-Fickian diffusion. Later on, alginate-chitosan based bioadhesive film was developed by Gilhotra and Mishra (2008) and variation in film properties due to cross-linking of surface was investigated. Formulation having 2% w/v and 1% w/v content of sodium alginate and chitosan respectively, and cross-linked surface showed maximum bioadhesive strength and capability to prolong the drug release up to 24 h. Table 10.5 gives an overview of the various bioadhesive formulations used for ocular delivery of various therapeutic agents.

10.7.6 Applications in gastrointestinal drug delivery

The oral route is well known for the administration of various bioactive molecules; however, many drugs administered through this route may show poor bioavailability problems and rapid first-pass metabolism. To avoid such problems new gastromucoadhesive

Table 10.4 Role of bioadhesive formulations for nasal drug delivery

Drug	Delivery system	Bioadhesive polymer	Sophisticated technique	Key findings	References
Insulin	Nasal insert	Hydroxypropyl methyl cellulose (HPMC)	Ultraviolet (UV) spectroscopy, gamma scintigraphy	Insert prepared with 2% HPMC increased residence time up to 5h in the nasal cavity, however enhancement of nasal absorption was not found	McInnes et al. (2007)
Domperidone	Microspheres	Starch	Ultraviolet (UV) spectroscopy, SEM	Prepared formulations showed good mucoadhesive property, swelling behavior, and drug release up to 73.11%–86.21%	Yadav and Mote (2008)
Ketorolac tromethamine	In-situ nasal gel	Chitosan, pectin, hydroxypropyl methylcellulose (HPMC)	Ultraviolet (UV) spectroscopy	Optimized formulations showed longer contact time, nonirritant behavior to nasal mucosa, and mucoadhesive power of gel containing chitosan and pectin got increased after addition of HPMC	Chelladurai et al. (2008)
Heat-inactivated influenza virus combined with LTR192G adjuvant	Spray-dried powder	Starch and cross-linked poly (acrylic acid), Carbopol 974P	Enzyme linked immunosorbent assay (ELISA), laser diffraction studies	The powder vaccine formulations induced systemic immune response and IgG titer volume was increased along with improvement in immune response due to the presence of Carbopol 974P	Coucke et al. (2009)

Δ⁹ - Tetra hydro cannabinol (THC)	Nasal solution and nasal gel	Chitosan	HPLC-Mass spectroscopy (HPLC-MS)	Optimized gel formulation showed higher bioadhesive strength in nasal mucosa of rabbit and higher absolute bioavailability compared to nasal solution	Al-Ghananeem et al. (2011)
Zolmitriptan	Nasal inserts	Chitosan–chondroitin sulfate	FT-IR, DSC, SEM	Results of in vitro and in situ showed that formulations having drug : polymer in 1:10 ratio gave 90% and 98% zolmitriptan release over a period of 8h	Kaur and Kaur (2013)
Zolmitriptan (ZT) & ketorolac tromethamine (KT)	Thermo reversible in-situ muco adhesive intranasal gel (TMISG)	Xyloglucan	HPLC	Inransal gel showed 21% increase in bioavailability of drug compared to oral formulation of same drug	Kumar et al. (2015)

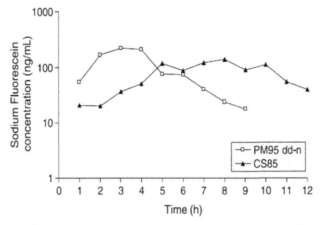

Figure 10.10 The tear film-cornea compartment concentrations of sodium fluorescein after application of the reference PM95dd-n and CS85 minitablets, sterilized at 25 kGy. CS (Cospray dried), PM (Physical mixture), dd (drum dried waxy maize starch), and n (native powder mixture). *Used with permission from Weyenberg, W., Bozdag, S., Foreman, P., Remon, J.P., Ludwig, A., 2006. Characterization and in vivo evaluation of ocular minitablets prepared with different bioadhesive Carbopolstarch components. Eur. J. Pharm. Biopharm. 62(2):202–209.*

formulations have been developed that show controlled drug release with improved pharmacokinetics (Lopes et al., 2016). Umamaheshwari et al. (2004) evaluated amoxicillin loaded mucoadhesive gliadin nanoparticles for eradication of *Helicobacter pylori*. Prepared nanoparticles showed higher GIT residence time because of greater mucoadhesion power, and effectively removed bacteria from GIT compared to free drug in the same dosage.

Furthermore, evaluation of mucoadhesive mini-tablets loaded with low molecular weight heparin (LMWH) to target stomach mucosa was carried out by Schmitz et al. (2005). Two different types of LMWH (LMWH-3 kDa (279 IU) and LMWH-6 kDa (300 IU)) were used to formulate tablets. Mini-tablets containing LMWH- 3 kDa (279 IU) showed relative bioavailability of 19.1%, which was higher compared to the control system, while (LMWH-6 kDa (300 IU)) loaded tablets showed relative bioavailability of 10.7% compared to control. Atyabi et al. (2007) prepared pectinate beads and studied the effect of trimethyl chitosan (TMC) on their mucoadhesive power. Furthermore, beads were also investigated to check mucoadhesion power for different parts of GIT. The highest mucoadhesion power was observed in the duodenum region and the least was found in the stomach area. Beads lacking TMC showed higher mucoadhesive power in the wet state, while in the dry state beads containing TMC showed higher mucoadhesive power. Various bioadhesive formulations used for gastrointestinal drug delivery are summarized in Table 10.6.

Table 10.5 Bioadhesive formulations used for ocular delivery of therapeutic agents

Drug	Delivery system	Bioadhesive polymers	Sophisticated technique	Key findings	References
5-Methoxy-carbonylamino-N-acetyltryptamine (5-MCA-NAT)	Solution	Propylene glycol, carboxymethyl cellulose (CMC) of low and medium viscosity, hydroxypropyl methyl cellulose	HPLC	Solution of drug containing CMC medium viscosity in 0.5% concentration effectively reduced intraocular pressure (IOP) for extended periods of 7h	Andrés-Guerrero et al. (2011)
Azithromycin (AZT)	Ocular inserts	Alginate, carbopol, and hydroxypropyl methylcellulose (HPMC)	Ultraviolet (UV) spectroscopy	Tensile strength and elasticity of the alginate based insert was higher compared to carbopol inserts; and insert containing carbopol and HPMC in the ratio 30:70 sustained the drug release for 6h	Gilhotra et al. (2011)
Moxifloxacin	In-situ gelling microparticle	Chondroitin sulfate-polyethylene glycol, poly(lactic-co-glycolic acid) PLGA	HPLC, SEM	In situ gelling microparticle showed controlled drug release higher than the minimum inhibitory concentration (MIC) of drug up to 10 days and high mucoadhesive strength in vitro	

(Continued)

Table 10.5 Bioadhesive formulations used for ocular delivery of therapeutic agents (Continued)

Drug	Delivery system	Bioadhesive polymers	Sophisticated technique	Key findings	References
Ketorolac tromethamine	In–situ gelling systems	Carbopol 980, HPMC K100LV, Poloxamer 407, Poloxamer 188	Ultraviolet (UV) spectroscopy, DSC	Optimized in situ gel showed high in vitro bioadhesion, long residence time, nonirritability and sustained drug release without affecting vision	Thakor et al. (2012)
Timolol maleate (TM)	Ocular film	Hyaluronic acid (HA), itaconic acid (IT), glutaraldehyde (GTA), polyethylene glycol diglycidyl ether (PEGDE)	Ultraviolet (UV) spectroscopy, Cell proliferation assay	Film containing PEGDE showed good in vivo mucoadhesive strength, reduced irritation, and high compatibility in rabbit eye compared to other films	Calles et al. (2013)
Brimonidine	Nano particles	Sodium alginate, chitosan	Ultraviolet (UV) spectroscopy	Nanoparticles prepared from both polymers showed sustained drug release for approximately 10h and effectively reduced intraocular pressure compared to marketed eye drops of the drug	Ibrahim et al. (2015)

Table 10.6 Bioadhesive formulations for gastrointestinal drug delivery

Drug	Delivery system	Bioadhesive polymer	Sophisticated technique	Key findings	References
Itraconazole	Tablets	Carbopol 934P (CP) and Methocel K4M (HPMC)	HPLC, DSC	Optimized formulation showed adequate mucoadhesive strength and C(max) equals to 1898 ± 75.23 ng/mL, T (max) of the formulation was 2h, and AUC was 28604.9 ng h/mL	Madgulkar et al. (2008)
Acyclovir	Microspheres	Sodium alginate	SEM, Gamma scintigraphy	Optimized formulation showed high mucoadhesive strength (66.42 ± 1.01%) and Gamma scintigraphy result showed gastric retention of formulation for more than 4h in vivo	Shadab et al. (2011)
Gliclazide	Microspheres	Tamarind seed polysaccharide (TSP)-alginate	FT-IR, SEM	Drug-loaded microspheres showed prolonged drug release in the stomach as well as intestinal pH for the periods of 12h and significant hypoglycemic activity in diabetic rats	Pal and Nayak (2012)
Lafutidine	Tablet	Sodium alginate, xanthan gum, karaya gum	FT-IR, DSC, X-ray imaging	Optimized formulation showed a mucoadhesive strength >35 g and X-ray analysis suggested that tablet was well adhered for >10h in rabbit's stomach	Patil and Talele (2015)

(Continued)

Table 10.6 Bioadhesive formulations for gastrointestinal drug delivery (Continued)

Drug	Delivery system	Bioadhesive polymer	Sophisticated technique	Key findings	References
Curcumin	Microspheres	Ethyl cellulose, carbopol 934P	FT-IR, DSC, SEM	The drug release from optimized formulation was also found to be slow and extended more than 8h due to prolonged stomach residence time of microspheres	Ali et al. (2014)
Ranitidine hydrochloride	Microspheres	Chitosan	SEM, FT-IR	Optimized formulation showed high drug encapsulation (70%), mucoadhesion (75%), and sustained drug release for 12h	Dhankar et al. (2014)
Puerarin	Microspheres	Sodium alginate, chitosan	DSC, SEM, Fluorescence Imaging	Significant protection of the stomach from ulceration was observed after administering the microparticles at different doses (150mg/kg, 300mg/kg, 450mg/kg, and 600mg/kg) in the case of ethanol induced ulcers	Hou et al. (2014)
Clarithromycin	Beads	Calcium alginate, chitosan	FT-IR, DSC, SEM	Prepared beads showed sustained drug release in mucin layer in vitro mimicking in vivo conditions in which *H. pylori* lies in gastric mucus	Adebisi et al. (2015)
Amoxicillin trihydrate	Beads	Sodium alginate, hydroxypropyl methylcellulose, chitosan	SEM, X-ray imaging	The optimized formulation showed 100% *Helicobacter pylori* growth inhibition in 15h in vitro culture and X-ray study in rabbit stomach confirmed the gastric retention of optimized formulation	Dey et al. (2016)

Table 10.7 List of intellectual property rights (IPR) related to bioadhesive polymeric formulation

Title of patent	Brief description	Patent number	References
Bioadhesive formulations for use in drug delivery	This invention deals with bioadhesive property for use in drug delivery to delivery single drug or drug combination to body cavities or body surfaces	WO2014027006 A1	Embil et al. (2014)
Bioadhesive drug formulations for oral transmucosal delivery	This invention deals with method of preparation of mucoadhesive hydrogel and its ability to deliver medicament in a sustained fashion to oral mucosa	US8252328 B2	Tzannis et al. (2012)
Bioadhesive microspheres and their use as drug delivery and imaging systems	This invention discloses method of development of bioadhesive microspheres using synthetic polymers and their capability to release medicament in GIT	US6235313 B1	Mathiowitz et al. (2001)
Bioadhesive rate controlled oral dosage formulations	This invention deals with a formulation in which a drug is encapsulated in a core which is coated with a bioadhesive rate controlling membrane and taken through oral route	WO2006026556 A2	Nangia et al. (2006)
Bioadhesive drug delivery system with enhanced gastric retention	This invention discloses a method of preparation of bioadhesive microspheres and their capability to adhere to stomach mucosa and sustained release of medicament there	WO2003051304 A2	Jacob et al. (2003)
Composition of a bioadhesive sustained delivery carrier for drug administration	This invention deals with development steps of a bioadhesive carrier for the treatment of systemic diseases or disease of oropharynx through sustained release of medicaments	EP0451433 B1	Bottenberg et al. (1996)

10.8 CONCLUSIONS

A wide range of APIs can be delivered in a targeted-controlled fashion by using the process of bioadhesion. The field of bioadhesion has had outstanding research growth and it mainly focuses on the development of new devices along with intelligent polymers. Bioadhesive polymeric-based systems may play an important role in delivering various bioactive molecules with the arrival of new large and small molecules in the field of drug research (Table 10.7). Taking various polymer properties and environmental factors into consideration, an appropriate bioadhesive polymer can be engineered which fulfills all needs of drug delivery. Although bioadhesive polymers are becoming trendy in the field of drug delivery, they are still ranked according to multiple available techniques due to lack of a universal technique for testing of their mucoadhesive strength. Hence, development of a universal bioadhesion evaluation technique is necessary for selection of polymers. Old generation polymers provide a strong bioadhesive character, but less effective targeting, while new generation polymers show targeted strong bioadhesion. However, toxicity concerns should be taken into consideration while using any of them. So, bioadhesive polymers may be utilized for drug delivery according to the needs of therapeutic challenge in various body parts like GIT, vagina, mouth, rectum, nose, and eyes.

REFERENCES

Adebisi, A.O., Laity, P.R., Conway, B.R., 2015. Formulation and evaluation of floating mucoadhesive alginate beads for targeting *Helicobacter pylori*. J. Pharm. Pharmacol. 67 (4), 511–524.

Ahagon, A., Gent, A.N., 1975. Effect of interfacial bonding on the strength of adhesion. J. Polym. Sci. Polym. Phys. 13, 1285–1300.

Albrecht, K., Greindl, M., Kremser, C., Wolf, C., Debbage, P., Bernkop-Schnürch, A., 2005. Comparative in vivo mucoadhesion studies of thiomer formulations using magnetic resonance imaging and fluorescence detection. J. Control. Rel. 115, 78–84.

Al-Ghananeem, A.M., Malkawi, A.H., Crooks, P.A., 2011. Bioavailability of Δ^9-tetrahydrocannabinol following intranasal administration of a mucoadhesive gel spray delivery system in conscious rabbits. Drug. Dev. Ind. Pharm. 37 (3), 329–334.

Ali, M.S., Pandit, V., Jain, M., Dhar, K.L., 2014. Mucoadhesive microparticulate drug delivery system of curcumin against *Helicobacter pylori* infection: design, development and optimization. J. Adv. Pharm. Technol. Res. 5 (1), 48–56.

Andrés-Guerrero, V., Molina-Martínez, I.T., Peral, A., de las Heras, B., Pintor, J., Herrero-Vanrell, R., 2011. The use of mucoadhesive polymers to enhance the hypotensive effect of a melatonin analogue, 5-MCA-NAT, in rabbit eyes. Invest. Ophthalmol. Vis. Sci. 52 (3), 1507–1515.

Andrew, B., 2010. Anorectal anatomy and physiology. Surg. Clin. N. Am. 90, 1–15.

Andrews, G.P., Laverty, T.P., Jones, D.S., 2009. Mucoadhesive polymeric platforms for controlled drug delivery. Eur. J. Pharm. Biopharm. 71, 505–518.

Asane, G.S., Nirmal, S.A., Rasal, K.B., Naik, A.A., Mahadik, M.S., Rao, Y.M., 2008. Polymers for mucoadhesive drug delivery system: a current status. Drug. Dev. Ind. Pharm. 34 (11), 1246–1266.

Atyabi, F., Majzoob, S., Dorkoosh, F., Sayyah, M., Ponchel, G., 2007. The impact of trimethyl chitosan on in vitro mucoadhesive properties of pectinate beads along different sections of gastrointestinal tract. Drug. Dev. Ind. Pharm. 33 (3), 291–300.

Avachat, A.M., Gujar, K.N., Wagh, K.V., 2013. Development and evaluation of tamarind seed xyloglucan-based mucoadhesive buccal films of rizatriptan benzoate. Carbohydr Polym. 91 (2), 537–542.

Baloğlu, E., Ozyazici, M., Yaprak Hizarcioğlu, S., Senyiğit, T., Ozyurt, D., Pekçetin, C., 2006. Bioadhesive controlled release systems of ornidazole for vaginal delivery. Pharm. Dev. Technol. 11 (4), 477–484.

Bansil, R., Turner, B., 2006. Mucin structure, aggregation, physiological functions and biomedical applications. Curr. Opin. Colloid Interf. Sci. 11, 164–170.

Barakat, N.S., 2009. In vitro and in vivo characteristics of a thermogelling rectal delivery system of etodolac. AAPS. Pharm. Sci. Tech. 10 (3), 724–731.

Bassi, P., Kaur, G., 2015. Bioadhesive vaginal drug delivery of nystatin using a derivatized polymer: development and characterization. Eur. J. Pharm. Biopharm. 96, 173–184.

Baudouin, C., 1996. Side effects of antiglaucomatous drugs on the ocular surface. Curr. Opin. Ophthalmol. 7, 80–86.

Bautista-Baños, S., Hernández-Lauzardo, A.N., Velázquez-delValle, M.G., Hernández-López, M., AitBarka, E., Bosquez-Molina, E., et al., 2006. Chitosan as apotential natural compound to control pre and post-harvest diseases of horticultural commodities. Crop. Protect. 25, 108–118.

Berginc, K., Suljaković, S., Škalko-Basnet, N., Kristl, A., 2014. Mucoadhesive liposomes as new formulation for vaginal delivery of curcumin. Eur. J. Pharm. Biopharm. 87 (1), 40–46.

Behl, C.R., Pimplaskar, H.K., Sileno, J., deMeireles, J., Romeo, V.D., 1998. Effects of physicochemical properties and other factors on systemic nasal drug delivery. Adv. Drug Deliv. Rev. 29, 89–116.

Bernkop-Schnürch, A., 2005. Mucoadhesive systems in oral drug delivery. Drug Discov. Today Tech. 2, 83–87.

Bernkop-Schnürch, A., Humenberger, C., Valenta, C., 1998. Basic studies on bioadhesive delivery systems for peptide and protein drugs. Int. J. Pharm. 165, 217–225.

Bernkop-Schnürch, A., Krauland, A., Leitner, V., Palmberger, T., 2004. Thiomers: potential excipients for non-invasive peptide delivery systems. Eur. J. Pharm. Biopharm. 58, 253–263.

Bottenberg, P., Remon J.P., De M.C., Slop, D., 1996. Composition of a bioadhesive sustained delivery carrier for drug administration. E.P. Patent # 0451433 B1.

Bravo-Osuna, I., Vauthier, C., Farabollini, A., Palmieri, G., Ponchel, G., 2007. Mucoadhesion mechanism of chitosan and thiolated chitosan-poly(isobutyl cyanoacrylate) core–shell nanoparticles. Biomaterials 28, 2233–2243.

Callens, C., Pringels, E., Remon, J.P., 2003. Influence of multiple nasal administrations of bioadhesive powders on the insulin bioavailability. Int. J. Pharm. 250 (2), 415–422.

Calles, J.A., Tártara, L.I., Lopez-García, A., Diebold, Y., Palma, S.D., Vallés, E.M., 2013. Novel bioadhesive hyaluronan-itaconic acid crosslinked films for ocular therapy. Int. J. Pharm. 455 (1-2), 48–56.

Capra, R., Baruzzi, A., Quinzani, L., Strumia, M., 2007. Rheological, dielectric and diffusion analysis of mucin/carbopol matrices used in amperometric biosensors. Sensors. Actuators. B. 124, 466–476.

Cevher, E., Sensoy, D., Zloh, M., Mülazimoğlu, L., 2008. Preparation and characterisation of natamycin: gamma-cyclodextrin inclusion complex and its evaluation invaginal mucoadhesive formulations. J. Pharm. Sci. 2008, 4319–4335.

Chelladurai, S., Mishra, M., Mishra, B., 2008. Design and evaluation of bioadhesive in-situ nasal gel of ketorolac tromethamine. Chem. Pharm. Bull. (Tokyo). 56 (11), 1596–1599.

Chen, G., Bunt, C., Wen, J., 2015. Mucoadhesive polymers-based film as a carrier system for sublingual delivery of glutathione. J. Pharm. Pharmacol. 67 (1), 26–34.

Chowdary, C.P.R., Rao, Y.S., 2004. Mucoadhesive microspheres for controlled drug delivery. Biol. Pharm. Bull. 27, 1717–1724.

Clark, M.A., Hirst, B., Jepson, M., 2000. Lectin-mediated mucosal delivery of drugs and microparticles. Adv. Drug Deliv. Rev. 43, 207–223.

Coucke, D., Schotsaert, M., Libert, C., Pringels, E., Vervaet, C., Foreman, P., et al., 2009. Spray-dried powders of starch and crosslinked poly(acrylic acid) as carriers for nasal delivery of inactivated influenza vaccine. Vaccine 27 (8), 1279–1286.

Davies, J., Viney, C., 1998. Water–mucin phases: conditions for mucus liquid crystallinity. Thermochim. Acta. 315, 39–49.

Dey, S.K., De, P.K., De, A., Ojha, S., De, R., Mukhopadhyay, A.K., et al., 2016. Floating mucoadhesive alginate beads of amoxicillin trihydrate: a facile approach for *H. pylori* eradication. Int. J. Biol. Macromol. 89, 622–631.

Dhankar, V., Garg, G., Dhamija, K., Awasthi, R., 2014. Preparation, characterization and evaluation of ranitidine hydrochloride-loaded mucoadhesive microspheres. Polim. Med. 44 (2), 75–81.

Dhiman, M., Yedurkar, P., Sawant, K.K., 2008. Formulation, characterization, and in vitro evaluation of bioadhesive gels containing 5-Fluorouracil. Pharm. Dev. Technol. 13 (1), 15–25.

Dobaria, N.B., Mashru, R.C., Badhan, A.C., Thakkar, A.R., 2009. A novel intravaginal delivery system for itraconazole: in vitro and in vivo evaluation. Curr. Drug. Deliv. 6 (2), 151–158.

Dodane, V., Vilivalam, V., 1998. Pharmaceutical applications of chitosan. Pharm. Sci. Technol. 1, 246–253.

Dodou, D., Breedveld, P., Wieringa, P., 2005. Mucoadhesives in the gastrointestinal tract: revisiting the literature for novel applications. Eur. J. Pharm. Biopharm. 60, 1–16.

Dudinski, O., Finnin, B.C., Reed, B.L., 1983. Acceptability of thickened eye drops to human subjects. Curr. Ther. Res. 33, 322–337.

Dyvik, K., Graffner, C., 1992. Investigation of the applicability of a tensile testing machine for measuring mucoadhesive strength. Acta. Pharm. Nord. 4, 79–84.

El-Kamel, A., Sokar, M., Naggar, V., Al-Gamal, S., 2002. Chitosan and sodium alginate based bioadhesive vaginal tablets. AAPS. Pharm. Sci. 4: article 44.

Embil, K., Figueroa, R., Dominguez, J.R., 2014. Bioadhesive formulations for use in drug delivery. W.O. Patent # 2014027006 A1.

Fefelova, N., Nurkeeva, Z., Mun, G., Khutoryanskiy, V., 2007. Mucoadhesive interactions of amphiphilic cationic copolymers based on [2-(methacryloyloxy)ethyl]trimethylammonium chloride. Int. J. Pharm. 339, 25–32.

Fiebrig, I., Harding, S., Rowe, A., Hyman, S., Davis, S., 1995. Transmission electron microscopy studies on pig gastric mucin and its interactions with chitosan. Carbohydr. Polym. 28, 239–244.

Gaikwad, S.S., Thombre, S.K., Kale, Y.K., Gondkar, S.B., Darekar, A.B., 2014. Design and *in-vitro* characterization of buccoadhesive tablets of timolol maleate. Drug. Dev. Ind. Pharm. 40 (5), 680–690.

Garg, S., Jambu, L., Vermani, K., 2007. Development of novel sustained release bioadhesive vaginal tablets of povidone iodine. Drug. Dev. Ind. Pharm. 33, 1340–1349.

Gilhotra, R.M., Mishra, D.N., 2008. Alginate-chitosan film for ocular drug delivery: effect of surface cross-linking on film properties and characterization. Pharmazie 63 (8), 576–579.

Gilhotra, R.M., Nagpal, K., Mishra, D.N., 2011. Azithromycin novel drug delivery system for ocular application. Int. J. Pharm. Investig. 1 (1), 22–28.

Gu, J.M., Robinson, J.R., Leung, S.H., 1988. Binding of acrylic polymers to mucin/epithelial surfaces: structure–property relationships. Crit. Rev. Ther. Drug. Carrier. Syst. 5, 21–67.

Guo, Q., Aly, A., Schein, O., Trexler, M.M., Elisseeff, J.H., 2012. Moxifloxacin in situ gelling microparticles-bioadhesive delivery system. Res. Pharma. Sci. 2, 66–71.

Hagesaether, E., Sande, S.A., 2007. *In-vitro* measurements of mucoadhesive properties of six types of pectin. Drug Dev. Ind. Pharm. 33, 417–425.

Hassan, N., Khar, R.K., Ali, M., Ali, J., 2009. Development and evaluation of buccal bioadhesive tablet of an anti-emetic agent ondansetron. AAPS Pharm. Sci. Tech. 10 (4), 1085–1092.

He, P., Davis, S., Illum, L., 1998. In-vitro evaluation of the mucoadhesive properties of chitosan microspheres. Int. J. Pharm. 166, 75–88.

Henriksen, I., Green, K., Smart, J., Smistad, G., Karlsen, J., 1996. Bioadhesion of hydrated chitosans: an in vitro and in vivo study. Int. J. Pharm. 145, 231–240.

Hiorth, M., Liereng, L., Reinertsen, R., Tho, I., 2013. Formulation of bioadhesive hexylaminolevulinate pellets intended for photodynamic therapy in the treatment of cervical cancer. Int. J. Pharm. 441 (1-2), 544–554.

Hou, J.Y., Gao, L.N., Meng, F.Y., Cui, Y.L., 2014. Mucoadhesive microparticles for gastroretentive delivery: preparation, biodistribution and targeting evaluation. Mar. Drugs 12 (12), 5764–5787.

Huang, Y., Leobandung, W., Foss, A., Peppas, N.A., 2000. Molecular aspects of mucoand bioadhesion: tethered structures and site-specific surfaces. J. Control. Rel. 65, 63–71.

Hussain, A., Ahsan, F., 2005. The vagina as a route for systemic drug delivery. J. Control. Rel. 103, 301–313.

Ibrahim, M.M., Abd-Elgawad, A.H., Soliman, O.A., Jablonski, M.M., 2015. Natural Bioadhesive Biodegradable Nanoparticle-Based Topical Ophthalmic Formulations for Management of Glaucoma. Transl. Vis. Sci. Technol. 4 (3), 12–15.

Ikinci, G., Senel, S., Tokgözoğlu, L., Wilson, C.G., Sumnu, M., 2006. Development and in vitro/in vivo evaluations of bioadhesive buccal tablets for nicotine replacement therapy. Pharmazie 61 (3), 203–207.

Illum, L., Fisher, A.N., Jabbal-Gill, I., Davis, S.S., 2001. Bioadhesive starch microspheres and absorption enhancing agents act synergistically to enhance the nasal absorption of polypeptides. Int. J. Pharm. 222 (1), 109–119.

Illum, L., 2003. Nasal drug delivery—possibilities, problems and solutions. J. Control. Rel. 87, 187–198.

Imam, M.E., Hornof, M., Valenta, C., Reznicek, G., Bernkop-Schnürch, A., 2003. Evidence for the interpenetration of mucoadhesive polymers into the mucous gel layer. STP Pharma. Sci. 13, 171–176.

Jabbari, E., Peppas, N.A., 1995. A model for interdiffusion at interfaces of polymers with dissimilar physical properties. Polymer 36, 575–586.

Jacob, J., Mathiowitz, E., Enscore, D., Schestopol, M., 2003. Bioadhesive drug delivery system with enhanced gastric retention. W.O. Patent # 2003051304 A2.

Jaipal, A., Pandey, M.M., Charde, S.Y., Raut, P.P., Prasanth, K.V., Prasad, R.G., 2015. Effect of HPMC and mannitol on drug release and bioadhesion behavior of buccal discs of buspirone hydrochloride: in-vitro and in-vivo pharmacokinetic studies. Saudi. Pharm. J. 23 (3), 315–326.

Jiménez-Castellanos, M.R., Zia, H., Rhodes, C.T., 1993. Mucoadhesive drug delivery systems. Drug. Dev. Ind. Pharm. 19, 143–194.

Kast, C.E., Valenta, C., Leopold, M., Bernkop-Schnürch, A., 2002. Design and in vitro evaluation of a novel bioadhesive vaginal drug delivery system for clotrimazole. J. Control. Rel. 81 (3), 347–354.

Kaur, K., Kaur, G., 2013. Formulation and evaluation of chitosan-chondroitin sulphate based nasal inserts for zolmitriptan. Biomed. Res. Int., 1–8.

Khutoryanskiy, V., 2007. Hydrogen-bonded interpolymer complexes as materials for pharmaceutical applications. Int. J. Pharm. 334, 15–26.

Kingshott, P., Griesser, H., 1999. Surfaces that resist bioadhesion. Curr. Opin. Solid State Mater. Sci. 4, 403–412.

Kinloch, A.J., 1980. The science of adhesion. J. Mater. Sci. 15, 2141–2166.

Kocevar-Nared, J., Kristl, J., Smid-Korbar, J., 1997. Comparative rheological investigation of crude gastric mucin and natural gastric mucus. Biomaterials 18, 677–681.

Koffi, A.A., Agnely, F., Besnard, M., Kablan Brou, J., Grossiord, J.L., Ponchel, G., 2008. In-vitro and in vivo characteristics of a thermogelling and bioadhesive delivery system intended for rectal administration of quinine in children. Eur. J. Pharm. Biopharm. 69 (1), 167–175.

Kumar, A., Garg, T., Sarma, G.S., Rath, G., Goyal, A.K., 2015. Optimization of combinational intranasal drug delivery system for the management of migraine by using statistical design. Eur. J. Pharm. Sci. 70, 140–151.

Kumar, K., Dhawan, N., Sharma, H., Vaidya, S., Vaidya, B., 2014. Bioadhesive polymers: novel tool for drug delivery. Artif. Cells Nanomed. Biotechnol. 42 (4), 274–283.

Lakshmi Prasanna, J., Deepthi, B., Rama Rao, N., 2012. Rectal drug delivery: a promising route for enhancing drug absorption. Asian J. Res. Pharm. Sci. 2 (4), 143–149.

Lee, J.W., Park, J.H., Robinson, J.R., 2000. Bioadhesive-based dosage forms: the next generation. J. Pharm. Sci. 89, 850–866.

Lehr, C., 2000. Lectin-mediated drug delivery: the second generation of bioadhesives. J. Control. Rel. 65, 19–29.

Lehr, C.M., Bouwstra, J.A., Schacht, E.H., Junginger, H.E., 1992. In-vitro evaluation of mucoadhesive properties of chitosan and some other natural polymers. Int. J. Pharm. 78, 43–48.

Lehr, C.M., Johanna, A.B., Tukker, J.J., Junginge, H.E., 1990. Intestinal transit of bioadhesive microspheres in an in situ loop in the rat-a comparative study with copolymers and blends based on poly (acrylic acid). J. Control. Rel. 13, 51–62.

Leitner, V., Walker, G., Bernkop-Schnürch, A., 2003. Thiolated polymers: evidence for the formation of disulphide bonds with mucus glycoproteins. Eur. J. Pharm. Biopharm. 56, 207–214.

Lemoine, D., Wauters, F., Bouchend'homme, S., Pre'at, V., 1998. Preparation and characterization of alginate microspheres containing a model antigen. Int. J. Pharm. 176, 9–19.

Lis, H., Sharon, N., 1986. Lectins as molecules and as tools. Annu. Rev. Biochem. 55, 35–67.

Lo, Y.L., Lin, Y., Lin, H.R., 2013. Evaluation of epirubicin in thermogelling and bioadhesive liquid and solid suppository formulations for rectal administration. Int. J. Mol. Sci. 15 (1), 342–360.

Lopes, C.M., Bettencourt, C., Rossi, A., Buttini, F., Barata, P., 2016. Overview on gastroretentive drug delivery systems for improving drug bioavailability. Int. J. Pharm. 510, 144–158.

Ludwig, A., 2005. The use of mucoadhesive polymers in ocular drug delivery. Adv. Drug. Deliv. Rev. 57, 1595–1639.

Madgulkar, A., Kadam, S., Pokharkar, V., 2008. Studies on formulation development of mucoadhesive sustained release itraconazole tablet using response surface methodology. AAPS. Pharm. Sci. Tech. 9 (3), 998–1005.

Madsen, F., Eberth, K., Smart, J., 1998. A rheological assessment of the nature of interactions between mucoadhesive polymers and a homogenised mucus gel. Biomaterials 1998, 1083–1092.

Mansuri, S., Kesharwani, P., Jain, K., Tekade, R.K., Jain, N.K., 2016. Mucoadhesion: a promising approach in drug delivery system. React. Funct. Polym. 100, 151–172.

Mathiowitz, E., Chickering, D., Jacob, J.S., 2001. Bioadhesive microspheres and their use as drug delivery and imaging systems. U.S. Patent # 6235313 B1.

McInnes, F.J., O'Mahony, B., Lindsay, B., Band, J., Wilson, C.G., Hodges, L.A., et al., 2007. Nasal residence of insulin containing lyophilised nasal insert formulations, using gamma scintigraphy. Eur. J. Pharm. Sci. 31 (1), 25–31.

Merabet, J., Thompson, D., Levinson, R.S., 2005. Advancing vaginal drug delivery. Expert. Opin. Drug. Deliv. 2, 769–777.

Meseguer, G., Gurny, R., Buri, P., 2004. In-vivo evaluation of dosage forms: application of gamma scintigraphy to non-enteral routes of administration. J. Drug. Rel. 2, 269–288.

Metia, P.K., Bandyopadhyay, A.K., 2008. In-vitro evaluation of novel mucoadhesive buccal tablet of oxytocin prepared with *Diospyros peregrina* fruits mucilages. Yakugaku. Zasshi. 128, 603–609.

Mortazavi, S.A., Smart, J., 1993. An investigation into the role of water movement and mucus gel dehydration in mucoadhesion. J. Control. Rel. 25, 197–203.

Mortazavi, S.A., Smart, J.D., 1994. Factors influencing gel-strengthening at the mucoadhesive-mucus interface. J. Pharm. Pharmacol. 46, 86–90.

Motwani, S.K., Chopra, S., Talegaonkar, S., Kohli, K., Ahmad, F.J., Khar, R.K., 2008. Chitosan-sodium alginate nanoparticles as submicroscopic reservoirs for ocular delivery: formulation, optimisation and in vitro characterisation. Eur. J. Pharm. Biopharm. 68 (3), 513–525.

Movassaghian, S., Barzegar-Jalali, M., Alaeddini, M., Hamedyazdan, S., Afzalifar, R., Zakeri-Milani, P., et al., 2011. Development of amitriptyline buccoadhesive tablets for management of pain in dental procedures. Drug. Dev. Ind. Pharm. 37 (7), 849–854.

Nangia, A., Jacob, J.S., Moslemy, P., 2006. Bioadhesive rate controlled oral dosage formulations. W.O. Patent # 2006026556 A2.

Ndesendo, V.M., Pillay, V., Choonara, Y.E., du Toit, L.C., Buchmann, E., Meyer, L.C., et al., 2010. Investigation of the physicochemical and physicomechanical properties of a novel intravaginalbioadhesive polymeric device in the pig model. AAPS. Pharm. Sci. Tech. 11 (2), 793–808.

Onishi, H., Machida, Y., 1999. Biodegradation and distribution of water-soluble chitosan in mice. Biomaterials 20, 175–182.

Pal, D., Nayak, A.K., 2012. Novel tamarind seed polysaccharide-alginate mucoadhesive microspheres for oral gliclazide delivery: in vitro-in vivo evaluation. Drug Deliv. 19 (3), 123–131.

Park, K., 1989. A new approach to study mucoadhesion colloidal gold staining. Int. J. Pharm. 53, 209–217.

Park, K., Robinson, J.R., 1984. Bioadhesive polymers as platforms for oral-controlled drug delivery: method to study bioadhesion. Int. J. Pharm. 19, 107–127.

Patel, P., Patel, P., 2015. Formulation and evaluation of clindamycin HCL in situ gel for vaginal application. Int. J. Pharm. Investig. 5 (1), 50–56.

Patil, S., Talele, G.S., 2015. Gastroretentive mucoadhesive tablet of lafutidine for controlled release and enhanced bioavailability. Drug. Deliv. 22 (3), 312–319.

Pavelić, Z., Skalko-Basnet, N., Jalsenjak, I., 2005. Characterisation and in vitro evaluation of bioadhesive liposome gels for local therapy of vaginitis. Int. J. Pharm. 301 (1-2), 140–148.

Pendekal, M.S., Tegginamat, P.K., 2012. Development and characterization of chitosan-polycarbophil inter-polyelectrolyte complex-based 5-fluorouracil formulations for buccal, vaginal and rectal application. Daru 20 (1), 67.

Peppas, N., Huang, Y., 2004. Nanoscale technology of mucoadhesive interactions. Adv. Drug Deliv. Rev. 56, 1675–1687.

Perioli, L., Ambrogi, V., Rubini, D., Giovagnoli, S., Ricci, M., Blasi, P., et al., 2004. Novel mucoadhesive buccal formulation containing metronidazole forthe treatment of periodontal disease. J. Control. Rel. 95 (3), 521–533.

Prasanna, R.I., Anitha, P., Chetty, C.M., 2011. Formulation and evaluation of bucco-adhesive tablets of sumatriptan succinate. Int. J. Pharm. Investig. 1 (3), 182–191.

Pund, S., Joshi, A., Vasu, K., Nivsarkar, M., Shishoo, C., 2011. Gastroretentive delivery of rifampicin: *in vitro* mucoadhesion and *in vivo* gamma scintigraphy. Int. J. Pharm. 411, 106–112.

Riley, R., Smart, J., Tsibouklis, J., Dettmar, P., Hampson, F., Davis, J.A., et al., 2001. An investigation of mucus/polymer rheological synergism using synthesised and characterised poly(acrylic acid)s. Int. J. Pharm. 217, 87–100.

Robinson, J., Mlynek, G., 1995. Bioadhesive and phase-change polymers for ocular drug delivery. Adv. Drug Deliv. Rev. 16, 45–50.

Roldo, M., Hornof, M., Caliceti, P., Bernkop-Schnürch, A., 2004. Mucoadhesive thiolated chitosans as platforms for oral controlled drug delivery: synthesis and in vitro evaluation. Eur. J. Pharm. Biopharm. 57, 115–121.

Schmitz, T., Leitner, V.M., Bernkop-Schnürch, A., 2005. Oral heparin delivery: design and in vivo evaluation of a stomach targeted mucoadhesive delivery system. J. Pharm. Sci. 94 (5), 966–973.

Senthil, V., Gopalakrishnan, S., Sureshkumar, R., Jawahar, N., Ganesh, G.N.K., Nagasamyvenkatesh, D., 2010. Mucoadhesive slowrelease tablets of theophylline: design and evaluation. Asian. J. Pharm. 4, 64–68.

Shadab, Ahuja, A., Khar, R.K., Baboota, S., Chuttani, K., Mishra, A.K., et al., 2011. Gastroretentive drug delivery system of acyclovir-loaded alginate mucoadhesive microspheres: formulation and evaluation. Drug. Deliv. 18 (4), 255–264.

Shahi, S.R., Tribhuwan, S.D., Tadwee, I.K., Gupta, S.K., Zadbuke, N.S., Shivanikar, S.S., 2011. Formulation of atenolol mucoadhesive microspheres for nasal delivery by spray drying technique: in vitro/ex vivo Evaluation. Der. Pharmacia. Sinica. 2, 54–63.

Shayeda, Gannu, R., Palem, C.R., Rao, Y.M., 2009. Development of novel bioadhesive buccal formulation of diltiazem: in vitro and in vivo characterization. PDA. J. Pharm. Sci. Technol. 63 (5), 401–408.

Shojaei, A., Li, X., 1997. Mechanisms of buccal mucoadhesion of novel copolymers of acrylic acid and polyethylene glycol monomethylether monomethacrylate. J. Control. Rel. 47, 151–161.

Sigurdsson, H., Loftsson, T., Lehr, C., 2006. Assessment of mucoadhesion by a resonant mirror biosensor. Int. J. Pharm. 325, 75–81.

Singh, S., Jain, S., Muthu, M.S., Tilak, R., 2008. Preparation and evaluation of buccal bioadhesive tablets containing clotrimazole. Curr. Drug. Deliv. 5 (2), 133–141.

Singla, A.K., Chawla, M., Singh, A., 2006. Potential applications of carbomer in oral mucoadhesive con-trolled drug delivery system: a review. Drug. Dev. Ind. Pharm. 26, 913–924.

Sudhakar, Y., Kuotsu, K., Bandyopadhyay, A.K., 2006. Buccal bioadhesive drug delivery – a promising option for orally less efficient drugs. J. Control. Rel. 114, 15–40.

Suzuki, Y., Makino, Y., 1999. Mucosal drug delivery using cellulose derivatives as a functional polymer. J. Control. Rel. 62, 101–107.

Thakor, S., Vhora, I., Desai, J., Thakkar, S., Thakkar, H., 2012. Physiologically activated phase transition systems for improved ocular retention of ketorolac tromethamine. J. Pharm. Bioallied. Sci. 4 (Suppl 1), S6–S7.

Topalkara, A., Guler, C., Arici, D.S., Arici, M.K., 2000. Adverse effects of topical antiglaucoma drugs on the ocular surface. Clin. Exp. Ophthalmol. 28, 113–117.

Tuğcu-Demiröz, F., Acartürk, F., Erdoğan, D., 2013. Development of long-acting bioadhesive vaginal gels of oxybutynin: formulation, in vitro and in vivo evaluations. Int. J. Pharm. 457 (1), 25–39.

Tzannis, S., Palmer, P., Schreck, T., Hamel, L., Poutiatine A.I., 2012. Bioadhesive drug formulations for oral transmucosal delivery. U.S. Patent # 8252328 B2.

Ugwoke, M., Sam, E., VanDenMooter, G., Verbeke, N., Kinget, R., 1999. Nasal mucoadhesive delivery systems of the anti-parkinsonian drug, apomorphine: influence of drug-loading on in vitro and in vivo release in rabbits. Int. J. Pharm. 181, 125–138.

Ugwoke, M.I., Agu, R.U., Verbeke, N., Kinget, R., 2005. Nasal mucoadhesive drug delivery: background, applications, trends and future perspectives. Adv. Drug. Deliv. Rev. 57, 1640–1665.

Umamaheshwari, R.B., Ramteke, S., Jain, N.K., 2004. Anti-Helicobacter pylori effect of mucoadhesive nanoparticles bearing amoxicillin in experimental gerbils model. AAPS. Pharm. Sci. Tech. 5 (2), e32.

Van Hoogdalem, E.J., de Boer, A.G., Breimer, D.D., 1991. Pharmacokinetics of rectal drug administration, part I: general considerations and clinical applications of centrally acting drugs. Clin. Pharmacokinet. 21 (1), 11–26.

Varshosaz, J., Sadrai, H., Heidari, A., 2006. Nasal delivery of insulin using bioadhesive chitosan gels. Drug Deliv. 13 (1), 31–38.

Vasir, J., Tambwekar, K., Garg, S., 2003. Bioadhesive microspheres as a controlled drug delivery system. Int. J. Pharm. 255, 13–32.

Vishnu, Y.V., Chandrasekhar, K., Ramesh, G., Rao, Y.M., 2007. Development of mucoadhesive patches for buccal administration of carvedilol. Curr. Drug. Deliv. 4 (1), 27–39.

Wang, L., Tang, X., 2008. A novel ketoconazole bioadhesive effervescent tablet for vaginal delivery: design, in vitro and in vivo. Int. J. Pharm. 350 (1-2), 181–187.

Wattanakorn, N., Asavapichayont, P., Nunthanid, J., Limmatvapirat, S., Sungthongjeen, S., Chantasart, D., et al., 2010. Pectin-based bioadhesive delivery of carbenoxolone sodium for aphthous ulcers in oral cavity. AAPS. Pharm. Sci. Tech. 11 (2), 743–751.

Weyenberg, W., Bozdag, S., Foreman, P., Remon, J.P., Ludwig, A., 2006. Characterization and in vivo evaluation of ocular minitablets prepared with different bioadhesive Carbopolstarch components. Eur. J. Pharm. Biopharm. 62 (2), 202–209.

Wilding, I.R., Davis, S.S., O'Hagan, D.T., 1994. Targeting of drugs and vaccines to the gut. Pharmacol. Ther. 62, 97–124.

Willits, R., Saltzman, W.M., 2001. Synthetic polymers alter the structure of cervical mucus. Biomaterials 22, 445–452.

Wirth, M., Gerhardt, K., Wurm, C., Gabor, F., 2002. Lectin-mediated drug delivery: influence of mucin on cytoadhesion of plant lectins in vitro. J. Control. Rel. 79, 183–191.

Woodley, J., 2001. Bioadhesion: new possibilities for drug administration? Clin. Pharmacokinet. 40, 77–84.

Yadav, A.V., Mote, H.H., 2008. Development of biodegradable starch microspheres for intranasal delivery. Ind. J. Pharm. Sci. 70 (2), 170–174.

Yang, X., Robinson, J.R., 1998. Bioadhesion in mucosal drug delivery. In: Okano, T. (Ed.), Biorelated Polymers and Gels Controlled Release Applications in Biomedicine. Academic Press, San Diego, pp. 135–192.

Yong, C.S., Sah, H., Jahng, Y., Chang, H.W., Son, J.K., Lee, S.H., et al., 2003. Physicochemical characterization of diclofenac sodium-loaded poloxamer gel as a rectal delivery system with fast absorption. Drug. Dev. Ind. Pharm. 29 (5), 545–553.

Zhou, X.H., 1994. Overcoming enzymatic and absorption barriers to nonparenterally administered protein and peptide drugs. J. Control. Rel. 29, 239–252.

CHAPTER 11

Dendrimers in Targeting and Delivery of Drugs

Prashant Kesharwani, Mohd Cairul Iqbal Mohd Amin, Namita Giri, Ashay Jain and Virendra Gajbhiye

Contents

Nanotechnology-Based Approaches for Targeting and Delivery of Drugs and Genes.
DOI: http://dx.doi.org/10.1016/B978-0-12-809717-5.00013-0

Disclosures: There are no conflict of interest and disclosures associated with this manuscript.

11.1 INTRODUCTION

In the present context of architectural chemistry, considerable efforts have been dedicated to the development of polymer systems. The 20th century has witnessed impressive innovations in polymer synthesis and advances in the design of biodegradable polymeric macromolecules. The dendrimers are the result of these advances and innovations in the field of polymer science. Dendrimers were, for the first time, synthesized during 1970–90 by two different groups, Buhleier et al. and Tomalia et al. In contrast to linear polymers, dendrimers developed by these two groups have precisely controlled architecture with tailor-made surface groups that could be finely tuned (Tomalia et al., 1985; Kesharwani et al., 2011, 2014a; Nanjwade et al., 2009).

The word dendrimer is derived from the Greek term dendron that means "tree," which is logical in view of their typical structure with a number of branching units. Dendrimers are defined as synthetic macromolecules characterized by high branching points, three-dimensional globular shape, monodispersity and nanometric size range. In literature, these are also popularly described as "Cascade molecules," "Arborols," "Dendritic molecules"; or as "nanometric architectures" because of their nanoscopic size and monodispersity (Nanjwade et al., 2009; Newkome et al., 1985; Birdhariya et al., 2014; Jain et al., 2015). The characteristic architecture of dendrimers provides a well-defined branched structure with globular shape, which renders a large number of surface groups that can be tailored to provide a template for drug delivery. Dendrimers are globular, nanosized (1–100 nm) macromolecules with a particular architecture constituted of three distinct domains: (1) a core at the center of dendrimer consisting of an atom or a molecule having at least two identical chemical functions; (2) branches, emanating from the core, constituted of repeat units having at least one branch junction, whose repetition is organized in a geometrical progression that results in a series of radially concentric layers called "generations"; and (3) many terminal functional groups, generally located at the surface of dendritic architecture (Fig. 11.1). These surface groups are vital in determining the properties of dendritic macromolecules (Kesharwani et al., 2012, 2014a,b,c; Mishra and Kesharwani, 2016).

11.2 PROPERTIES OF DENDRIMERS

Dendritic architecture holds immense potential over the other carrier systems, particularly in the field of drug delivery, because of their unique properties. As compared to traditional linear polymers, dendrimers exhibit significantly improved physical and chemical properties. Salient properties of dendrimers are discussed below.

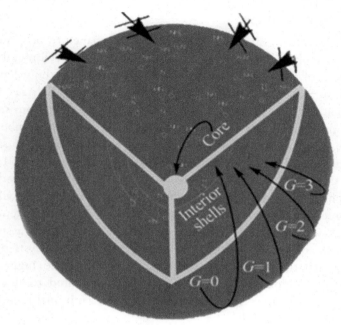

Figure 11.1 Basic structure of dendrimer.

11.2.1 Monodispersity

Dendrimers are the class of dendritic polymers that can be constructed with a well-defined molecular structure, i.e., monodisperse unlike linear polymers. Monodispersity offers researchers the possibility to work with a tool for well-defined and reproducible scalable size (Nanjwade et al., 2009; Patel et al., 2016; Kesharwani et al., 2011, 2015e; Mansuri et al., 2016). Monodispersity of dendrimers has been confirmed widely by mass spectroscopy, size exclusion chromatography, gel electrophoresis and transmission electron microscopy. Due to the purification at each step of synthesis, the convergent methods generally produce the most nearly isomolecular dendrimers (Mishra and Kesharwani, 2016; Miller and Neenan, 1990; Kesharwani et al., 2015a). Mass spectroscopy data have well established that PAMAM dendrimers produced by the divergent method are extremely monodisperse for earlier generation (1.0–5.0 G). Factors such as dendrimer bridging and incomplete removal of ethylenediamine at each of the generation sequences may affect the degree of monodispersity (Kesharwani et al., 2015b). This latter factor, at any point in dendrimer growth, will function as an initiator core to produce 0.5 G and subsequent generation dendrimers that lead to polydispersity (Nanjwade et al., 2009; Thakur et al., 2015; Dwivedi et al., 2016; Kesharwani et al., 2011; Gothwal et al., 2015).

11.2.2 Nanosize and shape

Dendrimers with uniform and well-defined size and shape are of prominent interest in biomedical applications because of their ability to cross cell membranes as well as reduce the risk of premature clearance from the body. The high level of control over the dendritic architecture makes them an ideal carrier. The size of dendrimers increases systematically, as does the generation number, ranging from several to tens of nanometers in diameter. Dendrimers are similar in size to a number of biological structures, e.g., 5.0 G PAMAM dendrimers are approximately the same size and shape as hemoglobin (Hb) (5.5 nm diameter) (Esfand and Tomalia, 2001; Singh et al., 2015). Several classes of dendrimers have been synthesized with a variety of core materials, branching units and surface modifications. Dendrimer size will also be relevant to the three-dimensional shape; lower-generation dendrimers tend to be open and amorphous structures whereas higher generations can adopt a spherical conformation, capable of incorporating drug molecules. X-ray analysis on dendrimer aggregates suggested that the molecular shape of the lower to higher generations becomes increasingly globular (i.e., more spherical compared to linear shape), in order to spread out the larger molecular structure with a minimal repulsion between the segments (Percec et al., 1998).

11.2.3 Biocompatibility

Regardless of their toxicity, dendrimers have been considered as "smart" carriers owing to their ability as an intracellular drug delivery vehicle, to cross biological barriers, to circulate in the body during time needed to exert a clinical effect, and to target specific structures. Toxicity of dendrimers is ascribed mainly to the end group present on its periphery. Generally amine-terminated PAMAM and PPI dendrimers display concentration-dependent toxicity and hemolysis (Nanjwade et al., 2009; Kesharwani et al., 2012, 2014c,f, 2015c,d; Mishra et al., 2010), whereas neutral or anionic groups terminated dendrimers have shown comparatively less toxicity and hemolysis (Fuchs et al., 2004; Malik et al., 2000; Mahor et al., 2016; Tiwari et al., 2015). Fortunately, the toxicity of cationic dendrimers can be overcome by partial or complete modification of their periphery with negatively charged or neutral groups (Malik et al., 2000; Kesharwani et al., 2014b; Choudhary et al., 2016). Although both PAMAM and PPI dendrimers have terminal amino groups, they display different patterns of toxicity. In the case of cationic PAMAM dendrimers, toxicity increases with each generation but unpredictably cationic PPI dendrimers do not follow this pattern of toxicity (Kesharwani et al., 2014a; Jain et al., 2010). The cytotoxicity behavior of cationic dendrimers is widely explained by the favored interactions between negatively charged cell membranes and the positively charged dendrimers surface, enabling these dendrimers to adhere to and damage the cell membrane, causing cell lysis. Masking of cationic end groups or conversion of end groups of dendrimers to neutral or anionic groups has resulted in dendrimers with decreased toxicity or even nontoxic dendrimers in both in vitro and in vivo studies, as observed in cases of neutral dendrimers like polyester, polyether and surface-engineered dendrimers, e.g., glycosylated, PEGylated dendrimers, etc. (Kesharwani et al., 2011).

11.2.4 Periphery charge

Dendrimers consist of three structural units, i.e., core, branching units, and a number of terminal end groups. End groups may possess positive, negative or neutral charges, which are vital in the exploration of dendrimers as drug delivery vehicles. This polyvalency can be exploited to play an important role in the application of dendrimers as gene carriers because cationic dendrimers like poly-l-lysine, PPI and PAMAM, etc., can form complexes with negatively charged DNA. Also the positive charges of dendrimers facilitate their interaction with negatively charged biological membranes leading to applicability of dendrimers for intracellular drug delivery. Challenging these advantages, the polyvalency of dendrimers also leads to the toxicities including cytotoxicity, hemolysis, etc. (Nanjwade et al., 2009; Patel et al., 2016; Jain et al., 2010; Mishra et al., 2014; Kesharwani et al., 2015b, 2015g; Luong et al., 2016). Fortunately, these toxicities can be overcome by surface modification (engineering) of dendrimers with different agents like carbohydrates, PEG, acetate etc. Thus, polyvalency has an important implication on the properties of dendrimers and provides a potential arena for scientists working in the field of dendrimers-mediated drug delivery (Mansuri et al., 2016; Kesharwani et al., 2015f; Kesharwani and Iyer, 2015).

11.2.5 Dendrimer–membrane interactions

Interaction of higher generation dendrimers having positively charged surface groups with negatively charged biological membrane results in formation of nanoscale holes and cell lysis. Two most widely used models to understand this mechanism include biological membranes as well as living cell membranes (Kesharwani et al., 2014a; Kesharwani et al., 2015d; Hong et al., 2006; Kesharwani and Iyer, 2015; Tekade et al., 2016). The interaction of cationic phosphorus-containing dendrimers (CPDs) (3.0 and 4.0 G) with model lipid membranes has been investigated. Model lipid bilayers consisting of 1,2-dimyristoyl-*sn*-glycero-3-phosphocholine and dispersed in aqueous HEPES buffer solution (10 mM, pH 7.4) were used, and then interactions were studied with differential scanning calorimetry and zeta-potential techniques. The result of calorimetric analysis displayed generation-dependent interaction. The presence of dendrimers attributed significant changes in the main transition enthalpy and phase transition temperature values. The rate of alteration of thermotropic behavior was found to be concentration-dependent. The fluidity rate of the lipid–dendrimer complexes was proportional to the dendrimer/lipid molar ratios (Leroueil et al., 2008).

11.2.6 End groups and toxicity

Dendrimers, being nanometric size, can nonspecifically interact with a variety of cells and cellular components menifesting toxic consequences. It has been reported that in the case of dendrimers the cell toxicity is associated with the number of end groups and surface charges. Cationic dendrimers like PAMAM, PPI and poly-l-lysine have shown toxicity in a dose-dependent manner; however, negatively charged dendrimers such as sulfonated, carboxylated, phosphonated, or neutral dendrimers, such as dendrimers with poly(ethylene oxide), acetyl, carboxyl, mannose, galactose end groups, revealed less toxicity compared to positively charged dendrimers. In light of these reports, modification of surface groups of cationic dendrimers with neutral molecules is preferred to possibly prevent toxicity of dendrimers (Leroueil et al., 2008; El-Sayed et al., 2002; Ionov et al., 2011) (Fig. 11.2). As discussed in the previous section dendrimers having positively charged end groups may interact with negatively charged membrane and may increase the permeability that facilitates the intracellular delivery of bioactives. But in the case of higher generation dendrimers, dendrimer–membrane interaction may result in disruption of membrane integrity followed by leakage of important intracellular components, which finally leads to cell death and toxicity. This toxicity is attributed to high charge density inherently associated with larger generations of positively charged dendrimers.

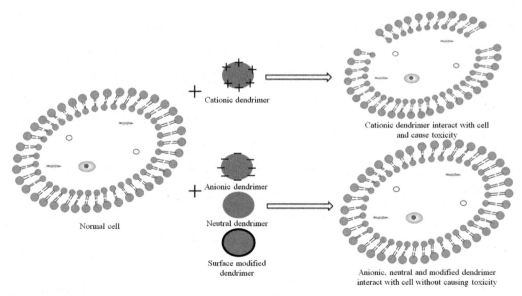

Figure 11.2 Biological cell interaction with different types of dendrimer.

11.3 SYNTHESIS OF DENDRIMERS

Dendrimers are symmetric, highly branched polymers with a compact spherical structure (diameter ranging from 1.1 nm for 1.0 G PAMAM to 9 nm for 8.0 G PAMAM dendrimer) (El-Sayed et al., 2002). They are normally synthesized from a central polyfunctional core by repetitive addition of monomers. The core is characterized by a number of functional groups. Addition of monomers to each functional group results into next dendrimer generation as well as expression of end groups for further reaction (Kesharwani et al., 2014a; Nanjwade et al., 2009; Kesharwani et al., 2015f; Stöckigt et al., 1996; Jain et al., 2012). The size of dendrimers increase as the generation number increases; a stage will soon be reached when the dendrimer attains its maximum size and becomes tightly packed looking like a ball. Divergent and convergent methods are most frequently used for dendrimer synthesis. In addition, other approaches like "hypercores" and "branched monomers" growth, "double exponential" growth, "lego" chemistry and "click" chemistry are also used.

11.3.1 Divergent approach

The divergent approach comprises two steps: first is the activation of functional surface groups, and second is the addition of branching monomer units. In this approach, the core is reacted with two or more moles of reagent containing at least two protecting/branching sites, followed by removal of the protecting groups. This will lead to the formation of first–generation dendrimer. This process is repeated several times until the dendrimer of the desired size is formed. PAMAM starburst dendrimers are prepared by this method. As compared to others methods, the divergent approach has some overriding advantages such as the ability to modify the surface of dendrimer molecules by changing the end groups at the outermost layer. Another advantage is that the overall chemical and physical properties of dendrimer can be configured to specific needs (Islam et al., 2005).

11.3.2 Convergent approach

This is an alternative method of dendrimer synthesis, first proposed by Hawker and Frechet in 1990. Only one kind of functional group on the outermost generation is the main constraint of the divergent growth method. Convergent growth would overcome such a weakness. The convergent method involves two stages: first a reiterative coupling of protected/deprotected branch to produce a focal point functionalized dendron, and second a divergent core anchoring step to produce various multidendron dendrimers. Precise control over molecular weight and production of dendrimers having functionalities in precise positions and number are some outstanding dividends of this method (Hawker and Frechet, 1990).

However, difficulty in synthesizing the dendrimer in large quantities, because of repeated reaction occurring during the convergent approach that necessitates the protection of active site, is a significant limitation of these methods. Presently dendrimers are commercially manufactured by companies like Dentritech (Midland, US), Dutch State Mines (DSM), Netherlands, Dow Chemicals (Michigan, US), Aldrich Chemical Company (Milwaukee, WI) and Weihai CY Dendrimer Technology (China).

11.3.3 Other approaches

11.3.3.1 Hypercores and branched monomers

This method involves the preassembly of oligomeric species to hasten the rate of dendrimer synthesis. In this method oligomeric species are linked together to yield dendrimers in fewer steps and/or higher yields. Essentially a hypercore having multiple attaching groups is grown from a core molecule and the surface units are linked to branched monomer with focal point activation leading to synthesis of blocks, which are then attached to the hypercore to generate a higher generation dendrimer (Kesharwani et al., 2014a; Tiwari et al., 2015; Mishra and Kesharwani, 2014).

11.3.3.2 Double exponential

This approach allows the preparation of monomers for both divergent and convergent growth from a single starting material, which is similar to a rapid growth technique for linear polymer. The resultant two products are then react to give an orthogonally protected trimer, which can be used to repeat the growth again. The advantage of the double exponential growth approach is rapid synthesis and applicability to either the divergent or convergent method (Kesharwani et al., 2014a).

11.3.3.3 Lego chemistry

In order to simplify the synthetic procedure for dendrimers, in terms of cost as well as duration of synthesis, various approaches have been explored by scientists; Lego chemistry is one of the outcomes of these explorations. Lego chemistry is based on the application of highly functionalized cores and branched monomers and has been utilized in the synthesis of phosphorus dendrimers. The basic synthetic scheme has undergone several modifications and has resulted in a refined scheme wherein a single step can amplify the number of terminal surface groups from 48 to 250. Apart from higher growth in the number of terminal surface groups in few reactions, this method also encompasses the advantage of utilizing minimum volume of solvent, allowing a simplified purification procedure with ecofriendly by–products like water and nitrogen (Kesharwani et al., 2014a; Svenson and Tomalia, 2005).

11.3.3.4 Click chemistry

Another approach for fast and reliable synthesis of dendrimers is based on click chemistry wherein small units are joined together. High chemical yield with innocuous by-products is the main characteristic of click chemistry reaction. Use of simple reaction conditions, easily available reagents, and benign solvent are the additional profits of click chemistry. Following the click chemistry strategy, dendrimers with various surface groups can be obtained in high purity and excellent yield. 2.0 and 3.0 G triazole dendrimers were synthesized using Cu (I)-catalyzed click chemistry reactions and obtained dendrimers were isolated as pure, solid sample with only sodium chloride as the major by-product using the chromatographic procedure (Nanjwade et al., 2009; Dwivedi et al., 2016; Svenson and Tomalia, 2005; Wang et al., 2012).

11.4 TYPES OF DENDRIMERS

A rapid development of dendritic novel carrier has been possible due to recent advances in synthetic chemistry and characterization techniques. Also a range of dendritic scaffolds has become available with defined nanoscopic size and ample numbers of functional end groups (Kesharwani et al., 2014a; Nanjwade et al., 2009) (Fig. 11.3).

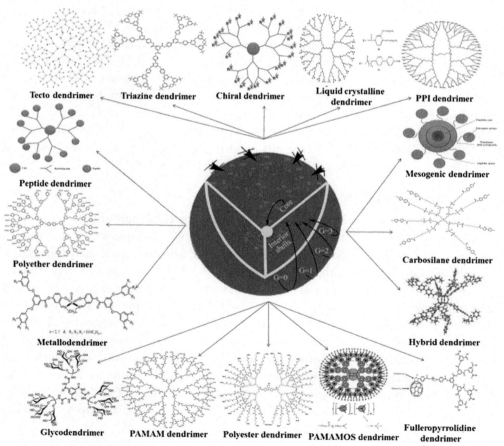

Figure 11.3 Types of dendrimers.

11.4.1 PAMAM dendrimer

The first dendritic structures that have been exhaustively investigated and have received widespread attention were Tomalia's PAMAM dendrimer. PAMAM dendrimers are synthesized by the divergent method starting from ammonia or ethylenediamine initiator core reagents. Products up to generation 10 (a molecular weight of over 930,000 g/mol) have been obtained. The polydispersity index of 5.0–10.0 G PAMAM dendrimers is less than 1.08, which means that the particle size distribution is very uniform for each generation. Due to the presence of positive charge on the surface, PAMAM dendrimers have the ability for condensation of DNA followed by transfection (Nanjwade et al., 2009; Kesharwani et al., 2015b).

11.4.2 PPI dendrimer

PPI dendrimers are amine terminated hyper-branched macromolecules, which are mainly synthesized by the divergent method. PPI dendrimer contains two types of nitrogen atoms: nitrogen of primary amine, and nitrogen of tertiary amine. As compared to tertiary nitrogen atoms, which are more acidic having a pKa around 6–9, primary nitrogen atoms are more basic having a pKa around 10. PPI dendrimers are synthesized by the divergent approach, in a sequence of repetition of double Michael addition of acrylonitrile to primary amines followed by heterogeneously catalyzed hydrogenation of nitriles. This repeated reaction results in a doubling of the number of primary amines. During the synthesis of PPI dendrimer 1, 4-diaminobutane is utilized as dendrimer core. A variety of molecules with primary or secondary amine groups can also be used as core in dendrimer synthesis (Gothwal et al., 2015; Kesharwani et al., 2012; Kesharwani, 2015). These PPI dendrimers are commercially available from DSM, Netherlands and Aldrich Chemical Company (Milwaukee, WI).

11.4.3 Liquid crystalline dendrimers

LC dendrimers consist of mesogenic liquid crystalline (LC) monomers, e.g., mesogen functionalized carbosilane dendrimers. Thermotropic LC phases or mesophases are usually formed by rod-like (calamitic) or disk-like (discotic) molecules (Nanjwade et al., 2009). Dendrimers with AB mesogens in the branches was first reported by Percec et al. in 1995 (Percec et al., 1995). Frey and coworkers also attached several mesogenic units to carbosilane dendrimers, such as cyanobiphenyl and cholesteryl (Frey et al., 1996). In a reported study, mesogenic 3,4-bis-(decyloxy)benzoyl groups functionalized PPI dendrimers of different generations (1.0–5.0 G) were investigated for mesogenic activity. It was found that apart from fifth-generation dendrimers, all other lower-generation dendrimers displayed a hexagonal columnar mesophase in which the dendrimers had a cylindrical conformation. The lack of mesomorphism for the fifth-generation dendrimer was due to its inability to reorganize into a cylindrical shape. In 2001, Boiko et al. reported the debut synthesis of photosensitive LC dendrimer with terminal cinnamoyl groups (Boiko et al., 2001). These LC dendrimers are being investigated by scientists for biomedical applications. Recently Pedziwiatr-Werbicka and coworkers suggested that amino-terminated carbosilane dendrimers have the potential to deliver short-chain siRNA and anti-HIV oligodeoxynucleotide to HIV-infected blood cells. Although these dendrimers had limited application in the delivery of long-chain double stranded nucleic acids, the dendriplexes of carbosilane dendrimers and anti-HIV nucleic acid were stable and less cytotoxic to blood cells than the plain dendrimers, suggesting their utility in the delivery of bioactives (Pedziwiatr-Werbicka et al., 2013).

11.4.4 Core-shell (tecto) dendrimers

Core-shell or tecto-dendrimers represent a polymeric architecture with a highly ordered structure obtained as a result of controlled covalent attachment of dendrimer building blocks (Schilrreff et al., 2012). Tecto-dendrimers are composed of a core dendrimer, which may or may not contain the therapeutic agent, surrounded by dendrimers. Synthesis of tecto-dendrimers has been reported with fluorescein in the core reagent for detection and folate as the targeting moiety (Betley et al., 2002). Such conjugates solved the solubility problems encountered in previous studies with aromatic fluorescein isothiocynate (FITC) moieties on dendrimeric surfaces. This conjugate was found to be overwhelmingly superior to those dendrimeric conjugates containing both FITC and folic acid attached to the surface. In contrast to simple dendrimers, the synthesis procedure for tecto-dendrimers is comparatively simple, and hence expected to inflate the application of dendrimers. Schilrreff et al. investigated the cytotoxicity of tecto-dendrimers to point out their application in the biomedical field. In this study tecto-dendrimers having amine-terminated 5.0 G PAMAM dendrimers as the core, surrounded by a shell composed of 2.5 G PAMAM dendrimers with surface carboxyl groups, were investigate for cytotoxicity towards SK-Mel-28 human melanoma cells. These tecto-dendrimers were found to inhibit growth of melanoma cells at a concentration which is safe to healthy keratinocytes epithelial cells (Schilrreff et al., 2012). Thus tecto-dendrimers could be explored for application in the field of nanomedicine including drug delivery.

11.4.5 Chiral dendrimers

Dendrimers based upon the construction of constitutionally different, but chemically similar branches to chiral core, are referred to as chiral dendrimers. Chiral, nonracemic dendrimers with well-defined stereochemistry are a particularly interesting subclass with potential applications in asymmetric catalysis and chiral molecular recognition. Ghorai et al. described the first molecules of anthracene-capped chiral dendrimers derived from a 1,3,5-trisubstituted aromatic core and carbohydrate units in the interior and periphery. These are claimed to be suitable for anchoring other useful functionalities aimed at applications as a drug delivery system and light harvesting materials (Ghorai et al., 2004). Evidence supporting the above claim is keenly awaited, particularly in the field of drug delivery application of chiral dendrimers.

11.4.6 Peptide dendrimers

Peptide dendrimers are radically branched macromolecules that contain a peptidyl branching core and/or peripheral peptide chains, and can be divided into three categories. The first category having peptides only as surface functionalities is referred to as "grafted" peptide dendrimers; the second category composed entirely of amino acids is known as peptide dendrimers; while the third one utilizes amino acids in the branching core and surface functional groups but having nonpeptide branching units. Divergent and convergent methods are frequently used for the synthesis of peptide dendrimers, and the availability of solid-phase combinatorial methods facilitates large libraries of peptide dendrimers to be produced and screened for desired properties. Peptide dendrimers have been used in industry as surfactants, and in the biomedical field as multiple antigen peptides, protein mimics and vehicles for drug and gene delivery (Sadler and Tam, 2002). Additionally, Darbre and Reymond have utilized peptide dendrimers as esterase catalysts (Darbre and Reymond, 2006).

11.4.7 Glycodendrimers

Dendrimers that encompass sugar moieties such as glucose, mannose, galactose (Cheng et al., 2007) and/or disaccharide into their structure are referred to as glycodendrimers. The vast majority of glycodendrimers have saccharide residues on their outer surface, but glycodendrimers containing a sugar unit as the central core, from which all branches emanate, have also been described. Generally, glycodendrimers can be divided into three categories: (1) carbohydrate-centered, (2) carbohydrate-based, and (3) carbohydrate-coated dendrimers (Choi et al., 2000; Woller and Cloninger, 2001). One anticipated application of these dendrimers is site-specific drug delivery to the lectin-rich organs. These dendrimers were anticipated to display better association with lectins anchored systems as compared to monocarbohydrate anchored systems (Kesharwani et al., 2011, 2014a; Roy and Baek, 2002).

11.4.8 Hybrid dendrimers

Hybrid dendrimers are a combination of dendritic and linear polymers in hybrid block or graft copolymer forms. The spherical shape and a large number of surface functional groups of dendrimers made the formation of dendritic hybrids possible. The small dendrimer segment coupled to multiple reactive chain ends provides an opportunity to use them as surface active agents, compatibilizers or adhesives, or hybrid dendritic linear polymers. The dendritic hybrids obtained from various polymers with dendrimers generated the compact, rigid, uniformly shaped globular dendritic hybrids, which have been explored for various aspects in the field of drug delivery (Pushechnikov et al., 2013).

11.4.9 PAMAM-organosilicon (PAMAMOS) dendrimers

Inverted unimolecular micelles that consist of hydrophilic, nucleophilic PAMAM interiors and hydrophobic organosilicon (OS) exteriors are known as radially layered PAMAMOS dendrimers (PAMAMOS). PAMAMOS dendrimers offer unique potential for novel application in electronics, chemical catalysis, nanolithography and photonics, etc., due to their unique properties such as constancy of structure and ability to form complex and encapsulate various guest species with nanoscopic topological precision (Agashe et al., 2007; Luong et al., 2016).

The continuous urge for optimum therapeutic delivery system for various medicaments used in treatment of infectious and noninfectious diseases leads to development and investigation of various types of dendrimers like polyether dendrimers, polyester dendrimers, triazine dendrimers, melamine dendrimers, citric acid dendrimers, etc., using different core and branching units. The arena of dendrimers is ever-expanding and applications of this versatile carrier system in the drug delivery are inflating day by day.

11.5 APPLICATIONS OF DENDRIMERS

11.5.1 Dendrimers as a carrier for drug delivery

Dendrimers have narrow polydispersity; nanometer size range of dendrimers can allow easier passage across biological barriers. All these properties make dendrimers suitable as a host either binding guest molecules in the interior of dendrimers or on the periphery of the dendrimers.

11.5.1.1 Dendrimers in transdermal drug delivery

Recently, dendrimers have found applications in transdermal drug-delivery systems. Generally, bioactive drugs have hydrophobic moieties in their structure, resulting in low water-solubility that inhibits efficient delivery into cells. Dendrimers designed to be highly water-soluble and biocompatible have been shown to be able to improve drug properties such as solubility and plasma circulation time via transdermal formulations and to deliver drugs efficiently. Nonsteroidal antiinflammatory drugs (NSAIDs) are very effective in the treatment of acute and chronic rheumatoid and osteoarthritis; however, clinical use of NSAIDs is often limited by adverse events such as gastrointestinal side effects (dyspepsia, gastrointestinal bleeding), or renal side effects when given orally. Transdermal drug delivery overcomes these adverse effects and also maintains therapeutic blood level for a longer period of time. Transdermal delivery suffers poor rates of transcutaneous delivery due to the barrier function of the skin. PAMAM dendrimer complex with NSAIDs (e.g., Ketoprofen, Diflunisal) could be improving the drug permeation through the skin as penetration enhancers (Cheng et al., 2007). The model drugs Ketoprofen and Diflunisal were conjugated with G5 PAMAM dendrimer and investigated for different studies. In vitro permeation studies on excised rat skin showed 3.4 times higher permeation of Ketoprofen from Ketoprofen–dendrimer complex than that from 2 mg/mL Ketoprofen suspended in normal saline. Similarly, a 3.2 times higher permeated amount was observed with Diflunisal–dendrimer complex. The antinociception effect of drugs was studied on mice; the results showed that Ketoprofen–dendrimer complex reduced writhing activity during the period of 1–8 hours after transdermal administration, while pure Ketoprofen suspension at the equivalent dose of Ketoprofen significantly decreased number of writhing between 4 and 6 hours after drug was transdermally given (Jain et al., 2010; Cheng et al., 2007).

Chauhan et al. investigated transdermal ability of PAMAM dendrimers by using indomethacin as the model drug for study. In vitro permeation studies showed increase in the steady-state flux as increase in concentration of all three types G4-NH$_2$, G4-OH and G-4.5 PAMAM dendrimers. For the in vivo pharmacokinetic and pharmacodynamic studies, indomethacin and dendrimer formulations were applied to the abdominal skin of the Wistar rats and blood collected from the tail vein at the scheduled time. The indomethacin concentration was significantly higher with PAMAM dendrimers when compared to the pure drug suspension. The results showed that effective concentration could be maintained for 24 hours in the blood with the G4 dendrimer–indomethacin formulation. Therefore, data suggested that the dendrimer–indomethacin based transdermal delivery system was effective and might be a safe and efficacious method for treating various diseases (Cheng et al., 2007).

11.5.1.2 Dendrimers in oral drug delivery

The oral drug-delivery system has been the dominant route for many years because of its significant advantages. It is by far the most convenient administration route with good patient compliance, especially in the patient's opinion. Along with these benefits, there are also some defects of oral delivery route, like low solubility in aqueous solutions and low penetration across intestinal membranes. D'Emanuele and his research group (Jevprasesphant et al., 2003) investigated the effect of dendrimer generation and conjugation on the cytotoxicity, permeation and transport mechanism of PAMAM dendrimer and surface-modified cationic PAMAM dendrimer using monolayers of the human colon adenocarcinoma cell line, Caco-2. As increase in the concentration and generation, there was increase in the cytotoxicity and permeation of dendrimers, while reduction in cytotoxicity was observed by conjugation with lauryl chloride. Modified dendrimers also reduced transepithelial electrical resistance (TEER) and significantly increased the apparent permeability coefficient (Papp).

In another study of transepithelial permeability of naproxen, a low solubility model drug was investigated (Najlah et al., 2007). The stability of these G0 PAMAM conjugates in 50% liver homogenate was compared to that in 80% human plasma, showing the lactate ester linker gave prodrug of high stability in plasma with slow hydrolysis in liver homogenate; such conjugates may have potential in controlled release systems, while using diethylene glycol as a linker gives conjugate that showed high chemical stability, but readily released drug in plasma and liver homogenate. So, these conjugations demonstrate potential as nanocarriers for the enhancement of oral bioavailability. P-glycoproteins (P-gp), an efflux transporter, are able to reduce the absorption of orally administered drugs and decrease their bioavailability, whereas propranolol is the substrate of the P-gp efflux transporter with poor water solubility.

D'Emanuele et al. synthesized the propranolol–PAMAM dendrimers conjugate and investigated transport of the conjugate across Caco-2 cell monolayer. The results showed that propranolol–dendrimer conjugate was able to bypass the P-gp efflux transporter. Investigators have finally concluded that dendrimers conjugate with drug could reduce the effect of intestinal P-gp on drug absorption of propranolol and many other orally administered drugs (D'Emanuele et al., 2004).

11.5.1.3 Dendrimers in ocular drug delivery

The topical application of active drugs to the eye is the most prescribed route of administration for the treatment of various ocular disorders. It is generally agreed that the intraocular bioavailability of topically applied drugs is extremely poor. These results are mainly due to drainage of the excess fluid via nasolacrimal duct and elimination of the solution by tear turnover. Several research advances have been made in ocular drug-delivery systems by using specialized delivery systems such as polymers, liposomes, or dendrimers to overcome some of these disadvantages. Ideal ocular drug-delivery systems should be nonirritating, sterile, isotonic, biocompatible, does not run out from the eye and biodegradable (Tolia, 2008). Dendrimers provide unique solutions to complex delivery problems for ocular drug delivery. Research efforts for improving residence time of pilocarpine in the eye were increased by using PAMAM dendrimers with carboxylic or hydroxyl surface groups. These surface-modified dendrimers were predicted to enhance pilocarpine bioavailability (Vandamme and Brobeck, 2005).

11.5.1.4 Dendrimers in pulmonary drug delivery

Dendrimers have been reported for pulmonary drug delivery. During one study, the efficacy of PAMAM dendrimers in enhancing pulmonary absorption of Enoxaparin was studied by measuring plasma antifactor Xa activity, and by observing prevention efficacy of deep vein thrombosis in a rodent model. G2 and G3 generation positively charged PAMAM dendrimers increased the relative bioavailability of Enoxaparin by 40%, while G2.5 PAMAM, half generation dendrimers, containing negatively charged carboxylic groups had no effect. Formulations did not adversely affect mucociliary transport rate or produce extensive damage to the lungs. So the positively charged dendrimers are suitable carriers for Enoxaparin pulmonary delivery (Bai et al., 2007).

11.6 DENDRIMERS IN TARGETED DRUG DELIVERY

Nowadays general cancer chemotherapeutics are less effective in their ability to cure tumors because of the nonselective action of these highly potent drugs, resulting in dose limiting side effects. The application of drug carrier systems for targeting tumor cells has gained credence as an alternative approach for treating cancer, and offers both increased therapeutic index and decreased drug resistance. An effective targeting drug-delivery system requires a base that is uniform and able to couple multiple components such as targeting molecule, drug and cancer imaging agent (Kesharwani et al., 2014a; Jain et al., 2010, 2012; Thomas et al., 2004).

Dendrimers have ideal properties that are useful in a targeted drug-delivery system. One of the most effective cell-specific targeting agents delivered by dendrimers is folic acid. Membrane associated high-affinity folate receptors are folate-binding proteins that are over expressed on the surface of different types of cancer cells (e.g., ovarian). PAMAM dendrimers conjugated with the folic acid and fluorescein isothiocyanate for targeting the tumor cells and imaging respectively. Further, these two molecules are linked with complementary oligonucleotides. DNA-assembled nanoclusters were evaluated in vitro, which helps in detecting tumor cell-specific binding and internalization. These DNA-assembled dendrimer conjugates may allow the combination of different drugs with different targeting and imaging agents so it is easy to develop combinatorial therapeutics (Choi et al., 2005). Antibodies specific to CD14 and PSMA (prostate-specific membrane antigen) were conjugated to a G5 PAMAM dendrimer bearing a fluorescein imaging tag. Cell binding and internalization of the antibody-conjugated dendrimers to antigen-expressing cells were evaluated by flow cytometry and confocal microscopy. It was found that the conjugates bound specifically to the antigen-expressing cells in a time- and dose-dependent fashion with an affinity similar to that of the free antibody. Confocal microscopy indicated that cellular internalization of the dendrimer conjugate did occur (Thomas et al., 2004). During another study, folic acid was conjugated to dendrimers as a targeting agent and then coupled with a model drug Methotrexate. These conjugates were injected in immune-deficient mice bearing Human KB tumors and evaluated. Biodistribution study shows that the percentage of the injected dose particularly in tumor cells after 1 day was around three times higher with targeted polymer conjugates (with folic acid) compare to nontargeted polymer conjugates (without folic acid). There was also observed the clear and rapid removal of conjugated dendrimers from the blood via kidney during the first day postinjection. Confocal microscopy images obtained of tumors after 15 hours of i.v. injection showed a significant number of fluorescent cells with targeted dye-conjugates compare to the nontargeted dye-conjugates. Flow cytometry analysis of a single-cell suspension isolated from the same tumors showed the same results as confocal microscopy images (Kukowska-Latallo et al., 2005).

Patri and coworkers have investigated that complexing a drug with dendrimers as an inclusion complex improves its solubility in water, a cleavable; while covalently linked dendrimer conjugate is better for targeted drug delivery because it does not release the drug prematurely in biological conditions. They reported less cytotoxic effect with the covalently linked dendrimer (Patri et al., 2005).

11.6.1 Dendrimers for controlled-release drug delivery

Fréchet and coworkers have prepared polyaryl ether dendrimers containing dual functionality on the surface. One is used to attach polyethylene glycol (PEG) units on the surface to improve water solubility, and the other one is utilized to attach hydrophobic drug molecules. They have also synthesized a series of dendritic unimolecular micelles with a hydrophobic polyether core surrounded by a hydrophilic PEG shell for drug encapsulation. A third-generation micelle with indomethacin entrapped as model drug gives slow and sustained in vitro release, as compared to cellulose membrane control (Liu et al., 1999). PEG-2000 was conjugated to generation G3 PAMAM dendrimers with varying degree of substitution. Methotrexate drug was encapsulated (loaded) to the prepared conjugates and investigated for drug release in a dialysis bag. The results found that PEG-dendrimers conjugated with encapsulated drug and sustained release of methotrexate as compare to unencapsulated drug.

Controlled release of the Flurbiprofen could be achieved by formation of complex with amine-terminated generation 4 (G4) PAMAM dendrimers. Prepared dendrimer complexes observed that loaded drug displayed initial rapid release (more than 40% until 3rd hour) followed by slow release. Pharmacodynamic study was performed using carrageenan induced paw edema model, revealed 75% inhibition at 4th hour that was maintained above 50% until 8th hour. The dendritic formulation showed twofold and threefold increase in mean residence time and terminal half-life, respectively, as compared to free drugs (Asthana et al., 2005).

11.6.2 Dendrimers in gene delivery

Dendrimer-based transfection agents have become routine tools for many molecular and cell biologists, but therapeutic delivery of nucleic acids remains a challenge. Because of their immunogenicity, dendrimers are extensively used as nonviral vector for gene delivery. The use of dendrimers as gene transfection agents and drug-delivery devices have been extensively reviewed (Broeren et al., 2004; Boas and Heegaard, 2004).

The ability of cationic dendrimers to deliver DNA or RNA has been reviewed previously. Besides that, some research recently indicated that the dendrimer-based gene delivery system also has significant potential in clinical trials. Kukowska-Latallo et al. reported that intravenous administration of G9 PAMAM dendrimer-complexed pCF1CAT plasmid could result in high level of gene expression in the lung tissues of rats. It enhances the transfection efficiency and expression pattern of dendrimer (Kukowska-Latallo et al., 2000). Amphiphilic dendrimers having a rigid diphenyl-ethyne core featured a variety of geometries and substitution patterns, all of which showed high transfection activity. The hydrophobic parameters influenced the DNA binding and transport more strongly than anticipated, exhibiting lower toxicity. In contrast to cationic dendrimers, these dendrimers did not have any size limitation for transfection (Joester et al., 2003). In another study, amphiphilic, Tomalia-type PAMAM dendrimers of generations 1 to 4, were synthesized utilizing di-n-dodecylamine as the core. It was anticipated that the hydrophobic components would mimic the membrane transfection ability of natural phospholipids such as dioleoylphosphatidyl-ethanolamine (DOPE) and enhance membrane penetration. These constructs form complexes with DNA and, in case of the $G = 2$–4 dendrimers, were able to cross cell membranes and efficiently deliver DNA. Polycationic dendrimers were also studied with different cell lines for the ability to transfect DNA and toxicity. In the result, some of polycation–DNA complexes were less toxic than lipid–DNA systems (Gebhart and Kabanov, 2001).

Glycoplexes are used to target the specific cells and/or to increase gene transfer activity. For example, a galactosylatedpolyethyleneimine (PEI) has high transfection efficiency to hepatocyte expressing asialoglycoprotein receptor. In addition, a man-nosylated PEI has high transfection efficiency to macrophages and dendritic cells, which were mediated by the mannose-specific receptor and DEC-205, respectively. In another approach, Wada et al. have reported that galactose-α-cyclodextrines conjuga-tions with degree of substitution of the galactose moiety was the most preferential car-rier among the prepared series because it provided good gene transfer activity in vitro with no cytotoxicity. Consequently, the potential use of Gal-α-CDE conjugate (DSG 4) could be expected as a nonviral vector to deliver gene and these data may be use-ful for design of α-cyclodextrins and galactose conjugates with other nonviral vectors (Wada et al., 2005). Glycosylation of polymer is one of the effective methods to deliver gene to target cells and/or to enhance gene transfer.

Cyclodextrins are cyclic (α-1,4)-linked oligosaccharides of α-D-glucopyranose containing a hydrophobic central cavity and hydrophilic outer surface, and they are known to function as novel drug carriers. The most common cyclodextrins are α-, β- and γ-cyclodextrins, which consist of six-, seven- and eight-D-glucopyra-nose units, respectively (Wada et al., 2005). PAMAM dendrimers functionalized with α-cyclodextrin showed luciferase gene expression about 100 times higher than for unfunctionalized PAMAM or for noncovalent mixtures of PAMAM and α-cyclodextrin (Arima et al., 2001). Arima et al. previously reported the potential use of polyamidoamine (PAMAM) starburst dendrimer (generation 3, G3) bearing α-cyclodextrins with the degree of substitution (DS) of 2.4 because of the highest transfection efficiency in vitro and in vivo with low cytotoxicity (Wada et al., 2005; Arima et al., 2001).

Unlike PAMAM dendrimers-based gene delivery systems, PPI dendrimers as gene complexes can lead to liver targeted gene expression. Dufes et al. indicated those different generations of PPI dendrimers as transfection agents, efficiently expressed gene in the liver rather than other organs. Furthermore, they demonstrated that intravenous administration of a gene medicine and G3 PPI dendrimer complex could result in intratumoral transgene expression and regression of the established tumors in all the experimental animals (Dufès et al., 2005).

11.7 DENDRIMERS AS IMAGING AGENTS

The first in vivo diagnostic imaging applications using dendrimer-based MRI contrast agents were demonstrated in the early 1990s by Lauterbur and colleagues (Wiener et al., 1994). In comparison with the commercially available small-molecule agent (Magnevist, Schering, AG), the dendrimer-based reagents exhibited blood pool properties and extraordinary relaxivity values when chelated gadolinium groups (Magnevist®). These generation dependent, dramatic enhancements of MRI contrast properties were some of the first examples of a "dendritic effect" (Tomalia et al., 2007). Gadolinium is an FDA-approved contrast agent for MRI which provides greater contrast between normal tissue and abnormal tissue in the brain and body. It is safer than the iodine type contrast used in CT scans and also nonradioactive and is rapidly cleared by kidneys (Patri et al., 2005). Different generation G1 ($n = 4$ Gd(III) ions per molecule), G3($n = 16$), G5($n = 64$) of Gd(III)DTPA-terminated poly(propylene imine) dendrimers and Gd(III)DTPA complex (G0 ($n = 10$)) reference were synthesized and investigated for relaxivities and concentration detection limits. Investigator revealed dendrimers G1, G3 and G5 have been shown an increase in both T1 and T2 relaxivities with increasing molecular weight. Relaxivity values, measured in blood plasma are in good agreement with earlier measurements in citrate buffers, it shows these dendritic contrast agents have fewer interactions with blood plasma proteins. The largest MRI contrast agent G5 with 64 Gd(III) ions gives lowest concentration detection limit, which would make G5 potentially the best dendritic MRI contrast agent (Langereis et al., 2006).

11.8 TOXICITY OF DENDRIMERS

The fields of medicine and drug delivery have particularly witnessed immense progress after the emergence of nanomaterials such as dendrimers. The well-defined structure, large number of surface groups, and nanometric size of dendrimers lead to tremendous advances in intracellular and targeted delivery of drugs. PAMAM dendrimers surface modified with lauryl chains and conjugated with paclitaxel have been found to increase permeability of drug across the cellular barriers and to show the higher cytotoxic potential against human colon adenocarcinoma cell line (Caco-2) (Nanjwade et al., 2009; Kesharwani et al., 2012, 2015e; Jain et al., 2010, 2014; Luong et al., 2016; Kesharwani, 2015; Teow et al., 2013; Thakur et al., 2013; Mishra et al., 2014; Giri et al., 2016). This increase in permeability and cytotoxicity is advantageous in the treatment of diseases like cancer. However, dendrimers have been found to exhibit toxic effects due to their size in the range of nanometers (1–100 nm) and presence of positively charged surface groups in case of cationic dendrimers. The nanometric size and cationic surface groups lead to nonspecific interaction of dendrimers with cellular components including cell membrane, nucleus, mitochondria, enzymes, endosomes, etc. Although nanomaterials have brought advances in drug delivery, the biological and toxicological potential of nanocarriers like nanoparticles, dendrimers, carbon nanotubes is one of the main concerns of nanotoxicology (Malik et al., 2000; Jain et al., 2010; Cancino et al., 2013).

Toxicity of dendrimers has been primarily explored ex vivo with cancer cell lines barring a few reports available on the in vivo toxicity potential of dendrimers. The existence of surface single bondNH2 groups and associated cationic charge on dendrimers confines their applications in drug delivery due to toxicity (Fig. 13.2). The toxicity of dendrimers is affected by concentration, surface charge, generation, and size as revealed through hemolytic toxicity, hematological toxicity, cytotoxicity, immunogenicity and in vivo toxicity (Malik et al., 2000; Jain et al., 2010).

Cancino and coworkers evaluated the toxicity of single-walled carbon nanotubes (SWCNT), PAMAM dendrimers and PAMAM–SWCNT complexes in mouse myoblast cell line (C2C12). In the results these nanomaterials were found to damage DNA and significantly toxic towards C2C12 cell. Finally, the authors concluded that the toxicity of nanomaterials is strongly correlated to their surface charge (Janaszewska et al., 2013). Surface charge of dendrimers plays an important role in in vivo exploration of dendrimers with regards to safety.

Toxicological issues pose most vital limitations in clinical application of dendrimers due to the presence and number of surface amine groups. Recent research has focused on the development of biocompatible dendrimers, which will hasten the era of dendrimer-mediated drug delivery. Apart from designing biocompatible dendrimers, surface engineering presents yet another attractive approach to diminish toxicity of dendrimers, in addition to other beneficial aspects like drug targeting, reduced drug leakage, increased stability, improved pharmacokinetic profile and biodistribution pattern, etc. Surface engineering masks the cationic charge of dendrimer surface either by neutralization of charge, e.g., PEGylation, acetylation, carbohydrate and peptide conjugation; or by introducing negative charge such as half-generations of dendrimers. The modification of PAMAM dendrimers with 4-carbomethoxypyrrolidone groups reduced the toxicity to a significant level. These pyrrolidone-modified dendrimers were found to elicit negligible toxicity against Chinese hamster fibroblasts (B14), embryonic mouse hippocampal cells (mHippoE-18), and rat liver derived cells (BRL-3A). The last two decades have witnessed significant development of biodegradable and/or biocompatible, and surface-engineered dendrimers resulting in improved therapeutic index. All these efforts resulted in a new class of dendrimer family comprising biocompatible, biodegradable and surface-engineered dendrimers (Jain et al., 2010; Janaszewska et al., 2013). The issues related to toxicity of dendrimers and the strategies to resolve these toxicities have been discussed in detail by various scientists including our group (Jain et al., 2010), which could be referred to for deeper insight.

11.9 CONCLUSION AND FUTURE PROGNOSIS

Although dendrimer drug delivery is in its infancy, it offers several attractive features. Dendrimers are expected to be a potential polymer for biomedical, pharmaceutical and biopharmaceutical fields in the 21st century. Easily controllable features of dendrimers such as their size, shape, branching length, and their surface functionality allow modification of the dendrimers as per the requirements, makes these compounds ideal carriers in many of the applications. In spite of this, their use is constrained due to their high manufacturing cost; further, they are not subject to GRAS status due to the inherent toxicity issues associated with them. However, many attempts have been made to resolve the problems pertaining to safety, toxicity and efficacy of dendrimers, and more research is going on to develop the functionalized moieties pertaining to their high potential as a nanocarrier. With improved synthesis, further understanding, of their unique characteristics and recognition of new applications, dendrimers will become promising candidates for further exploitation in drug discovery and clinical applications.

REFERENCES

Agashe, H.B., Babbar, A.K., Jain, S., Sharma, R.K., Mishra, A.K., Asthana, A., et al., 2007. Investigations on biodistribution of technetium-99m-labeled carbohydrate-coated poly(propylene imine) dendrimers. Nanomedicine 3, 120–127. http://dx.doi.org/10.1016/j.nano.2007.02.002.

Arima H., F. Kihara, F. Hirayama, K. Uekama, 2001. Enhancement of gene expression by polyamidoamine dendrimer conjugates with alpha-, beta-, and gamma-cyclodextrins. Bioconjug. Chem. 12 476–484. http://www.ncbi.nlm.nih.gov/pubmed/11459450 (accessed September 6, 2016).

Asthana, A., Chauhan, A.S., Diwan, P.V., Jain, N.K., 2005. Poly(amidoamine) (PAMAM) dendritic nanostructures for controlled site-specific delivery of acidic anti-inflammatory active ingredient. AAPS PharmSciTech 6, E536–E542. http://dx.doi.org/10.1208/pt060367.

Bai, S., Thomas, C., Ahsan, F., 2007. Dendrimers as a carrier for pulmonary delivery of enoxaparin, a low-molecular weight heparin. J. Pharm. Sci. 96, 2090–2106. http://dx.doi.org/10.1002/jps.20849.

Betley, T.A., Hessler, J.A., Mecke, A., Banaszak Holl, M.M., Orr, B.G., Uppuluri, S., et al., 2002. Tapping mode atomic force microscopy investigation of poly(amidoamine) core–Shell Tecto(dendrimers) using carbon nanoprobes. Langmuir 18, 3127–3133. http://dx.doi.org/10.1021/la025538s.

Birdhariya, B., Kesharwani, P., Jain, N.K., 2014. Effect of surface capping on targeting potential of folate decorated poly (propylene imine) dendrimers. Drug Dev. Ind. Pharm, 1–7. http://dx.doi.org/10.310 9/03639045.2014.954584.

Boas, U., Heegaard, P.M.H., 2004. Dendrimers in drug research. Chem. Soc. Rev 33, 43–63. http://dx.doi. org/10.1039/b309043b.

Boiko, N., Zhu, X., Bobrovsky, A., Shibaev, V., 2001. First photosensitive liquid crystalline dendrimer: synthesis, phase behavior, and photochemical properties. Chem. Mater. 13, 1447–1452. http://dx.doi. org/10.1021/cm001116x.

Broeren, M.A.C., van Dongen, J.L.J., Pittelkow, M., Christensen, J.B., van Genderen, M.H.P., Meijer, E.W., 2004. Multivalency in the gas phase: The study of dendritic aggregates by mass spectrometry. Angew. Chemie Int. Ed 43, 3557–3562. http://dx.doi.org/10.1002/anie.200453707.

Cancino, J., Paino, I.M.M., Micocci, K.C., Selistre-de-Araujo, H.S., Zucolotto, V., 2013. In vitro nanotoxicity of single-walled carbon nanotube-dendrimer nanocomplexes against murine myoblast cells. Toxicol. Lett. 219, 18–25. http://dx.doi.org/10.1016/j.toxlet.2013.02.009.

Cheng, Y., Man, N., Xu, T., Fu, R., Wang, X., Wang, X., et al., 2007. Transdermal delivery of nonsteroidal anti-inflammatory drugs mediated by polyamidoamine (PAMAM) dendrimers. J. Pharm. Sci. 96, 595–602. http://dx.doi.org/10.1002/jps.20745.

Choi, J.S., Joo, D.K., Kim, C.H., Kim, K., Park, J.S., 2000. Synthesis of a barbell-like triblock copolymer, poly(l -lysine) dendrimer- *block* -Poly(ethylene glycol)- *block* -Poly(l -lysine) dendrimer, and its self-assembly with plasmid DNA. J. Am. Chem. Soc. 122, 474–480. http://dx.doi.org/10.1021/ja9931473.

Choi, Y., Thomas, T., Kotlyar, A., Islam, M.T., Baker, J.R., 2005. Synthesis and functional evaluation of DNA-assembled polyamidoamine dendrimer clusters for cancer cell-specific targeting. Chem. Biol. 12, 35–43. http://dx.doi.org/10.1016/j.chembiol.2004.10.016.

Choudhary, S., Jain, A., Amin, M.C.I.M., Mishra, V., Agrawal, G.P., Kesharwani, P., 2016. Stomach specific polymeric low density microballoons as a vector for extended delivery of rabeprazole and amoxicillin for treatment of peptic ulcer. Colloids Surfaces B Biointerfaces 141. http://dx.doi.org/10.1016/j. colsurfb.2016.01.048.

D'Emanuele, A., Jevprasesphant, R., Penny, J., Attwood, D., 2004. The use of a dendrimer-propranolol prodrug to bypass efflux transporters and enhance oral bioavailability. J. Control. Release 95, 447–453. http://dx.doi.org/10.1016/j.jconrel.2003.12.006.

Darbre, T., Reymond, J.L., 2006. Peptide dendrimers as artificial enzymes, receptors, and drug-delivery agents. Acc Chem. Res. 39, 925–934. http://dx.doi.org/10.1021/ar050203y.

Dufès, C., Keith, W.N., Bilsland, A., Proutski, I., Uchegbu, I.F., Schätzlein, A.G., 2005. Synthetic anticancer gene medicine exploits intrinsic antitumor activity of cationic vector to cure established tumors. Cancer Res. 65, 8079–8084. http://dx.doi.org/10.1158/0008-5472.CAN-04-4402.

Dwivedi, N., Shah, J., Mishra, V., Mohd Amin, M.C.I., Iyer, A.K., Tekade, R.K., et al., 2016. Dendrimer-mediated approaches for the treatment of brain tumor. J. Biomater. Sci. Polym. Ed 27, 557–580. http:// dx.doi.org/10.1080/09205063.2015.1133155.

El-Sayed, M., Ginski, M., Rhodes, C., Ghandehari, H., 2002. Transepithelial transport of poly(amidoamine) dendrimers across Caco-2 cell monolayers. J. Control. Release 81, 355–365. http://dx.doi.org/ S0168365902000871.

Esfand, R., Tomalia, D.A., 2001. Poly(amidoamine) (PAMAM) dendrimers: from biomimicry to drug delivery and biomedical applications. Drug Discov. Today 6, 427–436. http://www.ncbi.nlm.nih.gov/ pubmed/11301287 (accessed September 5, 2016).

Frey, H., Lorenz, K., Mülhaupt, R., Rapp, U., Mayer-Posner, F.J., 1996. Dendritic polyols based on carbosilanes - lipophilic dendrimers with hydrophilic skin. Macromol. Symp. 102, 19–26. http://dx.doi. org/10.1002/masy.19961020105.

Fuchs, S., Kapp, T., Otto, H., Schöneberg, T., Franke, P., Gust, R., et al., 2004. A surface-modified dendrimer set for potential application as drug delivery vehicles: Synthesis, in vitro toxity, and intracellular localization. Chem. A Eur. J 10, 1167–1192. http://dx.doi.org/10.1002/chem.200305386.

Gebhart, C.L., Kabanov, A.V., 2001. Evaluation of polyplexes as gene transfer agents. J. Control. Release. 73, 401–416. http://www.ncbi.nlm.nih.gov/pubmed/11516515 (accessed September 6, 2016).

Ghorai, S., Bhattacharyya, D., Bhattacharjya, A., 2004. The first examples of anthracene capped chiral carbohydrate derived dendrimers: synthesis, fluorescence and chiroptical properties. Tetrahedron. Lett. 45, 6191–6194. http://dx.doi.org/10.1016/j.tetlet.2004.05.119.

Giri, N., Oh, B., Lee, C.H., 2016. Stimuli-sensitive nanoparticles for multiple anti-HIV microbicides. J. Nanoparticle Res 18, 140. http://dx.doi.org/10.1007/s11051-016-3449-3.

Gothwal, A., Kesharwani, P., Gupta, U., Khan, I., Amin, M.C.I.M., Banerjee, S., et al., 2015. Dendrimers as an effective nanocarrier in cardiovascular disease. Curr. Pharm. Des. 21, 4519–4526. http://www.ncbi.nlm.nih.gov/pubmed/26311317 (accessed May 27, 2016).

Hawker, C.J., Frechet, J.M.J., 1990. Preparation of polymers with controlled molecular architecture. A new convergent approach to dendritic macromolecules. J. Am. Chem. Soc. 112, 7638–7647. http://dx.doi.org/10.1021/ja00177a027.

Hong S., P.R. Leroueil, E.K. Janus, J.L. Peters, M.-M. Kober, M.T. Islam, et al. 2006. Interaction of polycationic polymers with supported lipid bilayers and cells: nanoscale hole formation and enhanced membrane permeability., Bioconjug. Chem. 17, 728–734. http://dx.doi.org/10.1021/bc060077y.

Ionov, M., Gardikis, K., Wróbel, D., Hatziantoniou, S., Mourelatou, H., Majoral, J.-P., et al., 2011. Interaction of cationic phosphorus dendrimers (CPD) with charged and neutral lipid membranes. Colloids Surf. B. Biointerfaces 82, 8–12. http://dx.doi.org/10.1016/j.colsurfb.2010.07.046.

Islam, M.T., Majoros, I.J., Baker, J.R., 2005. HPLC analysis of PAMAM dendrimer based multifunctional devices. J. Chromatogr. B. Analyt. Technol. Biomed. Life Sci 822, 21–26. http://dx.doi.org/10.1016/j.jchromb.2005.05.001.

Jain, A., Garg, N.K., Jain, A., Kesharwani, P., Jain, A.K., Nirbhavane, P., et al., 2015. A synergistic approach of adapalene-loaded nanostructured lipid carriers, and vitamin C co-administration for treating acne. Drug Dev. Ind. Pharm, 1–9. http://dx.doi.org/10.3109/03639045.2015.1104343.

Jain, K., Kesharwani, P., Gupta, U., Jain, N.K., 2012. A review of glycosylated carriers for drug delivery. Biomaterials 33, 4166–4186.

Jain, K., Kesharwani, P., Gupta, U., Jain, N.K., 2010. Dendrimer toxicity: Let's meet the challenge. Int. J. Pharm. 394, 122–142. http://dx.doi.org/10.1016/j.ijpharm.2010.04.027.

Jain, S., Kesharwani, P., Tekade, R.K., Jain, N.K., 2014. One platform comparison of solubilization potential of dendrimer with some solubilizing agents. Drug Dev. Ind. Pharm. http://dx.doi.org/10.3109/03639045.2014.900077.

Janaszewska, A., Ciolkowski, M., Wróbel, D., Petersen, J.F., Ficker, M., Christensen, J.B., et al., 2013. Modified PAMAM dendrimer with 4-carbomethoxypyrrolidone surface groups reveals negligible toxicity against three rodent cell-lines. Nanomedicine 9, 461–464. http://dx.doi.org/10.1016/j.nano.2013.01.010.

Jevprasesphant, R., Penny, J., Attwood, D., McKeown, N.B., D'Emanuele, A., 2003. Engineering of dendrimer surfaces to enhance transepithelial transport and reduce cytotoxicity. Pharm. Res. 20, 1543–1550. http://www.ncbi.nlm.nih.gov/pubmed/14620505 (accessed September 6, 2016).

Kesharwani, P., Iyer, A.K., 2015. Recent advances in dendrimer-based nanovectors for tumor-targeted drug and gene delivery. Drug Discov. Today 20, 536–547. http://dx.doi.org/10.1016/j.drudis.2014.12.012.

Kesharwani, P., Jain, K., Jain, N.K., 2014a. Dendrimer as nanocarrier for drug delivery. Progress Polymer Sci 39, 268–307. http://dx.doi.org/10.1016/j.progpolymsci.2013.07.005.

Kesharwani, P., Tekade, R.K., Jain, N.K., 2014b. Formulation development and in vitro-in vivo assessment of the fourth-generation PPI dendrimer as a cancer-targeting vector. Nanomedicine (Lond) http://www.ncbi.nlm.nih.gov/pubmed/24593000.

Kesharwani, P., Tekade, R.K., Jain, N.K., 2014c. Generation dependent cancer targeting potential of poly(propyleneimine) dendrimer. Biomaterials 35, 5539–5548. http://dx.doi.org/10.1016/j.biomaterials.2014.03.064.

Kesharwani, P., Ghanghoria, R., Jain, N.K., 2012. Carbon nanotube exploration in cancer cell lines. Drug Discov. Today 17, 1023–1030.

Kesharwani, P., Tekade, R.K., Gajbhiye, V., Jain, K., Jain, N.K., 2011. Cancer targeting potential of some ligand-anchored poly(propylene imine) dendrimers: a comparison. Nanomedicine 7, 295–304. http://dx.doi.org/10.1016/j.nano.2010.10.010.

Kesharwani, P., 2015. Relative study of cancer targeting potential of engineered dendrimer. https://www.morebooks.de/store/gb/book/relative-study-of-cancer-targeting-potential-of-engineered-dendrimer/isbn/978-3-659-51741-9 (accessed June 30, 2015).

Kesharwani, P., Banerjee, S., Padhye, S., Sarkar, F.H., Iyer, A.K., 2015a. Hyaluronic acid engineered nanomicelles loaded with 3, 4-difluorobenzylidene curcumin for targeted killing of CD44+ stem-like pancreatic cancer cells. Biomacromolecules. http://dx.doi.org/10.1021/acs.biomac.5b00941.

Kesharwani, P., Banerjee, S., Gupta, U., Mohd Amin, M.C.I., Padhye, S., Sarkar, F.H., et al., 2015b. PAMAM dendrimers as promising nanocarriers for RNAi therapeutics. Mater. Today. http://dx.doi.org/10.1016/j.mattod.2015.06.003.

Kesharwani, P., Mishra, V., Jain, N.K., 2015c. Generation dependent hemolytic profile of folate engineered poly(propyleneimine) dendrimer. J. Drug Deliv. Sci. Technol. 28, 1–6. http://dx.doi.org/10.1016/j.jddst.2015.04.006.

Kesharwani, P., Mishra, V., Jain, N.K., 2015d. Validating the anticancer potential of carbon nanotube-based therapeutics through cell line testing. Drug Discov. Today. http://dx.doi.org/10.1016/j.drudis.2015.05.004.

Kesharwani, P., Tekade, R.K., Jain, N.K., 2015e. Dendrimer generational nomenclature: the need to harmonize. Drug Discov. Today. http://dx.doi.org/10.1016/j.drudis.2014.12.015.

Kesharwani, P., Tekade, R.K., Jain, N.K., 2015f. Generation dependent safety and efficacy of folic acid conjugated dendrimer based anticancer drug formulations. Pharm. Res. 32, 1438–1450. http://dx.doi.org/10.1007/s11095-014-1549-2.

Kesharwani, P., Xie, L., Banerjee, S., Mao, G., Padhye, S., Sarkar, F.H., et al., 2015g. Hyaluronic acid-conjugated polyamidoamine dendrimers for targeted delivery of 3,4-difluorobenzylidene curcumin to CD44 overexpressing pancreatic cancer cells. Colloids Surf. B. Biointerfaces 136, 413–423. http://dx.doi.org/10.1016/j.colsurfb.2015.09.043.

Kesharwani, P., Gajbhiye, V., Jain, N.K., 2012. A review of nanocarriers for the delivery of small interfering RNA. Biomaterials 33, 7138–7150. http://dx.doi.org/10.1016/j.biomaterials.2012.06.068.

Kesharwani, P., Gajbhiye, V., Tekade, R.K., Jain, N.K., 2011. Evaluation of dendrimer safety and efficacy through cell line studies. Curr. Drug Targets 12, 1478–1497. http://dx.doi.org/10.2174/138945011796818135.

Kukowska-Latallo, J.F., Raczka, E., Quintana, A., Chen, C., Rymaszewski, M., Baker, J.R., 2000. Intravascular and endobronchial DNA delivery to murine lung tissue using a novel, nonviral vector. Hum. Gene Ther. 11, 1385–1395.

Kukowska-Latallo, J.F., Candido, K.A., Cao, Z., Nigavekar, S.S., Majoros, I.J., Thomas, T.P., et al., 2005. Nanoparticle targeting of anticancer drug improves therapeutic response in animal model of human epithelial cancer. Cancer Res 65, 5317–5324. http://dx.doi.org/10.1158/0008-5472.can-04-3921.

Langereis, S., de Lussanet, Q.G., van Genderen, M.H.P., Meijer, E.W., Beets-Tan, R.G.H., Griffioen, A.W., et al., 2006. Evaluation of Gd(III)DTPA-terminated poly(propylene imine) dendrimers as contrast agents for MR imaging. NMR Biomed. 19, 133–141. http://dx.doi.org/10.1002/nbm.1015.

Leroueil, P.R., Berry, S.A., Duthie, K., Han, G., Rotello, V.M., McNerny, D.Q., et al., 2008. Wide varieties of cationic nanoparticles induce defects in supported lipid bilayers. Nano Lett. 8, 420–424. http://dx.doi.org/10.1021/nl0722929.

Liu, M., Kono, K., Fréchet, J.M.J., 1999. Water-soluble dendrimer-poly(ethylene glycol) starlike conjugates as potential drug carriers. J. Polym. Sci. Part A Polym. Chem 37, 3492–3503. http://dx.doi.org/10.1002/(SICI)1099-0518(19990901)37:173492::AID-POLA73.0.CO;2-0.

Luong, D., Kesharwani, P., Killinger, B.A., Moszczynska, A., Sarkar, F.H., Padhye, S., et al., 2016. Solubility enhancement and targeted delivery of a potent anticancer flavonoid analogue to cancer cells using ligand decorated dendrimer nano-architectures. J. Colloid Interface Sci 484, 33–43. http://dx.doi.org/10.1016/j.jcis.2016.08.061.

Luong, D., Kesharwani, P., Deshmukh, R., Mohd Amin, M.C.I., Gupta, U., Greish, K., et al., 2016. PEGylated PAMAM dendrimers: Enhancing efficacy and mitigating toxicity for effective anticancer drug and gene delivery. Acta Biomater. http://dx.doi.org/10.1016/j.actbio.2016.07.015.

Mahor, A., Prajapati, S.K., Verma, A., Gupta, R., Iyer, A.K., Kesharwani, P., 2016. Moxifloxacin loaded gelatin nanoparticles for ocular delivery: Formulation and in-vitro, in-vivo evaluation. J. Colloid Interface Sci 483, 132–138. http://dx.doi.org/10.1016/j.jcis.2016.08.018.

Malik, N., Wiwattanapatapee, R., Klopsch, R., Lorenz, K., Frey, H., Weener, J.W., et al., 2000. Dendrimers: relationship between structure and biocompatibility in vitro, and preliminary studies on the biodistribution of 125I-labelled polyamidoamine dendrimers in vivo. J. Control. Release 65, 133–148. http://www.ncbi.nlm.nih.gov/pubmed/10699277 (accessed September 5, 2016).

Mansuri, S., Kesharwani, P., Tekade, R.K., Jain, N.K., 2016. Lyophilized mucoadhesive-dendrimer enclosed matrix tablet for extended oral delivery of albendazole. Eur. J. Pharm. Biopharm. Off. J.

Arbeitsgemeinschaft Für Pharm. Verfahrenstechnik e.V 102, 202–213. http://dx.doi.org/10.1016/j.ejpb.2015.10.015.

Miller, T.M., Neenan, T.X., 1990. Convergent synthesis of monodisperse dendrimers based upon 1,3,5-tri-substituted benzenes, Chem. Mater 2, 346–349. http://dx.doi.org/10.1021/cm00010a006.

Mishra, N.K.J.V., Kesharwani, P., 2014. Functionalized Polymeric Nanoparticles for Delivery of Bioactives. Nanobiomedicine, Publ. M/s Stud. Press LLC, USA, 91–123. https://scholar.google.co.in/citations?view_op=view_citation&hl=en&user=DJkvOAQAAAAJ&cstart=20&pagesize=80&citation_for_view=DJkvOAQAAAAJ:Se3iqnhoufwC (accessed November 29, 2015).

Mishra, V., Kesharwani, P., 2016. Dendrimer technologies for brain tumor. Drug Discov. Today 21, 766–778. http://dx.doi.org/10.1016/j.drudis.2016.02.006.

Mishra, V., Kesharwani, P., Jain, N.K., 2014. siRNA nanotherapeutics: a Trojan horse approach against HIV. Drug Discov. Today 19, 1913–1920. http://dx.doi.org/10.1016/j.drudis.2014.09.019.

Mishra, V., Gupta, U., Jain, N.K., 2010. Influence of different generations of poly(propylene imine) dendrimers on human erythrocytes. Pharmazie 65, 891–895. http://www.ncbi.nlm.nih.gov/pubmed/21284258 (accessed April 15, 2015).

Najlah, M., Freeman, S., Attwood, D., D'Emanuele, A., 2007. In vitro evaluation of dendrimer prodrugs for oral drug delivery. Int. J. Pharm. 336, 183–190. http://dx.doi.org/10.1016/j.ijpharm.2006.11.047.

Nanjwade, B.K., Bechra, H.M., Derkar, G.K., Manvi, F.V., Nanjwade, V.K., 2009. Dendrimers: emerging polymers for drug-delivery systems. Eur. J. Pharm. Sci. 38, 185–196. http://dx.doi.org/10.1016/j.ejps.2009.07.008.

Newkome, G.R., Yao, Z., Baker, G.R., Gupta, V.K., 1985. Micelles. Part 1. Cascade molecules: a new approach to micelles. A [27]-arborol, J. Org. Chem. 50, 2003–2004. http://dx.doi.org/10.1021/jo00211a052.

Patel, H.K., Gajbhiye, V., Kesharwani, P., Jain, N.K., 2016. Ligand anchored poly(propyleneimine) dendrimers for brain targeting: Comparative in vitro and in vivo assessment. J. Colloid Interface Sci 482, 142–150. http://dx.doi.org/10.1016/j.jcis.2016.07.047.

Patri, A.K., Kukowska-Latallo, J.F., Baker, J.R., 2005. Targeted drug delivery with dendrimers: comparison of the release kinetics of covalently conjugated drug and non-covalent drug inclusion complex. Adv. Drug Deliv. Rev 57, 2203–2214. http://dx.doi.org/10.1016/j.addr.2005.09.014.

Pedziwiatr-Werbicka, E., Fuentes, E., Dzmitruk, V., Sánchez-Nieves, J., Sudas, M., Drozd, E., et al., 2013. Novel "SiC" carbosilane dendrimers as carriers for anti-HIV nucleic acids: Studies on complexation and interaction with blood cells. Colloids Surf. B Biointerfaces 109, 183–189. http://dx.doi.org/10.1016/j.colsurfb.2013.03.045.

Percec, V., Chu, P., Ungar, G., Zhou, J., 1995. Rational design of the first nonspherical dendrimer which displays calamitic nematic and smectic thermotropic liquid crystalline phases. J. Am. Chem. Soc. 117, 11441–11454. http://dx.doi.org/10.1021/ja00151a008.

Percec, V., Cho, W.-D., Mosier, P.E., Ungar, G., Yeardley, D.J.P., 1998. Structural analysis of cylindrical and spherical supramolecular dendrimers quantifies the concept of monodendron shape control by generation number. J. Am. Chem. Soc. 120, 11061–11070. http://dx.doi.org/10.1021/ja9819007.

Pushechnikov, A., Jalisatgi, S.S., Hawthorne, M.F., Svenson, S., Medina, S.H., El-Sayed, M.E.H., et al., 2013. Dendritic closomers: novel spherical hybrid dendrimers. Chem. Commun. 49, 3579. http://dx.doi.org/10.1039/c3cc40597d.

Roy, R., Baek, M.-G., 2002. Glycodendrimers: novel glycotope isosteres unmasking sugar coding. Case study with T-antigen markers from breast cancer MUC1 glycoprotein. Rev. Mol. Biotechnol 90, 291–309. http://dx.doi.org/10.1016/S1389-0352(01)00065-4.

Sadler, K., Tam, J.P., 2002. Peptide dendrimers: applications and synthesis. Rev. Mol. Biotechnol 90, 195–229. http://dx.doi.org/10.1016/S1389-0352(01)00061-7.

Schilrreff, P., Mundiña-Weilenmann, C., Romero, E.L., Morilla, M.J., 2012. Selective cytotoxicity of PAMAM G5 core-PAMAM G2.5 shell tecto-dendrimers on melanoma cells. Int. J. Nanomed. 7, 4121–4133. http://dx.doi.org/10.2147/IJN.S32785.

Singh, R., Kesharwani, P., Mehra, N.K., Singh, S., Banerjee, S., Jain, N.K., 2015. Development and characterization of folate anchored Saquinavir entrapped PLGA nanoparticles for anti-tumor activity. Drug Dev. Ind. Pharm. 1–14. http://dx.doi.org/10.3109/03639045.2015.1019355.

Stöckigt, D., Lohmer, G., Belder, D., 1996. Separation and identification of basic dendrimers using capillary electrophoresis on-line coupled to a sector mass spectrometer. Rapid Commun. Mass Spectrom. 10, 521–526. http://dx.doi.org/10.1002/(SICI)1097-0231(19960331)10:5521::AID-RCM5183.0.CO;2-G.

Svenson, S., Tomalia, D.A., 2005. Dendrimers in biomedical applications--reflections on the field. Adv. Drug Deliv. Rev 57, 2106–2129. http://dx.doi.org/10.1016/j.addr.2005.09.018.

Tekade, R.K., Tekade, M., Kesharwani, P., D'Emanuele, A., 2016. RNAi-combined nano-chemotherapeutics to tackle resistant tumors. Drug Discov. Today. http://dx.doi.org/10.1016/j.drudis.2016.06.029.

Teow, H.M., Zhou, Z., Najlah, M., Yusof, S.R., Abbott, N.J., D'Emanuele, A., 2013. Delivery of paclitaxel across cellular barriers using a dendrimer-based nanocarrier. Int. J. Pharm. 441, 701–711. http://dx.doi.org/10.1016/j.ijpharm.2012.10.024.

Thakur, S., Kesharwani, P., Tekade, R.K., Jain, N.K., 2015. Impact of pegylation on biopharmaceutical properties of dendrimers. Polymer (Guildf) 59, 67–92. http://dx.doi.org/10.1016/j.polymer.2014.12.051.

Thakur, S., Tekade, R.K., Kesharwani, P., Jain, N.K., 2013. The effect of polyethylene glycol spacer chain length on the tumor-targeting potential of folate-modified PPI dendrimers. J. Nanoparticle Res. 15, 1625. http://dx.doi.org/10.1007/s11051-013-1625-2.

Thomas T.P., Patri, A.K., Myc, A., Myaing, M.T., Ye, J.Y., Norris, T.B., et al., 2004. In vitro targeting of synthesized antibody-conjugated dendrimer nanoparticles., Biomacromolecules. 5 2269–2274. http://dx.doi.org/10.1021/bm049704h.

Tiwari, A., Kesharwani, P., Gajbhiye, V., Jain, N.K., 2015. Synthesis and characterization of dendro-PLGA nanoconjugate for protein stabilization. Colloids Surf. B. Biointerfaces. 134, 279–286. http://dx.doi.org/10.1016/j.colsurfb.2015.06.064.

Tolia, G.T., Choi, H.H., 2008. The role of dendrimers in topical drug delivery. Pharm. Technol. 32, 88.

Tomalia, D.A., Baker, H., Dewald, J., Hall, M., Kallos, G., Martin, S., et al., 1985. A new class of polymers: starburst-dendritic macromolecules. Polym. J 17, 117–132. http://dx.doi.org/10.1295/polymj.17.117.

Tomalia, D.A., Reyna, L.A., Svenson, S., 2007. Dendrimers as multi-purpose nanodevices for oncology drug delivery and diagnostic imaging. Biochem. Soc. Trans 35, 61–67. http://dx.doi.org/10.1042/BST0350061.

Vandamme, T.F., Brobeck, L., 2005. Poly(amidoamine) dendrimers as ophthalmic vehicles for ocular delivery of pilocarpine nitrate and tropicamide. J. Controlled Release 102, 23–38. https://www.ncbi.nlm.nih.gov/pubmed/15653131.

Wada, K., Arima, H., Tsutsumi, T., Hirayama, F., Uekama, K., 2005. Enhancing effects of galactosylated dendrimer/alpha-cyclodextrin conjugates on gene transfer efficiency. Biol. Pharm. Bull. 28, 500–505. http://www.ncbi.nlm.nih.gov/pubmed/15744077 (accessed September 6, 2016).

Wang, F., Cai, X., Su, Y., Hu, J., Wu, Q., Zhang, H., et al., 2012. Reducing cytotoxicity while improving anticancer drug loading capacity of polypropylenimine dendrimers by surface acetylation. Acta Biomater 8, 4304–4313. http://dx.doi.org/10.1016/j.actbio.2012.07.031.

Wiener, E.C., Brechbiel, M.W., Brothers, H., Magin, R.L., Gansow, O.A., Tomalia, D.A., et al., 1994. Dendrimer-based metal chelates: a new class of magnetic resonance imaging contrast agents. Magn. Reson. Med. 31, 1–8. http://www.ncbi.nlm.nih.gov/pubmed/8121264 (accessed September 6, 2016).

Woller, E.K., Cloninger, M.J., 2001. Mannose Functionalization of a Sixth Generation Dendrimer. Biomacromolecules 2, 1052–1054. http://dx.doi.org/10.1021/bm015560k.

CHAPTER 12

Carbon Nanotubes in Targeting and Delivery of Drugs

Rakesh K. Tekade, Rahul Maheshwari, Namrata Soni and Muktika Tekade

Contents

Disclosures: There are no conflict of interest and disclosures associated with the manuscript.

Nanotechnology-Based Approaches for Targeting and Delivery of Drugs and Genes.
DOI: http://dx.doi.org/10.1016/B978-0-12-809717-5.00014-2

12.1 INTRODUCTION

In the array of booming nanotechnological evolutions, many researchers are coming up with new ideas involving the delivery of bio-actives and genetic materials through modified liposomes (Gorain et al., 2016; Tekade et al., 2016; Maheshwari et al., 2015a), dendrimers (Kesharwani et al., 2015; Tekade et al., 2008a,b), solid lipid nanoparticles (Soni et al., 2016) and a variety of other established vectors (Gorain et al., 2016; Tekade et al., 2015a; Sharma et al., 2015). In addition, the use of polymers in drug delivery further justifies and finds the opportunities to introduce these nanocarriers in clinic (Maheshwari et al., 2015b; Gandhi et al., 2014; Dua et al., 2016). The discovery of wide varieties of delivery systems have encouraged research focused on their applications in various biomedical fields (Choudhury et al., 2016; Maheshwari et al., 2012; Tekade et al., 2014b,c), with special emphasis on hybrid nanocarriers (Tekade and Chougule, 2013a; Mansuri et al., 2016; Moeendarbari et al., 2016; Prajapati et al., 2009). The applications of these delivery systems are more opportunistic when it comes to diseases like cancer. Many investigators have explored these novel delivery tools to provide platforms to fight against different variants of tumors (Youngren et al., 2013; Tekade et al., 2015b,c). In addition, investigators have also worked in the area to develop in vivo detection techniques for such novel devices (Tekade et al., 2013b; Jain and Tekade, 2013; Kurmi et al., 2010).

Carbon nanomaterials including carbon nanohorns (CNHs), graphenes (GRs), carbon nanorods (CNRs), polyhydroxy fullerenes (PHF), and carbon nanotubes (CNTs) represent safe and efficacious carrier systems for drug delivery and drug targeting because of their unique physicochemical properties. They hold promise for applications in medicine, gene and drug delivery areas. CNTs are tubular, hollow monolithic structures with a high surface-to-aspect ratio (length/diameter), rich functional surface chemistry and high drug-loading capacity. CNTs do not require any type of fluorescent labeling for detection because they can be detected directly due to their electron emission properties. They also represent a class of biocompatible, nonimmunogenic, and photo luminescent nanomaterials, making them smart nanocarriers for drug delivery and imaging applications (Mody et al., 2014).

CNTs are available as single-walled CNTs (SWCNTs), double-walled CNTs (DWCNTs), and multiwalled CNTs (MWCNTs), with cylindrical graphitic layers. CNTs are a versatile member of the carbon family (Zhang et al., 2013). Their crystal structures are very close to graphite, belonging to sp^2 bonded carbon, rather than sp^3-hybridized carbon as in the case of diamond. Topologically, SWCNTs can be constructed by rolling up a single layer of graphite or graphene along a certain direction into a tiny cylinder with a possible diameter from subnanometer to a few nanometers. Interestingly, the rolling-up direction and diameter or the chirality of SWCNTs determines their fundamental properties. Some SWCNTs have small energy bandgaps, showing semiconducting characteristics, whereas others do not have the bandgap and they are metallic ones. For CNTs with more than one cylindrical shell, the interactions between the shells must be taken into account (Zhang, 2012).

Several production techniques have been introduced for the synthesis, functionalization, filling, doping, and chemical modification of CNTs. With advancement in bioanalytical instrumentation technologies, the characterization, separation, and manipulation of individual CNTs are now possible with high precision. Parameters such as structure, surface area, surface charge, size distribution, surface chemistry, agglomeration state, as well as purity of the samples, have considerable impact on the reactivity of CNTs (Wang, 2013). The strength and flexibility of CNTs make them a potential candidate for use in controlling other nanoscale structures, suggesting their significant versatile role in nanotechnology and bioengineering (Eatemadi et al., 2014). This chapter highlights the current state of the art of CNTs in biomedical applications, their potential adverse human health effects, as well as the challenges posed by CNTs in the development of nanomedicines.

12.2 HISTORY

CNTs were discovered in 1991 by Sumio Ijima of the NEC laboratory in Tsukuba, Japan, during high-resolution transmission electron microscopy (TEM) observation of soot generated from the electrical discharge between two carbon electrodes. CNTs invariably contain at least two graphitic layers and generally have inner diameters of around 4 nm. The discovery was accidental, although it would not have been possible without Ijima's excellent microscopist skills and expertise (Iijima, 1991). After a few years, Ijima and Toshinari Ichihashi of NEC and Donald Bathune and Colleagues of IBM Almaden Research Center in California independently reported the synthesis of single-walled nanotubes (Iijima et al., 1999).

The C_{60} molecules, also known as Buckminster fullerenes, were previously discovered by Harold Kroto and Richard Smalley during the 1970s. Kroto and Smalley found that under the right arc-discharge conditions, carbon atoms would self-assemble spontaneously into molecules of specific shapes, such as the C_{60} molecule (Kroto et al., 1985). However, as shown by Ijima's discovery, under different experimental conditions, carbon atoms can instead self-assemble into CNTs. Although various carbon cages were studied, it was only in 1991 when Ijima observed for the first time tubular carbon structures. The nanotubes consisted of up to several tens of graphitic shells (so-called MWNT with adjacent shell separation of 0.34 nm, diameters of 1 nm and high length/diameter ratio. Two years later, Ijima et al. synthesized single-walled carbon nanotubes (SWNT) (Iijima et al., 1999). Currently, functionalization of CNTs has opened new perspectives in the study of their biological properties. Attachment of an organic moiety to nanosized tubes has made possible their use in diagnostics for imaging as well as for targeting purposes, especially in cancer therapy and infectious disease treatment.

12.3 METHODS OF PREPARATION

The production of CNTs involves the transformation of a carbon source into nano-tubular assembly, usually at high temperature and low pressure, wherein the synthesis conditions influence the characteristics of the final product. The synthesis of CNTs is usually associated with carbonaceous or metallic impurities; hence, purification is an essential step to be considered (Rastogi et al., 2014).

12.3.1 Arc-discharge

Ebbesen and Ajayan, for the first time, reported the synthetic methodology for the production of CNTs on a large scale (Ebbesen and Ajayan, 1992). They also optimized the yield of CNTs by varying the production parameters such as type of inert gas, pressure, the nature of current (A.C. or D.C.), the voltage, and the relative rod size (nm). A brief schematic representation of this method is given in Fig. 12.1A. The optimal yield of CNTs takes place at 500 torr pressure. In another study, Ohkohchi et al. studied the growth-promoting effect of scandium on nanotubes by using a carbon composite rod containing scandium oxide during the synthesis of CNTs by arc discharge evaporation method (1993). Some reports investigated that high yield of MWCNTs by electric arc discharges in liquid environments, particularly in liquid nitrogen and deionized water (Lee and Geckeler, 2010; Antisari et al., 2003).

Alternatively, Wang et al. used a sodium chloride solution as a liquid medium because of its significant cooling ability and excellent electrical conductivity rather than deionized water (Wang et al., 2005). The authors successfully synthesized MWCNTs with the formation of a single sheet of SWCNTs. Anazawa et al. demonstrated the introduction of magnetic field in arc discharge synthesis to obtain high-purity/defect-free (purity >95%) MWCNTs (Anazawa et al., 2002). Ando et al. modified this method by a newly developed direct current arc plasma jet method for the evaporation of metal doped carbon anode. They showed a high production rate (1.24 g/min) as compared to the conventional method (Ando et al., 2000). Cheng et al. devised a hydrogen arc discharge method for the production of SWCNTs with high hydrogen capacity using graphite powder, catalyst metal, and a growth promoter in an atmosphere containing hydrogen (Cheng et al., 2003).

12.3.2 Laser-ablation

In 2003, Smalley produced high yields (>75%) of SWNTs by laser ion (vaporization) of graphite rods with small amounts of Ni and Co at 1200°C. The brief schematic representation of this method is given in Fig. 12.1B. The tube grows until too many

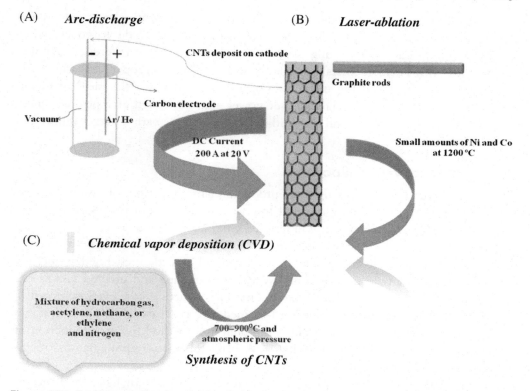

(A) *Arc-discharge*

CNTs deposit on cathode

Carbon electrode

Vacuum   Ar/He

DC Current
200 A at 20 V

(B) *Laser-ablation*

Graphite rods

Small amounts of Ni and Co
at 1200 °C

(C) *Chemical vapor deposition (CVD)*

Mixture of hydrocarbon gas,
acetylene, methane, or
ethylene
and nitrogen

700–900°C and
atmospheric pressure

Synthesis of CNTs

Figure 12.1 Exploring various methods of synthesis of CNTs such as (A) arc-discharge, (B) laser-ablation, and (C) chemical vapor deposition.

catalyst atoms aggregate on the end of the nanotube. The large particles either detach or become over-coated with sufficient carbon to poison the catalysis. This allows the tube to terminate with a fullerene-like tip or with a catalyst particle. Both arc-discharge and laser-ablation techniques have the advantage of high (>75%) yields of SWNTs and the drawback that they rely on evaporation of carbon atoms from solid targets at temperatures >3000°C. They also have one more drawback in regards to the entanglement of nanotubes, which makes the purification and application of the samples a tedious task (Smalley, 2003). In the laser-ablation technique reported by Thess et al., for producing CNTs, intense laser pulses were utilized to ablate a carbon target. The pulsed laser-ablation of graphite in the presence of an inert gas and catalyst formed SWNTs at 1250°C (Thess et al., 1996). Braidy et al., also synthesized SWNTs and other nanotubular structures (graphite nanocages and low aspect ratio nanotubules) by pulsed KrF laser-ablation of a graphite pellet at an argon pressure of 500 torr, a temperature of ~1200°C, and a laser intensity of 8–10W/cm. They observed that relatively high UV laser intensity was unfavorable towards the growth of SWNTs (Braidy et al., 2002).

By using high vacuum laser-ablation, Takahashi et al., synthesized multilayered MWNTs having a tip angle of 15–20 degrees. In the case of arc-discharge and laser-ablation techniques for synthesizing SWNTs, the by-products are fullerenes, graphitic polyhedrons with enclosed metal particles, and amorphous carbon (Takahashi et al., 2002). In general, some of the major parameters that determine the yield of CNTs are the amount and type of catalysts, laser power and wavelength, temperature, pressure, type of inert gas present and the fluid dynamics near the carbon target (Sinha and Yeow, 2005).

12.3.3 Chemical vapor deposition

The arc discharge technique is used for producing large quantities of raw nanotubes, but significant efforts are being directed towards the production processes that offer more controllable routes for the synthesis of CNTs. One such process is chemical vapor deposition (CVD) that seems to offer the best conditions to obtain a controllable process for the selective production of nanotubes with predefined properties. Apart from material scale-up, controlled synthesis of aligned and ordered CNTs can be achieved by using the CVD technique. The microstructure of CNTS tips synthesized by the CVD technique exhibits well-formed caps compared to other techniques (Rastogi et al., 2014). Therefore, CVD is the preferred method for the production of CNTs over other available methodologies. A brief schematic representation of this technique is presented in Fig. 12.1C.

In the CVD technique, a mixture of hydrocarbon gas (ethylene, methane, or acetylene) and a process gas (ammonia, nitrogen, and hydrogen) is made to react in a reaction chamber on heated metal substrate at a temperature of around 700–900°C, at atmospheric pressures. During this, the residual gas diffuses away, whereas free carbon atoms dissolve and then segregate on the catalyst surface to form CNTs. The defined parameters that must be critically considered during synthesis of CNTs by this technique are the nature of hydrocarbons, catalysts, and the growth temperature. Balbuena et al. demonstrated the role of the catalyst in the growth of SWCNTs by using model Co-Mo catalyst and also studying the role of catalyst/substrate interactions. They found that a strong cluster/substrate interaction increases the cluster melting point, modifying the initial stages of carbon dissolution and precipitation on the cluster surface (Balbuena et al., 2006).

Depending on the reaction conditions and catalyst preparations, this method may be applied to obtain either SWCNTs or MWCNTs. There are two possible mechanisms for the catalyst-assisted nanotube growth, namely, (1) tip growth and (2) root growth mechanism (Rastogi et al., 2014). In root growth mechanism, the tubes grow away from the metal particles, with carbon being continuously supplied to the base, whereas in tip growth metal particles are found at the tip.

12.4 FUNCTIONALIZATION OF CNTs

Functionalization of CNTs has been the most effective approach. It has been shown capable of decreasing cytotoxicity, improving biocompatibility, and giving opportunity to appendage molecules of drugs, proteins, or genes for the construction of delivery systems (Stoddart et al., 2007). Functionalization can be divided into two main sub-categories: noncovalent functionalization and covalent functionalization. The underlying process of functionalization involves the selective breaking of C=C bonds in the CNTs and is often done through oxidation (e.g., refluxing in nitric acid or through electrochemical modification) resulting in carboxyl groups that could then be used as subsidiary sites for addition reactions (Mehra et al., 2014; Yao et al., 2015). The schematic illustration of functionalization of CNTs is shown in Fig. 12.2.

12.4.1 Noncovalent functionalization

Many small as well as large polymeric anticancer agents can be adsorbed noncovalently onto the surface of pristine CNTs. Forces that governs such adsorption are the hydrophobic and π-π stacking interactions between the chains of the adsorbed molecules and the surface of CNTs. Since many anticancer drugs are hydrophobic in nature or have hydrophobic moieties, the hydrophobic forces are the main driving forces for the loading of such drugs into or onto CNTs (Mittal et al., 2015).

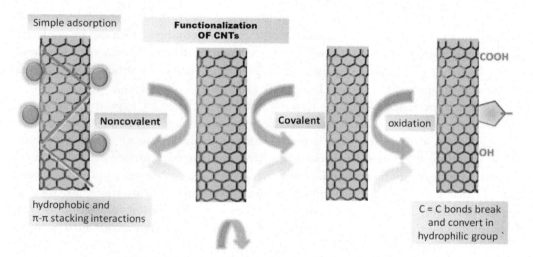

Figure 12.2 Different functionalization methods of CNTs.

The presence of charge on the nanotube surface due to chemical treatment can enable the adsorption of the charged molecules through ionic interactions. Aromatic molecules or the molecules with aromatic groups can be embarked on the solubilization of CNTs using nucleic acids and amphiphilic peptides based on the π-π stacking interactions between the CNTs surface and aromatic bases/amino acids in the structural backbone of these functional biomolecules (Ma et al., 2010; Ghosh et al., 2010).

Noncovalent functionalization of CNTs is particularly attractive because it offers the possibility of attaching chemical handles without affecting the electronic network of the tubes. Oxide surfaces modified with pyrene through π-π stacking interactions have been employed for the patterned assembly of single-walled carbon nanomaterials. The carbon graphitic structure can be recognized by pyrene functional groups with distinct molecular properties. The interactions between bifunctional molecules (with amino and silane groups) and the hydroxyl groups on an oxide substrate can generate an amine-covered surface. This was followed by a coupling step where molecules with pyrene groups were allowed to react with amines (Alpatova et al., 2010).

The patterned assembly of a single layer of SWCNTs could be achieved through π-π stacking with the area covered with pyrenyl groups. Alkyl-modified iron oxide nanoparticles have been attached onto CNTs by using a pyrene carboxylic acid derivative as a chemical cross-linker. The resulting material had an increased solubility in organic media due to the chemical functions of the inorganic nanoparticles. Surfactants were initially involved as dispersing agents in the purification protocols of raw carbon material. Then, surfactants were used to stabilize dispersions of CNTs for spectroscopic characterization, optical limiting property studies, and compatibility enhancement of composite materials. Functionalized nanotube surface can be achieved simply by exposing CNTs to vapors containing functionalization species that noncovalently bonds to the nanotube surface while providing chemically functional groups at the nanotube surface (Chang and Liu, 2010).

A stable functionalized nanotube surface can be obtained by exposing it to vapor stabilization species that reacts with the fictionalization layer to form a stabilization layer against desorption from the nanotube surface, while depositing chemically functional groups at the nanotube surface. The stabilized nanotube surface can be exposed further to at least another material layer precursor species that can deposit as a new layer of materials. A patent is pertinent to dispersions of CNTs in a host polymer or copolymer with delocalized electron orbitals, so that a dispersion interaction occurs between the host polymer or copolymer and the CNTs dispersed in that matrix. Such a dispersion interaction has advantageous results if the monomers of the host polymer/copolymer include an aromatic moiety, e.g., phenyl rings or their derivatives. It is claimed that dispersion force can be further enhanced if the aromatic moiety is naphthalenyl and anthracenyl (Wise et al., 2006).

A process registered by Stoddart et al. involves CNTs treated with poly {(5-alkoxy-*m*-phenylenevinylene)-co-[(2,5-dioctyloxy-*p*-phenylene)   vinyl-ene]}

(PAmPV) polymers and their derivatives for noncovalent fictionalization of the nanotubes, which increases solubility and enhances other properties of interest (Stoddart et al., 2007). Pseudorotaxanes are grafted along the walls of the nanotubes in a periodic fashion by wrapping of SWCNTs with these functionalized PAmPV polymers. Many biomolecules can interact with CNTs without producing covalent conjugates. Proteins are an important class of substrates that possess high affinity with the graphitic network (Zhao and Stoddart, 2009).

Nanotube walls can absorb proteins strongly on their external sides, and the products can be visualized clearly by microscopy techniques. Metallothione in proteins were adsorbed onto the surface of MWCNTs, as evidenced by high-resolution TEM. DNA strands have been reported by several groups to interact strongly with CNTs to form stable hybrids effectively dispersed in aqueous solutions. Kim et al. reported the solubilization of nanotubes with amylase by using dimethyl sulfoxide/water mixtures. The polysaccharide adopts an interrupted loose helix structure in these media. The studies of the same group on the dispersion capability of pullulan and carboxy methylamylase demonstrated that these substances could also solubilize CNTs but to a lesser extent than amylase (Kim et al., 2007).

12.4.2 Covalent functionalization/exohedral chemical functionalization

CNTs can be oxidized, giving CNTs hydrophilic groups as OH, COOH, and so on. Strong acid solution treatment can create defects in the side walls of CNTs, and the carboxylic acid groups are generated at the defect point, predominantly on the open ends. Excessive surface defects possibly change the electronic properties and cut longer CNTs into short ones, as drug carriers may need CNTs with different electronic properties (Karousis et al., 2010).

During the preparation of some drug delivery systems, CNTs were deliberately cut into smaller pieces. The functional groups on the oxidized CNTs can further react with $SOCl_2$ and carbodiimide to yield functional materials with great propensity for reacting with other compounds. Covalent functionalization of SWCNTs using addition chemistry is believed to be very promising for CNTs modification and derivatization. However, it is difficult to achieve complete control over the chemo and region selectivity of such additions and requires very special species such as arynes, carbenes, or halogens, and the reactions often occur only in extreme conditions for the formation of covalent bonds (Ma et al., 2010).

Furthermore, the characterization of functionalized SWCNTs and the determination of the precise location and mode of addition are also very difficult. The covalent chemistry of CNTs is not particularly rich with respect to variety chemical reactions to date. As regard to functionalization behavior of SWCNTs and MWCNTs, it has been reported that the functionalization percentage of MWCNTs is lower than that of SWCNTs with the similar process, which is assumingly attributed to the larger outer diameter and sheathed nature of MWCNTs that render many of their side walls inaccessible; nonetheless, a comparative study on functionalizing single-walled and MWCNTs is scarce hitherto in open literature (Zhang et al., 2011).

In comparison with noncovalent functionalization, there are new substances to develop and therefore most patents regarding functionalization of CNTs registered to date are based on covalent chemistry. Though covalent procedures are not highly diverse yet, the end products vary exceedingly in terms of characteristics depending upon the incorporated species. Methotrexate functionalization can be realized through 1,3-cyclo-addition reaction. Azomethine ylides consisting of a carbanion adjacent to an immonium ion are organic 1,3-dipoles, which give pyrrolidine intermediates upon cycloaddition to dipolarophiles. Through decarboxylation of immonium salts obtained from the condensa-tion of a-amino acids with aldehydes or ketones, azomethineylides can be easily produced. These compounds can make CNTs fused with pyrrolidine rings with varied substituents depending on the structure of used a-amino acids and aldehydes. Acyl peroxides can gen-erate carbon-centered free radicals for functionalization of CNTs. The promising method allows the chemical attachment of a variety of functional groups to the wall or end-cap of CNTs through covalent carbon bonds without destroying the wall or end-cap structure of CNTs, unlike in the case of treating with strong acid (Spitalsky et al., 2010).

Carbon-centered radicals generated from acyl peroxides can have terminal groups that render the modified sites capable of further reaction with other compounds. For example, organic groups with terminal carboxylic acid functionality can further react with acyl chloride and an amine to form an amide or with a diamine to form an amide with terminal amine (Su and Cheng, 2014).

12.5 PROPERTIES AND PHARMACOKINETICS OF CNTs

The discovery of wide varieties of CNTs (see Fig. 12.3 for more details) have stimu-lated intensive research pertaining to the potential application of CNTs due to the excellent mechanical and electronic properties. It is well established that CNTs have superior mechanical strength and low weight (tensile modules ~1 TPa) as well as good heat conductance (heat conductivity of MWCNTs bundles ~2000W/mK). They could be either metallic or semiconducting, depending on their diameter. Perhaps the most intriguing property of SWCNTs is the high room temperature mobility of semicon-ducting SWCNTs (Moghadam et al., 2015).

12.5.1 Electrical properties

To a large extent, the unique electrical properties of CNTs are derived from their one-dimensional character and the peculiar electronic structure that came from graphitic structure. They have extremely low electrical resistance, which occurs when an elec-tron collides with some defect in the crystal structure of the material through which it is passing. The defect could be an atom impurity, a defect in the crystal structure, or

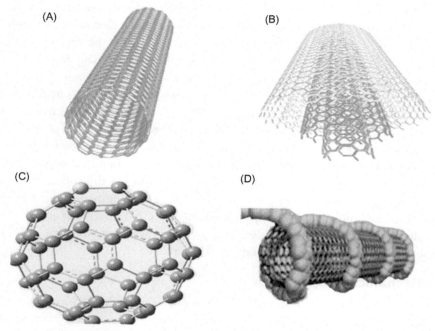

Figure 12.3 Structure of CNTs (A) SWCNTs, (B) MWCNTs, (C) Buckminster fullerene, and (D) functionalized CNTs.

an atom vibrating about its position in the crystal. Such collisions deflect the electron from its path. But the electrons inside a CNT are not so easily scattered due to their small diameter and huge length-to-diameter ratio or even higher. In a three-dimensional conductor, electrons have plenty of opportunity to scatter, since they can do so at any angle (Smalley, 2003; Chen et al., 2011).

However, in a one-dimensional conductor, electrons can travel only forward or backward. Under these circumstances, only backscattering (the change in electron motion from forward to backward) can lead to electrical resistance. But backscattering requires very strong collisions and is thus less likely to happen, thus, diminishing the possibilities to electron scattering. This reduced scattering gives CNTs very low resistance and at this state they can carry the current density measured as high as $109A/cm^2$. One use for nanotubes that has already been developed is as extremely fine electron guns, which could be used as miniature cathode ray tubes (CRTs) in thin high-brightness low-energy low-weight displays. This type of display would consist of a group of many tiny CRTs, each providing the electrons to hit the phosphor of one pixel, instead of having one giant CRT whose electrons are aimed at using electric and magnetic fields. These displays are known as field emission displays (Kaushik and Majumder, 2015; Chen et al., 1998).

A nanotube formed by joining nanotubes of two different diameters end-to-end can act as a diode, suggesting the possibility of constructing electronic computer circuits entirely out of nanotubes. Nanotubes have been shown to be superconducting at low temperatures. The electronic transport in metallic SWNTs and MWNTs occurs ballistically (without scattering) over long lengths owing to the nearly 1-d electronic structure. This property enables nanotubes to carry high currents with negligible heating. Distortions like bending and twisting affect the electrical and electronic properties of nanotubes. Bending introduces pentagon–heptagon pair in CNTs; this results in metal–metal and semiconductor–metal nanoscale junctions that can be used for nano-switches. The effect of bending becomes important when bending angles are more than 45°C. At this stage, bending appears in the structure of the tube. The ultimate effect of bending is reduction in conductivity of CNTs. Upon twisting, a band gap opens up that turns metallic CNTs to semiconducting CNTs. Twisting above a certain angle results in the collapse of CNTs structures. Also, superconductivity in SWNTs has been observed, but only at low temperatures (Li et al., 2012).

12.5.2 Mechanical properties

The CNTs possess high stiffness and axial strength as a result of the carbon–carbon sp^2 bonding. Mechanically, CNTs are one of the strongest known fibers available for biomedical applications. The practical application of the nanotubes requires the study of their elastic response, the inelastic behavior and buckling, yield strength and fracture aspects. Nanotubes are the stiffest known fiber, with a measured Young's modulus (a measure of stiffness) of 1.4 TPa. They have an expected elongation to failure of 20%–30%, which combined with the stiffness, projects to a tensile strength well above 100 GPa (possibly higher), by far the highest known (Esawi et al., 2010).

For comparison, the Young's modulus of high-strength steel is around 200 GPa, and its tensile strength is 1–2 GPa (Kaushik and Majumder, 2015). Elastic properties of CNTs can be obtained from experiment by assuming them structural members. Wong et al. determined the mechanical properties of MWNTs using atomic force microscopy (AFM) by pinning them at one end to molybdenum disulfide surfaces. They observed the average value of Young's modulus as 1280 GPa (Wong et al., 1997).

12.5.3 Thermal properties

CNTs have been shown to have twice the thermal conductivity as that of diamond. CNTs have the unique property of feeling cold to the touch, like metal, on the sides with the tube ends exposed, but similar to wood on the other sides. The specific heat and thermal conductivity of CNTs are determined primarily by phonons. The measurements yield linear specific heat and thermal conductivity above 1°K and below room temperature while a $T^{0.62}$ behavior of the specific heat was observed below 1°K. The linear temperature dependence can be explained with the linear k-vector dependence of the frequency of the longitudinal and twist acoustic phonons. The specific behavior of the specific heat below 1°K can be attributed to the transverse acoustic phonons with quadratic dependence. The measurements of the thermoelectric power (TEP) of nanotube systems give direct information for the type of carriers and conductivity mechanisms (Li et al., 2012).

The specific heat and thermal conductivity of CNTs are dominated by phonons, as the electronic contribution is negligible due to low density of free charge carriers. Experiments were performed by Yi et al. for measurements of specific heat of MWNTs. The measurements revealed linear dependence of the specific heat on the temperature over the entire temperature interval (Yi et al., 1999). MWNTs and bundles of SWNTs with an average diameter of 1.3 nm were used by Mizel et al. to measure the specific heat in the temperature range (Mizel et al., 1999). While MWNTs exhibited graphite-like behavior, SWNTs bundles had steep temperature dependence at low temperatures. The specific heat of SWNTs bundles was measured by Lasjaunias et al. down to a temperature of 0.1°K. The presence of sublinear temperature dependence of specific heat, dominant below 1.0°K, was observed. This dependence could not be understood and warrants further investigation into the thermal properties of CNTs. For measuring thermal conductivities of CNTs to isotopic and other atomic defects, the domination of phonons leads to unusual sensitivity of the experiments (Lasjaunias et al., 2002).

Zhang et al. found that the thermal conductivity of SWNTs depends on their length, radius, temperature, and chirality. Therefore, nanoscale devices with different heat conductivities can be made with nanotubes of different chirality. They also observed that the value of heat conductivity of zigzag nanotube is maximum, while the chiral nanotube has a minimum value (Zhang et al., 2011). Kim et al. determined that at room temperature the thermal conductivity for an individual MWNT (3000 W/mK) is greater than that of graphite (2000 W/mK; Kim et al., 2007).

12.5.4 Pharmacokinetics of CNTs

As a drug carrier system, the administration, absorption, and transportation of CNTs must be considered for obtaining desired treatment effects. The primary routes for CNTs administration include oral, subcutaneous injection, abdominal injection, and intravenous injection. There are different ways for the absorption and transportation when CNTs are administered in different ways. The absorbed CNTs are transported from the administration sites to the effective relevant sites by blood or lymphatic circulation (Mulvey et al., 2015).

12.5.4.1 Absorption

When subcutaneously and abdominally administered, a part of CNTs exist persistently in the local tissues while some may be absorbed through the lymphatic canal. Because endothelial cells of blood capillaries are 30 nm to −50 nm, approximately, while those in the endothelial cells of lymphatic capillaries are larger than 100 nm in diameter, the lymph absorption of bundled CNTs seemed to be easier than blood absorption. The lymphatically absorbed CNTs migrate along the lymph canal and are accumulated in the lymph node, which is in fact a lymphatic target effect. This is clinically important because lymphatic metastasis occurs extensively in cancers, resulting in frequent tumor recurrence, even after extended lymph node dissection. If anticancer drugs are loaded into CNTs, they will be delivered into the lymph system, where the drugs will be released to kill metastatic cancer cells (Kayat et al., 2011).

Zhang and coworkers delivered gemcitabine to lymph nodes with high efficiency by using a lymphatic-targeted drug delivery system based on magnetically guided MWCNTs (Zhang et al., 2014). When administered through veins, CNTs gets into the blood circulation and distributes in internal organs, such as liver, spleen, heart and kidneys. Some studies demonstrated that the blood clearance of intravenously injected CNTs largely depends upon the surface modification (Liu et al., 2005). Polyethylene glycol conjugation (PEGylation) is believed to be one of the most important strategies to prolong the circulation time of CNTs in blood. This is because the surface coverage with PEG lowers the immunogenicity of the carriers and prevents their nonspecific phagocytosis by the reticuloendothelial system (RES), thus, prolonging the half-life. It has been observed that PEGylated CNTs can persistently exist within liver and spleen macrophages for 4 months with excellent compatibility (Mulvey et al., 2015).

In a recent investigation, it was observed that fluorescein isothiocyanate (FITC) labeled PEGylated SWCNTs can penetrate through the nuclear membrane and gains entry into the nucleus in an energy-independent fashion. The presence of FITC-PEG-SWCNTs in the nucleus did not initiated any notable structural change in the nuclear organization and had no significant effects on the growth kinetics as well as cell cycle for up to 5 days. Surprisingly, upon removal of the FITC-PEG-SWCNTs from the culture medium, the internalized FITC-PEG-SWCNTs rapidly moved out of the nuclear compartment and was expelled out of the cells, suggesting that the internalization followed by excretion of CNTs from the cellular compartment is a bidirectional reversible process. These results well illustrated the successful use of SWCNTs as ideal nanovectors for biomedical and pharmaceutical applications, but they drive the concern about the excretion problems of CNTs (Zhang et al., 2011; Mulvey et al., 2015).

12.5.4.2 Distribution

Distribution indicates the sites at which the absorbed CNTs can distribute and localize, which are of great importance in clinical pharmacology and toxicology of drug carriers. There have been experiments to investigate in vivo and ex vivo biodistributions, as well as tumor targeting ability of radio-labeled SWCNTs (diameter, approximately 1–5 nm; length, approximately 100–300 nm) noncovalently functionalized with phospholipids (PL)-PEG in mice using positron emission tomography and Raman spectroscopy, respectively. It was interesting to note that the PEG chain lengths determine the biodistribution and circulation of CNTs. PEG-5400-modified SWCNTs have a circulation time ($t_{1/2} = 2$ h) much longer than that of PEG-2000-modified counterpart ($t_{1/2} = 0.5$ h), which may be attributed to its lower uptake of the former by the RES. By further functionalization of these PEGylated SWCNTs with arginine-glycine-aspartic acid (RGD) peptide, the accumulation in integrin-positive U87MG tumors was significantly improved from approximately 3%–4% to approximately 10%–15% of the total SWCNTs covalently modified with PEG, showing longer half-life in blood circulation in comparison with those noncovalently modified with PEG of similar chain lengths (Riviere, 2009).

SWCNTs covalently conjugated with branched chains of 7 kDa PEG effectively increased the half-life of SWCNTs up to 1 day, which is the longest among all of the tested PEGs. This PEG-modified SWCNTs had almost clear clearance from the main organs in approximately 2 months. There seemed to be a length limit in the relations between PEG chain lengths and their effects to increase the blood circulation time. Further increase in molecular weight from 7 to 12 kDa had no influence on the blood circulation time and RES uptake (Yang et al., 2006).

Nano-encapsulation of iodide within SWCNTs facilitated its biodistribution in tissues, and SWCNTs were completely redirected from tissue with intrinsic affinity (thyroid) to lungs. In this experiment, Na^{125}I filled glyco-SWCNTs were intravenously administered into mice and tracked in vivo by single photon emission computed tomography. Tissue-specific accumulation, coupled with high in vivo stability, prevented excretion or leakage of radionuclide to other high-affinity organs (thyroid and stomach), allowing ultrasensitive imaging and delivery of unprecedented radio dose densitylation (Sheng et al., 2010).

12.5.4.3 Metabolism and excretion

The nonbiodegradability in the body and noneliminability from the body interrogate on the possibility of their successful use in clinical practice, which has always been of much concerned in literature. Functionalized SWCNTs seem to be metabolizable in an animal body. For instance, SWCNTs with carboxylate surfaces have demonstrated their unique ability to undergo 90-day degradation in a phago–lysosomal simulant, resulting in shortening of length and accumulation of ultrafine solid carbonaceous debris. Unmodified, ozonolyzed, aryl-sulfonated SWCNTs exhibit no degradation under similar conditions. The observed metabolism phenomenon may be accredited to the unique chemistry of acid carboxylation, in addition to introducing the reactive, modifiable –COOH groups on the surface of CNTs. It also induces collateral damage to the tubular graphenic backbone in the form of neighboring active sites that provide points of attack for further oxidative degradation. Some experiments demonstrated that CNTs persisted inside cells for up to 5 months after administration and well-dispersed SWCNTs effectively managed to escape the RES and finally excreted through the kidneys and bile ducts. A very recent investigation reveals that the biodegradation of SWCNTs can be catalyzed by hypochlorite and reactive radical intermediates of the human neutrophil enzyme myeloperoxidase in neutrophils (Andon et al., 2013).

The phenomenon of CNTs metabolism can also be seen in macrophages to a lesser degree. Molecular modeling further reveals that the interaction between basic amino acid residues on the enzyme backbone and carboxyl acid groups of CNTs is favorable to orient the nanotubes close to the catalytic site. Notably, when aspirated into the lungs of mice, the biodegradation of the nanotubes does not engender an inflammatory response. These findings imply that the biodegradation of CNTs may be a key determinant of the degree and severity of the inflammatory responses in individuals exposed to them (Mulvey et al., 2015).

12.6 CLASSIFICATION OF CNTs

Nanotubes are mainly classified into two types depending upon the structure Fig. 12.3, namely: (1) SWNTs, and (2) MWNTs. In SWNTs only one sheet of graphene is arranged to give a cylindrical structure, which is one atom thick, having a radius of up to 1 nm. The SWNTs are closed at both ends with cap-like structures during the process of synthesis and the rings form ends by C–C bonds. MWNTs consist of a few layers of graphene sheets one atom thick, having an external diameter of >10 nm (Su and Cheng, 2014).

Also, the SWNTs are structurally different from MWNTs, having different basic arrangements of the carbon atoms to give three different structural configurations: armchair arrangement where the chiral vector is characterized by the presence of chairs perpendicular to the tube axis; zigzag arrangement where the tube is characterized by having a V-shape perpendicular to the tube axis; and chiral or helical arrangement, which is unidentical from the above two types of arrangements (Luo et al., 2012).

The degree of chirality in CNTs is a representative measure of their electrical as well as conductivity properties and helps in designing a wide variety of nanoelectronics instruments. Apart from this, chirality also affects the diameter of nanotubes as well as its metallic or semimetallic characteristics. MWNTs are further of two subtypes depending upon the pattern of arrangement of the graphitic sheets. One is a Russian–doll model structural arrangement where graphite sheets are arranged in concentric layers, e.g., a sheet of (0, 8) SWNT enclaved within a large diameter (0, 10) SWNTs. The second model is known as a parchment model in which a single sheet of graphite is rolled around itself, resembling a scroll of parchment or a rolled newspaper. Apart from these, there is another type of nanotube resembling SWNTs, known as double-walled nanotubes (DWNTs), having structural similarity with SWNTs; they are of considerable interest in the pharmaceutical field. Additionally, CNTs can be classified into three different prototypic structures depending upon their shapes. These include CNHs, nanobuds, and nanotorus (Beg et al., 2011).

12.7 APPLICATIONS OF CNTs IN DRUG DELIVERY

The concept of practicing safe and effective medicine with a high therapeutic potential is of great focus of the 21st century (Chopdey et al., 2015; Chougule et al., 2014). Recently CNTs have attracted much attention due to their ability to deliver drug molecules to a specific site in a controlled manner. They are used in the controlled release of drugs as well as in delivery of genetic material such as DNA, genes and antibodies. Drugs or biomolecules can be loaded inside the hollow tube or can be directly attached to the walls of CNTs (De-Volder et al., 2013; Dhakad et al., 2013; Dwivedi et al., 2013).

The special characteristic that makes nanotubes promising drug delivery carriers is their hollow monolithic structure having an outer and inner core, which can be modified by the method of functionalization with desired groups on the outer and inner areas. This helps in insertion of required drug molecules in the inner core environment while the outer surface can be modified for achieving biocompatibility and biodegradation (Gajbhiye et al., 2007, 2009). The functionalized CNTs can be used for the purpose of enhanced biocompatibility within the body, enhancement of the encapsulation tendency and solubility, and multimodal drug delivery and imaging (Zhang et al., 2011; Cirillo et al., 2014; Phan et al., 2014). The biomedical applications of CNTs are summarized in Table 12.1 and Fig. 12.4.

12.7.1 Delivery of small molecules

The ability of functionalized CNTs (f-CNTs) to penetrate into the cells offers the potential of using f-CNTs as vehicles for the delivery of small drug molecules. CNTs have gained tremendous attention as promising nanocarriers, owing to their distinct characteristics, such as high surface area, enhanced cellular uptake and the possibility to be easily conjugated with many therapeutics, including both small molecules and biologics, displaying superior efficacy, enhanced specificity and diminished side effects (Garmaroudi and Vahdati, 2010).

12.7.2 Delivery of anticancer molecules

Successful delivery of anticancer drug selectively to its site of action is still an unmet objective of pharmaceutical scientists (Tekade et al., 2008b, 2009). Vibrant research is going on to come up with versatile nanosystems that can efficiently load drugs with wide ranging physicochemical properties, viz: solubility, pKa, surface charge, etc. (Youngren et al., 2013; Tekade, 2014a; Tekade et al., 2014b,c). Most CNTs-based drug delivery systems had been engineered to combat different types of malignancies.

Table 12.1 Tabulation of different types of CNTs explored in the delivery of bioactive molecules

Indication	CNTS type	Functionalization of nanotubes	Bioactive molecule	Objective	Cell line	Inferences	References
Antifungal	MWCNTs	Amphotericin B attached on nanotube surface by conjugation with fluorescein molecules	Amphotericin B	Drug delivery of CNTS for antifungal effect	–	Reduced viable growth of fungus along with significant reduction of drug–related unwanted toxicity	Beg et al. (2011)
PDT	Graphine oxide	Folic acid conjugated Graphine	Ce6	Cancer treatment		Olic acid–conjugated GO loaded Ce6 had great potential as effective drug delivery system in targeting PDT	Huang et al. (2011)
Anticancer	SWCNTs	–	Curcumin	Prostate cancer treatment	PC–3 cells	Significant reduction in the size of prostate cancer	Beg et al. (2011)
Anticancer		Functionalized with ammonium groups and folic acid for site-specific targeting to folate receptors expressed on malignant cells	Cisplatin	Cancer of ovary, testis, lung	–	Better cellular uptake of drug due to increased circulation time in blood	Beg et al. (2011)

Anti-angiogenic agent	SWCNTs	–	Combretastatin	–		Considerable reduction in the growth of cancer cells	Beg et al. (2011)
Anticancer	MWCNTs	–	Carboplatin	Bladder cancer		Higher efficacy of drug-filled CNTs with reduced growth of bladder cancer cells as compared with nanotube-free drugs	Beg et al. (2011)
Excipient	MWCNTs		Cellulose	Cancer treatment		Better yield in productivity	Beg et al. (2011)
	MWCNTs	MWCNTs were functionalized with poly(acrylic acid) conjugated with a targeting ligand folic acid (FA)	Doxorubicin		U87 human glioblastoma cells	A potential candidate for targeted delivery of doxorubicin for cancer treatment	Lu et al. (2012)
Anticancer	SWCNTs	Chitosan and folic acid	Doxorubicin	Targeted DDS based on chitosan and folic acid modified SWCNTs for controllable loading/release of anticancer agent doxorubicin for liver cancer treatment	Hepato-cellular carcinoma SMMC-7721 cell lines	The chitosan–folic acid modified SWCNTs not only kill cancer cells efficiently, but also display much less in vivo toxicity than free doxorubicin	Rastogi et al. (2014)

(Continued)

Table 12.1 Tabulation of different types of CNTs explored in the delivery of bioactive molecules (Continued)

Indication	CNTS type	Functionalization of nanotubes	Bioactive molecule	Objective	Cell line	Inferences	References
Anticancer	SWCNTs	Bovine serum albumin–antibody, fluorescein	Doxorubicin	Triple function-alization of SWCNTs with doxorubicin, a monoclonal antibody and a fluorescent marker for targeted cancer therapy for colon	WiDr Human colon adeno-carcinoma cells	The triple functionalized complex was efficiently taken by cancer cells with subsequent intracellular release of doxorubicin	Rastogi et al. (2014)
Anticancer	SWCNTs	Polysaccharides (sodium alginate (ALG) and chitosan (CH))	Doxorubicin	Targeted delivery and controlled release by doxorubicin to cervical cancer cells using modified CNTs	HeLa cells	The doxorubicin released from the modified nanotubes has been shown to damage nuclear DNA and inhibit the cell proliferation	Rastogi et al. (2014)
Anticancer	SWCNTs	Bovine serum albumin–antibody, fluorescein	Doxorubicin	Triple function-alization of SWCNTs with doxorubicin, a monoclonal antibody and a fluorescent marker for targeted cancer therapy for colon	WiDr Human colon adeno-carcinoma cells	The triple fictionalized complex was efficiently taken by cancer cells with subsequent intracellular release of doxorubicin	Rastogi et al. (2014)

Cancer Sgc8c aptamer Reversible targeting and controlled release delivery of daunorubicin to cancer cells by aptamer-wrapped CNTs	SWCNTs		Daunorubicin	Human T cell leukemia cell MOLT-4 and U266 myeloma cells	The tertiary complex Dau–aptamer–SWCNTs was internalized effectively to MOLT-4 cells but not to U266 cells and is less toxic in U266 as compared to Dau–aptamer–SWCNTs complex is able to selectively target MOLT-4 cells	Rastogi et al. (2014)
Anti-inflammatory	SWCNTs	Drug loaded in oxidized single-walled Nanohorns	Dexamethasone		During in–vitro studies, sustained release profile observed in mouse bone marrow stromal ST2 cells due to induction of alkaline phosphatase level in mouse osteoblastic MC3T3-E1 cells	Beg et al. (2011)

(Continued)

Table 12.1 Tabulation of different types of CNTs explored in the delivery of bioactive molecules (Continued)

Indication	CNTS type	Functionalization of nanotubes	Bioactive molecule	Objective	Cell line	Inferences	References
Imaging		Selective functionalization	Fluorescein			Solubility increased due to partitioning	
Imaging	MWCNTs	Covalently decorating the surface of multiwalled carbon nanotubes (CNTs) by magnetite nanoparticles (Fe_3O_4), poly(ethylene glycol) (PEG)	FITC		Bioimaging	Fluorescent Fe_3O_4–PEG–FITC–CNTS nanosystem an ideal candidate for bioimaging, both in vitro and in vivo	Khandare et al. (2012)
Gene delivery	MWCNTs	Polyamidoamine dendrimer(d)	FITC-labeled antisense c-myc oligonucleotides (as ODN)	Synthesis and characterization of polyamidoamine dendrimer coated MWCNTs and their application in gene delivery systems for liver cancer	MCF-7 and MDA-MB-435 human breast cancer cell line HepG2 liver cancer cell	Laser confocal microscopy confirmed the entry of ODN into the tumor cell, within 15 min of incubation As ODN-dMNTs composites inhibit the cell growth and downregulated the expression of the c-myc gene and C-Myc protein	Rastogi et al. (2014)

Application	CNT type	Functionalization	Drug			Observation	Reference
Anticancer	MWCNTs	Magnetic functionalized CNTs as drug vehicles for cancer lymph node metastasis	Gemcitabine	Lymph node Metastatis	treatment BxPC-3 pancreatic cancer cells	mMWCNTs-GEM had high antitumor activity resulting in successful regression and inhibition of lymph node metastasis under the magnetic field	Rastogi et al. (2014)
Anticancer	MWCNTs	Folic acid (FA) conjugated multiwalled n CNTs	Gemcitabine	Breast cancer treatment	MCF-7 breast cancer cell line	Selectively at the tumor site while minimizing side effects and thus holds promise in chemotherapy	Singh et al. (2013)
Biosensor	SWCNTs	Galactose conjugated CNTs	Gelactin-3			Detection of cancer marker	Park et al. (2011)
Anti-inflammatory action		Functionalized by 3–amino-propyltrie-thoxysilane with CdS quantum dots	Ibuprofen (Drug molecules encapsulated in nanotubes made up of CaCO3 nanoparticles)			Comparative study of these functionalized nanotubes with functionalization-free ones showed a significantly slower release rate and thus a controlled drug delivery	Beg et al. (2011)

(Continued)

Table 12.1 Tabulation of different types of CNTs explored in the delivery of bioactive molecules (Continued)

Indication	CNTS type	Functionalization of nanotubes	Bioactive molecule	Objective	Cell line	Inferences	References
Anticancer		Double functionalization made on its side wall by amino groups by 1,3–dipolar cycloaddition reaction	Methotrexate			Better cellular uptake of drug [47] Methotrexate attached on nanotube surface by conjugation with fluorescein molecules	Rastogi et al. (2014)
Anticancer		Nanotubes pegylated with PEG (polyethylene glycol) moiety on its side wall surface in the form of small branches sprouting from center	Paclitaxel	Breast cell carcinoma		Bioavailability improved due to increased retention time of drug in blood vessels	Beg et al. (2011)
Photothermal therapy and gene therapy	SWCNTs	Polyethylenimine (PEI), Asn–Gly–Arg (NGR	Peptide SiRNA	Synergistic anticancer effect of RNAi and photothermal therapy mediated by functionalized SWCNTs	PC–3 cells Prostate cancer cell line	Human prostate carcinoma (GIV) cell PC–3 cells SWCNTS–PEI-SiRNA–NGR induces severe apoptosis and suppresses the proliferating of PC–3 cells without any level of toxicity	Rastogi et al. (2014)

Anticancer	SWCNTs	Phenosafranine (PS) and Nile blue (NB) dyes		SWCNTs modified with organic dyes: synthesis, characterization, and potential cytotoxic effects	Baby hamster kidney fibroblast cells BHK-21 cell line	Cytotoxicity of dye modified SWCNTS displayed low toxicity in the dark while being higher in the dark and higher in the presence of light	Rastogi et al. (2014)
Gene therapy	SWCNTs	Distearoyl-phosphatidyle-thanolamine (DSPE)-PEG-amine, mouse double minute 2 homolog (MDM2)	SiRNA	Functionalized SWCNTs enables efficient intracellular delivery of SiRNA targeting MDM2 to inhibit breast cancer cells growth	B-Cap-37 breast carcinoma cells	f-SWCNTs showed significant efficiency in carrying SiRNA and SiRNA–MDM2 complexes in B-Cap-37 cells and caused inhibition of proliferating cells	Rastogi et al. (2014)

(Continued)

Table 12.1 Tabulation of different types of CNTs explored in the delivery of bioactive molecules (Continued)

Indication	CNTS type	Functionalization of nanotubes	Bioactive molecule	Objective	Cell line	Inferences	References
Genetherapy	MWCNTs		SiRNA and DNA	Internalization of MWCNTs by microglia: possible application in immunotherapy of brain tumors	BV2 microglia and GL261 glioma cells	Uptake of MWCNTs by both BV2 and GL261 cells in vitro without any significant signs of cytotoxicity (MWNT-COOH)	Rastogi et al. (2014)
Asthma	MWCNTs	Nanohybrid gel mediated delivery system formed from hydrogen bond self-assembly of poly (methacrylic acid) (PMAA) networks and carboxyl-functionalized multiwalled carbon nanotubes	Theophylline			Controlled release profile of drug delivery to lungs	Beg et al. (2011)

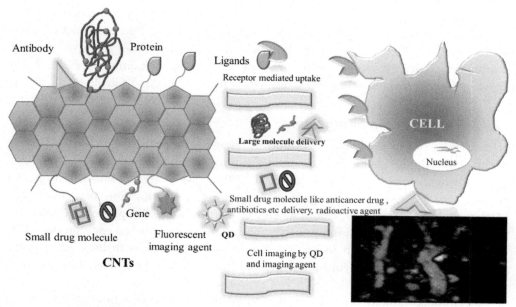

Figure 12.4 Bio-medicinal applications of CNTs.

Reports are available that employed CNTs as either the main carrier or adjunct material for the delivery of various nonanticancer drugs. The delivery of small molecule drugs is illustrated with special attention to the current progress of in vitro and in vivo research involving CNTs-based delivery system, before finally concluding with some consideration on inevitable complications that hamper successful disease intervention with CNTs. CNTs used for targeted delivery is widely accepted in treating malignant disorders such as choriocarcinoma, Burkitt's lymphoma, carcinoma of cervix, breast cancer, and testicular tumors (Sen et al., 2012).

For example, methotrexate in cancer chemotherapy usually showed a low level of cellular uptake due to improper absorption from the gastrointestinal tract. Welsher et al. reported that SWNTs with PEG functionalization and conjugated with the monoclonal antibody rituxan were able to target selectively the CD20 cell surface receptor on B-cells with little nonspecific binding to negative T-cells, and with Herceptin to recognize HER2/neu positive breast cancer cells. This antibody-mediated approach has thus proved to be an ideal alternative in targeting drug molecules to cancer cells (Welsher et al., 2008).

12.7.3 Delivery of antibacterial or antiviral agents

f-CNTs use for the delivery of anticancer, antibacterial or antiviral agents has not yet been fully ascertained. MWNTs were functionalized with amphotericin B and fluorescein. The antibiotic linked to the nanotubes was easily internalized into mammalian cells without toxic effects in comparison with the antibiotic incubated alone (Phan et al., 2014). In addition, amphotericin B bound to CNTs preserved its high antifungal activity against a broad range of pathogens, including *Candida albicans*, *Cryptococcus neoformans* and *Candida parapsilosis* (Zhang et al., 2014; Cirillo et al., 2014).

12.7.4 Delivery of imaging agent

Nanocarriers play a vital role in diagnostic imaging by assisting in the imaging of organs and help in identifying the site of action of drugs in targeted delivery systems (Ghanghoria et al., 2016; Huang et al., 2015). CNTs are shown to have greater potential to act as a contrast agent in imaging and identification of cancer cells. Methotrexate (an anticancer agent), when given with fluorescein probe functionalized nanotubes showed better visibility in the body due to the fluorescence produced by the drug-carrying probe on its surface. CNTs functionalized with fluorescent compounds are used as a radio-opaque substance that produces images of the desired in vivo organs.

Recent developments have shown that nano–imaging of several body structures can be done by administering nanotubes encapsulating miniaturized video systems in the form of a pill. In recent years, research revealed that CNTs can also generate Q-dots, or may behave like a Q-dot, due to their mechanism of coulomb blockade. Hence, the phenomena of electron quantization lead to formation of such types of light emitting nanoparticles. Q-dots emerged as a special tool in imaging because of their continuous light emitting characteristics and this does not fade when exposed to UV light. In addition, Q-dots also help in drug targeting and diagnosing the site of action of the drug molecules attached on it, e.g., the delivery of antisense agents and Q-dots attached on MWNTs (Huang et al., 2015).

Bhirde and coworkers delivered cisplatin and epidermal growth factor (EGF) attached on SWNTs specifically to target squamous cell cancer and compared them with the nontargeted control SWNT–cisplatin without EGF. Imaging studies in head and neck squamous carcinoma cells (HNSCC) overexpressing EGF receptors (EGFR) using Q-dot luminescence and confocal microscopy showed that SWNT-Qdot–EGF bioconjugates internalized rapidly into the cancer cells and vice-versa (Vittorio et al., 2011).

12.7.5 Delivery of neuropharmaceuticals

CNTs have the ability to cross the blood brain barrier and, hence, have attracted much attention for delivery of drug molecules to the brain. Kateb et al. reported that MWNTs are quite effective in delivering neuropharmaceutical agents to the inner environment of brain microglial cells (Kateb et al., 2007). Similar findings were reported by VanHandel et al. upon administering MWNTs by intratumoral injection to GL261 murine intracranial glioma cells for 24 h. Observation showed that 75% of MWNTs were taken up by macrophages of brain tumor cells. These findings suggested that the CNTs-loaded drug could invade the BBB effectively compared to existing techniques. Additionally, nanotubes also have a potential role in treating neurodegenerative disorders due to their magnetic properties (VanHandel et al., 2009).

12.7.6 Delivery of proteins and peptides

The peptides adopt the appropriate secondary structure around the CNTs that helps in its recognition by specific monoclonal and polyclonal antibodies. The immunogenic features of peptide–CNTs conjugates were subsequently assessed in vivo. Immunization of mice with foot and mouth disease virus (FMDV) peptide–nanotube conjugates elicited high antibody responses as compared to the free peptides. Authors studied the application of CNTs as a template for presenting bioactive peptides to the immune system. For this purpose, a B-cell epitope of the FMDV was covalently attached to the amine groups present on CNTs, using a bifunctional linker (Beg et al., 2011).

12.7.7 Vaccine delivery

The use of CNTs as potential novel vaccine delivery tools was validated by interaction with the complements. The complement is that portion of the human immune system composed of a series of proteins responsible for recognizing, opsonizing, clearing and killing pathogens, apoptotic or necrotic cells and foreign materials. Salvador-Morales et al. showed that pristine CNTs activate the complement following both the classical and the alternative way by selective adsorption of some of its proteins. Since complement activation is also involved in immune response to antigens, this might support the enhancement of antibody response following immunization with peptide–CNTs conjugates (Salvador-Morales et al., 2006).

12.7.8 Gene delivery

Gene therapy helps in the replacement of damaged or missing genes and can be improved prominently by using CNTs. The major problem associated with gene delivery is the complication of DNA passing through the cell membrane. In this regard, CNTs help in the transportation of DNA into cells. Researchers have made nanotubes with dendrimers grafted onto the surface for treating gene defects by delivering the grafted genes. Recently, the potential of CNTs as matrices to support and stimulate neural growth has been reported. Prato et al. reported ammonium-functionalized CNTs as a vector for gene-encoding nucleic acids and plasmid DNA to demonstrate the enhancement of gene therapeutic capacity in comparison with DNA alone (Prato et al., 2007).

Yang et al. delivered siRNA complexed with SWNTs for efficient gene delivery. The siRNA and SWNTs complex can be easily taken up by splenic immune recognizing cells such as CD11c$^+$ cells, CD11b$^+$ cells and Gr-1$^+$CD11b$^+$ cells to induce the immune response for a particular gene (Yang et al., 2006).

12.7.9 Nucleic acid and nucleotide delivery

Ammonium-functionalized CNTs were tested for their ability to form supramolecular complexes with nucleic acids via electrostatic interactions. Many cationic systems were being investigated for the delivery of nucleic acids to the cells. Recently, the efficiency of DNA transfer using f-CNTs was increased by covalent modification of the external walls of the tubes with polyethyleneimine (PEI). PEI-grafted MWNT complexed and delivered plasmid DNA to different cell types; however, the measured levels of luciferase expression were similar to that of PEI alone (Liu et al., 2005). CNTs were also used to deliver nonencoding RNA polymers into cells. SWNT condensed RNA by nonspecific binding. Translocation of the complexes between CNTs and poly(rU) RNA polymer into MCF-7 cells was assessed by radioisotope labeling and confocal fluorescence. The hybrids showed negligible toxicity as found by monitoring cell growth CNTs were also used to deliver nonencoding RNA polymers into cells. SWNTs condensed RNA by nonspecific binding. Translocation of the complexes between CNTs and poly(rU) RNA polymer into MCF-7 cells was assessed by radioisotope labeling and confocal fluorescence. The hybrids showed negligible toxicity as found by monitoring cell growth (Bianco et al., 2005).

12.8 BIOSAFETY/TOXICITY PROFILE OF CNTs

In nanomedicine, the most possible toxicological implications of CNTs is generally cytotoxicity and limited control over functionalized-CNTs behavior, both of which restrict predictability (Aschberger et al., 2010). Each CNTs could be intrinsically different due to limitations on the fabrication of structurally identical CNTs with minimal impurities. Subtle variations in local and overall charge, catalyst residue (typically Fe, Co, and Ni), and length of individual nanotubes are three representative issues that preclude precise use of CNTs in the biomedical sciences. It has been estimated that there is a 1 to 3% chance of finding nonhexagonal (seven- or five-membered) rings randomly distributed along a CNTs surface over a length of 4 µm. It has then been postulated that either deficit or excess charge may be present around the odd-membered rings, which causes deviations from neutrality (Garmaroudi and Vahdati, 2010).

Although charge modulation has been exploited for interesting applications, studies have shown that the type of charge and charge density on a functionalized nanotube can affect cellular interaction. For example, the amount of DNA, and the strength with which DNA strands bind onto CNTs depend on nanotube charge density, which varies with fabrication. Such effects can limit predictability (e.g., if CNTs are used for gene therapy). Extrinsic defects, such as catalyst residue, could also be harmful to biomedical application. As Fe and Ni catalysts used for CNTs production can constitute 25 to 40% of the CNTs by weight, these embedded metals can catalyze oxidative species in cells and tissue through free radical generation (Elhissi et al., 2011).

It is generally agreed that f-CNTs constitute a major improvement over unmodified, nonfunctionalized (pristine) CNTs, as the latter are often reported to cause adverse reactions from living tissue, whereas the former could be much less toxic due to more biocompatible functional groups that cytotoxicity was diminished as SWCNTs sidewall functionalization increased; e.g., using phenyl-SO_3H and phenyl-SO_3Na additives, and that even at high concentrations (2 mg/mL) there was insignificant damage to cells. The length and shape of the CNTs influence how well they traverse the membrane of macrophages and determine the resulting immunologic response cytotoxicity of pristine SWCNTs in the liver, spleen, and lungs; it was observed that indicators for oxidative stress due to SWCNTs (e.g., malondialdehyde (MDA) and glutathione (GSH) levels) increased in a dose-dependent manner in the liver and lung, whereas the stress remained relatively constant in the spleen as nanotube dosage increased (Luo et al., 2012).

If certain organs are sensitive to CNTs in different ways, this creates another facet to consider during the search for safe in vivo dosage; tumor cells interact differently with CNTs than do wild-type cells. It was seen that malignant mesothelial cells were able to maintain a dose-dependent increase of stress-response proteins (activator protein1 (AP-1) and nuclear factor kappa-light-chain-enhancer of activated B cells (NF-κB)) as a reaction to increasing toxicity due to longer incubation with CNTs in culture, whereas normal mesothelial cells had higher sensitivity (i.e., produced a higher level of similar proteins initially) but were unable to maintain production under longer incubation (Zhang et al., 2014).

Numerous studies suggest CNTs to be nontoxic in vivo outnumber those proposing otherwise (Johnston et al., 2010). For example, doses of 20 μg diethylene triamine pentaacetic acid (DTPA)-MWCNTS/μL phosphate buffer saline (PBS) and 20 μg DTPA-SWCNTS/μL PBS were administered in different mice intravenously with no acute toxicity observed. As yet another example, an intravenous injection of a 20 μg SWCNTS/kg body weight concentration into specimens confirmed safety of this dosage after a 24 h period. A recent update expanded our understanding of chronic toxicity of CNTs by asserting negligible toxicity in a sample of mice after 4 months of treatment (Pacurari et al., 2010).

New insight arises from the observation that the changes in neutrophil count for mice treated with PEGylated oxidized SWCNTs were larger than counts from those mice treated with PEGylated SWCNTs, which suggests that varying functionalization can modify toxicity. A recent in vivo cancer therapy study using CNTs originally designed as drug delivery enhancers was able to demonstrate that tumor cells respond to toxicity differently than do wild-type cells. SWCNTs conjugated with paclitaxel (a common chemotherapy drug) markedly decreased breast cancer tumors in mice, far more than by using paclitaxel alone (Foldbjerg et al., 2014; Kayat et al., 2011).

12.9 REGULATORY ASPECTS

Awareness of nanotechnology has dramatically risen in recent years among lawmakers, regulators and environmental activists. The major guideline involves the related environmental health and safety issues of existing nanomaterial. As per the reports of the Environmental Protection Agency (EPA), CNTs require major regulatory concern over toxicity as well as environmental safety. A number of recent developments at EPA obligates the manufacturers of CNTs to mandatorily meet their TSCA obligations (Hansen and Baun, 2012).

EPA officials have indicated that they plan to follow through on their previously announced plan to take enforcement action against companies manufacturing or importing CNTs that have not submitted premanufacture notices (PMNs) as required by the Toxic Substances Control Act (TSCA). Along with this threat of enforcement, the EPA has issued Significant New Use Rules (SNURs) for a single and a MWCNTs. The EPA has also indicated that it may issue a section 4(a) test rule for MWCNTs. In 2008, the EPA issued a policy document announcing that, for purposes of the TSCA Inventory, the EPA would classify a nanomaterial as a new chemical substance if the nanomaterial has a molecular identity that is not identical to the molecular identity of a chemical substance already on the TSCA Inventory. The EPA clarified that CNTs regulated under TSCA may be new chemicals with molecular identities distinct from graphite or other allotropes of carbon already listed on the TSCA Inventory (Jacobs et al., 2014).

In recent years, several companies have submitted PMNs to the EPA for various nanomaterials, including dendrimers, CNTs, and fullerenes. According to a recent statement by the Director of the EPA's Chemical Control Division, the EPA has received eleven PMNs and eight Low Release and Exposure (LoREX) exemption applications for CNTs. The director of the EPA indicated that when reviewing PMNs for CNTs, the EPA is focusing on the particular characteristics of each CNT, such as shape, length, and wall thickness. The EPA may consider a CNTs to be a "new chemical" if, for instance, a CNT has a different spatial arrangement of atoms than the CNTs on the TSCA Inventory. Despite the PMNs for nanomaterials that it has received, the EPA remains convinced that there are nanomaterials subject to the premanufacturing requirements for which PMNs have not been submitted. If the EPA issues a section 4(a) test rule for CNTs, the agency will specify whether manufacturers (including importers) or processors or both are subject to the rule. It remains an open question whether the EPA will require the participation of small-volume manufacturers and manufacturers solely for research and development (R&D) purposes (Canu et al., 2016).

Typically, entities that manufacture/import less than 500 kg of a chemical substance annually (i.e., a "small volume") must comply with a test rule only if the test rule specifically so states, or the EPA publishes a notice in the Federal Register that no entity has submitted a notice of intent to conduct a required test. The same limitations usually apply to the participation of R&D manufacturers. Since many manufacturers of CNTs manufacture less than 500 kg annually, and others manufacturer solely for R&D, the EPA may specifically require that such manufacturers comply with a test rule. Persons subject to a test rule must submit a notice of intent to conduct the required testing, or submit an application for an exemption, within 30 days of the effective date of the test rule. Another significant regulatory concern is the potential effects of bioaccumulation due to the high bioavailability of nanotubes.

A recent investigation reports that CNTs forms aggregates that lead to alteration of their basic physicochemical properties. It was suggested to judiciously consider the bulk behavior of these nanomaterials and their counterparts in developing regulatory norms. It is anticipated that definitive knowledge of the development and outline of regulatory guidelines for the production and handling of these nanomaterials will help in resolving the marketing concerns (Beg et al., 2011).

12.10 CONCLUSION AND FUTURE PERSPECTIVE

Since their discovery, research on CNTs has come a long way with sweeping technological innovations in electron microscopy and AFM; various aspects of CNTs have been studied extensively and reported in literature by various research groups across the world. Applications based on CNTs have been demonstrated in the field of sensors, solar cells, thin film transistors and medicine. However, this material is not yet able to become a utility material in our day-to-day life. In this context, the study of this material comes into prominence with issues pertaining to mass production, characterization and implementation in technological applications.

Engineered nanomaterials, particularly CNTs, hold great promise for a variety of industrial, consumer, and biomedical applications, due to their outstanding and novel properties. Over the last two decades many different types of CNTs have been produced at the industrial scale. Therefore, the exposure risk to humans associated with such a mass scale production has also increased substantially. This has led to increased concerns about the potential adverse health effects that may be associated with human exposure to CNTs, predominantly because of their size, their shape, and chemistry. CNTs are also intended for use in many biomedical applications; therefore their biocompatibility, biodistribution, and fate need to be carefully assessed.

REFERENCES

Alpatova, A.L., Shan, W., Babica, P., Upham, B.L., Rogensues, A.R., Masten, S.J., et al., 2010. Single-walled carbon nanotubes dispersed in aqueous media via non-covalent functionalization: effect of dispersant on the stability, cytotoxicity, and epigenetic toxicity of nanotube suspensions. Water Res. 44 (2), 505–520.

Anazawa, K., Shimotani, K., Manabe, C., Watanabe, H., Shimizu, M., 2002. High-purity carbon nanotubes synthesis method by an arc discharging in magnetic field. Appl. Phys. Lett. 81 (4), 739–741.

Ando, Y., Zhao, X., Hirahara, K., Suenaga, K., Bandow, S., Iijima, S., 2000. Mass production of single-wall carbon nanotubes by the arc plasma jet method. Chem. Phys. Lett. 323 (5), 580–585.

Andón, F.T., Kapralov, A.A., Yanamala, N., Feng, W., Baygan, A., Chambers, B.J., et al., 2013. Biodegradation of single walled carbon nanotubes by eosinophil peroxidase. Small 9 (16), 2721–2729.

Antisari, M.V., Marazzi, R., Krsmanovic, R., 2003. Synthesis of multiwall carbon nanotubes by electric arc discharge in liquid environments. Carbon 41 (12), 2393–2401.

Aschberger, K., Johnston, H.J., Stone, V., Aitken, R.J., Hankin, S.M., Peters, S.A., et al., 2010. Review of carbon nanotubes toxicity and exposure—Appraisal of human health risk assessment based on open literature. Crit. Rev. Toxicol. 40 (9), 759–790.

Balbuena, P.B., Zhao, J., Huang, S., Wang, Y., Sakulchaicharoen, N., Resasco, D.E., 2006. Role of the catalyst in the growth of single-wall carbon nanotubes. J. Nanosci. Nanotechnol. 6 (5), 1247–1258.

Beg, S., Rizwan, M., Sheikh, A.M., Hasnain, M.S., Anwer, K., Kohli, K., 2011. Advancement in carbon nanotubes: basics, biomedical applications and toxicity. J. Pharm. Pharmacol. 63 (2), 141–163.

Bianco, A., Hoebeke, J., Godefroy, S., Chaloin, O., Pantarotto, D., Briand, J.P., et al., 2005. Cationic carbon nanotubes bind to CpG oligodeoxynucleotides and enhance their immunostimulatory properties. J. Am. Chem. Soc. 127 (1), 58–59.

Braidy, N., El Khakani, M.M., Botton, G.G., 2002. Carbon nanotubular structures synthesis by means of ultraviolet laser ablation. J. Mater. Res. 17 (09), 2189–2192.

Canu, I.G., Bateson, T.F., Bouvard, V., Debia, M., Dion, C., Savolainen, K., et al., 2016. Human exposure to carbon-based fibrous nanomaterials: a review. Int. J. Hygiene Environ. Health 219 (2), 166–175.

Chang, C.M., Liu, Y.L., 2010. Functionalization of multi-walled carbon nanotubes with non-reactive polymers through an ozone-mediated process for the preparation of a wide range of high performance polymer/carbon nanotube composites. Carbon 48 (4), 1289–1297.

Chen, J., Hamon, M.A., Hu, H., Chen, Y., Rao, A.M., Eklund, P.C., et al., 1998. Solution properties of single-walled carbon nanotubes. Science 282 (5386), 95–98.

Chen, Z., Augustyn, V., Wen, J., Zhang, Y., Shen, M., Dunn, B., et al., 2011. High performance supercapacitors based on intertwined CNT/V2O5 nanowire nanocomposites. Adv. Mater 23 (6), 791–795.

Cheng, H., Liu, C., Cong, H., Liu, M., Fan, Y., Su, G., 2003. U.S. Patent No. 6,517,800. Washington, DC: U.S. Patent and Trademark Office.

Chopdey, P.K., Tekade, R.K., Mehra, N.K., Mody, N., Jain, N.K., 2015. Glycyrrhizin conjugated dendrimer and multi-walled carbon nanotubes for liver specific delivery of doxorubicin. J. Nanosci. Nanotechnol. 15 (2), 1088–1100.

Choudhury, H., Gorain, B., Chatterjee, B., Mandal, U.K., Sengupta, P., Tekade, R.K., 2016. Pharmacokinetic and pharmacodynamic features of nanoemulsion following oral, intravenous, topical and nasal route. Curr. Pharm. Des.

Chougule, M.B., Tekade, R.K., Hoffmann, P.R., Bhatia, D., Sutariya, V.B., Pathak, Y., 2014. Nanomaterial-based gene and drug delivery: pulmonary toxicity considerations. Bioint. Nanomater., 225–248.

Cirillo, G., Hampel, S., Spizzirri, U.G., Parisi, O.I., Picci, N., Iemma, F., 2014. Carbon nanotubes hybrid hydrogels in drug delivery: a perspective review. BioMed Res. Int. 2014, 825017.

De-Volder, M.F., Tawfick, S.H., Baughman, R.H., Hart, A.J., 2013. Carbon nanotubes: present and future commercial applications. Science 339 (6119), 535–539.

Dhakad, R.S., Tekade, R.K., Jain, N.K., 2013. Cancer targeting potential of folate targeted nanocarrier under comparative influence of tretinoin and dexamethasone. Curr. Drug Deliv. 10 (4), 477–491.

Dwivedi, P., Tekade, R.K., Jain, N.K., 2013. Nanoparticulate carrier mediated intranasal delivery of insulin for the restoration of memory signaling in Alzheimer's disease. Curr. Nanosci. 9 (1), 46–55.

Dua, K., Shukla, S.D., Tekade, R.K., Hansbro, P.M., 2016. Whether a novel drug delivery system can overcome the problem of biofilms in respiratory diseases? Drug Deliv. Trans. Res, 1–9.

Eatemadi, A., Daraee, H., Karimkhanloo, H., Kouhi, M., Zarghami, N., Akbarzadeh, A., et al., 2014. Carbon nanotubes: properties, synthesis, purification, and medical applications. Nanoscale Res. Lett. 9 (1), 393.

Ebbesen, T.W., Ajayan, P.M., 1992. Large-scale synthesis of carbon nanotubes. Nature 358 (6383), 220–222.

Elhissi, A., Ahmed, W., Hassan, I.U., Dhanak, V.R., D'Emanuele, A., 2011. Carbon nanotubes in cancer therapy and drug delivery. J. Drug Deliv. 2012, 837327.

Esawi, A.M.K., Morsi, K., Sayed, A., Taher, M., Lanka, S., 2010. Effect of carbon nanotube (CNT) content on the mechanical properties of CNT-reinforced aluminium composites. Comp. Sci. Technol. 70 (16), 2237–2241.

Foldbjerg, R., Irving, E.S., Wang, J., Thorsen, K., Sutherland, D.S., Autrup, H., et al., 2014. The toxic effects of single-walled carbon nanotubes are linked to the phagocytic ability of cells. Toxicol. Res. 3 (4), 228–241.

Gajbhiye, V., Vijayaraj Kumar, P., Tekade, R.K., Jain, N.K., 2007. Pharmaceutical and biomedical potential of PEGylated dendrimers. Curr. Pharm. Design 13 (4), 415–429.

Gajbhiye, V., Palanirajan, V.K., Tekade, R.K., Jain, N.K., 2009. Dendrimers as therapeutic agents: a systematic review. J. Pharm. Pharmacol. 61 (8), 989–1003.

Gandhi, N.S., Tekade, R.K., Chougule, M.B., 2014. Nanocarrier mediated delivery of siRNA/miRNA in combination with chemotherapeutic agents for cancer therapy: current progress and advances. J. Control. Release 194, 238–256.

Garmaroudi, F.S., Vahdati, R.A.R., 2010. Functionalized CNTs for delivery of therapeutics. Int. J.f Nano Dimens. 1 (2), 89–102.

Ghanghoria, R., Tekade, R.K., Mishra, A.K., Chuttani, K., Jain, N.K., 2016. Luteinizing hormone-releasing hormone peptide tethered nanoparticulate system for enhanced antitumoral efficacy of paclitaxel. Nanomedicine (Lond) 11 (7), 797–816.

Ghosh, A., Rao, K.V., Voggu, R., George, S.J., 2010. Non-covalent functionalization, solubilization of graphene and single-walled carbon nanotubes with aromatic donor and acceptor molecules. Chem. Phys. Lett. 488 (4), 198–201.

Gorain, B., Choudhury, H., Tekade, R.K., Karan, S., Jaisankar, P., Pal, T.K., 2016. Comparative biodistribution and safety profiling of olmesartan medoxomil oil-in-water oral nanoemulsion. Regul. Toxicol. Pharmacol. 82, 20–31.

Hansen, S.F., Baun, A., 2012. European regulation affecting nanomaterials-review of limitations and future recommendations. Dose-response 10 (3), 364–383.

Huang, P., Xu, C., Lin, J., Wang, C., Wang, X., Zhang, C., et al., 2011. Folic acid-conjugated graphene oxide loaded with photosensitizers for targeting photodynamic therapy. Theranostics 1 (13), 240–250.

Huang, C.W., Kearney, V., Moeendarbari, S., Jiang, R.Q., Christensen, P., Tekade, R.K., et al., 2015. Hollow gold nanoparticals as biocompatible radiosensitizer: an in vitro proof of concept study. J. Nano Res. 32, 106–112.

Iijima, S., 1991. Helical microtubules of graphitic carbon. Nature 354 (6348), 56–58.

Iijima, S., Yudasaka, M., Yamada, R., Bandow, S., Suenaga, K., Kokai, F., et al., 1999. Nano-aggregates of single-walled graphitic carbon nano-horns. Chem. Phys. Lett. 309 (3), 165–170.

Jacobs, M.M., Ellenbecker, M., Hoppin, P., Kriebel, D., Tickner, J., 2014. Precarious promise: a case study of engineered carbon nanotubes. UMASS, Massachusetts, 7–19.

Jain, N.K., Tekade, R.K., 2013. Dendrimers for enhanced drug solubilization. Drug Deliv. Strateg. Poorly Water-Soluble Drugs, 373–409.

Johnston, H.J., Hutchison, G.R., Christensen, F.M., Peters, S., Hankin, S., Aschberger, K., et al., 2010. A critical review of the biological mechanisms underlying the *in vivo* and *in vitro* toxicity of carbon nanotubes: the contribution of physico-chemical characteristics. Nanotoxicology 4 (2), 207–246.

Karousis, N., Tagmatarchis, N., Tasis, D., 2010. Current progress on the chemical modification of carbon nanotubes. Chem. Rev. 110 (9), 5366–5397.

Kateb, B., Van Handel, M., Zhang, L., Bronikowski, M.J., Manohara, H., Badie, B., 2007. Internalization of MWCNTs by microglia: possible application in immunotherapy of brain tumors. Neuroimage 37, S9–S17.

Kaushik, B.K., Majumder, M.K., 2015. Carbon nanotube: properties and applications. Carbon Nanotube Based VLSI Interconnects. Springer, India.17–37.

Kayat, J., Gajbhiye, V., Tekade, R.K., Jain, N.K., 2011. Pulmonary toxicity of carbon nanotubes: a systematic report. Nanomed. Nanotechnol. Biol. Med. 7 (1), 40–49.

Kesharwani, P., Tekade, R.K., Jain, N.K., 2015. Generation dependent safety and efficacy of folic acid conjugated dendrimer based anticancer drug formulations. Pharm. Res. 32 (4), 1438–1450.

Kim, T.H., Doe, C., Kline, S.R., Choi, S.M., 2007. Water-redispersible isolated single-walled carbon nanotubes fabricated by in situ polymerization of micelles. Adv. Mater. 19 (7), 929–933.

Khandare, J.J., Jalota-Badhwar, A., Satavalekar, S.D., Bhansali, S.G., Aher, N.D., Kharas, F., et al., 2012. PEG-conjugated highly dispersive multifunctional magnetic multi-walled carbon nanotubes for cellular imaging. Nanoscale 4 (3), 837–844.

Kroto, H.W., Heath, J.R., O'Brien, S.C., Curl, R.F., Smalley, R.E., 1985. C 60: buckminsterfullerene. Nature 318 (6042), 162–163.

Kurmi, B.D., Kayat, J., Gajbhiye, V., Tekade, R.K., Jain, N.K., 2010. Micro- and nanocarrier-mediated lung targeting. Exp. Opin. Drug Deliv. 7 (7), 781–794.

Lasjaunias, J.C., Biljaković, K., Benes, Z., Fischer, J.E., Monceau, P., 2002. Low-temperature specific heat of single-wall carbon nanotubes. Phys. Rev. B 65 (11), 113409.

Lee, Y., Geckeler, K.E., 2010. Carbon nanotubes in the biological interphase: the relevance of noncovalence. Adv. Mater. 22 (36), 4076–4083.

Li, H., Kang, Z., Liu, Y., Lee, S.T., 2012. Carbon nanodots: synthesis, properties and applications. J. Mater. Chem. 22 (46), 24230–24253.

Liu, Y., Wu, D.C., Zhang, W.D., Jiang, X., He, C.B., Chung, T.S., et al., 2005. Polyethylenimine-grafted multiwalled carbon nanotubes for secure noncovalent immobilization and efficient delivery of DNA. Angewandte Chemie 117 (30), 4860–4863.

Lu, C., Shao, C., Cobos, E., Singh, K.P., Gao, W., 2012. Chemotherapeutic sensitization of leptomycin B resistant lung cancer cells by pretreatment with doxorubicin. PLoS One 7 (3), 32895.

Luo, M., Deng, X., Shen, X., Dong, L., Liu, Y., 2012. Comparison of cytotoxicity of pristine and covalently functionalized multi-walled carbon nanotubes in RAW 264.7 macrophages. J. Nanosci. Nanotechnol. 12 (1), 274–283.

Ma, P.C., Siddiqui, N.A., Marom, G., Kim, J.K., 2010. Dispersion and functionalization of carbon nanotubes for polymer-based nanocomposites: a review. Compos. Part A: Appl. Sci. Manufact. 41 (10), 1345–1367.

Maheshwari, R., Tekade, R.K., Sharma, P.A., Gajanan, D., Tyagi, A., Patel, R.P., et al., 2012. Ethosomesand ultradeformable liposomes for transdermal delivery of clotrimazole: a comparative assessment. Saudi Pharm. J. 20, 161–170.

Maheshwari, R., Tekade, M., Sharma, P.A., Tekade, R.K., 2015a. Nanocarriers assisted siRNA gene therapy for the management of cardiovascular disorders. Curr. Pharm. Design 21 (30), 4427–4440.

Maheshwari, R., Thakur, S., Singhal, S., Patel, R.P., Tekade, M., Tekade, R.K., 2015b. Chitosan encrusted nonionic surfactant based vesicular formulation for topical administration of ofloxacin. Sci. Adv. Mater. 7 (6), 1163–1176.

Mansuri, S., Kesharwani, P., Tekade, R.K., Jain, N.K., 2016. Lyophilized mucoadhesive-dendrimer enclosed matrix tablet for extended oral delivery of albendazole. Eur. J. Pharm. Biopharm. 102, 202–213.

Mehra, N.K., Mishra, V., Jain, N.K., 2014. A review of ligand tethered surface engineered carbon nanotubes. Biomaterials 35 (4), 1267–1283.

Mittal, G., Dhand, V., Rhee, K.Y., Park, S.J., Lee, W.R., 2015. A review on carbon nanotubes and graphene as fillers in reinforced polymer nanocomposites. J. Ind. Eng. Chem. 21, 11–25.

Mizel, A., Benedict, L.X., Cohen, M.L., Louie, S.G., Zettl, A., Budraa, N.K., et al., 1999. Analysis of the low-temperature specific heat of multiwalled carbon nanotubes and carbon nanotube ropes. Phys. Rev. B 60 (5), 3264.

Mody, N., Tekade, R.K., Mehra, N.K., Chopdey, P., Jain, N.K., 2014. Dendrimer, liposomes, carbon nanotubes and PLGA nanoparticles: one platform assessment of drug delivery potential. Aaps Pharmscitech 15 (2), 388–399.

Moeendarbari, S., Tekade, R.K., Mulgaonkar, A., Christensen, P., Ramezani, S., Hassan, G., et al., 2016. Theranostic nanoseeds for efficacious internal radiation therapy of unresectable solid tumors. Scient. Rep. 6 (20614), 1–9.

Moghadam, A.D., Omrani, E., Menezes, P.L., Rohatgi, P.K., 2015. Mechanical and tribological properties of self-lubricating metal matrix nanocomposites reinforced by carbon nanotubes (CNTs) and graphene–a review. Compos. Part B: Eng. 77, 402–420.

Mulvey, J.J., Feinberg, E.N., Alidori, S., 2015. Synthesis, pharmacokinetics, and biological use of lysine-modified single-walled carbon nanotubes [Erratum]. Int. J. Nanomed. 10, 595.

Ohkohchi, M., Ando, Y., Bandow, S., SAIro, Y., 1993. Formation of carbon nanotubes by evaporation of carbon rod containing scandium oxide. Jpn. J. Appl. Phys. 32 (9A Pt 2) L107-L109.

Pacurari, M., Castranova, V., Vallyathan, V., 2010. Single-and multi-wall carbon nanotubes versus asbestos: are the carbon nanotubes a new health risk to humans? J. Toxicol. Environ. Health, Part A 73 (5-6), 378–395.

Park, Y.K., Bold, B., Lee, W.K., Jeon, M.H., An, K.H., Jeong, S.Y., et al., 2011. D-(+)-galactose-conjugated single-walled carbon nanotubes as new chemical probes for electrochemical biosensors for the cancer marker galectin-3. Int. J. Mol. Sci. 12 (5), 2946–2957.

Phan, N.M., Bui, H.T., Nguyen, M.H., Phan, H.K., 2014. Carbon-nanotube-based liquids: a new class of nanomaterials and their applications. Adv. Nat. Sci. Nanosci. Nanotechnol. 5 (1), 015014.

Prajapati, R.N., Tekade, R.K., Gupta, U., Gajbhiye, V., Jain, N.K., 2009. Dendrimer-mediated solubilization, formulation development and in vitro-in vivo assessment of piroxicam. Mol. Pharm. 6 (3), 940–950.

Prato, M., Kostarelos, K., Bianco, A., 2007. Functionalized carbon nanotubes in drug design and discovery. Accoun. Chem. Res. 41 (1), 60–68.

Rastogi, V., Yadav, P., Bhattacharya, S.S., Mishra, A.K., Verma, N., Verma, A., et al., 2014. Carbon nanotubes: an emerging drug carrier for targeting cancer cells. J. Drug Deliv. 2014, 670815.

Riviere, J.E., 2009. Pharmacokinetics of nanomaterials: an overview of carbon nanotubes, fullerenes and quantum dots. Wiley Interdiscipl. Rev. Nanomed. Nanobiotechnol. 1 (1), 26–34.

Salvador-Morales, C., Flahaut, E., Sim, E., Sloan, J., Green, M.L., Sim, R.B., 2006. Complement activation and protein adsorption by carbon nanotubes. Mol. Immunol. 43 (3), 193–201.

Sen, T., Sheppard, S.J., Mercer, T., Eizadi-Sharifabad, M., Mahmoudi, M., Elhissi, A., 2012. Simple one-pot fabrication of ultra-stable core-shell superparamagnetic nanoparticles for potential application in drug delivery. RSC Adv. 2 (12), 5221–5228.

Sharma, P., Maheshwari, R., Tekade, M., Kumar Tekade, R., 2015. Nanomaterial based approaches for the diagnosis and therapy of cardiovascular diseases. Curr. Pharm. Design 21 (30), 4465–4478.

Sheng, G.D., Shao, D.D., Ren, X.M., Wang, X.Q., Li, J.X., Chen, Y.X., et al., 2010. Kinetics and thermodynamics of adsorption of ionizable aromatic compounds from aqueous solutions by as-prepared and oxidized multiwalled carbon nanotubes. J. Hazardous Mater. 178 (1), 505–516.

Singh, R., Mo, Y.Y., 2013. Role of microRNAs in breast cancer. Cancer Biol. Ther. 14 (3), 201–212.

Sinha, N., Yeow, J.W., 2005. Carbon nanotubes for biomedical applications. IEEE Trans. Nanobiosci. 4 (2), 180–195.

Smalley, R.E., 2003. In: Dresselhaus, M.S. Dresselhaus, G. Avouris, P. (Eds.), Carbon Nanotubes: Synthesis, Structure, Properties, and Applications, Vol. 80. Springer Science & Business Media, New York.

Soni, N., Soni, N., Pandey, H., Maheshwari, R., Kesharwani, P., Tekade, R.K., 2016. Augmented delivery of gemcitabine in lung cancer cells exploring mannose anchored solid lipid nanoparticles. J. Colloid Interface Sci. 481, 107–116.

Spitalsky, Z., Tasis, D., Papagelis, K., Galiotis, C., 2010. Carbon nanotube–polymer composites: chemistry, processing, mechanical and electrical properties. Prog. Polymer Sci. 35 (3), 357–401.

Stoddart, J.F., Star, A., Liu, Y., Ridvan, L. Noncovalent functionalization of Nanotubes-Patent. 2007. US20077220818.

Su, W.C., Cheng, Y.S., 2014. Carbon nanotubes size classification, characterization and nasal airway deposition. Inhal. Toxicol. 26 (14), 843–852.

Takahashi, S., Ikuno, T., Oyama, T., Honda, S.I., Katayama, M., Hirao, T., et al., 2002. Synthesis and characterization of carbon nanotubes grown on carbon particles by using high vacuum laser ablation. J. Vac. Soc. Jpn. 45 (7), 609.

Tekade, R.K., 2014a. Editorial: contemporary siRNA therapeutics and the current state-of-art. Curr. Pharm. Design 21 (31), 4527–4528.

Tekade, R.K., Kumar, P.V., Jain, N.K., 2008a. Dendrimers in oncology: an expanding horizon. Chem. Rev. 109 (1), 49–87.

Tekade, R.K., Dutta, T., Tyagi, A., Bharti, A.C., Das, B.C., Jain, N.K., 2008b. Surface-engineered dendrimers for dual drug delivery: a receptor up-regulation and enhanced cancer targeting strategy. J. Drug Target 16 (10), 758–772.

Tekade, R.K., Dutta, T., Gajbhiye, V., Jain, N.K., 2009. Exploring dendrimer towards dual drug delivery: pH responsive simultaneous drug-release kinetics. J. Microencapsul. 26 (4), 287–296.

Tekade, R.K., Chougule, M.B., 2013a. Formulation development and evaluation of hybrid nanocarrier for cancer therapy: taguchi orthogonal array based design. BioMed Res. Int. 2013. Article ID 712678.

Tekade, R.K., D'Emanuele, A., Elhissi, A., Agrawal, A., Jain, A., Arafat, B.T., et al., 2013b. Extraction and RP-HPLC determination of taxol in rat plasma, cell culture and quality control samples. J. Biomed. Res. 27 (5), 394–405.

Tekade, R., Xu, L., Hao, G., Ramezani, S., Silvers, W., Christensen, P., et al., 2014b. A facile preparation of radioactive gold nanoplatforms for potential theranostic agents of cancer. J. Nuclear Med. 55 (1), 1047.

Tekade, R.K., Youngren-Ortiz, S.R., Yang, H., Haware, R., Chougule, M.B., 2014c. Designing hybrid onco-nase nanocarriers for mesothelioma therapy: a Taguchi orthogonal array and multivariate component driven analysis. Mol. Pharm. 11 (10), 3671–3683.

Tekade, R.K., Youngren-Ortiz, S.R., Yang, H., Haware, R., Chougule, M.B., 2015a. Albumin-chitosan hybrid onconase nanocarriers for mesothelioma therapy. Cancer Res. 75 (15 Suppl.), 3680.

Tekade, R., Maheshwari, Rahul, G.S., Sharma, P., Tekade, M., Singh Chauhan, A., 2015b. siRNA therapy, challenges and underlying perspectives of dendrimer as delivery vector. Curr. Pharm. Design 21 (31), 4614–4636.

Tekade, R.K., Tekade, M., Kumar, M., Chauhan, A.S., 2015c. Dendrimer-stabilized smart-nanoparticle (DSSN) platform for targeted delivery of hydrophobic antitumor therapeutics. Pharm. Res. 32 (3), 910–928.

Tekade, R.K., Tekade, M., Kesharwani, P., D'Emanuele, A., 2016. RNAi-combined nano-chemotherapeutics to tackle resistant tumors. Drug Discovery Today.

Thess, A., Lee, R., Nikolaev, P., Dai, H., 1996. Crystalline ropes of metallic carbon nanotubes. Science 273 (5274), 483.

VanHandel, M., Alizadeh, D., Zhang, L., Kateb, B., Bronikowski, M., Manohara, H., et al., 2009. Selective uptake of multi-walled carbon nanotubes by tumor macrophages in a murine glioma model. J. Neuroimmunol. 208 (1), 3–9.

Vittorio, O., Cuschieri, A., Duce, S.L., Raffa, V., 2011. Imaging and Biomedical Application of Magnetic Carbon Nanotubes. INTECH Open Access Publisher. 189–210

Wang, Z.L., (Ed.), 2013. Nanowires and nanobelts: materials, properties and devices, Metal and Semiconductor Nanowires, Vol. 1. Springer Science & Business Media, New York.

Wang, S.D., Chang, M.H., Lan, K.M.D., Wu, C.C., Cheng, J.J., Chang, H.K., 2005. Synthesis of carbon nanotubes by arc discharge in sodium chloride solution. Carbon 43 (8), 1792.

Welsher, K., Liu, Z., Daranciang, D., Dai, H., 2008. Selective probing and imaging of cells with single walled carbon nanotubes as near-infrared fluorescent molecules. Nano Lett. 8 (2), 586–590.

Wise, K.E., Park, C., Kang, J.H., Siochi, E.J., Harrison, J.S., 2006. U.S. Patent Application No. 11/644,019.

Wong, E.W., Sheehan, P.E., Lieber, C.M., 1997. Nanobeam mechanics: elasticity, strength, and toughness of nanorods and nanotubes. Science 277 (5334), 1971–1975.

Yang, R., Yang, X., Zhang, Z., Zhang, Y., Wang, S., Cai, Z., et al., 2006. Single-walled carbon nanotubes-mediated in vivo and in vitro delivery of siRNA into antigen-presenting cells. Gene Therapy 13 (24), 1714–1723.

Yao, X., Li, J., Kong, L., Wang, Y., 2015. Surface functionalization of carbon nanotubes by direct encapsulation with varying dosages of amphiphilic block copolymers. Nanotechnology 26 (32), 325601.

Yi, W., Lu, L., Dian-Lin, Z., Pan, Z.W., Xie, S.S., 1999. Linear specific heat of carbon nanotubes. Phys. Rev. B 59 (14), R9015.

Youngren, S.R., Tekade, R.K., Gustilo, B., Hoffmann, P.R., Chougule, M.B., 2013. STAT6 siRNA matrix-loaded gelatin nanocarriers: formulation, characterization, and ex vivo proof of concept using adeno-carcinoma cells. Biomed. Res. Int. 2013 (2013), 858946.

Zhang, Q. (Ed.), 2012. Carbon Nanotubes and Their Applications. CRC Press, Boca Raton, FL.

Zhang, W., Zhang, Z., Zhang, Y., 2011. The application of carbon nanotubes in target drug delivery systems for cancer therapies. Nanoscale Res. Lett. 6 (1), 1.

Zhang, Q., Huang, J.Q., Qian, W.Z., Zhang, Y.Y., Wei, F., 2013. The road for nanomaterials industry: a review of carbon nanotube production, post treatment, and bulk applications for composites and energy storage. Small 9 (8), 1237–1265.

Zhang, Y., Petibone, D., Xu, Y., Mahmood, M., Karmakar, A., Casciano, D., et al., 2014. Toxicity and efficacy of carbon nanotubes and graphene: the utility of carbon-based nanoparticles in nanomedicine. Drug Metabol. Rev. 46 (2), 232–246.

Zhao, Y.L., Stoddart, J.F., 2009. Noncovalent functionalization of single-walled carbon nanotubes. Accoun. Chem. Res. 42 (8), 1161–1171.

CHAPTER 13

Quantum Dots in Targeted Delivery of Bioactives and Imaging

Vijay Mishra, Ekta Gurnany and Mohammad H. Mansoori

Contents

Disclosures: There is no conflict of interest and disclosures associated with this manuscript.

13.1 INTRODUCTION

In the current era, nanotechnology is a rapidly expanding field owing to its multidisciplinary efforts from distinguished scientists and researchers. In the last two decades, outstanding progress has been noticed in the drug delivery and imaging areas using various nanovectors in the pharmaceutical domain. For example, polymer-based liposomes on active ingredients with fluorescence, micro, and nanoparticles have been the subject of intense research and development. In the present scenario, formulations

Nanotechnology-Based Approaches for Targeting and Delivery of Drugs and Genes.
DOI: http://dx.doi.org/10.1016/B978-0-12-809717-5.00015-4

427

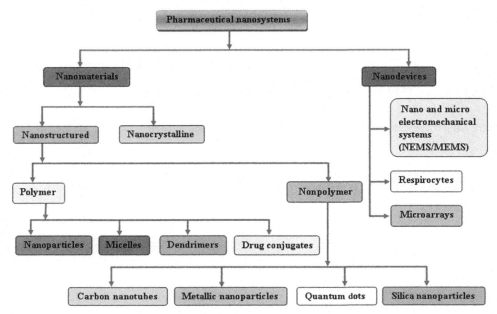

Figure 13.1 Various types of pharmaceutical nanosystems.

developed by the pharmaceutical industry are in abundance, though their efficacy is limited because of decreased bioavailability due to low aqueous solubility and cell membrane permeability. Conventional formulations have some limitations like unacceptable release pattern of drugs, poor solubility, and toxicity (Uehara et al., 2009).

The use of nanotechnology for developing various smart and intelligent nanocarriers and devices (Fig. 13.1) with well-defined shapes and sizes may resolve these limitations and help in the development of safe and effective nanomedicines (Duncan and Vicent, 2013). The solubility of the drug, its release at the site of disease, and reduced nonspecific toxicity can be easily altered through formulations using nanotechnology.

Various nanoparticles such as nanowires, quantum dots (QDs), nanorods, carbon nanotubes, dendrimers, or nanofilms have been intensively investigated due to their possible applications in several extremely important fields, e.g., in catalysis, coatings, textiles, data storage, biotechnology, health care, biomedical, and pharmaceutical industries. Concerning the medical applications, nanoparticles are one of the most widely developed and utilized carriers, which can be prepared using metallic, metal oxide-ceramic, polymer, carbon, core-shell, alloy, and biological components (Kumar, 2007).

QDs are nanosized radiant semiconductor crystals that have the ability to emit light when any source of energy excites their electrons (Abbasi et al., 2015; Bera et al., 2010). QDs constitute the part of technological future exhibiting various unique physicochemical properties due to their small size and highly condensed structure that are desirable for the imaging purposes including live-cell and whole-animal imaging as well as targeted drug delivery.

13.2 HISTORICAL BACKGROUND OF QUANTUM DOTS

QD is a conducting island of a size comparable to the Fermi wavelength in all spatial directions. Often called artificial atoms, however, the size is much bigger (100 nm for QDs vs 0.1 nm for atoms). In atoms the attractive forces are exerted by the nuclei, while in QDs by background charges. The number of electrons in atoms can be tuned by ionization, while in QDs by changing the confinement potential.

QDs were discovered in solids (glass crystals) in 1980 by the Russian physicist Alexei Ekimov; later, in 1982, the American chemist Louis E. Brus discovered the QDs in colloidal solutions (Zhu et al., 2013; Tokihiro and Hanamura, 1984). In the 1970s the first low-dimensional structures QW (quantum wells) were developed. 1D (quantum wires) and 0D (QDs) with continuous and discrete states of density, respectively, were subsequently developed. Canham (1990) reported the efficient light emission from silicon (Si). After that, QDs have been used in drug delivery and targeting as well as for diagnostic and imaging purposes. Additionally, QDs are used for sensing of DNA and oligonucleotides (Mitchell et al., 1999). Jaiswal et al. (2004) worked on the yield enhancement and photo-stability of QDs. Later on QD fluorescence quenching technique was employed for optical DNA and oligonucleotide sensors (Jin et al., 2005). High photochemical stability of core/shell QDs was proposed in 2008 as an alternative organic dye (Resch-Genger et al., 2008). In 2010, role of functional group for bio receptor immobilization and to anticipate toxicity issues was proposed (Biju et al., 2010).

13.3 DESIGN OF QUANTUM DOTS

QDs can be designed to have emission peaks at diverse wavelengths by adjusting their size (Misra, 2008). The conductivity of these semiconductors lies between that of distinct molecules and bulk semiconductors (Moriyama et al., 2005; Chakravarthy et al., 2011). On changing the size and shape of individual crystals, the conductive properties can be altered. The crystal size is inversely proportional to the band gap. The smaller the crystal size, the larger the band gap will be and hence the greater will be the difference in energy between the highest valence band and the lowest conduction band (Maksym and Chakraborty, 1990). Hence, the energy required to excite the dot will be more and therefore more energy is released when the crystal returns to its resting state (Minnich et al., 2009). As the crystal size gets smaller, there is a color shift from red to blue in the light emitted.

Broadly, QDs consist of three parts i.e., core, shell, and cap (Fig. 13.2). Core is made up of semiconductor material. Shell is the coat, which surrounds the semiconductor

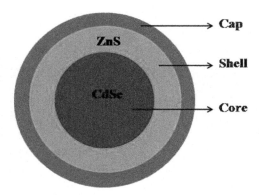

Figure 13.2 Typical components of quantum dot.

core for improving its optical properties and cap encapsulates the double layer of QDs by different materials like silica, which improves the solubility in aqueous buffers (Varshney and Shailender, 2012; Ghasemi et al., 2009). The QDs designed for biological applications have four basic steps: core synthesis, shell growth, aqueous solubilization, and biomolecular conjugation.

QDs cores are composed of semiconductors belonging to groups II–IV (Cadmium selenide (CdSe), Cadmium sulfide (CdS), Cadmium telluride (CdTe)), group IV–VI (Lead sulfide (PbS), Lead selenide (PbSe), Lead telluride (PbTe), Tin telluride (SnTe)), and group III–V (Indium phosphide (InP)) of the Periodic Table. The most common method for preparation of QDs core consists in the rapid injection of semiconductor precursors into a hot and vigorously stirred specific coordinating solvent like trioctylphosphine oxide (TOPO) or trioctylphosphine (TOP). Coordinating solvents stabilize the bulk semiconductors and prevent aggregation as the QDs grow (Talapin et al., 2001). The semiconductor core material must be protected from degradation and oxidation to optimize QDs performance.

Shell growth and surface modification enhance stability and increase photoluminescence of the core. The inorganic core-shell semiconductor nanoparticles are soluble in nonpolar solvents only. The most widely studied QDs consist of a core of CdSe or CdTe, because their quantum confinement region spans the entire optical spectrum (Gerion, 2006).

13.4 PROPERTIES OF QUANTUM DOTS

13.4.1 Physicochemical properties

QDs are semiconductor nanocrystals that possess unique optical properties including broad-range excitation, size-tunable narrow emission spectra, and high photo-stability, giving them considerable value in various applications (Kroutvar et al., 2004). The size and composition of QDs can be varied to obtain the desired emission properties and make them amenable to simultaneous detection of multiple targets. The general physicochemical properties of QDs are as follows:

1. QDs are more resistant to degradation than other optical imaging probes and hence allow tracking of the cellular process for longer period of time (Dabbousi et al., 1997).

2. They have a longer-lasting photo-stability than traditional inorganic dyes and fluorescence intensity (Ghasemi et al., 2009).

3. QDs have longer fluorescence and a significantly high photo resistance (Dubertret et al., 2002).

4. QDs are 10–20 times brighter than other organic dyes. QDs are stable fluorophores due to their inorganic composition, which reduces the effect of photo-bleaching as compared to organic dyes (Dubertret et al., 2002).

5. QDs have broader excitation spectra, large stokes shift, and sharp emission spectra with narrow emission peaks (Bera et al., 2010; Ghasemi et al., 2009).

6. QDs could be easily molded into any shape and coated with a variety of biomaterials (Kroutvar et al., 2004).

7. QDs are nanocrystals and provide better contrast with electron microscope as scattering increases (Hawrylak, 1999).

8. QDs have novel optical and electronic properties due to quantum confinement of electrons and photons in the nanostructure (Huang and Ren, 2011). Quantum confinement results in a widening of band gap (gap between valence and conduction band), which increases when the size of the nanostructure is decreased, hence QDs of the same material with different sizes emit different colors (Phillips, 2002).

9. QDs absorb photons when excitation energy exceeds band gap, and after absorbing that energy, electrons jump from the ground state to the excited state. The energy associated with optical absorption is directly related with the electronic structure of material. The excess energy results in recombination, and relaxation, which may be radiative or nonradiative (Maksym and Chakraborty, 1990; Kroutvar et al., 2004). Radiative relaxation causes spontaneous luminescence from QDs.

10. Optical properties can be influenced by varying different aspects of the QDs like core size and core composition, shell composition, and surface coating (Potasek et al., 2005). However, the core size and composition have the most influence over the range of the emission spectra (Bailey and Nie, 2003).

13.4.2 Merits of quantum dots

QDs demonstrate various meritorious attributes (Bakalova et al., 2004; Sattler, 2010; Resch-Genger et al., 2008), which are as follows:

1. *Physical stability*: QDs are more resistant to degradation than other optical imaging probes, allowing them to track cell processes for longer periods of time.

2. *Photo-stability*: QDs have greater photo-stability than traditional inorganic dyes and its fluorescence intensity do not diminish with time while organic dyes lose their intensities in 20 s.
3. *Signal-to-noise ratio*: QDs have high signal to noise (S/N) ratio as compared to organic dyes.
4. *Broader excitation and narrow emission*: QDs have broader excitation spectra and a narrow more sharply defined emission peak. Due to these properties, a single light source can be used to excite multicolor QDs simultaneously without overlap.
5. *Brightness*: The brightness of QDs compared to organic dyes is 10–20 times brighter.
6. *Fluorescent lifetime*: QDs are highly photo-resistant with significantly longer fluorescence lifetimes. Researchers can use their intense fluorescence to track individual molecules.
7. *Excitation by single or multiple sources*: QDs can be excited by the same source and multicolor QDs allow the use of many probes to track several targets in vivo simultaneously.
8. *Sensitive and precise*: Due to their large stokes shift and sharp emission spectra, QDs conjugates have a high signal intensity with minimal background interference.
9. *Shape flexibility*: QDs can be molded into different shapes and coated with a variety of biomaterials.
10. *Imaging agent*: QDs provide good contrast for imaging under electron microscope due to an increasing scattering nature.

13.4.3 Demerits of quantum dots

QDs may also show the following drawbacks (Pandurangan and Mounika, 2012):
1. QDs show a high reaction rate.
2. QDs demonstrated a poorly controlled growth rate and difficulties in high-throughput synthesis.
3. QDs, when positioned in live cells, may kill the cells due to aggregation.
4. The surface defects of QDs affect the recombination of electrons and holes by acting as temporary traps, which results in blinking and a decrease in the yield of QDs.
5. Biconjugation of QDs leads to delivery into the target difficult.
6. Building material of the QDs can be cytotoxic, e.g., CdSe.
7. The metabolism and excretion of QDs is unknown so the accumulation in body tissues can lead to toxicity.

13.5 METHODS OF QUANTUM DOTS FABRICATION

Originally, the QDs grown were semiconductors such as CdSe or CdTe, but now researchers have made QDs of nearly every semiconductor and also of many metals such as gold, silver, nickel, and cobalt, and insulators as well. However, the semiconductor dots are used in biological imaging applications while magnetic metal dots show promise as recording media. There are several methods for fabrication of QDs. The chemical synthesis of QDs represents a typical approach, which is generally divided into organic and water-phase approaches (Bandyopadhyay et al., 1999; Mishra et al., 2011).

13.5.1 Organic-phase method

13.5.1.1 Colloidal-synthesis method

There is a great interest in colloidal semiconductor QDs due to size-specific properties and flexible processing chemistry. This method is a potentially low-cost mass-production process that creates QDs in a liquid. This cheapest, less toxic method occurs at bench top condition. The size of QDs can be controlled in this process by the time duration that QDs remain in the solution.

Colloidal semiconductor nanocrystals are synthesized from precursor compounds dissolved in solutions. The temperature during the growth process is one of the critical factors in determining optimal conditions for the nanocrystal growth. It must be high enough to allow for rearrangement and annealing of atoms during the synthesis process while being low enough to promote crystal growth. Another critical factor that has to be stringently controlled during nanocrystal growth is the monomer concentration. The growth process of nanocrystals can occur in two different regimes, "focusing" and "defocusing." At high monomer concentrations, the critical size where nanocrystals neither grow nor shrink is relatively small, resulting in growth of nearly all crystals resulting in "focusing" of the size distribution to yield practically monodispersed particles. The size focusing is optimal when the monomer concentration is kept, such that the average nanocrystal size present is always slightly larger than the critical size. When the monomer concentration is depleted during growth, the critical size becomes larger than the average size present, and the distribution "defocuses" as a result of Ostwald ripening.

The synthesis of colloidal QDs is based on a three-component system composed of precursors, organic surfactants, and solvents. A reaction medium is heated to a sufficiently high temperature (e.g., 300°C), and under vigorous stirring the precursors are injected through with a syringe, which chemically transform into monomers. Once the monomers reach a high enough super-saturation level, the nanocrystal growth starts a nucleation process. The solution immediately begins to change from colorless to colors like yellow, orange, and red/brown, as the QDs increase in size by placing them under a black light. This reaction results in the formation of mono-dispersed QDs. Surfactant is used to avoid aggregation, which renders them water-soluble. Typically, these are made of binary alloys such as CdSe, CdS, InP, InAs (Indium arsenide). In addition, QDs may also be made from ternary alloys such as cadmium selenide sulfide (CdSeS). The reaction between cadmium oxide (CdO) dissolved in oleic acid (OA) and selenium (Se) dissolved in TOPO resulted in fabrication of

QDs, which contain as few as 100 to 100,000 atoms within the QD volume, with a diameter of 10–50 atoms, which corresponds to about 2–10 and 10 nm in diameter (Pandurangan and Mounika, 2012).

Li et al. described the raw materials and synthesizing processes required for fabrication of CdSe/CdS QDs (Li et al., 2003). Materials used to synthesize the CdSe/CdS core/shell QDs include CdO, Se powder, tributylphosphine (TBP), sulfur powder, TOPO, 1-octadecene (ODE), OA, octyldecylamine (ODA), and stearic acid (SA). For synthesis of CdSe core nanocrystals, mixtures of CdO, SA, and ODE are first heated up to 200°C to produce a colorless solution. After cool-down to room temperature, ODA and TOPO are added and then heated up to 280°C with argon (Ar) flow. A solution of dissolved Se in TBP and ODE is then added and CdSe nanocrystals of size 3.5 nm are precipitated. The second step is fabrication of CdS shell on the core surface. Solutions of CdO dissolved in oleic acid and sulfur dissolved in ODE are sequentially added to finish the core/shell structure (Li et al., 2003). Washing with methanol and hexane and centrifuging are the two methods used to purify and extract QDs (Li et al., 2003; Asokan et al., 2005).

13.5.1.2 Electron beam lithography method

Growth of the QDs in a semiconductor heterostructure refers to a plane of one semiconductor sandwiched between two other semiconductors. If this sandwiched layer is very thin, i.e., about 10 nm or less, then the electrons can no longer move vertically and thus are confined to a particular dimension. This is called the quantum well. When a thin slice of this material is taken to create a narrow strip, then it results in a quantum wire, as it gets trapped in a 2D area. Rotating this to 90 degrees and repeating the procedure results in the confinement of the electron in a 3D, which is called QD. According to quantum mechanics and Heisenberg's uncertainty principle, the more confined an electron is, the more uncertain its momentum is and hence the range of momentum is wider in addition to higher energy that is possessed by the electron. The electrons confined in an electron wire are free only in 1D. However, those confined in a QD are not free in any dimension (Mishra et al., 2012).

13.5.1.3 Molecular beam epitaxy method

This method involves the layer-by-layer deposition of crystals on a substrate. Pure elements are evaporated, then condensed and combined on the surface of the substrate under a vacuum condition. The QDs fabricated by this method are used for quantum computation, but the cost is high and the positioning of dots can't be controlled. For indium arsenic/gallium arsenic (InAs/GaAs) systems, In and As atoms are deposited on a GaAs substrate to form an epitaxial layer. A large-band gap material, such as InGaAs, is then deposited on top of the InAs epitaxial layer (Atkinson et al., 2006). Through postdeposition heat treatment, QDs are formed by self-assembly.

Self-assembled QDs can also be grown by depositing a semiconductor with larger lattice constant on a semiconductor with smaller lattice constant, e.g., Germanium on Si. These self-assembled QDs are then used to make QD lasers. Hence, the QDs are actually formed when very thin semiconductor films collapse due to stress on the lattice structure that's slightly different in size from those on which the films are grown.

13.5.1.4 Combined electron beam lithography and molecular beam epitaxy method

In this combined approach the electron beam is used to etch a pattern onto the substrate, and then molecular beam epitaxy is used to deposit QDs materials on top. Usually the pattern is an array of holes, and these holes act as preferential nucleation sites for dots. The cost is high, but the position of dots can be controlled (Atkinson et al., 2006).

The QDs, fabricated by an organic phase method, are generally capped with hydrophobic ligands like TOPO or TOP and hence cannot be directly employed in bioapplications (Mishra et al., 2012).

13.5.2 Water phase method

For the biological applications, QDs need to be soluble in aqueous solutions and require surface modifications to achieve biocompatibility and stability (Mishra et al., 2012). Water phase (aqueous synthesis) method is effective, less toxic, and a more reproducible method as compared to the organic approach. Furthermore, the products often show improved water solubility, biological compatibility as well as stability. The water phase method can be categorized as follows.

13.5.2.1 Cap exchange method

The hydrophobic layer of organic solvent can be replaced with bifunctional molecules containing a soft acidic group usually a thiol, e.g., mercaptoacetic acid, mercaptopropionic acid, mercaptoundecanoic acid or reduced glutathione (GSH), sodium thioglycolate, and hydrophilic groups, e.g., carboxylic or amino groups, which point outwards from the surfaces of QDs to bulk water molecules. In fact, substitution of monothiols by polythiols or phosphines usually improves stability. From these ligands, GSH seems to be a very perspective molecule, since it provides an additional functionality to the QDs due to its key function in detoxification of heavy metals in organisms.

13.5.2.2 Native surface modification

The addition of a silica shell to the nanoparticles using a silica precursor during the polycondensation rendered the QDs water soluble using several synthesis strategies, such as water-soluble ligands, organic dendrons, cysteines, dihydrolipoic acid, encapsulation with block-copolymer micelles, with amphiphilic polymers, amphiphilic polymers conjugated with polyethylene glycol (PEG), and surface coating with phytochelatin-related peptides. All these synthesis strategies have effectively solubilized CdSe or CdSe/ZnS QDs. In addition, QDs can be conjugated to biological molecules such as proteins and oligonucleotides, which are used to direct binding of the QDs to areas of interest for bio-labeling and bio-sensing (Ghasemi et al., 2009; Yuan et al., 2009; Zhang et al., 2012).

13.6 CHARACTERIZATION TECHNIQUES OF QUANTUM DOTS

QDs can be characterized by employing different sophisticated techniques like (1) spectroscopy techniques, e.g., ultra-violet (UV), infra-red (IR), nuclear magnetic resonance, X-ray diffraction, mass spectrometry, and Raman spectroscopy; (2) scattering techniques, e.g., small-angle neutron scattering, laser-light scattering; (3) microscopy, e.g., atomic-force microscopy, scanning electron microscopy, and transmission–electron microscopy; (4) electrical techniques, e.g., electrochemistry and electrophoresis; and (5) rheology, physical properties, e.g., dielectric spectroscopy and differential scanning calorimetry (Bajwa et al., 2016).

13.7 SURFACE ENGINEERING OF QUANTUM DOTS

For the biological applications, QDs should be biocompatible and soluble in aqueous solutions. In this regard, creating the hydrophilic QDs various methods that have been developed can be divided into two main categories (Medintz et al., 2008; Murcia et al., 2008).

1. *Cap exchange method*: In this method the hydrophobic organic solvent is replaced by bifunctional molecules containing a soft acidic group (usually a thiol, e.g., sodium thioglycolate) and hydrophilic groups (like carboxylic or amine groups) (Jorge et al., 2007; Wang et al., 2008). The substitution of monothiols by polythiols or phosphines usually improves stability.

2. *Native surface modification*: This can be achieved by the addition of a silica shell to the nanoparticles by using a silica precursor during the polycondensation (Koole et al., 2008). Amorphous silica shells can be further functionalized with other molecules or polymers. The method of QDs encapsulation into solid lipid nanoparticles (SLNs) composed of highly biocompatible lipids with long-term physical and chemical stability was also successfully tested (Liu et al., 2008). The SLNs were found to be more convenient than small molecules (e.g., mercapto-propionic acid) traditionally used for QDs surface modification, which were rather unstable since these could be easily degraded by hydrolysis or oxidation of the capping ligand.

QDs can be conjugated to biological molecules like proteins, targeting and imaging agents, oligonucleotides, and small molecules, which are used in the direct binding of QDs to areas of interest for bio-labeling, bio-sensing, imaging, and targeting (Reed, 1993; Alivisatos, 1996). Semiconductor QDs have attracted remarkable interest for bio-labeling and bio-imaging applications due to their considerable advantages over conventional organic dyes, such as high QY, size tunable emission, broad absorption with narrow photoluminescence spectra, photo-stability (low photo-bleaching), improved signal brightness, and resistance to chemical degradation. Surface modification of QDs developed a new generation of probes with integrated functionalities of labeling and drug/gene delivery (Guzelian et al., 1996; Reed, 1993).

Several functionalizations have been adapted to the ZnS layer coating and the CdSe core for making the CdSe/ZnS QDs as the most adaptable nanocarrier for biological applications. The methods for the functionalization of QDs include mercapto (–SH) exchange, adsorption, electrostatic attraction, and covalent linkage (Alivisatos et al., 2005; Lin et al., 2004).

Thiol-containing biomolecules can be conjugated to QDs via mercapto exchange. Under thiol-exchange reactions mercapto-coated QDs are mixed with thiolated biomolecules, and chemical equilibrium is reached between absorbed thiols and the free thiols through overnight incubation (Gao and Nie, 2001; Akerman et al., 2002).

However, simple small molecules such as oligonucleotides and various serum albumins were found to readily adsorb nonspecifically to the surface of water-soluble QDs. Another different strategy utilizes the phospholipid micelle to encapsulate QDs. The hydrophobic core of the micelle adsorbs to QDs through hydrophobic interactions, while the hydrophilic terminals on the outer side of the micelle interact with biomolecules (Dubertret et al., 2002). Electrostatic attraction can be used to conjugate proteins to QDs. The most common method involves covalent binding of biomolecules to the functional groups on the surface of QDs, e.g., the QD surface containing carboxyl groups can be conjugated to a biomolecule possessing the amine groups via 1-ethyl-3-(3-dimethyl aminopropyl) carbodiimide cross linker (Alivisatos et al., 2005; Lin et al., 2004). Functional group reaction strategy for designing the hydrophilic bio-conjugates of QDs involves primary amines, carboxylic acids, alcohols, and thiols as major reacting groups (Chan and Nie, 1998; Bruchez et al., 1998). A comparative account of different surface engineering methods is presented in Table 13.1.

13.8 QUANTUM DOTS RELATED TOXICITY AND ITS REMEDIAL STRATEGIES

QDs have been emerged as an alternative for organic dyes in the imaging of biological systems, due to their excellent fluorescent properties, good chemical stability, broad excitation ranges, and high photo-bleaching thresholds. However, the main shortcoming of QDs is their toxicity, which hampers their application, e.g., CdTe QDs are used as fluorescent probes for biological imaging as well as monitoring the targeted drug delivery,

Table 13.1 Comparative account of different surface engineering methods

Surface-engineering method	Conjugated biomolecules	Merits	Demerits	Reference
Mercapto (–SH) exchange	Thio–terminated DNA, thiolated peptides	Convenient	ZnS-thiol bond is not very strong. The conjugated biomolecules may break off and QDs may precipitate out of solution. It is time consuming	Gao and Nie (2001), Akerman et al. (2002)
Adsorption	Oligonucleotides and various serum albumins	Simple	Nonspecific	Dubertret et al. (2002)
Electrostatic attraction	DNA, recombinant fusion protein	Simple and fast, higher fluorescence yield of QDs	Prior modification of biomolecule is required, direct effect of environmental factors like pH on stability of QDs	Lin et al. (2004)
Covalent linkage between biomolecule to QD surface groups	IgG, tricosanthin, trypsin, bovine serum albumin	Most stable	Complex involving multi–steps modifications	Alivisatos et al. (2005), Lin et al. (2004)

but CdTe is toxic in nature. The toxicity aspects of QDs, their mechanism as well as remedies for decreasing/eliminating the toxicity of QDs have been discussed as follows:

13.8.1 Toxicity of quantum dots

Many of the QDs are cytotoxic to some extent. The cytotoxicity of QDs mainly depends on different parameters like size, shape, concentration, capping material used, dose, surface chemistry, coating bioactivity, residual organic molecules, charge, redox activity, mechanical stability, and QD exposure route (Derfus et al., 2004). QDs toxicity is directly related to oxidation of the core/shell material, which leads to the release of free Cd. Oxidation of the QD surface due to their exposure to air before solubilization or catalysis by UV light causes the oxidation of selenium and/or sulfur, rendering the release of Cd (Alivisatos, 1996).

Two major mechanisms involved in the toxicological effects of the QDs (Chen et al., 2012; Hoshino et al., 2011; Zheng et al., 2012; Singh et al., 2012) are as follows:

1. Cd^{2+} ions existing in QDs structure: These metal ions exhibit toxic effects through several routes such as interference with DNA repair and substitution for physiologic Zn. Cd^{2+} ions increase oxidative stress, however; they cannot directly generate free radicals.

2. Free radical formation: Since QDs of CdSe and CdTe are highly reactive hence photo-activation of these QDs via visible or UV light leads to their oxidation. The photo-activation of QDs results in the conversion of molecular oxygen into singlet oxygen, which reacts with water or other biological molecules and produces the free radicals.

Kauffer et al. recently demonstrated that variation in core compositions and surface chemistries of QDs lead to their different cytotoxicity. The CdSe QDs produced ˙OH radicals immediately after light activation, while the CdS QDs required extensive irradiation to generate an equivalent amount of radicals. Therefore, the toxicity of CdSe QDs could be directly related to ˙OH radicals produced. Hence, cytotoxicity of QDs can be controlled by selecting the appropriate materials for QD core and suitable irradiation condition (Kauffer et al., 2014).

Walling et al. (2009) reported that under in vivo QDs application, Cd^{2+} ions were determined as the primary cause of cytotoxicity, because of their ability to bind with thiol groups on critical molecules in the mitochondria and caused enough stress and damage resulting in significant cell death. However, additional surface coatings could reduce or eliminate the release of Cd^{2+}. The use of gelatin during the production of CdTe QDs reduced their toxicity (Byrne et al., 2007). In another report, Qian et al. (2006) used tripeptide glutathione as a stabilizer, which detoxified the Cd^{2+} ions due to its chelating capability.

The toxicity of QDs is dependent on the type of core material used. The groups III–IV QDs displayed less cytotoxicity (Shiohara et al., 2004).

13.8.2 Remedies for reducing toxicity of quantum dots

Oxidation of the shell material leads to toxicity due to resultant oxidation of the core and release of cadmium. Oxidation due to air exposure before solubilization is nearly eliminated when QDs are coated with a ZnS shell; however, shell oxidation occurs only with prolonged UV light exposure. Coatings on QDs reduce cytotoxicity because these additional layers act as a physical barrier to the core, preventing access, with different coatings having varying levels of passivation (Hoshino et al., 2004).

Ligands with terminal carboxylic acid, hydroxyl, or amine groups are used for solubilization. QDs with amine-terminal ligands demonstrated lesser cytotoxicity as compared to other ligands (Ryman-Rasmussen et al., 2006). The added protection of the PEG layer, as PEG-modified silica, has been proven exceptionally stable and almost entirely eliminated cytotoxicity (Zhang et al., 2006). However, PEG-modified QDs demonstrated the reduced uptake by cells as compared to other surface coatings.

The biomolecules like bovine serum albumin were also used as coatings for shielding the environment from the toxic cores. There is a direct relationship between the cytotoxicity of QD and the number of coating applied on the surface of QD. The cytotoxicity decreases as the number of surface layers increases until or unless the coating should remain intact. However, the addition of large number of layers is not a perfect strategy as size can become an issue. The fabrication of QDs with optimal clearance features minimizes the toxicity risks by reducing the exposure time. Hence, well established QDs characteristics associated with optimal clearance from the body will be helpful in the design of optimal QDs (Longmire et al., 2008).

Choi et al. (2010) reported that tumor-targeting QDs functionalized with high-affinity, small-molecule ligands with a hydrodynamic diameter less than 5.5 nm and 5–10 ligands per QD could be cleared from the body through the kidneys under renal clearance. The study suggested a set of design rules for the clinical translation of targeted QDs eliminating through the kidneys.

13.9 QUANTUM DOTS IN DRUG DELIVERY

The applications of QDs in drug delivery are well established. In context to the effective bioactive delivery, QDs exhibit various vital features like sufficient blood circulation, protection of the cargo from degradation, large drug pay load, controlled drug release profile as well as incorporation of multiple targeting ligands on their surface.

Development of single QD as in vivo drug delivery vehicle is venerable because the nanosize range (about 10–20 nm) of QD reduces the renal clearance as well as the uptake by reticulo-endothelial system, which results in the rising of blood circulation time and improving the delivery efficiency. Further, QD core offers a structural scaffold for the loading of different types of drug molecules (Fig. 13.3). Flexibility in the shell material facilitates the engineering of drug carriers with various physical properties

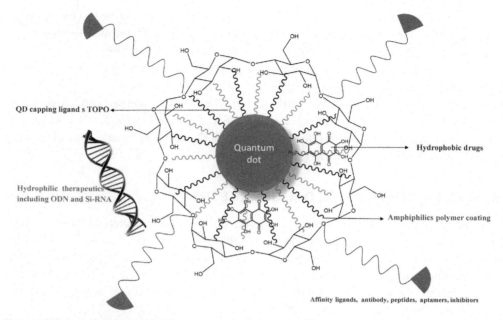

Figure 13.3 Drug loading and delivery aspects of quantum dot.

like size, charge, biodegradability, etc., which can be utilized for different specific applications (Jain 2005; Qi and Gao, 2008).

Chen et al. cotransfected the QDs and siRNA using Lipofectamine 2000 for monitoring the transfection efficiency via QD fluorescence and demonstrated that 1:1 mass ratio of QDs and transfection reagent showed better QD signal intensity and the degree of gene silencing. However, more cotransfection of different siRNA molecules with various QD colors facilitated the multiplexed monitoring of gene silencing (Chen et al., 2005).

Tan et al. reported more accurate quantitative data about the number of siRNA molecules delivered into cells using the QD doped chitosan nanobeads. The intracellular delivery of siRNA molecules present on the surface of nanobeads could be monitored by the nanobead fluorescence (Tan et al., 2007).

Jia et al. demonstrated the intracellular delivery of QDs tagged antisense oligodeoxynucleotides by functionalized multiwalled carbon nanotubes. The straight tagging of plasmid DNA with QDs followed by Lipofectamine-mediated transfection enabled long-term intracellular and intranuclear localization and transport of plasmid DNA, while preserving the ability of expressing reporter protein encoded by the plasmid (Jia et al., 2007; Srinivasan et al., 2006).

Manabe et al. established the in vivo delivery of captopril, an antihypertensive drug conjugated on the surface of QD with the therapeutic effect similar to that of the free drug and monitored the QD-drug biodistribution over a period of 96 h. The design of biocompatible QD surface coatings presents a crucial method for modeling pharmaco-kinetics and pharmacodynamics of QD (Fig. 13.3) (Manabe et al., 2006).

Surface modified QDs can be utilized for controlled drug delivery (Alivisatos et al., 2005). Internalization of peptide-coated QDs established the worth of QDs in drug delivery (Akerman et al., 2002; Rozenzhak et al., 2005). Lai et al. used surface-modified CdS QDs to retain the drug molecules as well as neurotransmitters. The CdS cap kept the drug inside the system until released by splitting the disulfide bonds, thus preventing the premature leakage of the drug (Lai et al., 2003).

13.10 QUANTUM DOTS IN DRUG TARGETING

Accurate recognition of key molecular targets distinguishing unhealthy from healthy cells enables targeted drug delivery with negligible side effects (Fig. 13.4). The QDs has a long blood circulation time, protection, large drug-loading capacity, controlled drug release profile, and integration of multiple targeting ligands on surface (Cheki et al., 2013). Suitably functionalized QDs can target specific subcellular targets (Alivisatos et al., 2005; Hoshino et al., 2004).

The determination of the extent of folate receptor expression can be used as a possible diagnostic tool for the cancer diagnosis as any significant intensity increase than the normal expression levels is an indication of over expression, which is generally observed in cancerous cells. Singh et al. synthesized two different types of QDs, i.e., CdSe/CdS/ZnS and CdTe QDs, by adopting a successive ion layer adsorption and reaction

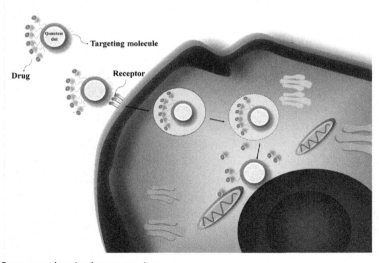

Figure 13.4 Quantum dots in drug targeting.

technique and a direct aqueous method, respectively. The 3-mercaptopropionic acid (MPA) and mercaptosuccinic acid (MSA) were used as stabilizer. Photoluminescence quantum yield (QY) of FA conjugated CdSe/CdS/ZnS-MPA and CdTe-MSA QDs was found to be 59% and 77% as compared to that nonfolated hydrophilic QDs and showed higher cellular internalization through folate receptor-mediated delivery. Folated and nonfolated CdTe-MSA QDs were found highly toxic (only 10% cell viability) as compared to CdSe/CdS/ZnS-MPA QDs (more than 80% cell viability). The anti-HER2 antibody conjugated CdSe/CdS/ZnS-MPA QDs demonstrated positive results with better sensitivity and specificity under immunohistochemistry of human breast cancer tissue samples (Singh et al., 2016).

13.11 QUANTUM DOTS IN DIAGNOSTIC IMAGING

For an optimal performance in biological imaging, semiconductor QDs are being developed in order to optimize their luminescent, surface and chemical stability properties, which require a very complex multilayered chemical assembly where the nanocrystal core determines its emission color, the passivation shell determines its brightness, and photo-stability and the organic capping layer determines its stability and functionality. For biological application/analysis (Fig. 13.5), QDs must meet the following criteria:

1. Biomolecules like protein, DNA, peptides must exist in an aqueous environment. Modifying the surface of QDs to be hydrophilic and compatible to varieties of biomolecules is vital.
2. Designing the technique for specific labeling cells and biomolecules with QDs is necessary.
3. The nontoxic nature of QDs is important for in vivo applications.

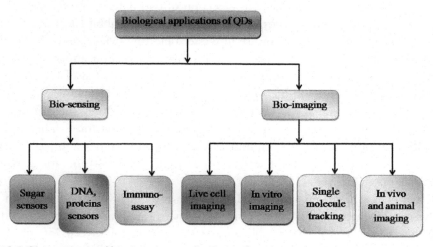

Figure 13.5 Bio-sensing and bio-imaging applications of quantum dots.

A sensitive and widely applicable method for detecting biomolecules and for scrutinizing biomolecular processes can be developed by attaching the biomolecules to the nanosized semiconductors. Molecules labeled with the QDs remain active for biochemical reactions and brightly colored products are produced by the tagged species. This methodology represents a new class of biological dyes by taking advantage of the efficient fluorescence and high photo-stability of the semiconductor QDs (Bruchez et al., 1998). For the first time Bruchez et al. (1998) used CdSe QDs coated with silica and mercaptoacetic acid layers for specific labeling.

The QDs are exceptionally suitable for immunolabeling, cell motility assays, in situ hybridization, and as live cell markers as well as nonspecific fluorescent stain (Alivisatos et al., 2005; Bruchez, 2005). Various sophisticated techniques including confocal microscopy, total internal reflection microscopy, wide-field epifluorescence microscopy, fluorometry, and high-throughput screening are available for the tracking and detection of QDs (Jain, 2005; Michalet et al., 2005; Bruchez et al., 1998; Chan and Nie, 1998; Smith et al., 2006).

Several QDs can be excited by the same wavelength of light that opens up several multiplexing potentials including the use of multiple QD labels for optical bar coding of targets by selecting the QD labels with isolated emission spectra (Jain, 2003; Bruchez et al., 1998). The multicolor optical label can be used as a specific bar code for a particular analyte or cell type for collecting the information about the location, abundance, and distribution of multiple proteins in living cells. The multicolor encoding of cells or beads can be used for obtaining data on differential response of cell types to the common stimuli in cancer diagnosis (Smith et al., 2006).

QDs are also employed in the field of proteomics for the detection of multiple proteins on Western blots (Bruchez, 2005). Makrides et al. successfully demonstrated the utility of streptavidin-coated QDs in simultaneous detection of two different proteins in a Western blot assay. The diagnostic kits employing QD conjugated anti-rabbit and anti-mouse antibodies are available in the market for use in Western blotting (Makrides et al., 2005).

Photo-stability, a specific advantage of QDs for in vivo applications, allows images to be recorded over a longer period of time as compared to that available with the use of fluorescent dyes or proteins due to their resistance to photobleaching.

Jiang et al. demonstrated the ability of QDs with near-IR region emission wavelengths for in vivo analysis of deep tissue or noninvasive applications. The application of near-IR emission wavelength minimizes the absorption and scattering of light by native tissues, allowing the use of longer wavelengths (650–900 nm) for a diagnostic emission window (Jiang et al., 2008).

Zimmer et al. synthesized a series of InAs/ZnSe core/shell QDs with less than 10 nm hydrodynamic diameter. The small core size along with variable shell thickness and composition offered a range of size tunable emission wavelengths between 750 and 920 nm. The conjugation of dihydrolipoic acid (DHLA) to the QDs allowed the observations within the interstitial fluid in rats and showed the potential to interrogate the delivery mechanism of QDs to tumor cells. No extravasation was observed in QDs without the DHLA coating (Zimmer et al., 2006).

QDs are used for diagnostic purposes in neuroscience manifestation. Antibody-functionalized QDs followed lateral diffusion of glycine receptor in a culture of primary spinal cord neurons (Ballou et al., 2004). Biocompatible water-soluble QD micelles demonstrated the uptake and intracellular dispersion in cultured neurons (Chan et al., 2002). QDs–ligand interaction is used in DNA defection, protein detection, and cellular labeling (Walling et al., 2009; Misra, 2008).

The use of QDs for cancer imaging is one of the most promising applications of QDs. Wu et al. (2003) developed an assay for detection of the receptor Her2 (hairy-related 2 protein) on SK–BR–3 breast cancer cells by employing humanized anti-Her2 antibody and a biotinylated goat anti-human IgG secondary antibody. Her2 was detected by streptavidin-coated QDs.

A prostate-specific membrane antigen coupled with QDs was successfully employed for the detection of prostate cancer xenografts using antibodies (Gao et al., 2004). Akerman et al. (2002) conjugated three different peptides to QDs for targeting the lung endothelial cells, brain endothelial cells, and breast carcinoma cells, both in vitro and in vivo. Microinjection of QDs conjugated to suitable peptides can also target cellular organelles such as the nucleus or mitochondria (Michalet et al., 2005).

Larson et al. injected water-soluble QDs into mice for the imaging the skin and adipose tissues. The vasculatures containing QDs at about $1\,\mu M$ were clearly visible using multiphoton microscopy. Fluorescence correlation spectroscopy indicated the stability of water-soluble QDs for over 9 months (Larson et al., 2003). In another study, Stroh et al. (2005) utilized QDs for imaging the vasculature within a tumor environment.

13.12 CONCLUDING REMARKS AND FUTURE PERSPECTIVES

The QD-based–targeted drug delivery offers several advantages including drug dosage reduction and pharmaceutical action in the malignant tissue. Moreover, the impact of QD application as fluorescent labels in drug delivery systems must also be considered. Thorough studies should be allocated for examination of QDs influence on pharmacokinetics of the original drugs. The toxicity of QDs may result from their interactions with blood components, immune response, or accumulation in organs. However, despite heavy metal contents in many QD cores, the long-term toxicity and pharmacokinetic profile, like evaluating degradation, excretion, and persistence of QDs, should always be a concern.

QDs can be made very effective for cellular imaging and drug delivery by coupling with specific ligands including vitamins, proteins, peptides, or antibodies. The QD based bio-imaging in small animals cannot be directly scaled up to bio-imaging in humans.

The commonly used graphical analysis function set in fluorescence imaging system, especially the in vivo imaging system, can only offer semi-quantitative results, which cannot fulfill the quantitative requirement of biopharmaceutical analysis in living animals.

The exploration of full capabilities of QDs in the field of drug delivery and imaging requires various advantageous optical properties of QDs, but also necessitates biocompatibility, low toxicity, appropriate attachment of biomolecules, and navigation of the cascade of events involved in the immune response. Hence, for the next few years, there are various promising areas of research, which need more attention particularly (1) development of novel type of QDs and surface coating with high biosafety; (2) Influence of labeling QDs on the inherent property of nanomedicines; and (3) quantitative analysis of QDs fluorescence imaging. Probably future targets can be achieved by meticulous studies undertaken by multidisciplinary scientific groups.

REFERENCES

Abbasi, E., Kafshdooz, T., Bakhtiary, M., Nikzamir, N., Nikzamir, N., Nikzamir, M., et al., 2015. Biomedical and biological applications of quantum dots. Artif. Cells Nanomed. Biotechnol. 23, 1–7.

Akerman, M.E., Chan, W.C., Laakkonen, P., Bhatia, S.N., Rouslahti, E., 2002. Nanocrystal targeting in vivo. Proc. Natl. Acad. Sci. U.S.A. 99, 12617–12621.

Alivisatos, A.P., 1996. Semiconductor clusters, nanocrystals, and quantum dots. Science 271, 933–937.

Alivisatos, A.P., Gu, W., Larabell, C., 2005. Quantum dots as cellular probes. Annu. Rev. Biomed. Eng. 7, 55–76.

Asokan, S., Krueger, K.M., Alkhawaldeh, A., Carreon, A.R., Mu, Z., Colvin, V.L., et al., 2005. The use of heat transfer fluid in the synthesis of high quality CdSe QDs, core/shell QDs and quantum rods. Nanotechnology 16, 2000–2011.

Atkinson, P., Bremner, S.P., Aderson, D., Jones, G.A.C., Ritchie, D.A., 2006. . Molecular beam epitaxial growth of site-controlled InAs quantum dots on pre-patterned GaAs substrates. Microelectron. J. 37 (12), 1436–1439.

Bailey, R.E., Nie, S., 2003. Alloyed semiconductor quantum dots: tuning the optical properties without changing the particle size. J. Am. Chem. Soc. 125, 7100–7106.

Bajwa, N., Mehra, N.K., Jain, K., Jain, N.K., 2016. Pharmaceutical and biomedical applications of quantum dots. Artif. Cells Nanomed. Biotechnol. 44 (3), 758–768.

Bakalova, R., Ohba, H., Zhelev, Z., Ishikawa, M., Baba, Y., 2004. Quantum dots as photosensitizers. Nat. Biotechnol. 22, 1360–1361.

Ballou, B., Lagerholm, B.C., Ernst, L.A., Bruchez, M.P., Waggoner, A.S., 2004. Noninvasive imaging of quantum dots in mice. Bioconjug. Chem. 15, 79–86.

Bandyopadhyay, S., Menon, L., Kouklin, N., Zeng, H., Sellmyer, D.J., 1999. Electrochemically self-assembled quantum dot arrays. J. Elect. Mater. 28 (5), 515–519.

Bera, D., Qian, L., Tseng, T.K., Holloway, P.H., 2010. Quantum dots and their multimodal applications: a review. Materials 3, 2260–2345.

Biju, V., Mundayoor, S., Omkumar, R.V., Aaas, A., Ishikawa, M., 2010. Bioconjugated quantum dots for cancer research: present status, prospects and remaining issues. Biotechnol. Adv. 28, 199–213.

Bruchez, M.P., 2005. Turning all the lights on: quantum dots in cellular assays. Curr. Opin. Chem. Biol. 9, 533–537.

Bruchez Jr, M., Moronne, M., Gin, P., Weiss, S., Alivisatos, A.P., 1998. Semiconductor nanocrystals as fluorescent biological labels. Science 281, 2013–2016.

Byrne, S.J., Williams, Y., Davies, A., Corr, S.A., Rakovich, A., Yurii, K., et al., 2007. Jelly dots: synthesis and cytotoxicity studies of CdTe quantum dot-gelatin nanocomposites. Small 3 (7), 1152–1156.

Canham, L., 1990. Silicon quantum wire array fabrication by electrochemical and chemical dissolution of wafers. Appl. Phys. Lett. 57, 1046–1048.

Chakravarthy, K.V., Davidson, B.A., Helinski, J.D., Ding, H., Law, W.C., Yong, K.T., et al., 2011. Doxorubicin-conjugated quantum dots to target alveolar macrophages and inflammation. Nanomedicine 7, 88–96.

Chan, W.C., Nie, S., 1998. Quantum dot bioconjugates for ultrasensitive nonisotopic detection. Science 281, 2016–2018.

Chan, W.C., Maxwell, D.J., Gao, X., Bailey, R.E., Han, M., Nie, S., 2002. Luminescent quantum dots for multiplexed biological detection and imaging. Curr. Opin. Biotechnol. 13, 40–46.

Cheki, M., Moslehi, M., Assadi, M., 2013. Marvelous applications of quantum dots. Eur. Rev. Med. Pharmacol. Sci. 17, 1141–1148.

Chen, A.A., Derfus, A.M., Khetani, S.R., Bhatia, S.N., 2005. Quantum dots to monitor RNAi delivery and improve gene silencing. Nucl. Acids Res. 33, 190.

Chen, N., He, Y., Su, Y., Li, X., Huang, Q., Wang, H., et al., 2012. The cytotoxicity of cadmium-based quantum dots. Biomaterials 33 (5), 1238–1244.

Choi, S.H., Liu, W., Liu, F., Nasr, K., Misra, P., Bawendi, M.G., et al., 2010. Design considerations for tumor-targeted nanoparticles. Nat. Nanotechnol. 5, 42–47.

Dabbousi, B.O., Rodriguez-Viejo, J., Mikulec, F.V., Heine, J.R., Mattoussi, H., Ober, R., et al., 1997. (CdSe) ZnS core-shell quantum dots: synthesis and characterization of a size series of highly luminescent nanocrystallites. J. Phys. Chem. B 101, 9463–9475.

Derfus, A.M., Chan, W.C.W., Bhatia, S.N., 2004. Probing the cytotoxicity of semiconductor quantum dots. Nano Lett. 4, 11–18.

Dubertret, B., Skourides, P., Norris, D.J., Noireaux, V., Brivanlou, A.H., Libchaber, A., 2002. In vivo imaging of quantum dots encapsulated in phospholipid micelles. Science 298, 1759–1762.

Duncan, R., Vicent, M.J., 2013. Polymer therapeutics-prospects for 21st century: the end of the beginning. Adv. Drug Deliv. Rev. 65, 60–70.

Gao, X., Nie, S., 2001. Biologists join the dots. Nature 13, 450–452.

Gao, X., Cui, Y., Levenson, R.M., Chung, L.W., Nie, S., 2004. In vivo cancer targeting and imaging with semiconductor quantum dots. Nat. Biotechnol. 22, 969–976.

Gerion, D., 2006., first ed. Fluorescence Imaging in Biology Using Nanoprobes, vol. 3 Wiley-VCH, Weinheim.

Ghasemi, Y., Peymani, P., Afifi, S., 2009. Quantum dot: magic nanoparticle for imaging, detection and targeting. Acta Biomed. 80, 156–165.

Guzelian, A.A., Banin, U., Kadavanich, A.V., Peng, X., Alivisatos, A.P., 1996. Colloidal chemical synthesis and characterization of InAs nanocrystal quantum dots. Appl. Phys. Lett. 69, 1432–1434.

Hawrylak, P., 1999. Excitonic artificial atoms: engineering optical properties of quantum dots. Phys. Rev. B 60, 5597–5608.

Hoshino, A., Fujioka, K., Oku, T., Suga, M., Sasaki, Y.F., Ohta, T., et al., 2004. Physicochemical properties and cellular toxicity of nanocrystal quantum dots depend on their surface modification. Nano Lett. 4, 2163–2169.

Hoshino, A., Hanada, S., Yamamoto, K., 2011. Toxicity of nanocrystal quantum dots: the relevance of surface modifications. Arch. Toxicol. 85 (7), 707–720.

Huang, X., Ren, J., 2011. Gold nanoparticles based chemiluminescent resonance energy transfer for immunoassay of alpha fetoprotein cancer marker. Anal. Chim. Acta 686, 115–120.

Jain, K.K., 2003. Nanodiagnostics: application of nanotechnology in molecular diagnostics. Expert Rev. Mol. Diagn. 3, 153–161.

Jain, K.K., 2005. Nanotechnology in clinical laboratory diagnostics. Clin. Chim. Acta 358, 37–54.

Jaiswal, J.K., Goldman, E.R., Mattoussi, H., Simon, S.M., 2004. Use of quantum dots for live cell imaging. Nat. Methods 1, 73–78.

Jia, N., Lian, Q., Shen, H., Wang, C., Li, X., Yang, Z., 2007. Intracellular delivery of quantum dots tagged antisense oligodeoxynucleotides by functionalized multiwalled carbon nanotubes. Nano Lett. 7 (10), 2976–2980.

Jiang, W., Singhal, A., Kim, B.Y.S., Zheng, J., Rutka, J.T., Wang, C., et al., 2008. Assessing near-infrared quantum dots for deep tissue, organ, and animal imaging applications. J. Assoc. Lab. Autom. 13, 6–12.

Jin, W.J., Fernandezz-Arguelles, M.T., Costa-Fernandez, J.M., Pereriro, R., Sanz-Medel, A., 2005. Photoactivated luminescent CdSe quantum dots as sensitive cyanide probes in aqueous solutions. Chem. Commun., 883–885.

Jorge, P., Martins, M.A., Trindade, T., Santos, J.L., Farahi, F., 2007. Optical fiber sensing using quantum dots. Sensors 7 (12), 3489–3534.

Kauffer, F.A., Merlin, C., Balan, L., Schneider, R., 2014. Incidence of the core composition on the stability, the ROS production and the toxicity of CdSe quantum dots. J. Hazard. Mater. 268, 246–255.

Koole, R., Schooneveld, V.M.M., Hilhorst, J., Donega, C.M., Dannis, C., Blaaderen, A., et al., 2008. On the incorporation mechanism of hydrophobic quantum dots in silica spheres by a reverse microemulsion method. Chem. Mater. 20 (7), 2503–2512.

Kroutvar, M., Ducommun, Y., Heiss, D., Bichler, M., Schuh, D., Abstreiter, G., et al., 2004. Optically programmable electron spin memory using semiconductor quantum dots. Nature 432, 81–84.

Kumar, C.S.S.R., 2007. Nanomaterials for medical applications Kirk-Othmer Encyclopedia of Chemical Technology. Wiley, New York, NY.

Lai, C.Y., Trewyn, B.G., Jeftinija, D.M., Jeftinija, K., Xu, S., Jeftinija, S., et al., 2003. A mesoporous silica nanosphere-based carrier system with chemically removable CdS nanoparticle caps for stimuli-responsive controlled release of neurotransmitters and drug molecules. J. Am. Chem. Soc. 125, 4451–4459.

Larson, D.R., Zipfel, W.R., Williams, R.M., Clark, S.W., Bruchez, M.P., Wise, F.W., et al., 2003. Water-soluble quantum dots for multiphoton fluorescence imaging in vivo. Science 300, 1434–1436.

Li, J.J., Wang, Y.A., Guo, W., Keay, J.C., Mishima, T.D., Johnson, M.B., et al., 2003. Large scale synthesis of nearly monodisperse CdSe/CdS core/shell nanocrystals using air-stable reagents via successive ion layer and reaction. J. Am. Chem. Soc. 125, 12567–12575.

Lin, Z., Su, X., Mu, Y., Jin, Q., 2004. Methods for labeling quantum dots to biomolecules. J. Nanosci. Nanotechnol. 4, 641–645.

Liu, W., He, Z., Liang, J., Zhu, Y., Xu, H., Yang, X., 2008. Preparation and characterization of novel fluorescent nanocomposite particles: CdSe/ZnS core-shell quantum dots loaded solid lipid nanoparticles. J. Biomed. Mater. Res. Part A 84A (4), 1018–1025.

Longmire, M., Choyke, P.L., Kobayashi, H., 2008. Clearance properties of nano-sized particles and molecules as imaging agents: considerations and caveats. Nanomedicine 3, 703–717.

Makrides, S.C., Gasbarro, C., Bello, J.M., 2005. Bioconjugation of quantum dot luminescent probes for Western blot analysis. Biotechniques 39, 501–506.

Maksym, P., Chakraborty, T., 1990. Quantum dots in a magnetic field: role of electron-electron interactions. Phys. Rev. Lett. 65, 108–111.

Manabe, N., Hoshino, A., Liang, Y.Q., Goto, T., Kato, N., Yamamoto, K., 2006. Quantum dot as a drug tracer in vivo. IEEE Trans. Nanobiosci. 5, 263–267.

Medintz, I.L., Mattoussi, H., Clapp, A.R., 2008. Potential clinical applications of quantum dots. Int. J. Nanomedicine 3 (2), 151–167.

Michalet, X., Pinau, F.F., Bentolila, L.A., Tsay, J.M., Doose, S., Li, J.J., et al., 2005. Quantum dots for live cells, in vivo imaging and diagnostics. Science 307, 538–544.

Minnich, A., Dresselhaus, M., Ren, Z.F., Chen, G., 2009. Bulk nanostructured thermoelectric materials: current research and future prospects. Energy Environ. Sci. 2, 466–479.

Mishra, P., Vyas, G., Harsoliya, M.S., Pathan, J.K., Raghuvanshi, D., Sharma, P., et al., 2011. Quantum dot probes in disease diagnosis. Int. J. Pharm. Pharm. Sci. Res. 1 (2), 42–46.

Mishra, S., Tripathy, P., Sinha, S.P., 2012. Advancements in the field of quantum dots. Int. J. Adv. Res. Technol. 1 (3), 1–5.

Misra, R.D., 2008. Quantum dots for tumor-targeted drug delivery and cell imaging. Nanomedicine (Lond.) 3, 271–274.

Mitchell, G.P., Mirkin, C.A., Letsinger, R.L., 1999. Programmed assembly of DNA functionalized quantum dots. J. Am. Chem. Soc. 121, 8122–8123.

Moriyama, S., Fuse, T., Suzuki, M., Aoyagi, Y., Ishibashi, K., 2005. Four-electron shell structures and an interacting two-electron system in carbon nanotube quantum dots. Phys. Rev. Lett. 94 (18), 186806.

Murcia, M.J., Shaw, D.L., Long, E.C., Naumann, C.A., 2008. Fluorescence correlation spectroscopy of CdSe/ZnS quantum dot optical bioimaging probes with ultra-thin biocompatible coatings. Opt. Commun. 281 (7), 1771–1780.

Pandurangan, D.K., Mounika, K.S., 2012. Quantum dot aptamers-an emerging technology with wide scope in pharmacy. Int. J. Pharm. Pharm. Sci. 4 (3), 24–31.

Phillips, J., 2002. Evaluation of the fundamental properties of quantum dot infrared detectors. J. Appl. Phys. 91, 4590–4594.

Potasek, M., Gersten, B., Zaitsev, A., 2005. Organic-inorganic nanostructured composites with large optical nonlinearity for optical applications. Front. Optics Opt. Soc. Am., 16–25.

Qi, L., Gao, X., 2008. Emerging application of quantum dots for drug delivery and therapy. Expert Opin. Drug Deliv. 5 (3), 263–267.

Qian, H.F., Dong, C., Weng, J., Ren, J., 2006. Facile one-pot synthesis of luminescent, water-soluble, and biocompatible glutathione-coated CdTe nanocrystals. Small 2 (6), 747–751.

Reed, M.A., 1993. Quantum dots. Sci. Am. 268, 118–123.

Resch-Genger, U., Grabolle, M., Cavaliere-JaricotS, Nitschke, R., Nann, T., 2008. Quantum dots versus organic dyes as fluorescent labels. Nat. Methods 5 (9), 763–775.

Rozenzhak, S.M., Kadakia, M.P., Caserta, T.M., Westbrook, T.R., Stone, M.O., Naik, R.R., 2005. Cellular internalization and targeting of semiconductor quantum dots. Chem. Commun. (Camb.), 2217–2219.

Ryman-Rasmussen, J.P., Riviere, J.E., Monteiro-Riviere, N.A., 2006. Surface coatings determine cytotoxicity and irritation potential of quantum dot nanoparticles in epidermal keratinocytes. J. Invest. Dermatol. 127, 143–153.

Sattler, K.D., 2010. Handbook of Nanophysics: Nanomedicine and Nanorobotics. CRC Press, Boca Raton, FL. 1–5

Shiohara, A., Hoshino, A., Hanaki, K., Suzuki, K., Yamamoto, K., 2004. On the cytotoxicity caused by quantum dots. Microbiol. Immunol. 48, 669–675.

Singh, B.R., Singh, B.N., Khan, W., Singh, H.B., Naqvi, A.H., 2012. ROS-mediated apoptotic cell death in prostate cancer LNCaP cells induced by biosurfactant stabilized CdS quantum dots. Biomaterials 33 (23), 5753–5767.

Singh, G., Kumar, M., Soni, U., Arora, V., Bansal, V., Gupta, D., et al., 2016. Cancer cell targeting using folic acid/anti-HER2 antibody conjugated fluorescent CdSe/CdS/ZnS-mercaptopropionic acid and CdTe-mercaptosuccinic acid quantum dots. J. Nanosci. Nanotechnol. 16 (1), 130–143.

Smith, A.M., Dave, S., Nie, S., True, L., Gao, X., 2006. Multicolor quantum dots for molecular diagnostics of cancer. Expert Rev. Mol. Diagn. 6, 231–244.

Srinivasan, C., Lee, J., Papadimitrakopoulos, F., Silbart, L.K., Zhao, M., Burgess, D.J., 2006. Labeling and intracellular tracking of functionally active plasmid DNA with semiconductor quantum dots. Mol. Ther. 14 (2), 192–201.

Stroh, M., Zimmer, J.P., Duda, D.G., Levchenko, T.S., Cohen, K.S., Brown, E.B., et al., 2005. Quantum dots spectrally distinguish multiple species within the tumor milieu in vivo. Nat. Med. 11, 678–682.

Talapin, D.V., Rogach, A.L., Kornowski, A., Haase, M., Weller, H., 2001. Highly luminescent monodisperse CdSe and CdSe/ZnS nanocrystals synthesized in a hexadecylamine-trioctylphosphine oxide-trioctylphosphine mixture. Nano Lett. 1 (4), 207–211.

Tan, W.B., Jiang, S., Zhang, Y., 2007. Quantum-dot based nanoparticles for targeted silencing of HER2/neu gene via RNA interference. Biomaterials 28 (8), 1565–1571.

Tokihiro, T., Hanamura, E., 1984. Multi-polariton scattering via excitonic molecules. Solid State Commun. 52, 771–774.

Uehara, T., Ishii, D., Uemura, T., Suzuki, H., Kanei, T., Takagi, K., et al., 2009. gamma-Glutamyl PAMAM dendrimer as versatile precursor for dendrimer-based targeting devices. Bioconjug. Chem. 21, 175–181.

Varshney, H.M., Shailender, M., 2012. Nanotechnology current status in pharmaceutical science: a review. Int. J. Ther. Appl. 6, 14–24.

Walling, M.A., Novak, J.A., Shepard, J.R.E., 2009. Quantum dots for live cell and in vivo imaging. Int. J. Mol. Sci. 10, 441–491.

Wang, H.Q., Zhang, H.L., Li, X.Q., Wang, J.H., Huang, Z.L., Zhao, Y.D., 2008. Solubilization and bioconjugation of QDs and their application in cell imaging. J. Biomed. Mater. Res. Part A 86A (3), 833–841.

Wu, X., Liu, H., Liu, J., Haley, K.N., Treadway, J.A., Larson, J.P., et al., 2003. Immunofluorescent labeling of cancer marker Her2 and other cellular targets with semiconductor quantum dots. Nat. Biotechnol. 21, 41–46.

Yuan, J., Guo, W., Yin, J., Wang, E., 2009. Glutathione-capped CdTe quantum dots for the sensitive detection of glucose. Talanta 77 (5), 1858–1863.

Zhang, J., Zhao, S.-Q., Zhang, K., Zhou, J.-Q., Cai, Y.-F., 2012. A study of photoluminescence properties and performance improvement of Cd-doped ZnO quantum dots prepared by the sol-gel method. Nanoscale Res. Lett. 7, 1–7.

Zhang, T., Stilwell, J.L., Gerion, D., Ding, L., Elboudwarej, O., Cooke, P.A., et al., 2006. Cellular effect of high doses of silica-coated quantum dot profiled with high throughput gene expression analysis and high content cellomics measurements. Nano Lett. 6, 800–808.

Zheng, X., Tian, J., Weng, L., Wu, L., Jin, Q., Zhao, J., et al., 2012. Cytotoxicity of cadmium-containing quantum dots based on a study using a microfluidic chip. Nanotechnology 23 (5), 055102.

Zhu, J.J., Li, J.J., Huang, H.P., Cheng, F.F., 2013. Quantum dots for DNA biosensing. SpringerBriefs Mol. Sci. 165, 341–346.

Zimmer, J.P., Kim, S.W., Ohnishi, S., Tanaka, E., Frangioni, J.V., Bawendi, M.G., 2006. Size series of small indium arsenide-zinc selenide core-shell nanocrystals and their application to in vivo imaging. J. Am. Chem. Soc. 128, 2526–2527.

Future Developments and Challenges in Nanomedicine

CHAPTER 14

Toxicity Concerns of Nanocarriers

Shima Tavakol, Vali Kiani, Behnaz Tavakol, Mohammad A. Derakhshan, Mohammad Taghi Joghataei and Seyed Mahdi Rezayat

Contents

Nanotechnology-Based Approaches for Targeting and Delivery of Drugs and Genes.
DOI: http://dx.doi.org/10.1016/B978-0-12-809717-5.00016-6

14.1 NANOEMULSION

Nanoemulsion is composed of two immiscible liquids, usually oil and water. The nano-droplet particle sizes (Weissig et al., 2014) are prepared with different techniques consisting of phase inversion (Shinoda and Saito, 1968), sonication (Onodera et al., 2015; Tadros et al., 1983), high-pressure homogenizer (Floury et al., 2003), and other nonenergetic methods. The physical appearances of nanoemulsions are clear, translucent, and turbid even in sizes less than 100 nm (Weissig et al., 2014; Onodera et al., 2015; Sari et al., 2015). To form a nanoemulsion of oil and water, a surfactant and co-surfactant are necessary. With regard to this subject of nanoemulsion toxicology, in addition to shape, size, and other physical characteristics of nanoemulsions, a consideration of chemical substitutes is worthy of value especially when applying a surfactant and cosurfactant.

The ionic surfactant as compared to nonionic surfactant has considerably significant toxicity, so the usage of ionic surfactants has been limited. The commonly used nonionic surfactants and cosurfactants are Tween (20, 21, 40, 60, 61, 65, 80, 81, 85) and Span families (20, 40, 60, 65, 80, and 85) (Meaning of HLB Advantages and Limitations, 1980), ethyl alcohol, propyl alcohol, and polyethylene glycol 400 (PEG). To increase the stability of the system, it has been suggested that two or more surfactants be used with low and high HLB. It is better to use Tween and Span from the same families. However, in some cases, ideal results will obtain in blending of two different chemical class families, e.g., Tween and Span "20" (laurate esters), "40" (palmitate esters), "60" (stearates), and "80" (oleates).

It is interesting to note that the food industries may apply one surfactant in nanoemulsion, and that will be sufficient to stabilize the system due to natural surfactants in the food ingredient that resulted in the blending of complex surfactants in the final formulation.

In the following section toxicology related to nanoemulsion and ingredients will be discussed.

14.2 GENOTOXICITY

With regard to the impressive progress and development in the field of nanotechnology especially in the nanopharmaceutics industries, hazard identification in genotoxicology has a critical importance. Nanomaterial via two pathways will induce genotoxicity: direct and indirect that resulted in a reproductive impact in germ cells and induction or a progression of carcinogenesis. In a direct pathway, nanomaterials directly interact with genetic material and in the indirect pathway, nanomaterials, via reactive oxygen species and other mechanisms interact with intermediate biomolecules such as cell division machinery and finally influence genetic material. Since the toxicity mechanisms of nanomaterial have a difference in some levels with bulk material, a genotoxicity assay refinement is necessary. Organisation for Economic Co-operation and Development (OECD) has defined some tests for the genotoxicity evaluation of material in in vitro consisting of a bacterial reverse mutation test (Ames) (OECD 471): in vitro micronucleus assay (MNvit OECD 487) and HPRT forward mutation assay (OECD 476) (Doak et al., 2012). However, findings have disclosed that among these assays, the Ames assay due to false negative results in face to the nanomaterial is not valuable in a genotoxicity investigation of nanomaterial. In fact, nanomaterials larger than 20 nm or in the agglomerated form cannot diffuse to prokaryotic cells. Besides, uptake rate of nanomaterial in prokaryotic is less than eukaryotic cells due to lack of endocytosis mechanism and existence of the cell wall in bacteria (Doak et al., 2012).

The Food Safety Commission indicated that there is no mutagenic potential regarded Tween 20 (Commission, 2007) while Eskandani disclosed that Tween 20 induces DNA cleavage by comet assay. Interestingly, the DNA cleavage potential of $2\,\mu L/mL$ of Tween 20 is in the range of $200\,\mu M$ H_2O_2 (Eskandani et al., 2013). As for Tween 60, Kada et al. (2013) disclosed reverse mutation by rec-assay. However, other researchers reported negative results by Ames and other assays (Kawachi et al., 1980; Morita et al., 1981). The micronucleus assay showed no sign of micronucleus in the face of Tween 65 with a dose of 2000 mg/kg (A micronucleus test on Tween65 using mice) and were negative for the reverse mutation and clastogenic assays (A reverse mutation study on Tween65 using bacteria; A chromosomal aberration study on Tween65 using Chinese hamster cultured cells). Treatment with Tween 80 disclosed negative results derived from rec-assay (Kawachi et al., 1980; Magee, 1976), reverse mutation (Kawachi et al., 1980; Morita et al., 1981), micronucleus assays (Jenssen and Ramel, 1980), and chromosome aberration investigation (Ishidate and Odashima, 1977). In conclusion, in 2015, based on OECD guidlines the Panel on Food Additives and Nutrient Sources added to Food (ANS) released a document that disclosed that Tween 20, 40, 60, 65, and 80 (75 mg/kg bw/day) induced no DNA genotoxicity (EFSA, 2015). The safe dose of Tween 60 and 80 as a view of DNA damage were 5 mg/disk DMSO. However, there is some controversial data in the case of DNA damage to Tween 20 ($2\,\mu L/mL$, 2 mg/assay) by comet, and Ames assays (Eskandani et al., 2013; Odunola et al., 1998).

Another ingredient used in a nanoemulsion system is ethanol as a cosurfactant. There are a lot of reports that investigated genotoxicity of ethanol. However, data based on OECD guidelines disclose nongenotoxicity of ethanol in dermal and inhalation exposure. The OECD accepted (guideline no 471) dose of ethanol is 5 mg/plate. Meanwhile ethanol at a high dose as an aneugen induces Sister Chromatid Exchange (SCE). It is notable that the safepharm concentration of ethanol without chromosome aberration and mutagenic effect is 8 mg/mL, equivalent to 174 mM. Vedmaurty et al. investigated the mutagenic potential of DMSO and ethanol. Based on OECD guideline of 471, it was not seen mutagenic potential of acetone and acetonitril at the concentrations of 100 ul/plate (50 ul/ml) by ames assay (2982 pdf). However, it is mentioned that metabolic products of ethanol is mutagenic. The results of comet assay on *Drosophila melanogaster* showed that ethanol at a higher concentration of 0.625% is genotoxic, and this effect is dose dependent.

In conclusion, ethanol is not mutagenic and clastogenic in the in vitro assessments and micronucleus test in in vivo based on EC B17/OECD 476, EC B14/OECD 473 (5 mg/mL or 10 mM), and EC B11/OECD 474 (2 g/kg), respectively (Phillips and Jenkinson, 2001). However, its effect on drosophila by comet assay was the reverse. To the best of our knowledge and documents surveyed, there is no paper that investigates genotoxicity of nanoemulsions.

14.3 HEMOCOMPATIBILITY AND IMMUNOTOXICITY IN MICELLES AND NANOEMULSION

The aim of hemocompatibility assays is to investigate the effect of nanomaterial on blood cells and proteins. ISO 10993-4 has defined a list of recommended tests consisting evaluation of thrombosis, coagulation, platelet activation, blood cells changes, and complement activation. The nanomaterial via immune stimulation can induce an infusion reaction or hypersensitivity with specific IgE formation that influences whole organs and may result in death. However, there is another similar outcome without specific IgE formation and with a complement activation system that is named C activation-related pseudoallergy (CARPA). The reason that a nanocarrier is recognized by immune cells and complement systems is that pattern recognition receptors on immune cells recognizes repetitive elements on the surface of a nanocarrier that are similar to human pathogenic viruses (40–300 nm) (Szebeni, 2012). Besides, a resemblance of nanocarriers to viruses owing to lack of proteins in the membrane for protecting cells from a complement attack and lack of surface coating to avoid complement escape make them susceptible to a complement attack.

In the subject of nanoemulsion with regards to increase circulation time, when they enter the blood stream they can interact with every component of blood and if they lyse RBCs or if some opsonin is absorbed in their surface they will be engulfed by macrophages and be cleared from the body (Desai, 2012; Fukui et al., 2003; Wehrung et al., 2012; Moghimi and Szebeni, 2003). However, as mentioned earlier they can induce complement system (Moghimi and Szebeni, 2003).

Apart from nanostructures and the charge of micelles and nanoemulsions, ingredients such as surfactants and cosurfactants play a critical role in immunotoxicity and hemotoxicity. Hemolysis percent of span 20, 40, 60, and 80 were investigated by red blood cells treated at different concentrations of surfactants for 30 min in 37°C. Results showed that a percent of hemolysis gradually increases in span 80 (7%), 60 (9%), 40

(12%), and 20 (17%) at the concentration of 1 mM. In fact, attachment of surfactants to RBC influences molecular structure of cell membrane, increases membrane permeability, and afterwards an imbalance of osmotic-colloids resulted in swelling and RBCs' rupture (Noudeh et al., 2009). Schott et al. demonstrated that in one family of nonionic surfactants that have less HLB and higher hydrophobic contents induces less hemolysis. In fact, span 80 as a surfactant in nanoemulsion preparation with the least hemolysis percentage, higher emulsifying index and the potential of foaming production is favored to the other Span members (Schott, 1995). Bulky surfactants with slow lipid solubilization potential that form spheroechinocytes induce less hemolysis as compared to linear surfactants with a spherocyte formation (Ohnishi and Sagitani, 1993).

Bruxel et al. investigated the effect of adsorbed antisense oligonucleotides on cationic nanoemulsion against *Plasmodium falciparum* topoisomerase II. They disclosed that unloaded nanoemulsions (142–200 nm) induce hemolysis (40%) due to cationic lipid surrounding them, while nanoemulsion with antisense oligonucleotides complex (182–343 nm, 9%–12% hemolysis) on their self were uptaken by RBS and attack parasites. However, an increase of charge-influenced hemolytic potential resulted in a hemolysis increment in infected and noninfected RBCs (Bruxel et al., 2011). Apart from positive charge of nanoemulsion that influences cell uptake and hemolysis, it seems that to decrease macrophase and phagositic uptake of nanoemulsion, its stealth with some polymers such as PEG (Desai, 2012), poloxamer (Stolnik et al., 2001), and poloxamine (Alvarez-Lorenzo et al., 2010) via the formation of steric shield around it, will be useful. Also, PEGylation of nanoemulsion has dual effects on complement activation. It seems that although addition of a concentration about 50 mol % decreases particle size from 121 to 98 nm Nevertheless, it will activate the complement system by over-expression of CD11b gene, and in vivo circulation half-lives decrement while 5 mol % does not (Hak et al., 2015).

Khurana et al. synthesized nanoemulsion containing meloxicam as a nonsteroidal anti-inflammatory nanodrug via transdermal delivery route. The composition of nanoemulsion was consisted of caprylic acid, Tween 80 (0.2 mL), water (0.7 mL), Carbopol 940, propylene glycol, and meloxicam (5 mg). Interestingly, nanoemulsion (125 nm) induced less hemolysis percentage as compared to the drug solution, and Tween 80 had a noncomparable difference of hemolysis as compared to water, while hemolysis derived from propylene glycol was less. In fact, meloxicam nanoemulsion induces less hemolysis as compared to its entire component alone, and via lipid modification and extraction of stratum corneum in the face of nanoemulsion, permeability increases (Khurana et al., 2013).

There are some marketed drug micelles that induce a hypersensitivity reaction. Fasturec, a poloxmaric micelle form (15 nm) of Rasburicase that is prescribed for hyperuricemia induces anaphylaxis and bronchospasm and in the patients with glucose-6-phosphate dehydrogenase deficiency induces hemolysis. Taxol, Vumon, and Cyclosporine injections that are a Cremophor EL micelle form of Paclitaxel, Teniposide, and Cyclosporine (8–20 nm), respectively, induce anaphylaxis, angioedema, arrhythmias, and bronchospasm. Other micelle formulations that have the potential to induce severe hypersensitivity are Etoposide and Taxotere with Tween 80 as an emulsifier and a micelle with average diameter of 8–16 nm. These nanodrugs are prescribed for different cancers, and their active ingredients are Podophyllotoxin and Docetaxel, respectively (Szebeni, 2012).

14.4 PULMONARY TOXICITY CONSIDERATIONS

In the case of pulmonary diseases such as asthma, cystic fibrosis, and chronic obstructive pulmonary disease (Zhang et al., 2011), it seems that inhaled drugs are a convenient and efficacious method with decrement of dose and first pass metabolism due to a thin epithelial layer (0.2–1 μm), extended vascularization, and a large surface area (>100 m^2) (Zhang et al., 2011; Weber et al., 2014) that resulted in a high concentration of nanomedication in the target site (Todoroff and Vanbever, 2011). Although, an investigation on lipid-based nanoparticles with pulmonary route administration is still in the initial steps (Weber et al., 2014) in which its application has generated negligible toxicological concerns (Garcia, 2014). For example, the administration of a higher dose of intravenous lipid emulsion (ILE) (20 and 40 mg/kg) as compared to the typical dose had no adverse effect as a view of LD50 and histological observation on pulmonary system in rats (Hiller et al., 2010).

14.5 CARDIOVASCULAR TOXICITY

Cardiac toxicity is one of the most threatening complications of some drugs and to overcome this obstacle drug encapsulation into lipid-based carriers has been suggested. One of the novel methodologies to overcome the cardiac toxicity of lipophilic drugs that is the same as a local anesthetic-induced cardiovascular collapse is ILE (Arora et al., 2013; Bourne et al., 2010; Choi et al., 2004; Garrett et al., 2013; Mir and Rasool, 2014; Rothschild et al., 2010). It is important to mention that this methodology has moved to clinical practice. Its discovery backs to investigation on the cardiac toxicity of bupivacaine along with lipid emulsion, although they hypothesized that it had a synergism potential in cardiac toxicity, but the experimental results were different. The main composition usually are soy bean, egg phospholipids and triglyceride, and lipid sink has been assumed as one important mechanism of a drug scavenger (Bourne et al., 2010). However, the 100% long chain triglyceride instead of long and short triglyceride will be more effective for binding of an overdose drug to ILE (Bariskaner et al., 2003).

14.6 LIPOSOME

The first reported investigation for the preparation of liposome containing anthracycline backs to 1970s, when anthracycline cardiotoxicity push researchers to encapsulate them into liposomes. So, three liposome–dedicated start-up companies in the United States tried to industrialize it (Weissig et al., 2014). It is noticeable that there is no document that refers term of "nano" to liposome structure by 2000. However, the first FDA approval was issued for Doxil, a liposomal formulation of doxorubicin in 1995 (Weissig et al., 2014).

Liposome as a drug nanocarrier is a vesicular structure with a bilayer membrane surrounding a hydrophilic core. Its segregated chemical formulation enables it to carry hydrophobic in a bilayer membrane and hydrophilic compartments in interior. Its main component is phosphatidylcholine derived from soya bean lecithin or egg lecithin (Parnham and Wetzig, 1993). At the outset researchers attempted to decrement the drug's cardiotoxicity by using liposomes, but there were some obstacles related to it that in the present section will be discussed.

14.7 CYTOTOXICITY AND GENOTOXICITY

One of the major components of liposomes as mentioned above is lipid; soya lecithin, phosphatidylcholine, serine, inositol, etc. Soya lecithin and phosphatidylcholine are frequently used in food and in the pharmaceuticals industry, respectively (Parnham and Wetzig, 1993). No genotoxicity is reported by phosphatidylcholine (Gundermann and Schumacher, 1990). The chemical composition of lipids influence cell proliferation and viability. Reports indicated that cardiolipin and stearylamine induce significantly higher cell mortality as compared to the phosphatidylglycerol and phosphatidylserine while liposomes consisting of phosphatidylcholine or dipalmitoylphosphatidylcholine were nontoxic even at 3000–4000 µM concentration (Mayhew et al., 1987). Lysolecithin has been known as a powerful cytotoxic and hemolytic agent in medicine (Jeannet and Hässig, 1966; Bergfeld and Belsito, 2014) and exhibits wheal and erythema reactions through intracutaneous injection. However, lecithin and hydrogenated lecithin in the free and liposomal formulation are not irritants and sensitizers for human skin (Lecithin, 2001).

The charge of liposomes is another important factor in liposome's cytotoxicity. The positively charged liposomes such as stearylamine due to fast amine desorption rate by cells have a significantly higher cytotoxic potential as compared to the amphipathic amines with two chains (Panzner and Jansons, 1979). Over all, it might be said that some positively charged liposomes have higher cytotoxic potential as compared to the negatively and neutral ones (Rustum et al., 1979). The positively charged micelles and liposomes induce strand breaks and the DNA repair enzyme of OGG1 in rats' lung and liver, respectively. They do not alter HMOX1 gene over expression in tissues at single dose that confirmed the toxic mechanism of positively charged liposomes are independent to oxidative stress (Knudsen et al., 2015). Increase of surface charge increases genotoxicity by sensitive analysis of comet assay (Shah et al., 2013). In fact, the positive charge of liposomes induces oxidative stress and intracellular calcium increment in cells and damages to the cell organelles, the same as mitochondria or bimolecules of DNA (Knudsen et al., 2015). There is no sign of genotoxicity in mouse lymphoma L5178Y cells and human lymphocytes by phosphatidylserine (Heywood et al., 1987). Shafaei et al., investigated the genotoxicity of liposomes containing *Orthosiphon stamineus* ethanolic extract in rats by Ames assay in acute and subchronic studies. The results showed no genotoxicity derived from liposomes (Shafaei et al., 2015). In another investigation to decrease the mutagenic effects of cyclophosphamide, a liposomal formulation containing of phosphatidylcholine, cholesterol, and tuftsin (80 $\pm$ 10 nm) was prepared. The data disclosed that not only the pretreatment of nanoliposomal formulation has been inhibited mutagenic effects of cyclophosphamide but also increased mitogenic index in mice (Arif et al., 2009). However, Abdella demonstrated that encapsulation of cyclophosphamide into multilamellar liposomes (Hydrogenated Soy Phosphatidylcholine (HSPC) and cholesterol), which resulted in chromosomal aberrations, SCEs frequencies increment in the same mice as a free drug with a mitogenic index decrement and prolonged cell cycle kinetics. These effects are eventually due to a fusion of liposomes with cell membrane and afterwards accumulation of drug in cells to cause genotoxicity (Abdella, 2012). Overall, it seems that positively charged liposomes are genotoxic even at a noncytotoxic concentration while neutral ones were not from micronuclei in cells (Shah et al., 2013). As a view of comparative investigation between cationic emulsion and liposomal-based gene carriers, findings showed that cationic emulsion (DC-Chol/DOPE, Caster oil and Tween 80) induced higher gene transfection efficacy and prolonged circulation time as compared to the cationic liposome (DC-Chol/DOPE) at the same particle size range (150–230 nm) (Hiller et al., 2010). Although the liposome (L3) at the same size range had induced higher cell viability as compared to the emulsion ones (E4), both of them had a worsened cell viability as compared to naked ones (Choi et al., 2004).

14.8 HEMOCOMPATIBILITY AND IMMUNOTOXICITY OF LIPOSOMES

Liposomes are famous for inducing CARPA in humans (Szebeni, 2012). For the first time, in 1999, scientists observed a pseudo-allergy reaction in pigs that had been infused by doxorubicin liposome. The abnormalities were related to the bronchopulmonary, hemodynamic, cardiac airtimes, and skin changes in patients (Szebeni, 2012). In fact, Doxil is not distinguishable from HIV-1 virus and might be due to this similarity in size to the viruses capables to provoke an immune response. Another suggested reason is related to the absence of cell membrane proteins that hide cells from complement (Szebeni, 2012).

To the subject of hemocompatibility derived from ingredient of liposome, it is worthy of value to mention that hydrolysis of phosphatidylcholine is capable to induce hemolysis (30%) as compared to the phosphatidylcholine in in vitro (Parnham and Wetzig, 1993). However, liposomes containing *O. stamineus* ethanolic extract did not show any sign of hematological changes in rats in the acute and sub-chronic toxicity investigations (Shafaei et al., 2015). It is valuable to mention that the charge of phospholipids affects their hemocompatibility. Negatively charged phospholipids induce platelet aggregation in pigs and addition of stearylamine or phosphatidic acid to phosphatidylcholine exhibit human platelet aggregation, as well (Parnham and Wetzig, 1993). The positively charged liposomes induce over expression of proinflammatory cytokines IL6, CCL2 (Knudsen et al., 2015) in liver and IL12, TNF and IFNγ (Tan et al., 1999; Cui et al., 2005; Elsabahy and Wooley, 2013) while the positively micelles over expressed oxidative stress response gene HMOX1, as well (Knudsen et al., 2015). The lung's pro-inflammatory gene expression was unchanged. In fact cationic head group in cationic liposomes induces inflammation not the liposomes itself (Dokka et al., 2000). It is interesting to mention that a combination of cationic liposomes with bacterial DNA induces over expression of CD80/CD86 as a cell surface marker of dendritic cells with an unaltered level of cytokine production (Cui et al., 2005). However, in some cases, immuno-responsiveness of liposomes is unfavorable for the biosystems while it is helpful in cancer treatment due to its antitumor potential (Dileo et al., 2003; Whitmore et al., 2001). Not only polycation plasmid DNA's liposomes provoke pro-inflammatory responses but are also involved in TLR-dependent signaling, which is important to the maturation of phagosome and response thereof (Cui et al., 2005; Blander and Medzhitov, 2006; Dobrovolskaia and McNeil, 2007).

In addition to charges, the degree of phospholipid saturation and liposome stability are important factors because saturated or more stable liposomes especially in multi-lamellar vesicles overuptake in macrophages and interfere with phagocytic function. Conversely, saturated phospholipids are nontoxic to erythrocytes while unsaturated ones have a bit toxicity (Parnham and Wetzig, 1993). Liposomes that have cholesterol as an ingredient owing to opsonization with circulation cholesterol antibody activate complement and led to lethal consequence in pigs (Wassef et al., 1989). Others demonstrated that cardiolipin liposomes containing doxorubicin with cardiotoxicity decrement and spleen overuptake did not show immunotoxicity as a view of alloantigens and mitogenic responsiveness as compared to the free drug in single and multiple doses (Rahman et al., 1986).

14.9 HISTOCOMPATIBILITY

To the subject of the effect of liposomes in whole organs, it is worthy to note that the administration of positively charged micelles and liposomes had no significantly effect in rats' body weight and hematology parameters. However, liposomes increase albumin and triglycerides while micelles had increased albumin in rats (Knudsen et al., 2015).

Although it seems that phosphatidylcholine has no adverse effect except of hypercholesterolemia in dose administration, lysophosphatidylserine's metabolite derived from phosphatidylserine exhibits endocrine and central effects owing to brain catecholamine metabolism interference (Toffano and Bruni, 1980). Dicetylphosphate and stearylamine with a negative and positive charge, respectively induce epileptic seizures and death in mice (Parnham and Wetzig, 1993). Sterol liposomes have less lethal dose in mice and higher cell mortality as compared to the solid and liquid ones, respectively. However, the multilamellar vesicles with larger particle size than unilamellar vesicles more influence mortality of mice and were more toxic in three types of liposomes (Szoka et al., 1987). There are no changes in lung histology and alveolar macrophage phagocytic function in face to inhalation of liposomes consisting of HSPC (Bergfeld and Belsito, 2014). The oral administration of phosphatidylserine in human (300–600 mg) for 12 weeks resulted in no changes in blood chemistry or hematology (Bergfeld and Belsito, 2014).

Based on a meta analysis report in 2012 that compared the conventional anthracycline with liposomal ones, it was disclosed that liposomal anthracycline is more safe for the heart, and induces less myelo suppression, vomiting, nausea, and alopecia in patients (Rafiyath et al., 2012).

14.10 SOLID LIPID NANOPARTICLES

Polymeric nanoparticles, liposomes, and solid lipid nanoparticles (SLNs) have received particular attention in drug delivery for encapsulation of drugs (Durán et al., 2011). SLNs were introduced in 1991 and shown better functions than traditional colloidal carriers such as emulsions, liposomes, and polymeric micro emulsions (Ekambaram et al., 2012). They are solid at room temperature and are dispersed in a mixture of surfactant/cosurfactant solution (Kaur et al., 2012). SLNs have properties such as small size (50–1000 nm), large surface area, high drug loading and the interaction of phases at the interface and are attractive for their potential to improve performance of pharmaceuticals (Rudolph et al., 2004). They are also used as a carrier to nucleic acid delivery for therapeutic purpose (Jin et al., 2012).

SLNs composed of physiological lipid, dispersed in aqueous surfactant solution represent the most interesting class of nanocarriers, as they offer advantages of the ability to readily incorporating lipophilic candidates, improved drug stability, possibility of controlled release, and a higher safety threshold due to avoidance of organic solvents (Bondì and Craparo, 2010; MuÈller et al., 2000; Mehnert and Mäder, 2001). SLNs have some limitations as low drug loading, particle growth, unexpected dynamics of polymeric transitions, risk of gelation, and drug leakage during storage owning to lipid polymorphism (Müller et al., 2002). The small particle size of SLNs ensures the nanoparticles are in close contact to the stratum corneum, thus can increase the amount of encapsulated agents penetrating into the viable skin (Mei et al., 2003). Formulation of cationic solid lipid nanoparticles (CSLNs) for delivering drugs into endothelial and neoplastic tissue is an emerging subject in pharmaceutical science (Kim et al., 2008). This nanosized biomaterial as a drug delivery system combines the advantages of lipid matrix and hydrophilic layer (Kuo and Chen, 2009). In fact, the lipid core of CSLNs serves as a reservoir to load hydrophobic drugs, and the cationic surface favors endocytosis (Kuo and Chen, 2010). In addition, the positive surface charge can modify drug entrapment and release to produce an interesting pharmacokinetic distribution (Xiao et al., 2011; Kuo et al., 2011).

SLNs also have been used for the encapsulation of different active agents (Schubert and Müller-Goymann, 2005). This type of carrier offers the same benefits as traditional ones such as liposomes and polymeric nanoparticles, without their disadvantages (such as physical instability, scale-up problems, leakage of the active agent, and in some cases, cytotoxicity). SLNs are attractive to the pharmaceutical and cosmetic industries because they are stable, versatile, use safe excipients, and are readily scalable (Puglia et al., 2008). Nanostructured lipid carriers (NLCs) are considered to be the second generation of lipid-based particles (SLNs being the first generation). They present at least one liquid lipid at room temperature to form their inner core (Battaglia and Gallarate, 2012). Although both types of lipid nanoparticles, SLNs and NLCs, are able to encapsulate high amounts of hydrophobic drugs. NLCs have not shown the limitations of SLN, regarding drug expulsion during storage caused by the crystalline rearrangements of the solid lipid (Wissing et al., 2004).

A crucial aspect is that SLNs safeguard drugs to a greater extent against chemical degradation as compared to liposomes because there is little access of water to the inner core of lipid particles (Soppimath et al., 2001). At room temperature the particles are in the solid state so the mobility of incorporated drugs, a prerequisite for controlled drug release, is greatly reduced. They can be stabilized by nontoxic surfactants, such as poloxamers and lecithin (Alex et al., 2011). In vitro tests show that temozolomide-loaded SLNs may be suitable as a drug-carrier system for potential i.v. use because of its relatively low cytotoxicity compared to polymeric nanoparticles (Müller et al., 1997). A SLN characteristic of no–burst release was obtained with fat emulsions (Müller et al., 1995).

In recent years, tremendous work has been conducted on the development of SLNs as delivery systems for cytotoxic anticancer agents. Serpe et al. investigated the improved efficacy achieved with doxorubicin-loaded SLNs, and the 50% inhibitory concentration (i.e., IC50) values for HT-28 cell growth of SLN drug formulation were lower than the corresponding conventional drug solutions (doxorubicin: 81.87 vs 126.57 nm) indicating that the SLN-encapsulated formulation can significantly increase bio-activity of cytotoxic drugs (Serpe et al., 2004).

SLNs can be produced without using organic solvents, minimizing the toxicological risks and are easy to scale up and sterilize, thus fulfilling the requirements for an optimum particulate carrier system (Cipolla et al., 2014). The advantages of using SLNs over the use of liposomes are their long-term stability as well as superior drug incorporation efficiency. However, SLNs have not yet been fully explored for the respiratory delivery of antitubercular drugs. In fact, the nebulization of SLNs is a new and upcoming area of research (Pandey and Khuller, 2005). SLNs also are drug delivery systems that have shown promising results of delivering anticancer agents across the BBB to tumor tissues (Battaglia et al., 2014). A good explanation for the increasing interest in these systems was that lipids enhance oral bioavailability and reduce plasma profile variability.

14.11 PREPARATION

Components in the SLNs are water, surfactant, and lipid that is solid in the temperature of body and room. One of the important factors in making SLNs is a surfactant. Lipids that are used in SLNs commonly are triglycerides, partial glycerides, fatty acids, and steroids (Marcato, 2008). Many methods has been used for making SLNs as a solvent diffusion, emulsion and solvent evaporation, heated microemulsion, and high-pressure homogenization.

14.12 SOLVENT EMULSIFICATION–DIFFUSION

A preparation method is based on butyl lactate or benzyl alcohol emulsification of a solid lipid in an aqueous solution in different emulsifiers, followed by dilution of the emulsion with water. It was used to prepare glyceryl monostearate nanodispersions with a narrow size distribution. To increase the loading the process was conducted at $47 \pm 2°C$, and in order to reach the submicron size a high-shear homogenizer was used. The particle size of the SLNs was affected by using different emulsifiers and different lipid loads. By using lecithin and taurodeoxycholic acid sodium salt, an increase of the GMS percentage from 2.5 to 10 and an increase of the mean diameter from 205 to 695 nm and from 320 to 368 nm was observed for the SLNs prepared using benzyl alcohol and butyl lactate, respectively. Transmission electron micrographs of SLNs reveal nanospheres with a smooth surface (Mukherjee et al., 2009).

14.13 HEATED MICROEMULSION

It is consisting of 15% of surfactant (e.g., polysorbate 20 or 60), more than 10% of cosurfactant (e.g., Poloxamer), and water. The procedure is started with the lipid melted (5–10°C above the fusion temperature of lipid), which is added into a hot surfactant and cosurfactant solution and afterwards, under agitation is formed oil/water microemulsion. Under agitation, it is cold at 2–3°C and led to solidification of the particles. This method has a disadvantage due to the excess water, which is necessary to be removed at a high surfactant and cosurfactant concentration. This excess of water can be removed via ultracentrifugation, lyophilization or dialysis. SLN dispersion can be used as granulation fluid for transferring into solid product (tablets, pellets) by a granulation process. High-temperature gradients facilitate rapid lipid crystallization and prevent aggregation (Mathur et al., 2010).

14.14 CHARACTERIZATION

Many methods have been used for this aim, e.g., transmission electron microscopy (TEM), differential scanning calorimetry, X-ray powder diffraction, Fourier infrared spectroscopy, near–infrared spectroscopy (NIR), NIR-chemical imaging, proton nuclear magnetic resonance (NMR), and microscopy techniques (Rahman et al., 2010).

14.15 PROTON NUCLEAR MAGNETIC RESONANCE

NMR technique characterizes the liquid lipid domains within SLN, e.g., the mobility and arrangement of the molecules of oil, through the technique of ^{1}H NMR spectroscopy. NMR data can verify whether a lipid is in the liquid or semisolid/solid state through the difference in the relaxation time of protons in the liquid state and the semisolid/solid state, since the protons in the liquid state have a narrower signal of higher amplitude than those in the semisolid/solid state (Pedersen et al., 2006).

14.16 MICROSCOPY TECHNIQUES

One of the most appropriated microscopic techniques is cryogenic transmission. Electron microscopy (Cryo-TEM) is used in order to obtain direct images of SLNs with high-resolution. TEM allowed to determine the size (diameter) distribution and crystalline structure of nanoparticles. Atomic force microscopy is also used in the characterization of SLNs (Chen et al., 2006).

14.17 TOXIC EFFECTS OF SOLID LIPID NANOPARTICLES IN VIVO

In 1991 SLNs were introduced as an alternative colloidal drug carrier for oral, parenteral, and topical application (Müller and Lucks, 1996). However, prior to its use as a carrier system, there was a need for the quantification of in vivo toxicity of intravenously administered SLNs. For some years toxicity of SLNs were evaluated first in vitro then in vivo. One of the first studies that was done in this area was investigation the viability of human granulocytes incubated with magnetite-loaded SLN that it disclosed an ED50 over 10% (Müller et al., 1996b). In another study was evaluated the toxicity of intravenous administration of SLNs when they were injected in multiple doses. For example, 400 mL SLN dispersion (lipid content 10% [m/m]) were administered to mice via a bolus injection for six times within a period of 20 days and hepatic and splenic tissues were analyzed histologically. High-dosed Compritol containing formulations led to an accumulation of the lipid in liver and spleen and subsequently to pathological alterations. Lipid accumulation and pathological alterations of high-dosed Compritol SLNs were attributed to the slow degradation of the Compritol matrix, which could be shown by performing in vitro studies in human plasma (Weyhers et al., 2006).

In another study with human granulocytes, SLNs were found to be ten times less cytotoxic than PLA nanoparticles and 100-fold less cytotoxic than butylcyanoacrylate nanoparticles (Müller et al., 1996a). Some investigations show the direct or indirect effect of SLNs on peritoneal macrophage however, there were not seen immunomodulatory effects (Schöler et al., 2002). In some cases, it is applied stabilizers in which they have some cytotoxicity effects on cells (Kristl et al., 2008).

Inhalation is a noninvasive approach for both local and systemic drug delivery. To estimate the toxic dose of SLNs in vitro, A549 cells and murine precision-cut lung slices were exposed to increasing concentrations of SLN. Cytotoxic effects of this nanoparticle were assessed with MTT and NRU assays. Inflammation was assessed by measuring chemokine KC and TNF-a levels; results show SLNs induced no significant signs of inflammation. No consistent increase in LDH release, protein levels, or other signs of inflammation such as chemokine KC, IL-6, or neutrophilia were observed in this study. In contrast, the particle control (carbon black) caused inflammatory and cytotoxic effects at corresponding concentrations. These results confirm that repeated inhalation exposure to SLN30 at concentrations lower than a 200-lg deposit dose is safe in a murine inhalation model (Nassimi et al., 2010).

Today's use of SLNs as a carrier for the pulmonary administration has increased. SLNs were prepared by rotro-stator homogenizations using from Compritol SLM as a surfactant. The particles showed a spherical shape and smooth surface. In vivo assessment has been done for the rats with two groups of placebo and SLN. Total cell counts showed no significant difference between placebo and SLN groups. These results suggest that a single intratracheal administration of SLNs do not induce a significant inflammatory airway response in rats and that the SLNs might be a potential carrier for encapsulated drug by the pulmonary route (Sanna et al., 2004). SLNs using cetyl palmitate (CP) (189.0 ± 1.8 nm and zeta potential 34.7 ± 4.1 mV), myristyl myristate (MM) (185.4 ± 6.3 nm and zeta potential −31.9 ± 2.1 mV), and cetyl esters (CE) (197.5 ± 3.6 nm and zeta potential 30.5 ± 1.5 mV) were produced by hot HPH. The solid lipid was heated to around 10°C above its melting point. Afterwards, the mixture was added to a hot aqueous solution of Pluronic F68 under high agitation in an Ultra-Turrax to form a pre-emulsion, which was homogenized using a high pressure homogenizer, applying three homogenization cycles at 600 bar and cooled to form the SLN. The results obtained in cytotoxicity by MTT assays on BALB/c 3T3, and HaCaT cell lines showed that the cytotoxic effect influenced by the lipid matrix. The lipid that showed smaller reduction in cell viability in the tested concentration range was the CE. Although in some concentrations of SLNs, cell viability reduced at the highest concentration (500 µg/mL) (Maia et al., 2000).

SLN composites with lipid cholesteryl-butyrate (Chol-but) (12%), Epikuron 200 (containing about 95% of soy phosphatidylcholine), and sodium taurocholate (3%) were prepared by microemulsion method. The average diameter of SLNs was around 100–150 nm with zeta potential of 29 mV and their shape was spherical. Chol-but SLNs were tested in vitro and proved to be an effective and suitable pro drug of butyrate in cancer treatment. Cytotoxic activity analysis (MTT test at 72 h) performed on four human glioma cell cultures (U87, U373, Lipari, DF) showed that Chol-but SLNs exhibited an IC_{50} of around 0.18–0.25 mM for all cell cultures, indicating that Chol-but SLNs are able to induce a great cell growth inhibition (Brioschi et al., 2008).

Cationic SLNs were prepared by a modified solvent-emulsification method using cholesteryl oleate, glyceryl trioleate, L-α-dioleoyl phosphatidylethanolamine (DOPE), and 3-β-[N-(N0,N0-dimethylaminoethane)-carbamoyl] cholesterol hydrochloride (DC-cholesterol). The cytotoxicity of SLNs (117 nm size and zeta potential of +41.8 mV) was performed using human breast adenocarcinoma cells (MDAMB435 cells) by Cell Counting Kit-8 assay. The concentration of SLNs was varied from 3 to 72 µg/mL, the same range for transfection experiments. SLNs did not induce any damage to cells up to 48 µg/mL and had 20 % cell lethality at the concentration of 72 µg/mL. The cytotoxicity study clearly revealed that SLNs were not toxic so might be used as non cytotoxic core for molecule transfection (Marcato and Durán, 2014).

Cationic SLNs were prepared by phase inversion temperature (PIT) method using three cationic lipids (tetradecyltrimethyl ammonium bromide (CTAB), dimethyl-dioctadecylammonium bromide (DDAB), and N-[1-(2,3-dioleoyloxy)propyl]-N,N,N-trimethylammonium chloride (DOTAP)). The SLNs with DOTAP exhibited a higher zeta potential and smaller particle size (462.9 nm) than the other particles (SLN-CTAB, SLN-DDAB). Thus, the cytotoxicity of these particles on two models of cell cultures (human prostate cancer androgen-non-responsive DU-145 cells and primary cultures of rat astrocytes) was evaluated. DU-145 cells proved to be more sensible, exhibiting an IC_{50} of 125 µg/mL, whereas in primary cultures of rat astrocytes, the IC_{50} was 500 µg/mL. Probably this difference was due to different proliferative capacity of the two cellular types as well as to their different cellular membrane composition (Carbone et al., 2012).

14.18 EFFECT OF THE SIZE ON THE CYTOTOXICITY AND GENOTOXICITY

In one study SLNs consisting of stearic acid, polysorbate 80, and stearylamine has been produced with melt-emulsification. The cytotoxicity and physical properties of SLNs for dermal applications showed a significant dependency to the formulation process. The viability of J774 macrophages, mouse 3T3 fibroblasts, and HaCaT keratinocytes was significantly reduced in the presence of stearylamine. Survival of macrophages was highly affected by stearic acid and stearylamine. In general a viability of more than 90% was observed when semisynthetic glycerides or hard fat was employed to formulate nanoparticles (Weyenberg et al., 2007).

SLNs were prepared by HPH technique using a lipid phase hydrogenated palm oil (S154) and hydrogenated soybean lecithin (Lipoid S100) as an aqueous phase water, oleyl alcohol, thimerosal, and Sorbitol. Both phases were mixed and homogenized at 1000 bar by 20 cycles, obtaining particles with 145.00_3.39 nm size and zeta potential of −19.50 ± 1.80 mV. The MTT assay was conducted as a cytotoxicity evaluation on human breast cancer cell lines, MCF-7 and MDA-MB231. The IC_{50} of SLNs was time-dependent and was different between the tested breast cell lines (287.5 ± 17.7 µg/mL for MCF-7 and 290.2 µg/mL for MDA-MB231). The IC50 values indicated that SLNs had low cytotoxicity effect in these cells. The low cytotoxicity of the SLNs can be attributed to lecithin and components of the aqueous phase, especially the nonionic emulsifier (Abbasalipourkabirreh et al., 2011).

Another study reported an in vivo exposure to two different SLNs in mice. Overall, animals survived without any observational evidence of toxic effect, except for an inflammatory response in adipose tissue eventually due to the effect of SLN components at the adipose tissue sites. Though the exact mechanism responsible for this response is not clear at present. This kind of nanoparticle should be theoretically biocompatible by the nature of the lipids, but surprisingly we faced an inflammation process, allowing us to conclude that this study deserves attention since it provides a basis for further toxicological studies of SLNs in long-term in vivo exposure as well as to investigate the profile of the inflammatory response (Silva et al., 2014).

14.19 COMMENTS AND FUTURE PERSPECTIVES

Administration of lipophilic drugs by oral route is a challenging task due to multiple factors that include poor solubilization of a drug in the gastrointestinal tract, P-glycoprotein efflux, and significant first pass elimination attributed to predisposition through cytochrome P450 enzymes, all of which resulted in poor in-vivo bioavailability. It is better to use a lipid-based system, and SLNs can be one of the best choices for this aim (Porter et al., 2007). It might be used liquid emulsions (LE), but LE have some problems: it has a poor drug-loading capacity and the restriction on usage level of formulation excipients, e.g., surfactants and cosolvents led to its failure in bioavailability improvement in LE (Narang et al., 2007).

Advancements in formulation of lipid nanoceuticals led to the advent of novel lipid nanocarriers, commonly referred to second generation of SLNs which unlike LEs and SLNs are composed of a mixture of incompatible liquid lipids and solid lipids inappropriate and permissible proportions (Muchow et al., 2008).

SLNs and NLS are attractive carriers of actives used as cosmetic and pharmaceutical products. An advantage of NLS is its easy production in large scale and sterilization that led to an increased industry interest in this carrier. Different methods are used for their preparations with the absence of organic solvents or only reagents previously approved by the Food and Drug Administration (FDA) through the GRAS rules. A disadvantage is sometimes the low loading efficiency; however, this can be possible to circumvent, building different kinds of SLNs or NLC that exhibits different capacities for oral, nasal, parenteral, dermic, and ophthalmic administrations. This chapter showed that SLNs or NLC, in general, is nontoxic to cell cultures or in vivo assays. However, it depends on many factors, such as size, chemical components, capped surface, or surface charge. In order to maintain the safety condition it must be studied case by case. It is not possible to generalize their behavior in cells or in vivo studies. Curiously, very few studies related to their genotoxicity were found in the actual literature. In the next years a larger emphasis must be done in this direction.

14.20 POLYMERIC NANOPARTICLES

14.20.1 Chitosan polymer and particles

Chitosan (CS) is a natural polysaccharide that has gained the largest utilization and distribution after cellulose in the biomaterial field (Mincea et al., 2012). This natural polymer is provided from chitin and is the structural element of crustacean exoskeleton, such as in shrimps, lobsters, and crabs.

Today, there is a special interest in utilizing chitosan in biomedical/pharmaceutical applications. A variety of chitosan's therapeutic properties, i.e., in pain relief, suppression of microorganism growth, hemostasis promotion, and cell growth (Badawy et al.,

2004; Balakrishnan et al., 2005; Howling et al., 2001) have been studied. Specially, chitosan has shown a notable toxicity to several bacteria, fungi, and parasites (Jumaa et al., 2002; Guo et al., 2006; Pujals et al., 2008). Therefore, this property would be promising in controlling infectious diseases. Also, different delivery systems based on chitosan, including micro- or nanoparticles and films, have been prepared (Elgadir et al., 2015).

With the emergence of novel nanotechnology-based systems utilizing polymers, a number of studies that address the biocompatibility/toxicity of chitosan carriers are now available. In fact, these works would shed light on the way towards the fabrication of smarter and safer delivery systems. In the literature, chitosan is generally mentioned to be a nontoxic, biocompatible material and therefore appropriate for preparing drug delivery systems. The American FDA has already approved chitosan as a wound dressing material (Wedmore et al., 2006) and also, in some countries, it is approved as a dietary component (Ilium, 1998). According to this statement, biocompatibility/toxicity of chitosan requires to be evaluated for each special case (any of biological parts targeted by chitosan-based systems) beyond the FDA's approval for wound-dressing applications.

A variety of prominent factors that affect chitosan biocompatibility/toxicity includes the polymer characteristics (MW and Degree of Deacetylation), formulation procedures, and the administration routes (Garcia-Fuentes and Alonso, 2012). The cytotoxicity of CS is reported to be totally concentration-dependent, with regard to the deacetylation degree and molecular weight (MW) of CS. To explore the in vivo toxicity, studies conducted on dogs and rabbits exhibited subcutaneous cytotoxicity (for doses of 5–50 mg/kg/day) (Carreño-Gómez and Duncan, 1997; Minami et al., 1996). However, neither of the mouse and rat models has demonstrated the specific clinical toxicity after oral administration of the CS (dose range of 1–15 g/kg/day) for up to 3 months. In accordance with this data, oral intake of CS (up to 6.75 g/day) also has not indicated clinical toxicity in healthy human volunteers (Garcia-Fuentes and Alonso, 2012; Tapola et al., 2008).

[99]mTc labeled nanoparticles (<100 nm) injected in Swiss albino mice have shown an evasion from reticuloendothelial system (RES) (Banerjee et al., 2002). In a contradictory finding, after intravenous administration in Sprague–Dawley rats, a significant uptake of paclitaxel from *N*-octyl-*O*-sulfate chitosan micelles by RES was indicated (Zhang et al., 2008). These variations in scientific findings would be related to the particle diameters in which an augmented size of particles over 200 nm would facilitate uptake by spleen and liver RES (Gupta and Gupta, 2005). In this regard, it is well demonstrated that the liver is a special organ affecting distribution of chitosan after intravenous administration (Alexis et al., 2008). At the cell level, studies illustrated that chitosan nanoparticles (less than 100 nm) are able to be internalized into cells through endocytosis (Mao et al., 2010).

14.21 GENOTOXICITY

A probable risk of toxicity concerns the damages to cellular DNA and genes. In this regard, providing a genotoxicity assay is required to determine the level of damage upon contact of chitosan particles with the cells. The comet assay is routinely utilized to illustrate the DNA damage in individual cells through gel electrophoresis (Lewinski et al., 2008). Yet there are few available studies on genotoxic evaluation of chitosan. Neither of low MW chitosan and chitosan oligosaccharides have elicited genotoxicity in lymphocytes (Fernandes et al., 2011). Furthermore, the comet assay has demonstrated that chitosan-coated silver nanoparticles in a 3 ppm concentration do not show a genotoxicity on mouse macrophages (RAW264.7 cells) despite what is seen in the case of 20 ppm of the nanoparticles (Jena et al., 2012). While, Tavakol et al. (2016) showed that uncoated silver nanoparticles, even when the concentration of cell viability is 100%, have the potential to induce chromosomal breakage. Additionally, CS nanoparticles utilized in food packaging have not indicated any genotoxicity (De Lima et al., 2010).

14.22 HEMOCOMPATIBILITY

Hemocompatibility of chitosan-based structures have to be considered precisely, as according to the literature, it has the potential to induce thrombogenic or/ and hemolytic responses. A composition is blood compatible if it would not lead to hemolysis or activation of blood coagulation system. Chitosan has shown to be highly thrombogenic by the ability to activate complement and blood coagulation systems (Hirano et al., 2000; Brandenberg et al., 1984; Suzuki et al., 2003; Balan and Verestiuc, 2014).

Primary adsorption of plasma proteins onto the chitosan-based particles and then adhesion and activation of platelets are involved in the formation of thrombus (Chou et al., 2003). The induction of conformational changes in fibrinogen (a prominent part of plasma proteins) when interacting with positive charges of free amino groups on chitosan particles resulted in the adhesion of platelets and monocytes. Consequently, these cells will release a number of cleaving molecules that break fibrinogens to form a fibrin network (clot) (Butnaru et al., 2012; Furie and Furie, 2008).

Indeed, hemagglutination is the result of interactions between chitosan (free amino groups) and acidic groups of blood cells (Rao and Sharma, 1997; Shelma and Sharma, 2011).

In another study, it is demonstrated that high concentrations of chitosan oligosaccharides induces human erythrocytes adhesion, aggregation, or even both. Accordingly, it is concluded that the toxic effects of chitosan on erythrocytes was dependent on its concentration and MW so the higher MW or an augmented concentration causes more significant damage on the cells (Fernandes et al., 2008). It is also reported that the association of complement components (C3, C5, and factor B) with chitosan has not activated the complement cascade (Marchand et al., 2010).

To improve the blood compatibility of the chitosan, a variety of strategies has been introduced (Balan and Verestiuc, 2014). In some of these approaches chemical substitution of chitosan polymer is considered and other ones have attempted to utilize a modification on the surface of the prepared particles. In the first attempts, acylation of chitosan was followed and the prepared chitosan derivatives that in addition to good water solubility illustrated significant hemocompatibility (Zhu et al., 2009).

Recently, the antithrombotic, hemolytic, and anticoagulant properties of two water-soluble chitosan derivatives including N-succinyl-chitosan and N,O-succinyl-chitosan (N,O-SCs) have been studied (Xiong et al., 2011). According to the results, excellent blood compatibility of the mentioned chitosan derivatives has been reported and it was concluded that the induction of anticoagulation activity is related to the carboxyl groups attached to the C-6 or C-2 hydroxyl function.

Modification of the chitosan polymer with sulfates has been another approach to provide water-soluble chitosan and increase the hemocompatibility of this polymer (Jayakumar et al., 2007; Vongchan et al., 2002; Xue et al., 2013). In this regard, the studies on the sulfated derivatives of chitosan have also demonstrated anticoagulant activity, hemagglutination inhibition, and antimicrobial and antioxidant activities.

PEGylation of chitosan has been another strategy to provide more biocompatible biopolymeric systems. It is well known that PEGylation of polymeric delivery systems or nanoparticles would increase the blood stability of these systems, through increasing hydrophilicity and rendering particles with an adequate steric hindrance to reduce the undesirable uptake and removal by RES. Therefore, PEGylation results in a prolonged circulation in the blood (Chan et al., 2007; Pasut and Veronese, 2007; Knop et al., 2010). Indeed, this kind of chitosan modification minimizes the unspecific interactions of chitosan delivery systems with the components of blood, including plasma proteins and cells.

14.23 POLY (ETHYLENE IMINE)

Poly (ethylene imine) (PEI) is one of the most utilized nonviral polymeric vectors for gene delivery applications. Indeed, owing to its pH buffering capability, PEI facilitates endosomal escape of gene carrier to support an efficient transduction of desirable genes (Jere et al., 2009; Rekha and Sharma, 2011). PEI-based carriers have been used to deliver DNA or siRNA to the targeted cells. However, lack of degradability is one of the problems that may confine the applicability of PEI in therapeutic purposes. It could be concluded from the literature that PEI-mediated toxicity would depend on its MW and the structure of PEI. In this respect, although LMW-PEIs would not support an efficient gene transfection, their cytotoxicity (LMW linear or branched PEI) is lower compared to the high MW counterparts. Furthermore, the cytotoxicity of linear form of PEIs is lower than branched ones (Cho, 2012).

The increased cytotoxicity of high MW PEIs could be attributed to its more unspecific interactions of cationic charges with cell membranes.

In one study, it was prepared a novel composition containing of siRNA to improve the intracellular siRNA delivery. In this report, chemical conjugation of 1,2-dioleoyl-sn-glycero-3-phosphoethanolamine (DOPE) phospholipid to low MW PEI significantly improved the siRNA delivery while keeping low cytotoxicity effects (Essex et al., 2015).

As mentioned earlier, the lack of degradability of PEIs requires to be considered when applied in biological systems. To prepare degradable PEIs, a number of crosslinking processes were investigated to help intracellular degradation through simple hydrolysis, hydrolysis in acidic pH of endosomes, enzymatic cleaving function, and cytosolic reduction in the cytosol by glutathione (Cho, 2012).

14.24 ALBUMIN NANOPARTICLES

Albumin is one the most important plasma proteins and is a versatile carrier for small molecules in the blood, and consequently provides an efficient delivery system. Human serum albumin (HSA) is now utilized for a wide range of therapeutic applications such as treating hypoalbuminemia, burns, cardiopulmonary bypass, and hemodialysis (Kratz, 2008). ABI 007 (Abraxane), as the first HSA-based nanoparticle formulation, was approved by the FDA in 2005 (Iglesias, 2009). These nanoparticles (130 nm) carry the anticancer drug Paclitaxel. To improve the solubility of paclitaxel in water, polyethylated castor oil (CremophorEL) and ethanol are involved in the conventional drug formulation (Taxol) (Kratz, 2008). The abraxane fabrication procedure removes the need for organic solvents to formulate paclitaxel. However, some consideration is required when administering abraxane. It is reported that bone marrow suppression by abraxane is dose-dependent and there is a dose-limiting toxicity. According to the manufacturer's guideline, abraxane therapy should not be administered to patients with baseline neutrophil counts of less than $1.5 \times 10^9 \mathrm{L}^{-1}$. Also, the peripheral neuropathy of abraxane is a dose-dependent and dose-limiting toxicity and it occurs frequently with abraxane. Also, the occurrence of cardiac events is possible when receiving abraxane and thus monitoring of these patients by physicians is recommended (Rafiyath et al., 2012).

In addition to HSA protein, bovine serum albumin (BSA) has been successfully utilized in the formulation process of therapeutics. In one study, ciprofloxacin-loaded

BSA microspheres were fabricated by a spray drying technique for lung delivery (Li et al., 2001). In another investigation, BSA was employed to deliver and maintain tetrandrine in the lung in order to reduce any antisilicotic effects to other body organs (Wang et al., 2007).

Although albumin as a natural protein is considered to be biocompatible, an investigation of its biodistribution and potential routs of toxicity and metabolism of the aggregated albumin when using it as a drug delivery carrier is needed.

Recently, Elblbesy (2016) has prepared bovine albumin nanoparticles (BANs) with a simple coacervation approach to evaluate their hemocompatibility. In vitro hemocompatibility tests demonstrated that the hemolysis percentage of erythrocytes was reduced due to the exposure to the BAN (250–350 nm). Also, a normal range of parameters including hemoglobin, mean corpuscle hemoglobin, and mean corpuscle hemoglobin concentration was observed. The results of this study supports the idea of using BAN in biomedical applications specifically as there are no notable blood toxicity effects.

In another study, the in vivo administration of albumin nanoparticles (2 and 20 μg/mouse) illustrated no inflammatory responses (TNF-α, IL-6, cellular infiltration and protein concentration) compared to vehicle controls. However, a high dose administration (390 μg/mouse) revealed an increased mononucleocytes and a mild inflammatory effect (Woods et al., 2015).

The animal study of pulmonary administration of albumin suspension particles showed initial pulmonary retention and then a clearance from the lungs and translocation to the liver and spleen. Particle suspensions with smaller diameters illustrated a faster clearance from the lungs. It was also shown that in the case of smaller particles and lower proportion of larger ones in the suspension, the incidence of capillary embolization was greatly reduced (Taplin et al., 1964). One limitation in the administration of formulations prepared using albumin nanoparticles is their poor in vivo colloidal stability.

As mentioned earlier, investigation of the fate and biodistribution of nanoparticulate systems consisting of albumin would help design and fabricate more efficient delivery systems. It was shown that the clearance of ^{111}In labeled albumin solution was faster than ^{111}In labeled albumin nanoparticles over 48 h following a single pulmonary administration to mice (Woods et al., 2015). The results showed a significantly higher level of radioactivity for albumin nanoparticles located within the lung tissue compared to the lung fluid. Detection of low amounts of ^{111}In activity in other organs including the liver, kidneys, and intestine indicates that some of the administered formulation is cleared from the lungs and translocated to other organs. This study provides important information on the fate of albumin vehicles in the lungs, which may be used to direct future formulation design of inhaled nanomedicines.

14.25 NANOFIBERS AS DRUG DELIVERY SYSTEMS

The delivery of pharmaceutical agents or drugs to the desirable tissue and cells utilizing nanofibrous structures has attracted a special consideration from researchers. In such a delivery system, biodegradable and nondegradable polymers could be used. Three main approaches including self-assembly (Tavakol et al., 2016 a,b and c), phase-separation (Salehi et al, 2015), and electrospinning (Hoveizi, 2016 and 2017) have been employed to prepare nanofibers (Rathinamoorthy, 2012). Drugs would be sequestered inside or in the matrix of the comprised materials.

The number of studies that concern the toxicity of fibrous scaffolds capable of drug release is limited. However, it seems that their toxicity is related to the type of materials used and also the characteristics of the nanofibers (e.g., diameter of the fibers). In this respect, the chemical composition of the polymers controls the interactions around molecules and cells. There are some reports on the cytotoxic investigations of self-assembling peptide nanofibers that disclosed its cell protection is related to the biological motif and the core of self-assembling section. It is demonstrated that based on the cell membrane damage derived from the acidic pH of self-assembling's core. the reactive acterocyte and dark neurons will be increased in in-vivo model of spinal cord injury in rat. However the nature of biological motif will be important, as well and sensitivness of cells are different together in face to acidic pH or proton concentration and neuroblastoma cells are more resistant to cell membrane damage and viability as compared to the endometrial stromal cells (Tavakol, 2016 a,b and c).

An in vitro study revealed that a decrease in mean diameter of fibers results in a reduction of proinflammatory cytokine secretion by macrophages (Saino et al., 2011). As the scaffolds are mostly used to regenerate the lost tissues and to support cell functions, there may also be interactions of macrophages with the fibers and, consequently, a release of proinflammatory molecules. In a recent and contradictory study, the interactions of macrophages with thicker fibers containing larger pores and also thinner fiber scaffolds with smaller pores have been investigated (Wang et al., 2014). The results illustrated that thicker-fiber scaffolds with larger pores may help macrophage polarization into M2 phenotype. This polarized phenotype secretes wound-healing cytokines and helps cell infiltration and vascularization. However, the thinner-fiber ones with smaller pores play a negative role in vascular regeneration by preserving the macrophages in M1 phenotype, which secrete proinflammatory factors.

In another interesting study, the in vivo stability, distribution, and toxicity of supramolecular nanofibers formed by Nap-GFFYGRGD (L-amino acid-based, L-fibers) and Nap-G^{D}F^{D}F^{D}YGRGD (D-amino acid−based, D–fibers), respectively, were investigated (Yang et al., 2015). These nanofibers were formed by self-assembly method. The results demonstrated that D-fibers possess a better in vitro and in vivo bio-stabilities compared to L-fibers. Also, D-fibers show an appropriate integrity in plasma during 24 h, while half of the L-fibers are broken down upon plasma incubation for 6 h. The distribution studies illustrated that L-fibers mostly accumulate in the stomach, whereas D-fibers are well distributed in the liver. The toxicity investigations revealed that consecutive administrations of both L- and D-fibers with the dose of 30 mg/kg/dose provoke no significant inflammation, liver and kidney dysfunction, immune responses, or impairment of the hematopoietic system.

REFERENCES

Abbasalipourkabirreh, R., Salehzadeh, A., Abdullah, R., 2011. Cytotoxicity effect of solid lipid nanoparticle on human breast cancer cell lines. Biotechnology 10, 528–533.

Abdella, E.M., 2012. Short-term comparative study of the cyclophosphamide genotoxicity administered free and liposome-encapsulated in mice. Iran. J. Cancer Prev. 5, 51.

A chromosomal aberration study on Tween65 using Chinese hamster cultured cells (sponsored by the National Institute of Health Sciences) Hatano Research Institute, Food and Drug Safety Center.

Alex, M.A., Chacko, A., Jose, S., Souto, E., 2011. Lopinavir loaded solid lipid nanoparticles (SLN) for intestinal lymphatic targeting. Eur. J. Pharm. Sci. 42, 11–18.

Alexis, F., Pridgen, E., Molnar, L.K., Farokhzad, O.C., 2008. Factors affecting the clearance and biodistribution of polymeric nanoparticles. Mol. Pharm. 5, 505–515.

Alvarez-Lorenzo, C., Rey-Rico, A., Sosnik, A., Taboada, P., Concheiro, A., 2010. Poloxamine-based nanomaterials for drug delivery. Front. Biosci. 2, 424–440.

A micronucleus test on Tween65 using mice (sponsored by the National Institute of Health Sciences) Hatano Research Institute, Food and Drug Safety Center.

A reverse mutation study on Tween65 using bacteria (sponsored by the National Institute of Health Sciences) Hatano Research Institute, Food and Drug Safety Center.

Arif, K., Ejaj, A., Maroof, A., Azmat, A., Arun, C., et al., 2009. Protective effect of liposomal formulation of tuftsin (a naturally occurring tetrapeptide) against cyclophosphamide-induced genotoxicity and oxidative stress in mice. Ind. J. Biochem. Biophys. 46, 45.

Arora, N.P., Berk, W.A., Aaron, C.K., Williams, K.A., 2013. Usefulness of intravenous lipid emulsion for cardiac toxicity from cocaine overdose. Am. J. Cardiol. 111, 445–447.

Badawy, M.E., Rabea, E.I., Rogge, T.M., Stevens, C.V., Smagghe, G., et al., 2004. Synthesis and fungicidal activity of new N, O-acyl chitosan derivatives. Biomacromolecules 5, 589–595.

Balakrishnan, B., Mohanty, M., Umashankar, P., Jayakrishnan, A., 2005. Evaluation of an in situ forming hydrogel wound dressing based on oxidized alginate and gelatin. Biomaterials 26, 6335–6342.

Balan, V., Verestiuc, L., 2014. Strategies to improve chitosan hemocompatibility: a review. Eur. Polym. J. 53, 171–188.

Banerjee, T., Mitra, S., Singh, A.K., Sharma, R.K., Maitra, A., 2002. Preparation, characterization and biodistribution of ultrafine chitosan nanoparticles. Int. J. Pharm. 243, 93–105.

Bariskaner, H., Tuncer, S., Taner, A., Dogan, N., 2003. Effects of bupivacaine and ropivacaine on the isolated human umbilical artery. Int. J. Obstet. Anesth. 12, 261–265.

Battaglia, L., Gallarate, M., 2012. Lipid nanoparticles: state of the art, new preparation methods and challenges in drug delivery. Expert Opin. Drug Deliv. 9, 497–508.

Battaglia, L., Gallarate, M., Peira, E., Chirio, D., Muntoni, E., et al., 2014. Solid lipid nanoparticles for potential doxorubicin delivery in glioblastoma treatment: preliminary in vitro studies. J. Pharm. Sci. 103, 2157–2165.

Bergfeld, W.F., Belsito, D.V., 2014. Safety Assessment of Lecithin and Other Phosphoglycerides as Used in Cosmetics. http://www.cir-safety.org/sites/default/files/lecithl122014tent.pdf.

Blander, J.M., Medzhitov, R., 2006. On regulation of phagosome maturation and antigen presentation. Nat. Immunol. 7, 1029–1035.

Bondì, M.L., Craparo, E.F., 2010. Solid lipid nanoparticles for applications in gene therapy: a review of the state of the art. Expert Opin. Drug Deliv. 7, 7–18.

Bourne, E., Wright, C., Royse, C., 2010. A review of local anesthetic cardiotoxicity and treatment with lipid emulsion. Local Reg. Anesth. 3, 11.

Brandenberg, G., Leibrock, L.G., Shuman, R., Malette, W.G., Quigley, H., 1984. Chitosan: a new topical hemostatic agent for diffuse capillary bleeding in brain tissue. Neurosurgery 15, 9–13.

Brioschi, A., Zara, G.P., Calderoni, S., Gasco, M.R., Mauro, A., 2008. Cholesterylbutyrate solid lipid nanoparticles as a butyric acid prodrug. Molecules 13, 230–254.

Bruxel, F., Cojean, S., Bochot, A., Teixeira, H., Bories, C., et al., 2011. Cationic nanoemulsion as a delivery system for oligonucleotides targeting malarial topoisomerase II. Int. J. Pharm. 416, 402–409.

Butnaru, M., Dimitriu, C.D., Macocinschi, D., Knieling, L., Bredetean, O., et al., 2012. Biocompatibility and Biological Performance of the Improved Polyurethane Membranes for Medical Applications. INTECH Open Access Publisher, Croatia.

Carbone, C., Tomasello, B., Ruozi, B., Renis, M., Puglisi, G., 2012. Preparation and optimization of PIT solid lipid nanoparticles via statistical factorial design. Eur. J. Med. Chem. 49, 110–117.

Carreño-Gómez, B., Duncan, R., 1997. Evaluation of the biological properties of soluble chitosan and chitosan microspheres. Int. J. Pharm. 148, 231–240.

Chan, P., Kurisawa, M., Chung, J.E., Yang, Y.-Y., 2007. Synthesis and characterization of chitosan-g-poly (ethylene glycol)-folate as a non-viral carrier for tumor-targeted gene delivery. Biomaterials 28, 540–549.

Chen, H., Chang, X., Du, D., Liu, W., Liu, J., et al., 2006. Podophyllotoxin-loaded solid lipid nanoparticles for epidermal targeting. J. Control. Release 110, 296–306.

Cho, C.-S., 2012. Design and development of degradable polyethylenimines for delivery of DNA and small interfering RNA: an updated review. ISRN Mater. Sci. 2012, 798247.

Choi, W.-J., Kim, J.-K., Choi, S.-H., Park, J.-S., Ahn, W.S., et al., 2004. Low toxicity of cationic lipid-based emulsion for gene transfer. Biomaterials 25, 5893–5903.

Chou, T.-C., Fu, E., Wu, C.-J., Yeh, J.-H., 2003. Chitosan enhances platelet adhesion and aggregation. Biochem. Biophys. Res. Commun. 302, 480–483.

Cipolla, D., Shekunov, B., Blanchard, J., Hickey, A., 2014. Lipid-based carriers for pulmonary products: preclinical development and case studies in humans. Adv. Drug Deliv. Rev. 75, 53–80.

Commission FS, 2007. Evaluation Report of Food Additives Polysorbates (Polysorbates 20, 60, 65 and 80).

Cui, Z., Han, S.-J., Vangasseri, D.P., Huang, L., 2005. Immunostimulation mechanism of LPD nanoparticle as a vaccine carrier. Mol. Pharm. 2, 22–28.

De Lima, R., Feitosa, L., Pereira, Ad.E.S., De Moura, M.R., Aouada, F.A., et al., 2010. Evaluation of the genotoxicity of chitosan nanoparticles for use in food packaging films. J. Food Sci. 75, N89–N96.

Desai, N., 2012. Challenges in development of nanoparticle-based therapeutics. AAPS J. 14, 282–295.

Dileo, J., Banerjee, R., Whitmore, M., Nayak, J.V., Falo, L.D., et al., 2003. Lipid–protamine–DNA-mediated antigen delivery to antigen-presenting cells results in enhanced anti-tumor immune responses. Mol. Ther. 7, 640–648.

Doak, S., Manshian, B., Jenkins, G., Singh, N., 2012. In vitro genotoxicity testing strategy for nanomaterials and the adaptation of current OECD guidelines. Mutat. Res. 745, 104–111.

Dobrovolskaia, M.A., McNeil, S.E., 2007. Immunological properties of engineered nanomaterials. Nat. Nanotechnol. 2, 469–478.

Dokka, S., Toledo, D., Shi, X., Castranova, V., Rojanasakul, Y., 2000. Oxygen radical-mediated pulmonary toxicity induced by some cationic liposomes. Pharm. Res. 17, 521–525.

Durán, N., Teixeira, Z., Marcato, P.D., 2011. Topical application of nanostructures: solid lipid, polymeric and metallic nanoparticles Nanocosmetics and Nanomedicines. Springer, Germany. 69–99

EFSA, 2015. Scientific Opinion on the re-evaluation of polyoxyethylene sorbitan monolaurate (E 432), polyoxyethylene sorbitan monooleate (E 433), polyoxyethylene sorbitan monopalmitate (E 434), polyoxyethylene sorbitan monostearate (E 435) and polyoxyethylene sorbitan tristearate (E 436) as food additives1. Parma, Italy: EFSA Panel on Food Additives and Nutrient Sources added to Food (ANS).

Ekambaram, P., Sathali, A.A.H., Priyanka, K., 2012. Solid lipid nanoparticles: a review. Sci. Rev. Chem. Commun. 2, 80–102.

Elblbesy, M.A., 2016. Hemocompatibility of albumin nanoparticles as a drug delivery system—an in vitro study. J. Biomater. Nanobiotechnol. 7, 64.

Elgadir, M.A., Uddin, M.S., Ferdosh, S., Adam, A., Chowdhury, A.J.K., et al., 2015. Impact of chitosan composites and chitosan nanoparticle composites on various drug delivery systems: a review. J. Food Drug Anal. 23, 619–629.

Elsabahy, M., Wooley, K.L., 2013. Cytokines as biomarkers of nanoparticle immunotoxicity. Chem. Soc. Rev. 42, 5552–5576.

Eskandani, M., Hamishehkar, H., Ezzati Nazhad Dolatabadi, J., 2013. Cyto/genotoxicity study of polyoxyethylene (20) sorbitan monolaurate (Tween 20). DNA Cell Biol. 32, 498–503.

Essex, S., Navarro, G., Sabhachandani, P., Chordia, A., Trivedi, M., et al., 2015. Phospholipid-modified PEI-based nanocarriers for in vivo siRNA therapeutics against multidrug-resistant tumors. Gene Ther. 22, 41–50.

Fernandes, J.C., Eaton, P., Nascimento, H., Belo, L., Rocha, S., et al., 2008. Effects of chitooligosaccharides on human red blood cell morphology and membrane protein structure. Biomacromolecules 9, 3346–3352.

Fernandes, J.C., Borges, M., Nascimento, H., Bronze-da-Rocha, E., Ramos, O.S., et al., 2011. Cytotoxicity and genotoxicity of chitooligosaccharides upon lymphocytes. Int. J. Biol. Macromol. 49, 433–438.

Floury, J., Desrumaux, A., Axelos, M.A., Legrand, J., 2003. Effect of high pressure homogenisation on methylcellulose as food emulsifier. J. Food Eng. 58, 227–238.

Fukui, H., Koike, T., Nakagawa, T., Saheki, A., Sonoke, S., et al., 2003. Comparison of LNS-AmB, a novel low-dose formulation of amphotericin B with lipid nano-sphere (LNS®), with commercial lipid-based formulations. Int. J. Pharm. 267, 101–112.

Furie, B., Furie, B.C., 2008. Mechanisms of thrombus formation. New Engl. J. Med. 359, 938–949.

Garcia, Fd.M., 2014. Nanomedicine and therapy of lung diseases. Einstein (São Paulo) 12, 531–533.

Garcia-Fuentes, M., Alonso, M.J., 2012. Chitosan-based drug nanocarriers: where do we stand? J. Control. Release 161, 496–504.

Garrett, R., Kaura, V., Kathawaroo, S., 2013. Intravenous lipid emulsion therapy–the fat of the land. Trends Anaesth. Criti. Care 3, 336–341.

Gundermann, K.-J., Schumacher, R., 1990. 50th Anniversary of Phospholipid Research (EPL). wbn-Verlag, Bingen.

Guo, Z., Chen, R., Xing, R., Liu, S., Yu, H., et al., 2006. Novel derivatives of chitosan and their antifungal activities in vitro. Carbohydr. Res. 341, 351–354.

Gupta, A.K., Gupta, M., 2005. Synthesis and surface engineering of iron oxide nanoparticles for biomedical applications. Biomaterials 26, 3995–4021.

Hak, S., Garaiova, Z., Olsen, L.T., Nilsen, A.M., de Lange Davies, C., 2015. The effects of oil-in-water nanoemulsion polyethylene glycol surface density on intracellular stability, pharmacokinetics, and biodistribution in tumor bearing mice. Pharm. Res. 32, 1475–1485.

Heywood, R., Cozens, D., Richold, M., 1987. Toxicology of a phosphatidylserine preparation from bovine brain (BC-PS). Clin. Trials J. 24, 25–32.

Hiller, D.B., Di Gregorio, G., Kelly, K., Ripper, R., Edelman, L., et al., 2010. Safety of high volume lipid emulsion infusion: a first approximation of LD50 in rats. Reg. Anesth. Pain Med. 35, 140–144.

Hirano, S., Zhang, M., Nakagawa, M., Miyata, T., 2000. Wet spun chitosan–collagen fibers, their chemical N-modifications, and blood compatibility. Biomaterials 21, 997–1003.

Hoveizi, E., Massumi, M., Ebrahimi-barough, S., Tavakol, S., Ai, J., 2015. Differential effect of Activin A and WNT3a on definitive endoderm differentiation on electrospun nanofibrous PCL scaffold. Cell Biol. Int. 39 (5), 591–599.

Hoveizi, E., Ebrahimi-Barough, S., Tavakol, S., Sanamiri, K., 2017. In vitro differentiation of human iPS cells into neural like cells on a biomimetic polyurea. Mol. Neurobiol. 54 (1), 601–607.

Howling, G.I., Dettmar, P.W., Goddard, P.A., Hampson, F.C., Dornish, M., et al., 2001. The effect of chitin and chitosan on the proliferation of human skin fibroblasts and keratinocytes in vitro. Biomaterials 22, 2959–2966.

Iglesias, J., 2009. nab-Paclitaxel (Abraxane®): an albumin-bound cytotoxic exploiting natural delivery mechanisms into tumors. Breast Cancer Res. 11, S21.

Ilium, L., 1998. Chitosan and its use as a pharmaceutical excipient. Pharm. Res. 15, 1326–1331.

Ishidate, M., Odashima, S., 1977. Chromosome tests with 134 compounds on Chinese hamster cells in vitro—a screening for chemical carcinogens. Mutat. Res. 48, 337–353.

Jayakumar, R., Nwe, N., Tokura, S., Tamura, H., 2007. Sulfated chitin and chitosan as novel biomaterials. Int. J. Biol. Macromol. 40, 175–181.

Jeannet, M., Hässig, A., 1966. [The role of lysophosphatides and fatty acids in hemolysis]. Helvet. Med. Acta 30, 756–795.

Jena, P., Mohanty, S., Mallick, R., Jacob, B., Sonawane, A., 2012. Toxicity and antibacterial assessment of chitosan-coated silver nanoparticles on human pathogens and macrophage cells. Int. J. Nanomedicine 7, 1805–1818.

Jenssen, D., Ramel, C., 1980. The micronucleus test as part of a short-term mutagenicity test program for the prediction of carcinogenicity evaluated by 143 agents tested. Mutat. Res. 75, 191–202.

Jere, D., Jiang, H., Arote, R., Kim, Y., Choi, Y., et al., 2009. Degradable polyethylenimines as DNA and small interfering RNA carriers. Expert Opin. Drug Deliv. 6, 827–834.

Jin, S.-E., Kim, C.-K., Kim, Y.-B., 2012. Cellular delivery of cationic lipid nanoparticle-based SMAD3 antisense oligonucleotides for the inhibition of collagen production in keloid fibroblasts. Eur. J. Pharm. Biopharm. 82, 19–26.

Jumaa, M., Furkert, F.H., Müller, B.W., 2002. A new lipid emulsion formulation with high antimicrobial efficacy using chitosan. Eur. J. Pharm. Biopharm. 53, 115–123.

Kada, T., Hirano, K., Shirasu, Y., 2013. Screening of environmental chemical mutagens by the Rec-assay system with Bacillus subtilis Chemical Mutagens. Springer, Berlin. 149–373.

Kaur, I., Singh, M., Verma, M., 2012. Need to Reformulate Streptomycin for Unconventional Routes of Administration-An Overview. New Delhi.

Kawachi, T., Yahagi, T., Kada, T., Tazima, Y., Ishidate, M., et al., 1980. Cooperative programme on short-term assays for carcinogenicity in Japan. IARC (Int. Agency Res. Cancer) 27, 323–330.

Khurana, S., Jain, N., Bedi, P., 2013. Nanoemulsion based gel for transdermal delivery of meloxicam: physico-chemical, mechanistic investigation. Life Sci. 92, 383–392.

Kim, H.R., Kim, I.K., Bae, K.H., Lee, S.H., Lee, Y., et al., 2008. Cationic solid lipid nanoparticles reconstituted from low density lipoprotein components for delivery of siRNA. Mol. Pharm. 5, 622–631.

Knop, K., Hoogenboom, R., Fischer, D., Schubert, U.S., 2010. Poly (ethylene glycol) in drug delivery: pros and cons as well as potential alternatives. Angew. Chem. Int. Ed. 49, 6288–6308.

Knudsen, K.B., Northeved, H., Ek, P.K., Permin, A., Gjetting, T., et al., 2015. In vivo toxicity of cationic micelles and liposomes. Nanomedicine 11, 467–477.

Kratz, F., 2008. Albumin as a drug carrier: design of prodrugs, drug conjugates and nanoparticles. J. Control. Release 132, 171–183.

Kristl, J., Teskac, K., Milek, M., Mlinaric-Rascan, I., 2008. Surface active stabilizer tyloxapol in colloidal dispersions exerts cytostatic effects and apoptotic dismissal of cells. Toxicol. Appl. Pharmacol. 232, 218–225.

Kuo, Y.-C., Chen, H.-H., 2009. Entrapment and release of saquinavir using novel cationic solid lipid nanoparticles. Int. J. Pharm. 365, 206–213.

Kuo, Y.-C., Chen, H.-H., 2010. Effect of electromagnetic field on endocytosis of cationic solid lipid nanoparticles by human brain-microvascular endothelial cells. J. Drug Target. 18, 447–456.

Kuo, Y.-C., Lin, P.-I., Wang, C.-C., 2011. Targeting nevirapine delivery across human brain microvascular endothelial cells using transferrin-grafted poly (lactide-co-glycolide) nanoparticles. Nanomedicine 6, 1011–1026.

Lecithin, H., 2001. Final report on the safety assessment of lecithin and hydrogenated lecithin. Int. J. Toxicol. 20, 21–45.

Lewinski, N., Colvin, V., Drezek, R., 2008. Cytotoxicity of nanoparticles. Small 4, 26–49.

Li, F.Q., Hu, J.H., Lu, B., Yao, H., Zhang, W.G., 2001. Ciprofloxacin-loaded bovine serum albumin microspheres: preparation and drug-release in vitro. J. Microencapsul. 18, 825–829.

Magee, P.N., 1976. Fundamentals in Cancer Prevention. University Park Press, Baltimore, MD.

Maia, C.S., Mehnert, W., Schäfer-Korting, M., 2000. Solid lipid nanoparticles as drug carriers for topical glucocorticoids. Int. J. Pharm. 196, 165–167.

Mao, S., Sun, W., Kissel, T., 2010. Chitosan-based formulations for delivery of DNA and siRNA. Adv. Drug Deliv. Rev. 62, 12–27.

Marcato, P., 2008. Preparation, characterization and application in drugs and cosmetics of solid lipid nanoparticles. Rev. Eletron. Fárm. 6, 1–37.

Marcato, P.D., Durán, N., 2014. Cytotoxicity and Genotoxicity of Solid Lipid Nanoparticles. Nanotoxicology. Springer, New York, NY. 229–244.

Marchand, C., Bachand, J., Perinet, J., Baraghis, E., Lamarre, M., et al., 2010. C3, C5, and factor B bind to chitosan without complement activation. J. Biomed. Mater. Res. Part A 93, 1429–1441.

Mathur, V., Satrawala, Y., Rajput, M.S., Kumar, P., Shrivastava, P., et al., 2010. Solid lipid nanoparticles in cancer therapy. Int. J. Drug Deliv. 2, 192–199.

Mayhew, E., Ito, M., Lazo, R., 1987. Toxicity of non-drug-containing liposomes for cultured human cells. Exp. Cell Res. 171, 195–202.

Meaning of HLB Advantages and Limitations, 1980. pp. 1–22.

Mehnert, W., Mäder, K., 2001. Solid lipid nanoparticles: production, characterization and applications. Adv. Drug Deliv. Rev. 47, 165–196.

Mei, Z., Chen, H., Weng, T., Yang, Y., Yang, X., 2003. Solid lipid nanoparticle and microemulsion for topical delivery of triptolide. Eur. J. Pharm. Biopharm. 56, 189–196.

Minami, S., Oh-Oka, M., Okamoto, Y., Miyatake, K., Matsuhashi, A., et al., 1996. Chitosan-inducing hemorrhagic pneumonia in dogs. Carbohydr. Polym. 29, 241–246.

Mincea, M., Negrulescu, A., Ostafe, V., 2012. Preparation, modification, and applications of chitin nanowhiskers: a review. Rev. Adv. Mater. Sci. 30, 225–242.

Mir, S.A., Rasool, R., 2014. Reversal of cardiovascular toxicity in severe organophosphate poisoning with 20% intralipid emulsion therapy: case report and review of literature. Asia Pacific J. Med. Toxicol. 3, 169–172.

Moghimi, S.M., Szebeni, J., 2003. Stealth liposomes and long circulating nanoparticles: critical issues in pharmacokinetics, opsonization and protein-binding properties. Prog. Lipid Res. 42, 463–478.

Morita, K., Ishigaki, I., Abe, T., 1981. Mutagenicity of cosmetic-related substances. J. Soc. Cosmet. Chem. Japan 15, 243–253.

Muchow, M., Maincent, P., Müller, R.H., 2008. Lipid nanoparticles with a solid matrix (SLN®, NLC®, LDC®) for oral drug delivery. Drug Dev. Ind. Pharm. 34, 1394–1405.

MuÈller, R.H., MaÈder, K., Gohla, S., 2000. Solid lipid nanoparticles (SLN) for controlled drug delivery—a review of the state of the art. Eur. J. Pharm. Biopharm. 50, 161–171.

Mukherjee, S., Ray, S., Thakur, R., 2009. Solid lipid nanoparticles: a modern formulation approach in drug delivery system. Ind. J. Pharm. Sci. 71, 349.

Müller, R., Lucks, J., 1996. Arzneistoffträger aus festen lipidteilchen, feste lipidnanosphären (sln). European Patent 605497.

Müller, R., Mehnert, W., Lucks, J.-S., Schwarz, C., Zur Mühlen, A., et al., 1995. Solid lipid nanoparticles (SLN): an alternative colloidal carrier system for controlled drug delivery. Eur. J. Pharm. Biopharm. 41, 62–69.

Müller, R., Maaben, S., Weyhers, H., Mehnert, W., 1996a. Phagocytic uptake and cytotoxicity of solid lipid nanoparticles (SLN) sterically stabilized with poloxamine 908 and poloxamer 407. J. Drug Target. 4, 161–170.

Müller, R., Maaβen, S., Weyhers, H., Specht, F., Lucks, J., 1996b. Cytotoxicity of magnetite-loaded polylactide, polylactide/glycolide particles and solid lipid nanoparticles. Int. J. Pharm. 138, 85–94.

Müller, R., Maassen, S., Schwarz, C., 1997. Solid lipid nanoparticles (SLN) as potential carrier for human use: interaction with human granulocytes. J. Control. Release 47, 261–269.

Müller, R.H., Radtke, M., Wissing, S.A., 2002. Solid lipid nanoparticles (SLN) and nanostructured lipid carriers (NLC) in cosmetic and dermatological preparations. Adv. Drug Deliv. Rev. 54, S131–S155.

Narang, A.S., Delmarre, D., Gao, D., 2007. Stable drug encapsulation in micelles and microemulsions. Int. J. Pharm. 345, 9–25.

Nassimi, M., Schleh, C., Lauenstein, H., Hussein, R., Hoymann, H., et al., 2010. A toxicological evaluation of inhaled solid lipid nanoparticles used as a potential drug delivery system for the lung. Eur. J. Pharm. Biopharm. 75, 107–116.

Noudeh, D., Khazaeli, P., Mirzaei, S., Sharififar, F., Nasrollahosaiani, S., 2009. Determination of the toxicity effect of sorbitan esters surfactants group on biological membrane. J. Biol. Sci. 9, 423–430.

Odunola, O., Uwaifo, A., Olorunsogo, O., 1998. Comparative genotoxicities of Tween 20 and Tween 80 in Escherichia coli PQ37. Biokemistri 123–120.

Ohnishi, M., Sagitani, H., 1993. The effect of nonionic surfactant structure on hemolysis. J. Am. Oil Chem. Soc. 70, 679–684.

Onodera, T., Kuriyama, I., Andoh, T., Ichikawa, H., Sakamoto, Y., et al., 2015. Influence of particle size on the in vitro and in vivo anti-inflammatory and anti-allergic activities of a curcumin lipid nanoemulsion. Int. J. Mol. Med. 35, 1720–1728.

Pandey, R., Khuller, G., 2005. Solid lipid particle-based inhalable sustained drug delivery system against experimental tuberculosis. Tuberculosis 85, 227–234.

Panzner, E.A., Jansons, V.K., 1979. Control of in vitro cytotoxicity of positively charged liposomes. J. Cancer Res. Clin. Oncol. 95, 29–37.

Parnham, M.J., Wetzig, H., 1993. Toxicity screening of liposomes. Chem. Phys. Lipids 64, 263–274.

Pasut, G., Veronese, F., 2007. Polymer–drug conjugation, recent achievements and general strategies. Prog. Polym. Sci. 32, 933–961.

Pedersen, N., Hansen, S., Heydenreich, A.V., Kristensen, H.G., Poulsen, H.S., 2006. Solid lipid nanoparticles can effectively bind DNA, streptavidin and biotinylated ligands. Eur. J. Pharm. Biopharm. 62, 155–162.

Phillips, B.J., Jenkinson, P., 2001. Is ethanol genotoxic? A review of the published data. Mutagenesis 16, 91–101.

Porter, C.J., Trevaskis, N.L., Charman, W.N., 2007. Lipids and lipid-based formulations: optimizing the oral delivery of lipophilic drugs. Nat. Rev. Drug Discov. 6, 231–248.

Puglia, C., Blasi, P., Rizza, L., Schoubben, A., Bonina, F., et al., 2008. Lipid nanoparticles for prolonged topical delivery: an in vitro and in vivo investigation. Int. J. Pharm. 357, 295–304.

Pujals, G., Suñé-Negre, J., Pérez, P., García, E., Portus, M., et al., 2008. In vitro evaluation of the effectiveness and cytotoxicity of meglumine antimoniate microspheres produced by spray drying against *Leishmania infantum*. Parasitol. Res. 102, 1243–1247.

Rafiyath, S.M., Rasul, M., Lee, B., Wei, G., Lamba, G., et al., 2012. Comparison of safety and toxicity of liposomal doxorubicin vs. conventional anthracyclines: a meta-analysis. Exp. Hematol. Oncol. 1, 10.

Rahman, A., Joher, A., Neefe, J., 1986. Immunotoxicity of multiple dosing regimens of free doxorubicin and doxorubicin entrapped in cardiolipin liposomes. Br. J. Cancer 54, 401–408.

Rahman, Z., Zidan, A.S., Khan, M.A., 2010. Non-destructive methods of characterization of risperidone solid lipid nanoparticles. Eur. J. Pharm. Biopharm. 76, 127–137.

Rao, S.B., Sharma, C.P., 1997. Use of chitosan as a biomaterial: studies on its safety and hemostatic potential. J. Biomed. Mater. Res. 34, 21–28.

Rathinamoorthy, R., 2012. Nanofiber for drug delivery system–principle and application. Pak. Text J. 61, 45–48.

Rekha, M., Sharma, C.P., 2011. Hemocompatible pullulan–polyethyleneimine conjugates for liver cell gene delivery: in vitro evaluation of cellular uptake, intracellular trafficking and transfection efficiency. Acta Biomater. 7, 370–379.

Rothschild, L., Bern, S., Oswald, S., Weinberg, G., 2010. Intravenous lipid emulsion in clinical toxicology. Scand. J. Trauma Resusc. Emerg. Med. 18, 51.

Rudolph, C., Schillinger, U., Ortiz, A., Tabatt, K., Plank, C., et al., 2004. Application of novel solid lipid nanoparticle (SLN)-gene vector formulations based on a dimeric HIV-1 TAT-peptide in vitro and in vivo. Pharm. Res. 21, 1662–1669.

Rustum, Y.M., Dave, C., Mayhew, E., Papahadjopoulos, D., 1979. Role of liposome type and route of administration in the antitumor activity of liposome-entrapped 1-β-D-arabinofuranosylcytosine against mouse L1210 leukemia. Cancer Res. 39, 1390–1395.

Saino, E., Focarete, M.L., Gualandi, C., Emanuele, E., Cornaglia, A.I., et al., 2011. Effect of electrospun fiber diameter and alignment on macrophage activation and secretion of proinflammatory cytokines and chemokines. Biomacromolecules 12, 1900–1911.

Sanna, V., Kirschvink, N., Gustin, P., Gavini, E., Roland, I., et al., 2004. Preparation and in vivo toxicity study of solid lipid microparticles as carrier for pulmonary administration. AAPS PharmSciTech 5, 17–23.

Salehi, M., Naseri Nosar, M., Amani, A., Azami, M., Tavakol, S., Ghanbari, H., 2015. Preparation of pure PLLA, pure chitosan, and PLLA/chitosan blend porous tissue engineering scaffolds by thermally induced phase separation method and evaluation of the corresponding mechanical and biological properties. Int. J. Polym. Mater. Polym. Biomater. 64 (13).

Sari, T., Mann, B., Kumar, R., Singh, R., Sharma, R., et al., 2015. Preparation and characterization of nano-emulsion encapsulating curcumin. Food Hydrocolloids 43, 540–546.

Schöler, N., Hahn, H., Müller, R., Liesenfeld, O., 2002. Effect of lipid matrix and size of solid lipid nanoparticles (SLN) on the viability and cytokine production of macrophages. Int. J. Pharm. 231, 167–176.

Schott, H., 1995. Hydrophilic–lipophilic balance, solubility parameter, and oil–water partition coefficient as universal parameters of nonionic surfactants. J. Pharm. Sci. 84, 1215–1222.

Schubert, M., Müller-Goymann, C., 2005. Characterisation of surface-modified solid lipid nanoparticles (SLN): influence of lecithin and nonionic emulsifier. Eur. J. Pharm. Biopharm. 61, 77–86.

Serpe, L., Catalano, M., Cavalli, R., Ugazio, E., Bosco, O., et al., 2004. Cytotoxicity of anticancer drugs incorporated in solid lipid nanoparticles on HT-29 colorectal cancer cell line. Eur. J. Pharm. Biopharm. 58, 673–680.

Shafaei, A., Esmailli, K., Farsi, E., Aisha, A.F., Majid, A.M.S.A., et al., 2015. Genotoxicity, acute and sub-chronic toxicity studies of nano liposomes of *Orthosiphon stamineus* ethanolic extract in Sprague Dawley rats. BMC Complement. Altern. Med. 15, 360.

Shah, V., Taratula, O., Garbuzenko, O.B., Patil, M.L., Savla, R., et al., 2013. Genotoxicity of different nanocarriers: possible modifications for the delivery of nucleic acids. Curr. Drug Discov. Technol. 10, 8–15.

Shelma, R., Sharma, C.P., 2011. Development of lauroyl sulfated chitosan for enhancing hemocompatibility of chitosan. Colloids Surf. B Biointerfaces 84, 561–570.

Shinoda, K., Saito, H., 1968. The effect of temperature on the phase equilibria and the types of dispersions of the ternary system composed of water, cyclohexane, and nonionic surfactant. J. Colloid Interface Sci. 26, 70–74.

Silva, A.H., Locatelli, C., Filippin-Monteiro, F.B., Zanetti-Ramos, B.G., Conte, A., et al., 2014. Solid lipid nanoparticles induced hematological changes and inflammatory response in mice. Nanotoxicology 8, 212–219.

Soppimath, K.S., Aminabhavi, T.M., Kulkarni, A.R., Rudzinski, W.E., 2001. Biodegradable polymeric nanoparticles as drug delivery devices. J. Control. Release 70, 1–20.

Stolnik, S., Daudali, B., Arien, A., Whetstone, J., Heald, C., et al., 2001. The effect of surface coverage and conformation of poly (ethylene oxide)(PEO) chains of poloxamer 407 on the biological fate of model colloidal drug carriers. Biochim. Biophys. Acta 1514, 261–279.

Suzuki, Y., Miyatake, K., Okamoto, Y., Muraki, E., Minami, S., 2003. Influence of the chain length of chitosan on complement activation. Carbohydr. Polym. 54, 465–469.

Szebeni, J., 2012. Hemocompatibility testing for nanomedicines and biologicals: predictive assays for complement mediated infusion reactions. Eur. J. Nanomed. 4, 33–53.

Szoka, F., Milholland, D., Barza, M., 1987. Effect of lipid composition and liposome size on toxicity and in vitro fungicidal activity of liposome-intercalated amphotericin B. Antimicrob. Agents Chemother. 31, 421–429.

Tadros, T.F., Vincent, B., Becher, P., 1983. In: Becher, P. (Ed.), Encyclopedia of emulsion technology. Dekker, New York, NY.

Tan, Y., Li, S., Pitt, B.R., Huang, L., 1999. The inhibitory role of CpG immunostimulatory motifs in cationic lipid vector-mediated transgene expression in vivo. Human Gene Ther. 10, 2153–2161.

Taplin, G., Johnson, D., Dore, E., Kaplan, H., 1964. Suspensions of radioalbumin aggregates for photoscanning the liver, spleen, lung and other organs. J. Nucl. Med. 5, 259–275.

Tapola, N.S., Lyyra, M.L., Kolehmainen, R.M., Sarkkinen, E.S., Schauss, A.G., 2008. Safety aspects and cholesterol-lowering efficacy of chitosan tablets. J. Am. Coll. Nutr. 27, 22–30.

Tavakol, S., Hoveizi, E., Kharrazi, S., Tavakol, B., Karimi, S., et al., 2016. Organelles and chromatin fragmentation of human umbilical vein endothelial cell influence by the effects of Zeta potential and size of silver nanoparticles in different manners. Artif. Cells Nanomed. Biotechnol., 1–7.

Tavakol, S., Mousavi, S.M.M., Tavakol, B., Hoveizi, E., Ai, J., Sorkhabadi, S.M.R., 2016a. Mechano-transduction signals derived from self-assembling peptide nanofibers containing long motif of laminin influence neurogenesis in in-vitro and in-vivo. Mol. Neurobiol., 1–14.

Tavakol, S., Saber, R., Hoveizi, E., Aligholi, H., Ai, J., Rezayat, S.M., 2016b. Chimeric self-assembling nanofiber containing bone marrow homing peptide's motif induces motor neuron recovery in animal model of chronic spinal cord injury; an in vitro and in vivo investigation. Mol. Neurobiol. 53 (5), 3298–3308.

Tavakol, S., Saber, R., Hoveizi, E., Tavakol, B., Aligholi, H., Ai, J., et al., 2016c. Self-assembling peptide nanofiber containing long motif of laminin induces neural differentiation, tubulin polymerization, and neurogenesis: in vitro, ex vivo, and in vivo studies. Mol. Neurobiol. 53 (8), 5288–5299.

Todoroff, J., Vanbever, R., 2011. Fate of nanomedicines in the lungs. Curr. Opin. Colloid Interface Sci. 16, 246–254.

Toffano, G., Bruni, A., 1980. Pharmacological properties of phospholipid liposomes. Pharmacol. Res. Commun. 12, 829–845.

Vongchan, P., Sajomsang, W., Subyen, D., Kongtawelert, P., 2002. Anticoagulant activity of a sulfated chitosan. Carbohydr. Res. 337, 1239–1242.

Wang, C., Wu, Q.-H., Li, C.-R., Wang, Z., Ma, J.-J., et al., 2007. Interaction of tetrandrine with human serum albumin: a fluorescence quenching study. Anal. Sci. 23, 429–433.

Wang, Z., Cui, Y., Wang, J., Yang, X., Wu, Y., et al., 2014. The effect of thick fibers and large pores of electrospun poly (ε-caprolactone) vascular grafts on macrophage polarization and arterial regeneration. Biomaterials 35, 5700–5710.

Wassef, N.M., Johnson, S.H., Graeber, G.M., Swartz, G., Schultz, C.L., et al., 1989. Anaphylactoid reactions mediated by autoantibodies to cholesterol in miniature pigs. J. Immunol. 143, 2990–2995.

Weber, S., Zimmer, A., Pardeike, J., 2014. Solid lipid nanoparticles (SLN) and nanostructured lipid carriers (NLC) for pulmonary application: a review of the state of the art. Eur. J. Pharm. Biopharm. 86, 7–22.

Wedmore, I., McManus, J.G., Pusateri, A.E., Holcomb, J.B., 2006. A special report on the chitosan-based hemostatic dressing: experience in current combat operations. J. Trauma Acute Care Surg. 60, 655–658.

Wehrung, D., Geldenhuys, W.J., Oyewumi, M.O., 2012. Effects of gelucire content on stability, macrophage interaction and blood circulation of nanoparticles engineered from nanoemulsions. Colloids Surf. B Biointerfaces 94, 259–265.

Weissig, V., Pettinger, T.K., Murdock, N., 2014. Nanopharmaceuticals (part 1): products on the market. Int. J. Nanomedicine 9, 4357–4373.

Weyenberg, W., Filev, P., Van den Plas, D., Vandervoort, J., De Smet, K., et al., 2007. Cytotoxicity of submicron emulsions and solid lipid nanoparticles for dermal application. Int. J. Pharm. 337, 291–298.

Weyhers, H., Ehlers, S., Hahn, H., Souto, E., Müller, R., 2006. Solid lipid nanoparticles (SLN)–effects of lipid composition on in vitro degradation and in vivo toxicity. Pharmazie 61, 539–544.

Whitmore, M.M., Li, S., Falo, L., Huang, L., 2001. Systemic administration of LPD prepared with CpG oligonucleotides inhibits the growth of established pulmonary metastases by stimulating innate and acquired antitumor immune responses. Cancer Immunol. Immunother. 50, 503–514.

Wissing, S., Kayser, O., Müller, R., 2004. Solid lipid nanoparticles for parenteral drug delivery. Adv. Drug Deliv. Rev. 56, 1257–1272.

Woods, A., Patel, A., Spina, D., Riffo-Vasquez, Y., Babin-Morgan, A., et al., 2015. In vivo biocompatibility, clearance, and biodistribution of albumin vehicles for pulmonary drug delivery. J. Control. Release 210, 1–9.

Xiao, K., Li, Y., Luo, J., Lee, J.S., Xiao, W., et al., 2011. The effect of surface charge on in vivo biodistribution of PEG-oligocholic acid based micellar nanoparticles. Biomaterials 32, 3435–3446.

Xiong, W.-Y., Yi, Y., Liu, H.-Z., Wang, H., Liu, J.-h, et al., 2011. Selective carboxypropionylation of chitosan: synthesis, characterization, blood compatibility, and degradation. Carbohydr. Res. 346, 1217–1223.

Xue, J., Zhao, W., Nie, S., Sun, S., Zhao, C., 2013. Blood compatibility of polyethersulfone membrane by blending a sulfated derivative of chitosan. Carbohydr. Polym. 95, 64–71.

Yang, C., Chu, L., Zhang, Y., Shi, Y., Liu, J., et al., 2015. Dynamic biostability, biodistribution, and toxicity of L/D-peptide-based supramolecular nanofibers. ACS Appl. Mater. Interfaces 7, 2735–2744.

Zhang, C., Qu, G., Sun, Y., Wu, X., Yao, Z., et al., 2008. Pharmacokinetics, biodistribution, efficacy and safety of N-octyl-O-sulfate chitosan micelles loaded with paclitaxel. Biomaterials 29, 1233–1241.

Zhang, J., Wu, L., Chan, H.-K., Watanabe, W., 2011. Formation, characterization, and fate of inhaled drug nanoparticles. Adv. Drug Deliv. Rev. 63, 441–455.

Zhu, A., Zhao, F., Ma, T., 2009. Photo-initiated grafting of gelatin/N-maleic acyl-chitosan to enhance endothelial cell adhesion, proliferation and function on PLA surface. Acta Biomater. 5, 2033–2044.

CHAPTER 15

Microporation and Nanoporation for Effective Delivery of Drugs and Genes

Bhupinder Singh, Rajneet K. Khurana, Atul Jain, Ripandeep Kaur and Rajendra Kumar

Contents

15.1 INTRODUCTION

Transport of a drug across the skin is quite challenging and demanding, which is primarily due to the physiology of the skin. The farthest epidermal layer of the skin, i.e., the stratum corneum, is calloused and generally remains intact, and therefore, constitutes the main barrier to the penetration and absorption of drugs into the skin. The proteinaceous corneocytes of the stratum corneum are flat, anuclear, arranged as a bilayer, and jam-packed within the extracellular lipid matrix. The architectural collection of the stratum corneum is often referred as a "bricks and mortar" model

Nanotechnology-Based Approaches for Targeting and Delivery of Drugs and Genes.
DOI: http://dx.doi.org/10.1016/B978-0-12-809717-5.00004-X

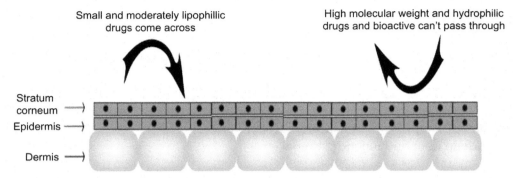

Figure 15.1 Stratum corneum: A physiological barrier of skin.

(Prow et al., 2011). Such a structure provides a remarkable property to the stratum corneum in that it poses a diffusion barrier and restricts the molecular diffusion of drugs (Gratieri et al., 2013). The lipophilicity also imparts an additional barrier and restricts the permeation of therapeutically active hydrophilic, high molecular weight and charged compounds, including peptides, DNA, and small interfering RNA into the systemic circulation (Singh et al., 2010).

Apart from the above-mentioned reasons, the optimal and adequate physiochemical properties of a transdermal-delivered drug also constitute some of the vital factors concerning the delivery of drugs via a transdermal route. Transdermal drug delivery offers stellar advantages like improved patient compliance and a high surface area of skin over which the drug could be delivered, which enables quick termination of drug delivery. It also helps in evading a major factor associated with the oral delivery of biological, i.e., poor oral bioavailability due to first-pass hepatic metabolism. Also, it helps in obliterating the pain allied with the use of hypodermic needles and injection (Thomas and Finnin, 2004). Fig. 15.1 illustrates the physiology of stratum corneum.

15.2 MICROPORATION AS AN UPCOMING APPROACH TO TRANSFER DRUG VIA SKIN

Of late, a technique called microporation has evolved that configure the skin by forming micron-sized micropores or microchannels in the skin to deliver the drugs and macromolecules through the top most layer of the skin. Mechanical microneedles, thermal ablation, radiofrequency, electroporation, ultrasound, laser, and high-pressure jet are some of the microchannels creating technologies that are commonly employed for this purpose. These technologies have provided new frontiers for the delivery of various biopharmaceuticals and neutraceuticals, as these cannot be delivered via skin passively (Banga, 2009). These physical methods are used to quickly perturb the skin barrier and endow it with additional driving forces to facilitate the transportation of drug molecules

(Gratieri et al., 2013). Another vital application of the above-mentioned techniques is the minimally invasive sampling and monitoring of the biological fluids.

15.2.1 Merits of microporation

Microporation has been categorized as the third generation of enhancement strategies, the usage of which would significantly impact the medical field. The micropores created in the skin are superficial, and breach the stratum corneum in localized areas of the skin, and thereby restrict the epidermis. Nerve endings reside in the dermis alongside capillaries and lymphatic vessels, where microneedles seldom reach. Therefore, these technologies are painless, provided they are properly designed and constructed (Banga, 2011). Being a simple mechanical device, these systems are quite inexpensive (Paudel et al., 2010).

15.3 MICROPORATOR

Microporator is a device capable of creating microchannels in the skin. The probe may contain an optically heated topical dye or absorber layer, electro-mechanical actuator, a microlancet, or an array of microneedles. Other principles to create micropores are sonic energy ablator, laser ablation system, or a high-pressure fluid jet puncture (Eppstein and Hatch, 2000).

The micropores of 1–1000 µm size in diameter can be created by the above-mentioned principles that may expand deep into the biological membrane and impair the skin barrier without unfavorably distressing or harming the underlying tissues. The term "micropore" is generally used in the singular form for simplicity, but the devices frequently used may form multiple artificial openings.

15.3.1 Microneedles

Microneedles are thin, short, solid or hollow cannulae, graded on 0.2–3.00 micrometer-scale which, upon the application of pressure, can pierce the stratum corneum (about 10–20 µm thick). This is probably due to their smaller size and sharper tips which reduces the odds of encountering a nerve. Microneedles with a length of few hundreds of microns are able to cross this barrier. Nevertheless, in comparison with classic hypodermic needles, these provide a minimally invasive, safe, and painless disruption of the skin barrier (Gratieri et al., 2013; Vandervoort and Ludwig, 2008).

A commercially available microneedle system consists of hundreds of microfabricated microneedles located at the bottom substrate, and could aid the transdermal drug delivery. These systems could pierce the stratum corneum to generate temporary pathways and assist the delivery of therapeutic molecules (Prausnitz, 1999). However, employing microneedles as tool for drug delivery causes little skin frustration without any bleeding. The physical and chemical properties of these sharp needles could be altered to achieve its desired application. Fig. 15.2A and B shows a dermaroller available in the market for the said purpose.

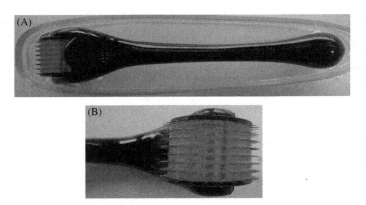

Figure 15.2 Photographic image of a dermaroller. (A) Overall view and (B) close-up view.

15.3.1.1 Microfabrication of microneedles

Recent developments in microfabrication technologies have enabled the fabrication of microneedles with diverse geometries using a wide range of materials, thereby creating miniature mechanical structures employing metal, polymer, silicon, and other materials. These microfabrication processes involve the usage of integrated circuits, electronic packages, and other microelectronic devices, augmented by additional methods used in the field of micro matching. The microneedles may have dimensions as minute as a few nanometers and can be bulk-produced economically. The methods employed for fabrication of microneedles are described as follows (Prausnitz, 1999).

15.3.1.1.1 Electrochemical etching of silicon

Electrochemical engraving of solid silicon is employed to generate extremely fine $0.01\,\mu m$ silicon networks to form porous silicon, which can be applied as piercing agents. On the application of light in aqueous hydrofluoric acid, electrolytic anodization of silicon occurs, which etch channels into the silicon. Concentration of electrolyte and intensity of incident light decide the pore size of the silicon. The remaining material that is not engraved (i.e., the silicon residual) forms the microneedles. The limitation of this method is that it may lead to the production of irregular needle-type structures (Donnelly et al., 2010).

15.3.1.1.2 Plasma etching

Microneedles formed by deep plasma etching of silicon are a diameter of $0.1\,\mu m$ or larger. A silicon wafer substrate is covered by a befitting masking material (e.g., metal) in the form of dots for producing microneedles of a requisite diameter. Needles may be patterned directly using photolithography, i.e., transferring geometric shapes on a mask to the surface of a silicon wafer, thereby providing a greater control over the final microneedle geometry (Prausnitz, 1999).

15.3.1.1.3 Electroplating

In electroporation, a top of planar substrate is employed for casting a metallic layer by an evaporation process. Then a photoresist layer is deposited onto the metal for the formation of patterned mold that results into metal-exposed region in the shape of needles. Further, the mold enclosed by photoresist layer can be packed with electroplated material. This process generally leads to the formation of microneedles having diameters of 1 μm or larger (Zahn et al., 2009).

15.3.2 Design of microneedles

Plausible factors that may influence the design of microneedles are discussed as follows.

15.3.2.1 Microneedle fabrication

The microneedles can be prepared with the help of a wide range of natural and synthetic materials, including ceramics, metals, polymers, composites, semiconductors, and organics, while both the flexible as well as the rigid portions of microneedles could be fabricated from metal, plastic, ceramic, silicone rubber, latex, vinyl, polyurethane, plastic, polyethylene, etc. The preferred materials for the preparation of microneedles includes pharmaceutical-grade stainless steel, copper, iron, tin, gold, titanium, nickel, chromium, and/or alloys of these metals, silicon, silicon dioxide, and many biodegradable and non-biodegradable polymers (Gonnelli et al., 2002).

Commonly employed biodegradable polymers used while fabricating microneedles are hydroxy acids such as lactic and glycolic acid, polylactide, polyglycolide, polylactide-*co*-glycolide, and copolymers with poly-ethylene-glycol (PEG), polyanhydrides, poly(ortho)esters, polyurethanes, poly(butyric acid), poly(valeric acid), and poly(lactide-*co*-caprolactone), while synthetic nonbiodegradable polymers include polycarbonate, polymethacrylic acid, ethylene vinyl acetate, polytetrafluoroethylene (TEFLON), polyesters, etc.

15.3.2.2 Microneedle geometry

Typical microneedle dimensions may diverge from 150 to 1500 μm in length, 50 to 250 μm in width, and 1 to 25 μm in tip diameter. When inserted into the skin, microneedle tips can break off; these tips therefore should preferably be biodegradable (Prausnitz, 2003). Fabrication of biodegradable microneedles, in this regard, can provide an increased level of safety while comparing with nonbiodegradable ones (Arora et al., 2008). Other prospective design parameters like needle height, shape, density, tip dimensions, and radius of curvature also influence their penetration performance and the force required to pierce the skin.

15.3.2.3 Mechanical strength

Ideally, the microneedles must remain intact and could tolerate the pressure exerted on it while being inserted into the skin or at the time of removal after application. Nevertheless, apt strength of these microneedles enables to perform important function of collecting biological fluid (Davis et al., 2004). Although metals are strong enough, yet the biodegradable polymers of sufficient mechanical strength are preferably used, as they may remain inside the body without any potential risk such as dissolution of microneedles. The sharpness of the tip of microneedles strongly affects the force required for microneedles to insert the skin tissue (Gill et al., 2010).

15.3.3 Mechanism of drug delivery through microneedles

Microneedles differ with respect to their composition and design. These can be broadly classified among four major categories as follows.

15.3.3.1 Solid microneedles

Solid microneedles pierce the skin to make it more permeable. Applications include either pressing or scraping of microneedles onto the skin ultimately leading to the creation of microscopic holes, therefore, increasing the skin permeability by four folds of magnitude. This is followed by the topical application of drugs or vaccines. Residual holes that are created on the removal of microneedles measure a few microns in size and have a lifetime of more than a day when kept under occlusion, but less than 2 h when left uncovered (Fig. 15.3A) (Arora et al., 2008).

15.3.3.2 Drug-coated microneedles

Direct coating on microneedles is generally used for vaccines or highly potent drugs. Hence, a very small quantity of active material can be coated onto the microneedles. Since microneedle arrays do not penetrate the skin to their full length, coatings should be applied to the tip or upper part of the microneedles. Higher loading of dose could be obtained by multiple coatings by repeatedly immersing the microneedles into the coating solution with a drying time of 5 s in-between each coating. Coatings of calcein, vitamin B, bovine serum albumin, plasmid DNA, proteins, and viruses have been demonstrated by the researchers (Fig. 15.3B) (Banga, 2011).

15.3.3.3 Dissolving or biodegradable microneedles

Biodegradable microneedles are generally referred to as "dissolvable microneedles" (Xiaoyun et al., 2013). These are prepared by coating solid microneedles with an active pharmaceutical ingredient before insertion. Replacing inert materials with biocompatible material or readily dissolvable sugars is another novel approach. In this case, the drug is loaded into the microneedle itself. After incorporation into skin, the microneedles degrade within the membrane and releases the drug therein (Fig. 15.3C).

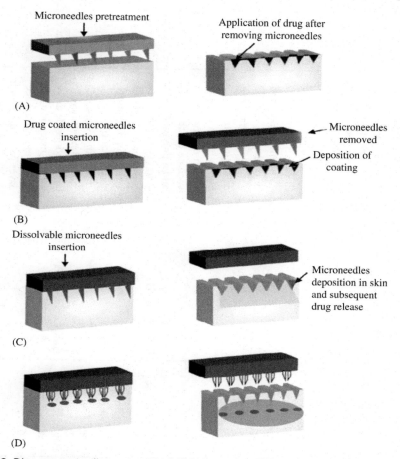

Figure 15.3 Diagrammatic illustration depicting the working mechanism of different types of microneedles. (A) Solid microneedles; (B) coated microneedles; (C) biodegradable microneedles; (D) hollow microneedles.

15.3.3.4 Hollow microneedles

Drug transportation through microneedles is the easiest approach to enhance drug permeability. These have a conduit within the needle that contains a drug solution. Its delivery is via hollow microneedles using a driving force such as pressure or iontophoresis. Hollow silicon microneedles with a pore diameter of about 3–20 μm have been reported in the literature (Banga, 2011). Micromolding and selective electrode position produces side-opened, sharp-tipped, hollow metallic microneedles. This approach enables cost-effective large-scale production of microneedles, which is an advantage over prior methodologies like direct fabrication of microneedles (Fig. 15.3D). With further research and development, this microneedle design could facilitate simple, reliable intradermal injections (Norman et al., 2013). Table 15.1 consists of some of the patented technologies on microneedles.

Table 15.1 Key patented technologies on microneedles

Patent No.	Assignee(s)	Type	Drug(s)	Therapeutic indication(s)	Reference(s)
US 8696637 B2	Kimberly–Clark Worldwide	Transdermal patch containing microneedles	Pharmaceutical bioactive agents	The activation of the patch includes at least partially separating the release member from the rate control membrane and the support	Ross (2014)
US 8911749 B2	Corium International, Inc.	Vaccine delivery via microneedle arrays	Vaccine	Method of preventing disease is provided, comprising insertion into the skin of a patient an array of microprojections consisting of a layer which comprises a vaccine for that disease and a polymer	Ghartey-Tagoe et al. (2014)
US20130253446	Daniel C Duan, Stanley Rendon	Thermotropic liquid crystalline polymeric microneedles	Rhodamine B	Liquid crystalline polymer microneedles consist of 80% more penetration power into skin compared to normal microneedles	Mir et al. (2012)
US20130296790	Koichi Masaoka, Keizo Ikari, Takashi Oda, Katsunori Kobayashi, Hidetoshi Hamamoto, Masaki Ishibashi, Kiyotsuna Toyohara	Polymer microneedle	Antiplatelet drugs	The microneedle array of this invention runs into the skin surface layer of a patient only by pressing them with fingers have both aspects of safety and easiness to use	Mir et al. (2012)

Patent number	Applicant/Inventor	Type	Active agent	Indication	Reference
US8328720B2	Infotonics Tech Center, Inc.	Coated microneedle	Histamine, serotonin, norepinephrine, and EDTA	Determination of prothrombin time	Mir et al. (2012)
EP2328530A2	Al-Ghananeem Abeer M	Patch microneedle hybrid	Apomorphine	Erectile dysfunction, Parkinson's disease	Al-Ghananeem et al. (2011)
US20110052694A1	Alltranz Inc.	Solid microneedle	Prodrugs of cannabidiol	Pancreatitis, Pancreatic cancer	Stinchcomb et al. (2011)
US20110118189A1	Farr Stephen J, Roger Hawley, Schuster Jeffrey A, John Turanin	Velocity-based microporation	5-HT antagonist	Migraine	Farr Stephen et al. (2011)
US20090062752A1	Gonnelli Robert R	Microneedle array	Insulin	Diabetes	Gonnelli (2009)
US20080255034A1	Pantec Biosolutions	Laser microporation	Follicle stimulating and luteinizing hormone	Infertility	Bohler et al. (2008)
US20080008745AI	University of Kentucky Research Foundation	Solid microneedle	Naltrexone and naltrexol hydrochloride salt, bis(hydroxymethyl) propionyl-3–O–ester naltrexone	Narcotic dependence, alcohol abuse, and/or alcoholism	Stinchcomb et al. (2008)
US20060030838A1	Gonnelli Robert R	Pump driven microneedle	Glucagon-like peptide-1	Diverse disease conditions	Gonnelli (2006)
US 6908453 B2	3M Innovative Properties Company	Microneedle devices and methods of manufacture	The microneedle devices include microneedles protruding from a substrate, with the microneedles piercing a cover placed over the substrate surface from which the microneedles protrude	Manufacturing methods may include simultaneous application of pressure and ultrasonic energy when piercing the cover with the microneedles	Allen et al. (2005)

(Continued)

Table 15.1 Key patented technologies on microneedles (Continued)

Patent No.	Assignee(s)	Type	Drug(s)	Therapeutic indication(s)	Reference(s)
US 6611707 B1	Georgia Tech Research Corporation	Microneedle drug delivery device	Peptides, proteins, carbohydrates, nucleic acid molecules, lipids, organic molecules, biologically active inorganic molecules, and combinations thereof	Microneedle device consist of rate control parameter as a semi-permeable membrane, to regulate the delivery rate of drug through the microneedles	Prausnitz et al. (2003)
EP1038016A2	Baylor College of Medicine, Valentis Inc.	Velocity-based microporation	Human growth hormone	Immuno-stimulant	Michael Barry and Smith (2000)
US006022316A	Altea Tech Inc., Spectrx, Inc.	Electroporation coupled to microporation	Hydrogel	—	Eppstein and Hatch (2000)
US 6142939A	Altea Technologies, Inc., Spectrx, Inc.	Thermal microporation	Hydrogel	—	Jonathan et al. (2000)

15.4 OTHER TECHNIQUES TO CREATE MICROPORES

15.4.1 Laser microporation

The prime objective of creating micropores is to create channels through the upper skin layers that allocate the permeation of topically applied therapeutic agents. Laser microporation uses laser energy for the formation of micropore, and categories as a reproducible technique. Laser microporation has not yet been considered as a well-established technique for the application of therapeutic agents. It has, however, dermatological benefits when employed for the local handling of medical conditions like keloids or warts. Also, it is frequently employed for aesthetic applications like wrinkle reduction or removal of hair and age-spots (Weiss et al., 2012).

Transdermal vaccination via fractional laser-generated micropores is yet another application in the field of skin vaccination. The advantage of this method is not only to deliver low-molecular weight drugs through the skin, but also shows the transport of recombinant proteins and functional antibodies into deeper skin layers, and subsequently into the circulation (Scheiblhofer et al., 2013).

The P.L.E.A.S.E. (Precise Laser EpidermAl SystEm) device primarily forms aqueous micropores in the skin. It does so by coupling the precision of its laser scanning technology with the flexibility to vary the number, density and the depth of the micropores (Weiss et al., 2012). Therefore, the rate and extent of drug delivery into the skin can be controlled or modified. Limitation of this technology is the higher cost of the laser devices presently available.

15.4.2 Radiofrequency and thermally induced microporation

This is yet another alternative technique that mechanically induces perforation to the outer layer of the epidermis on applying some form of physical energy. Viador technology, formerly called Viaderm, is involved in the application of radiofrequency to thermally ablated stratum corneum. The device consists of an array of micro-electrodes that are placed in contact with the prior warmed skin. Application of the alternating current causes fractional heating, as ions present within the tissue attempt to realign in response to the changing electric field. The local heating lead to cell ablation and thereby, results into the formation of number of microchannels in the stratum corneum that function as transport condits. Exposure of the skin to short but high temperature pulses can cause structural disruption of the stratum corneum without significantly damaging the deeper tissues. The highlighting applications are skin rejuvenation, resurfacing, and tightening. Recently, Gratieri and Alberti have reported delivery of α-bisabolol and epidermal growth factor into murine skin with a custom-built radiofrequency device (Gratieri et al., 2013).

15.4.3 Jet injectors: Velocity-based devices

Since the past four decades, the gun-shaped apparatus has been designed for the deliverance of vaccines through the stratum corneum without employing needles.

Of late, such devices have been redesigned such as to enable them to deliver small molecules and proteins either in the form of a liquid or powder (Gratieri et al., 2013).

Liquid jet injections are devoid of needles, these injections are based upon the utilization of a high-speed jet to prick the skin and thereby transfer the drugs. The basic design of commercial liquid jet injectors consists of a power source (consisting of compressed gas or spring), piston, drug-loaded compartment and a nozzle, with an orifice size typically ranging between 150 and 300 μm. Upon actuation, the power source pushes the piston, which influences the drug-loaded compartment, thereby leading to a quick increase in the pressure. This forces the drug solution through the nozzle orifice as a liquid jet with velocity ranging between 100 and 200 m/s. As the jet device progresses deeper into the skin, velocity decreases until it does not have sufficient energy to create a minute hole. This completes the first phase of injection, i.e., unidirectional skin puncture, followed by the second phase, i.e., multidirectional jet dispersion from the end point of penetration. These could be primarily employed for mass immunization programs and even for diseases like measles, smallpox, cholera, hepatitis B, influenza, and polio (Arora et al., 2008).

Another key concern is the time period during which the micropores remain open. Studies for the pores resealing kinetics in humans under nonocclusive conditions have confirmed that skin barrier properties recovered within 2 h after insertion of either stainless steel solid microneedles using a 26-gauge hypodermic needle. However, under occlusion, resealing took a longer time and was strongly dependent on microneedle length, number, and the cross-sectional area. It can be supposed that even if the time during which the skin breached is short, pathogens can still gain access. Therefore, an in vitro study showed that no microorganisms crossed the viable epidermis when skin was microneedle-punctured, in contrast to when it was macropunctured (Donnelly et al., 2009).

Further, Tables 15.2 and 15.3 enlist the various applications of microporation. While, Table 15.4 provides a glimpse of the current status of the ongoing clinical trials on the said technologies and devices.

15.5 NANOPORATION

Nanopores are the minuscule pores formed on the cell surfaces when electric fields of appropriate or threshold magnitudes are applied for nanoseconds, called nanosecond pulsed electric fields (nsPEFs). It is basically an electroporation-based technology for site-specific treatments that came into picture very recently in which primarily pores are created on the surface of cells and permeability of cells is altered for introducing drugs and/or biologics into the cells which are normally not diffusible across the cell membrane for experimental or therapeutic purposes. Therefore, in the virtue of these altered properties of the cell, local and effective treatment of solid tumors or

Table 15.2 Applications of microporation technique in the delivery of active medicinal agents

Drug(s)	Route	Technique	Conclusions	Reference(s)
Active medicinal agents				
Diclofenac	Transdermal	Effect of microporation on the passive and iontophoretic delivery	Microneedle pretreatment increased significantly the systemic exposure of diclofenac from either passive or iontophoretic delivery, whereas the effect in skin was less pronounced	Patel et al. (2015)
Nile red	Dermal	Bioerodible polymeric microneedles	Incorporation of drug-loaded nanostructured lipidic carriers into microneedles could represent a promising strategy for controlled transdermal delivery of lipophilic drugs	Lee et al. (2014)
Triamcinolone acetonide (TA)	Suprachoroidal space (SCS)	Microneedles	Delivery of TA (from dose 0.2 to 2 mg) to the SCS using microneedles was simple and effective in reducing acute inflammation in the ocular posterior segment as high-dose IVT injection and not associated with adverse effects or toxicity low-dose SCS TA was also effective in reducing inflammation; however, low-dose IVT TA was not	Gilger et al. (2013)
Vitamins A and B3	Topical	Elongate microparticles	Successfully employed high aspect ratio elongate microparticles for enhanced topical delivery to the skin	Raphael et al. (2013)
Rhodamine 110 (as model drug)	Transdermal	Polymer-based microneedle	An innovative fabrication process to make biocompatible SU-8 microtubes integrated with biodissolvable maltose tips as novel microneedles for the transdermal drug delivery applications	Xiang et al. (2013)
TiO_2 nanoparticles and Al_2O_3 microparticles	Transdermal	Fractional laser ablation	In in vivo experiments, reflectance spectroscopy, optical coherence tomography, and clinical photography were used to monitor the skin status during 1 month after suspension administering	Genina et al. (2013)

(*Continued*)

Table 15.2 Applications of microporation technique in the delivery of active medicinal agents (Continued)

Drug(s)	Route	Technique	Conclusions	Reference(s)
Active medicinal agents				
Drugs, protein antigens, or antibodies	Transdermal	Laserporation	Needle–free and painless vaccination approaches have the potential to replace standard methods due to their improved safety and optimal patient compliance. The use of fractional laser devices for stepwise ablation of skin layers might be advantageous for both vaccination against microbial pathogens, as well as immunotherapeutic approaches, such as allergen-specific immunotherapy	Scheiblhofer et al. (2013)
Iron FPP	Transdermal	Microneedle and iontophoresis	Iontophoresis (0.15 mA/cm^2 for 4 h when was combined with microneedle pretreatment (for 2 min), therapeutically adequate amount of FPP was delivered and there was significant recovery of rats from anemia	Modepalli et al. (2013)
α-Bisabolol and epidermal growth factor (EGF)	Topical	Radiofrequency microporation	Radiofrequency microporation enhances the topical delivery of active ingredients with high molecular weight or of small hydrophilic or lipophilic molecules. Thus, this technology can effectively improve photo–induced hyperpigmentation and wrinkle formation by enhancing topical delivery of active agents	Kim et al. (2012)
Recombinant human erythropoietin alfa	Transdermal	Erythropoietin coated titanium microneedle patch system	Preclinical studies in rats showed the EPO microneedle patches, coated with 750–22,000 IU, delivered with high efficiency (75%–90%) with a linear dose response	Peters et al. (2012)
Diclofenac	Transdermal	Laser microporation P.L.E.A.S.E. (Precise Laser EpidermAl SystEm) technology	After 24 h, cumulative drug permeated across skin with 150, 300, 450, and 900 shallow pores (50–80 μm) was 3.7-, 7.5-, 9.2-, and 13-fold superior to that permeated across untreated skin	Bachhav et al. (2011)

Phenylephrine (PE)	Rectal (anal sphincter muscle)	Polymeric microneedle	For rats pretreated with microneedles, topical application of 30% PE gel rapidly increased the mean resting anal sphincter pressure from $7 \pm 2\,cm\ H_2O$ to a peak value of $43 \pm 17\,cm\ H_2O$ after 1 h	Baek et al. (2011)
Lidocaine	Transdermal	Coated microneedle	Lidocaine dissolved rapidly off the microneedles into skin such that in 1 min wear time it achieved apt tissue levels needed to cause analgesia	Zhang et al. (2011)
Parathyroid hormone (PTH)	Bone	PTH coated microneedle patch system	Rapid PTH plasma profile was achieved with T_{max} 3 times shorter and apparent $T_{1/2}$ 2 times shorter than FORTEO	Daddona et al. (2011)
Diclofenac	Transdermal	Laser microporation	Laser microporation significantly increased diclofenac transport from both simple and semi-solid formulations through porcine and human skin. It was also concluded that pore depth and pore number could modulate delivery kinetics	Bachhav et al. (2011)
Maltose microneedles	Transdermal	Microneedles	Maltose microneedles penetrated the stratum corneum barrier and created microchannels in skin which completely close within 15 h after poration	Kalluri and Banga (2011)
Prednisone	Transdermal	Laser microporation	Transport of the drug across the membrane was controlled by pore number and depth. Increasing pore depth so that micropores reached the epidermis produced corresponding increases in prednisone transport and it was substantially improved over delivery through intact skin	Yu et al. (2010)
Lidocaine	Transdermal	Laser microporation (P.L.E.A.S.E. Painless Laser Epidermal System)	P.L.E.A.S.E. are well known to create well-defined conduits in the skin, to provide a controlled enhancement of transdermal transport and, to enable improvement in both the rate and extent of drug delivery	Bachhav et al. (2010)

(Continued)

Table 15.2 Applications of microporation technique in the delivery of active medicinal agents (Continued)

Drug(s)	Route	Technique	Conclusions	Reference(s)
Active medicinal agents				
Carboxyfluorescein and radiolabeled Mannitol	Skin	Dermaroller microneedle device	Dermarollers being already commercially available for cosmetic purposes appeared also promising for drug delivery purposes particularly those with medium (500 m) and shorter (150 m) needle lengths	Badran et al. (2009)
Immunoglobulin G Human monoclonal antibody IgG	Transdermal	Maltose microneedles	Delivery of 0.72 mg over 24 h, using a 10 cm^2 patch of IgG. Flux of 269.53 ng/(cm^{2}h) of the monoclonal antibody enhanced following treatment with microneedles	Banga (2011)
Sumatriptan	Transdermal	Velocity-based needle free system	A successful abort of migraine attack was seen in out-patients with the needle-free system	Brandes et al. (2009)
Chondroitin sulfate	Skin	Dermaroller microneedle device	Microneedle treatment enhanced the permeation of chondroitin sulfate into skin	Kim et al. (2009)
Pilocarpine hydrochloride	Ocular	Coated microneedle	Pupil constricted from 8 to 5.5 mm in diameter within 15 min with microneedles while with topical application constriction was just 7 mm with slower kinetics	Jiang et al. (2007)
Insulin	Transdermal	Solid metal microneedles	Microneedles reduced blood glucose levels to an extent similar to 0.05–0.5 units insulin injected subcutaneously	Martanto et al. (2004)
Granisteron hydrochloride Diclofenac sodium	Transdermal	Radiofrequency microporation	Plasma plateau levels of granisetron microneedle patches are about 30 times higher than levels obtained by 24-h passive diffusion of the applied drug diclofenac	Sintov et al. (2003)
Human growth hormone	Transdermal	Medi-Jector (velocity-based, needle-free device)	Medi-Jector specially used in growth hormone therapy tends to lead to fewer adverse psychological responses than multidose injection	Verrips et al. (1998)

Table 15.3 Applications of microporation technique in protein, peptide, and nucleic acid delivery

Gene	Cell types	Technique	Conclusions	Reference(s)
Protein and peptide delivery				
A mannosylated cell-penetrating peptide-grafted-polyethylenimine	Antigen-presenting cells	Coated microneedles	Man-PEI1800-CPP was a potential APCs targeted of nonvirus vector for gene therapy in the epidermis and dermis of skin and targeted on splenocytes after percutaneous coating based on microneedles in vivo	Hu et al. (2014)
DNA vaccination (cutaneous delivery)	Antigen-presenting cells	Polyplex microneedles	The resulting microneedle-based polyplex delivery systems for enhanced DNA vaccination was able to successfully induce a robust humoral immune response compared to conventional subcutaneous injection with hypodermal needles	Kim et al. (2014)
Protein (Alexa Fluor 555 bovine serum albumin conjugate (AF-BSA))	Transdermal route	Combination of iontophoresis and microporation	Combination of microneedles and iontophoresis significantly increased skin's penetration depth of AF-BSA (300 vs 110 μm) and achieved 23.7-fold (8.2-fold, in vivo) higher for combination as compared with microporation alone, in vitro	Bai et al. (2014)
Peptide (leuprolide acetate)	Transdermal delivery	Iontophoresis and/or microneedles	Combination treatment resulted in faster and increased drug delivery (3.54 ± 0.08 ng/mL) due to propulsion of the drug through the preformed micropores as compare	Sachdeva and Zhou (2013)
Novel DNA vectors, minicircles (mC)	Embryonic stem cell-derived neural stem cells (NSC)	Novel electroporation method (Neon Transfection System)	Microporation when combined with mC, 75% of NSC expressing a transgene was achieved. While comparing mC with their plasmid DNA (pDNA) counterparts, cells harboring mC showed 10% higher cell viability, maintaining 90% of survival at least for 10 days	Madeira et al. (2013)
Vaccine delivery	Transcutaneous	Transcutaneous P.L.E.A.S.E. (Precise Laser Epidermal System)	The potential of transcutaneous immunization via (P.L.E.A.S.E. Device) laser-generated micropores for induction of specific immune responses and compared the outcomes to conventional subcutaneous injection	Weiss et al. (2012)

(Continued)

Table 15.3 Applications of microporation technique in protein, peptide, and nucleic acid delivery (Continued)

Gene	Cell types	Technique	Conclusions	Reference(s)
Protein and peptide delivery				
Plasmid DNA encoding green fluorescence protein (GFP)	Human bone marrow mesenchymal stem cells (BM-MSC)	Novel electroporation method (Neon Transfection System)	Microporation demonstrated to be a reliable and efficient method to genetically modify hard-to-transfect cells giving rise to the highest levels of cell survival reported so far along with superior gene-delivery efficiencies	Madeira et al. (2011)
Enhanced green fluorescent protein (EGFP); brain-derived neurotropic factor (BDNF) plasmid DNA	Human umbilical cord blood-derived mesenchymal stem cells (hUCB-MSCs)	Electroporation-based gene transfer technique of microporation	hUCB-MSCs were transfected with EGFP with higher efficiency (83%) in comparison to conventional liposome-based reagent (<20%) or established electroporation methods (30%–40%). BDNF expression remained fairly constant for the first 2 weeks for both in vitro and in vivo	Lee et al. (2013)
Enhanced green fluorescent protein (EGFP) and luciferase	Human Adipose tissue-derived stem cells (hADSCs)	Electroporation	hADSCs retained their multipotency and reporter gene expression was maintained for >2 weeks in vitro and in vivo	Wang et al. (2009)
Interferon alpha-2b (IFNalpha2b)	Transdermal	Combination of aqueous microchannels and iontophoresis	Delivery of IFNalpha2b is not possible through unharmed skin by itself or during iontophoresis. IFNalpha2b can be delivered In vivo by micropores created in the outer layer of the skin. Beside this, Iontophoresis enhanced delivery by twofold (722 ± 169 ng) in the hairless rat	Badkar et al. (2007)
Lipid: Polycation: pDNA (LPD) nonviral gene therapy vectors	Human keratinocytes (HaCaT cells)	Silicon-based microneedles	In vitro cell culture was used to confirm that LPD complexes mediated efficient reporter gene expression in human keratinocytes in culture when formulated at the appropriate surface charge	Chabri et al. (2004)

Antigen	Vaccine	Technique	Conclusions	Reference
Nucleic acid delivery				
A/PR8 influenza hemagglutinin DNA and A/PR8 inactivated virus	Influenza	Coated microneedles	Study showed DNA solution as a microneedle coating agent and demonstrating cross-protection by coimmunization with inactivated virus and DNA vaccine using coated microneedles	Kim et al. (2013)
Ovalbumin and HIV	Oral mucosal vaccination	Coated microneedles	Coated microneedle is a novel method to induce systemic IgG and secretory IgA in saliva, and could offer a versatile technique for oral mucosal vaccination	Ma et al. (2014)
Attenuated live tuberculosis bacillus	Bacillus Calmette– Guérin (BCG)	BCG coated microneedle	Characteristic IFN-γ levels with high frequencies of CD4$^+$IFN-γ^+, CD4$^+$TNF-α^+, and CD4$^+$IFN-γ^+TNF-α^+ T cells	Hiraishi et al. (2011)
N-Trimethyl chitosan (TMC) adjuvanted diphtheria toxoid (DT)	Diphtheria	Microneedle array of different types	Transcutaneous immunization with the TMC/DT mixture elicited eightfold higher IgG titers compared to the TMC nanoparticles or DT solution	Bal et al. (2010)
Plasmid encoding hepatitis C virus NS3/4A protein	Hepatitis C	Coated microneedle	Delivery of 8 μg plasmid encoding hepatitis C virus NS3/4A protein using microneedles induced in vitro functional NS3/4A-specific cytotoxic T lymphocytes comparable to 4 μg DNA delivered using complex gene gun technology	Gill et al. (2010)
Inactivated A/ Aichi/2/68 (H3N2) influenza virus	Influenza	Vaccine coated microneedle array	100% protection was assured against lethal viral challenge	Koutsonanos et al. (2009)

(Continued)

Table 15.3 Applications of microporation technique in protein, peptide, and nucleic acid delivery (Continued)

Antigen	Vaccine	Technique	Conclusions	Reference
Tetanus toxoid	Tetanus	Bioneedle	Bioneedles with adjuvanted tetanus toxoid showed a comparable functional antibody response in mice to the group receiving conventional liquid injections. The response achieved was with four times lower antigen concentration using the bioneedles compared to the regular injections	Hirschberg et al. (2008)
Avian H5N1 influenza virus	Influenza	Novel needle-free transdermal patch	Micropores were created by electrical current Induction of robust serum antibody responses	Garg et al. (2007)
Recombinant protective antigen (rPA) of *Bacillus anthracis*	Anthrax	Stainless steel microneedle	Rabbit's immunized intradermally with 10μg of rPA displayed 100% protection from aerosol spore challenge, while intramuscularly injection of the same dose provided slightly lower protection, i.e., 71%	Mikszta et al. (2005)
Live, attenuated, chimeric JE Flavivirus vaccine (ChimeriVax-JE)	Japanese encephalitis	Stainless steel microneedle	The viremia induced was longer (5–7 days)	Dean et al. (2005)

Nucleic acid delivery

Table 15.4 Ongoing clinical trials involving microporation technique for transdermal delivery

Methods of microporation	Sponsor (collaborator)	Outcome	Disease state
Electroporation	Karolinska University Hospital (Karolinska Institutet, Swedish Institute for Infection Disease and Control Cyto Pulse Science, Inc.)	DNA vaccine delivered by intradermal electroporation (device: DermaVax)	Colorectal cancer
	Uppsala University (Karolinska Institutet and Cyto Pulse Science Inc.)		Prostate cancer
Microneedle	Emory University	Microneedle versus subcutaneous catheter for insulin delivery	Type 1 diabetes
	The University of Hong Kong (Hospital Authority, Hong Kong)	Micronjet versus intramuscular injection H1N1 vaccine	Vaccination and influenza
	NanoPass Technologies Ltd	Conventional subcutaneous injection versus Micronjet for insulin delivery	Intradermal injection and healthy volunteers
	Hadassah Medical Organization	Intradermal unadjuvant Pandemrix(R) via a microneedle device, compared with intramuscular adjuvated Pandemrix	Healthy subject
	NanoPass Technologies Ltd	Delivery of influenza with microneedle versus intramuscular injection	Influenza
	Becton, Dickinson and Company	Extended microneedle delivery of insulin	Diabetes
Radiofrequency ablation	Eli Lilly Company (TransPharma Medical)	Transdermal versus subcutaneous teriparatide (device: ViaDerm drug delivery system)	Osteoporosis

Source: www.clinicaltrials.gov (accessed 20.02.17).

gene electrotransfer for DNA vaccination and gene therapy can be amplified (www.nanoporation.eu).

Current chemotherapy for cancer treatment suffers from pitfalls of dangerous side effects due to its permeation to healthy cells and tissues besides cancer cells. To overcome the current chemotherapy-related challenges, Institute for Medical Science and

Technology (IMSaT) at the University of Dundee, the EU-funded Nanoporation project is developing a new treatment method that solves some chemotherapy-related problems, by concentrating drug delivery only, where it is needed while avoiding side effects.

Trials of the new treatment are planned for 2017 and industrial partners are keen to commercialize the biotechnologies resulting from the project. Nowadays, many experiments are being conducted with mice, and clinical trials are likely to be launched in successive years. According to this technique, anticancer drugs are loaded with advanced nanocapsules that will open or deliver the drug only when magnetic resonance-guided, focused ultrasound is applied to the tumor itself, followed by opening of cell membranes. The drug gets released from the nanocapsules and kills the cancer cells. While the remnant of nanocapsules remain inactive in the rest of the body and that after metabolism get extracted via the kidney, thus preventing healthy cells from being exposed to active chemotherapeutic agent (www.nanoporation.eu).

Rapid influx of Ca^{2+} was observed through nanopores created upon cellular exposure to nsPEF for modulation of neuronal function. It was observed that influx of Ca^{2+} still occurred in the presence of Cadmium, a nonspecific Ca^{2+} channel blocker, suggesting that observed influx is likely due to nanoporation (Roth et al., 2013). The hypothesis that such "nanopores" exist has been confirmed through various studies using diverse techniques such as fluorescence microscopy, electrophysiology, and direct ion measurement in bulk solution for longer pulse widths. Nanoporation has been tried by different methods, some of which are illustrated as follows.

15.5.1 Electric pulses induced electroporation

Electroporation is the phenomenon in which the cell membrane is exposed to a sufficiently high electric field thereby augmenting cell membrane permeability (Rebersek et al., 2007). Electroporation has now been well established as a tool to increase the cell permeability or membrane flow of solutes down to the concentration and voltage gradients. Pakhomov et al. (2009) reported that electric impulses of nanosecond duration (nsEP) may trigger the formation of voltage-sensitive and inward-rectifying membrane pores of size equivalent to 1 nm. nsEP-treated cells remain generally impermeable to propidium, signifying that the utmost pore size is about 1 nm, although nanopores can be stable for many minutes and therefore could significantly impact cell electrolyte and water balance. The rise in time of conventional electroporation pulses alone is often longer than the entire pulse duration of nsPEFs. For long pulse durations, ions accumulate along the plasma membrane, and a counter field develops inside the cell that successfully protects the organelles from additional exposure to the applied electric field. As a result, prolonged exposure only affects the outer membrane and pores tend to form once a cell-characteristic threshold voltage gets reached (Neumann et al., 1999). On the contrary, the short duration of exposures affects organelle membranes in a similar manner as the outer cell membrane (Scarlett et al., 2012).

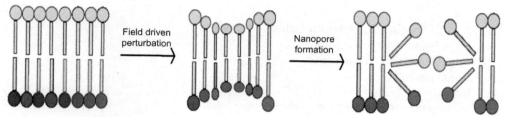

Figure 15.4 Field-driven nanopore formation.

15.5.2 Laser illumination

A nanoplasmonics-based opto-nanoporation technique has been adopted for creating nanopores in the membrane of red blood cells (RBCs) in which laser illumination is used to induce diffusion and trigger the release of both small and large molecules (Delcea et al., 2012). The method was implemented by absorbing gold nanoparticles (Au-NP) aggregates on the membrane of loaded RBCs. Localized heating of gold nanoparticles with near-IR laser light resulted in increased temperature that eventually impacted the membrane permeability by affecting the lipid bilayer or transmembrane proteins. This work seemed to be quite appropriate for generating nanopores for bio-molecule trafficking through polymeric and lipidic membranes. The probable sequence of steps has been illustrated as Fig. 15.4.

15.5.3 Microfluidic biochip

Dalmay et al. (2011) reported a microfluidic biochip for the exposure of living cells to nsPEF of 3–10 ns, wherein the cells get exposed to drugs or transgene vectors because of disturbances on the plasma membrane and on the intracellular components. The biochip was made of thick gold electrodes in order to convey the maximum energy to the biological medium containing cells. Biological experiments were also performed with this biochip for characterization of the effect of nanoporation on living cells, especially on permeabilization of plasma membrane. The results demonstrated that the nanoporation of live cells was valuable while keeping a sensible level of viability.

15.5.4 Nanoneedles

To perforate the cell's membrane, another method was reported employing nanoneedles. Nanoneedles-based intracellular delivery system consists of a nanoneedle (a nanotube or a nanowire) on a macroscopic handle (an etched metallic wire or simply a pulled glass micropipette) and a micromanipulator to penetrate into a target cell under the observation of an optical microscope. Through this arrangement, the whole process of nanoneedle-based delivery can be directly monitored visually. However, it is restricted by the imperfect resolution of the microscope (Yum et al., 2010).

> ## BOX 15.1 Features of Nanoneedles
> Important Features of Nanoneedles
> - Composed of silicon materials
> - Biocompatible
> - Sharp needle 7–10 μm in height
> - Tip diameter corresponds to few hundred nanometer
> - A range of 4–40 nanoneedle per 100 μm^2

Denoual and his collaborators (2006) presented a nanostructured support, constituted of nanoneedles with controlled height and densities having tip diameter below 200 nm for the culture of cells. Important features of nanoneedles are enumerated in Box 15.1. The careful observation of the narrow interface between the membranes of cells and nanoneedles concluded perforation of the cell membranes. Such nanostructured chips allowed specific interaction and may lead to the opening of doors to a large number of exciting and valuable applications such as nanoporation for transfection or internal cell potential recording.

A highly efficient gene delivery of over 70% was accomplished in human mesenchymal stem cells because of the fact that nanoneedles could be inserted into the nucleus directly without causing significant cell damage (Han et al., 2008). Similarly, these nanoneedles could be easily utilized for intracellular site-specific delivery of normally nondiffusible or nonpermeable drugs by somehow conjugating the drugs on the surface of the nanoneedles or inside the nanoneedles.

15.5.5 Potential applications of nanoporation

Nanoporation could create large number of nanopores in all cell membranes, a phenomenon called supra-electroporation. These distinctive characteristics anticipated for apoptosis induction as well as other cell death mechanisms, which have been revealed with nanopulse electroporation in several cell types and tumor tissues (Beebe and Schoenbach, 2005). Under a nanoporation technique, anticancer substances are placed inside advanced nanocapsules that deliver the substances only when a focused ultrasound is applied, as mentioned previously. The nanocapsules are destroyed and afterwards metabolized. The released drug kills the cancer cells, while nanocapsules remain inactive in the rest of the body and are excreted via the kidney, preventing healthy cells from being exposed.

Electroporation approach has enabled the successful transfection of plasmid DNA, gene silencing by siRNA, and insertion of exogenous proteins into various cell types, serving as vital tool for biological and clinical investigators. Introduction of cytotoxic substances, such as bleomycin and cisplatin, into cells by electroporation,

Table 15.5 Differences between microporation and nanoporation

Characteristics	Microporation (intradermal delivery)	Nanoporation (intracellular delivery)
Pore size	Micro meter	Nano meter
Mechanism	Micronsized pores created on the surface of skin	Nanosized pores created on the surface of cell
Technologies Employed	Mechanical microneedles, thermal ablation, radiofrequency, laser, electroporation, ultrasound and high-pressure/velocity jet	Electric, laser, microfluidic biochip, nanochips
Duration of applied frequency	200–500 μs	100–300 ns
Applications	Mostly used for hydrophilic macromolecules that cannot be delivered via skin passively	Mostly used in cases of dyes employed for diagnostic purpose, plasmids, oncology, genetics, cell biology, cancer therapy, and immune stimulation
Width of tip of microneedle	7–10 μm	1 nm

termed electrochemotherapy, has been documented as a proficient and safe approach of tumor ablation (Ibey et al., 2010). Tissue ablation and tumor devastation are probably the most promising medical applications of high-voltage nanopulses (Andre et al., 2010). Table 15.5 summarizes the salient differences between microporation and nanoporation.

15.6 EMERGING AND FUTURE PROSPECTS

Transdermal delivery has traditionally been used as a method for the administration of a low-molecular weight drug, predominantly lipophilic drugs with high potency. Albeit the latter property remains an obvious advantage in the selection of drug candidates, the technology under development enables many other small molecules, peptides, and proteins with very different characteristics to be considered as candidates for percutaneous delivery. Furthermore, in addition to challenge the previous notions about feasibility, they can also allow improvement in bioavailability and modification in the delivery kinetics. Sustained release of drug from dissolvable microneedles can be a potential system for a skin delivery because of long-term delivery of the drug and reduction in the number of microneedle (Baek et al., 2011).

nsPEFs, if applied for longer time serves as irreversible electroporation technique developed as new modalities to cure tumors. These promising techniques still require further studies before entering the clinics. On the contrary, during the last decade, electrochemotherapy has been established as a safe, cheap, efficient, and highly selective treatment for tumor nodules, DNA or RNA vaccination, genetic diseases, and other pathologies (Breton et al., 2012).

Many studies have shown the feasibility of using microneedle-arrays to deliver a wide variety of drugs. The increasing knowledge of production processes, natural and supramolecular materials, coating techniques, injection dynamics, and long-term effects on skin conditions will eventually lead to the commercial production of low cost, safe, and convenient-to-use systems. Development of sensors in combination with patches with hollow microneedles offers exciting potential for the systems which release drugs as the need occurs, for instance, for the management of diabetes. The goal for the future is to translate this potential into a new generation of pharmaceutical products through innovative formulations and device designs.

ACKNOWLEDGMENTS

Financial grants received from University Grant Commission (UGC) under the "Center for Potential Excellence in Application of Nanoparticles, Nanomaterials and Nanocomposites" for research assistance is deeply acknowledged.

REFERENCES

Al-Ghananeem, A.M., inventor; US Worldmeds LLC, assignee, 2011. Transdermal Delivery of Apomorphine Using Microneedles. EP2328530 A2.

Allen, M., Prausnitz, M., McAllister, D., Cros, F.; inventors. Allen Mark, G., Prausnitz Mark, R., Mcallister Devin, V., Cros Florent, P.M., assignee, 2005. Microneedle Devices and Methods of Manufacture Use Thereof. Google Patents. CA2330207A1.

Andre, P., Men'shchikov, A., Bontemps, S., Konyve, V., Motte, F., Schneider, N., et al., 2010. From filamentary clouds to prestellar cores to the stellar IMF: Initial highlights from the Herschel Gould belt survey. Astron. Astrophys. 518, 1–5.

Arora, A., Prausnitz, M.R., Mitragotri, S., 2008. Micro-scale devices for transdermal drug delivery. Int. J. Pharm. 364 (2), 227–236.

Bachhav, Y.G., Summer, S., Heinrich, A., Bragagna, T., Böhler, C., 2010. Effect of controlled laser microporation on drug transport kinetics into and across the skin. J. Control. Release 146 (1), 31–36.

Bachhav, Y.G., Heinrich, A., Kalia, Y.N., 2011. Using laser microporation to improve transdermal delivery of diclofenac: Increasing bioavailability and the range of therapeutic applications. Eur. J. Pharm. Biopharm. 78 (3), 408–414.

Badkar, A.V., Smith, A.M., Eppstein, J.A., 2007. Transdermal delivery of interferon alpha-2B using microporation and iontophoresis in hairless rats. Pharm. Res. 24 (7), 1389–1395.

Badran, M.M., Kuntsche, J.J., Fahr, A.A., 2009. Skin penetration enhancement by a microneedle device (Dermaroller®) in vitro: dependency on needle size and applied formulation. Eur. J. Pharm. Sci. 36, 511–523.

Baek, C., Han, M., Min, J., Prausnitz, M.R., Park, J.H., 2011. Local transdermal delivery of phenylephrine to the anal sphincter muscle using microneedles. J. Control. Release 154 (2), 138–147.

Bai, Y., Sachdeva, V., Kim, H., Friden, P.M., 2014. Transdermal delivery of proteins using a combination of iontophoresis and microporation. Ther. Deliv. 5 (5), 525–536.

Bal, S.M., Ding, Z., Kersten, G.F., Jiskoot, W., Bouwstra, J.A., 2010. Microneedle-based transcutaneous immunisation in mice with N-trimethyl chitosan adjuvanted diphtheria toxoid formulations. Pharm. Res. 27 (9), 1837–1847.

Banga, A.K., 2009. Microporation applications for enhancing drug delivery. Expert Opin. Drug Deliv. 6 (4), 343–354.

Banga, A.K., 2011. Microporation-mediated transdermal drug delivery Transdermal and Intradermal Delivery of Therapeutic Agents: Applications of Physical Technologies. Taylor and Francis Group, LLC, New York, NY p. 73.

Beebe, S.J., Schoenbach, K.H., 2005. Nanosecond pulsed electric fields: a new stimulus to activate intracellular signaling. J. Biomed. Biotechnol. 4, 297–300.

Bohler, C., Bragagna, T., Braun, R., Braun, W., Zech, H., inventors; Pantec Biosolutions AG, assignee, 2008. Transdermal Delivery System for Treating Infertility. EP1945193A1.

Brandes, J.L., Cady, R.K., Freitag, F.G., Smith, T.R., Chandler, P., Fox, A.W., et al., 2009. Needle-free subcutaneous sumatriptan (Sumavel DosePro): bioequivalence and ease of use. Headache 49 (10), 1435–1444.

Breton, M., Delemotte, L., Silve, A., Mir, L.M., Tarek, M., 2012. Transport of siRNA through lipid membranes driven by nanosecond electric pulses: an experimental and computational study. J. Am. Chem. Soc. 134 (34), 13938–13941.

Chabri, F., Bouris, K., Jones, T., Barrow, D., Hann, A., Allender, C., et al., 2004. Microfabricated silicon microneedles for nonviral cutaneous gene delivery. Br. J. Dermatol. 150, 869–877.

Daddona, P.E., Matriano, J.A., Mandema, J., Maa, Y.F., 2011. Parathyroid hormone (1-34)-coated microneedle patch system: clinical pharmacokinetics and pharmacodynamics for treatment of osteoporosis. Pharm. Res. 28 (1), 159–165.

Dalmay, C., Villemejane, J., Joubert, V., Silve, A., Arnaud-Cormos, D., Français, O., et al., 2011. A microfluidic biochip for the nanoporation of living cells. Biosens. Bioelectron. 26 (12), 4649–4655.

Davis, S.P., Landis, B.J., Adams, Z.H., Allen, M.G., Prausnitz, M.R., 2004. Insertion of microneedles into skin: measurement and prediction of insertion force and needle fracture force. J. Biomech. 37 (8), 1155–1163.

Dean, R.A., Talbot, N.J., Ebbole, D.J., Farman, M.L., Mitchell, T.K., Orbach, M.J., et al., 2005. The genome sequence of the rice blast fungus *Magnaporthe grisea*. Nature 434 (7036), 980–986.

Delcea, M., Sternberg, N., Yashchenok, A.M., Georgieva, R., Bäumler, H., Möhwald, H., et al., 2012. Nanoplasmonics for dual-molecule release through nanopores in the membrane of red blood cells. ACS Nano 6 (5), 4169–4180.

Denoual, M., Macé, Y., Pioufle, B., Mognol, P., Castel, D., Gidrol, X., 2006. Vacuum casting to manufacture a plastic biochip for highly parallel cell transfection. Meas. Sci. Technol. 17 (12), 3134–3140.

Donnelly, R.F., Singh, T.R.R., Tunney, M.M., Morrow, D.J., McCarron, P.A., O'Mahony, C., et al., 2009. Microneedle arrays allow lower microbial penetration than hypodermic needles in vitro. Pharm. Res. 26 (11), 2513–2522.

Donnelly, R.F., Singha, T.R.R., David, A., 2010. Microneedle-based drug delivery systems: microfabrication, drug delivery, and safety. Drug Deliv. 17 (4), 187–207.

Eppstein, J.A., Hatch, M.R., inventors; Altea Technologies, Inc., Spectrx, Inc., assignee, 2000. Apparatus and Method for Electroporation of Microporated Tissue for Enhancing Flux Rates for Monitoring and Delivery Applications. US6022316 A.

Farr Stephen, J.R.H., Schuster Jeffrey, A., Turanin, J., inventors; Farr Stephen, J., Hawley, R., Schuster Jeffrey, A., Turanin, J., assignee, 2011. Novel Formulations for Treatment of Migraine. EP2756756 A1.

Garg, S., Hoelscher, M., Belser, J.A., Wang, C., Jayashankar, L., Guo, Z., et al., 2007. Needle-free skin patch delivery of a vaccine for a potentially pandemic influenza virus provides protection against lethal challenge in mice. Clin. Vaccine Immunol. 14 (7), 926–928.

Genina, E.A., Bashkatov, A.N., Dolotov, L.E., Maslyakova, G.N., Kochubey, V.I., Yaroslavsky, I.V., et al., 2013. Transcutaneous delivery of micro- and nanoparticles with laser microporation. J. Biomed. Opt. 18 (11), 111406.

Ghartey-Tagoe, E., Wendorf, J., Williams, S., Singh, P., Worsham, R.W., Trautman, J.C., et al., inventors; Corium International, Inc., assignee, 2014. Vaccine Delivery via Microneedle Arrays. Google Patents. US8911749 B2.

Gilger, B.C., Abarca, E.M., Salmon, J.H., Patel, S., 2013. Treatment of acute posterior uveitis in a porcine model by injection of triamcinolone acetonide into the suprachoroidal space using microneedles. Invest. Ophthalmol. Vis. Sci. 54 (4), 2483–2492.

Gill, H.S., Soderholm, J., Prausnitz, M.R., Sallberg, M., 2010. Cutaneous vaccination using microneedles coated with hepatitis C DNA vaccine. Gene Ther. 17 (6), 811–814.

Gonnelli, R.R., inventor; Gonnelli, R.R., assignee, 2006. Methods and Devices for Delivering GLP-1 and Uses Thereof. US20060030838.

Gonnelli, R.R., inventor; Gonnelli, R.R., assignee, 2009. Switchable Microneedle Arrays and Systems and Methods Relating to Same. US20030135166 A1.

Gonnelli, R.S., Daghero, D., Ummarino, G.A., 2002. Direct evidence for two-band superconductivity in MgB_2 single crystals from directional point-contact spectroscopy in magnetic fields. Phys. Rev. Lett. 89 (24), 247004.

Gratieri, T., Alberti, I., Lapteva, M., Kalia, Y.N., 2013. Next generation intra- and transdermal therapeutic systems: using non- and minimally-invasive technologies to increase drug delivery into and across the skin. Eur. J. Pharm. Sci. 50 (5), 609–622.

Han, S.W., Nakamura, C., Kotobuki, N., Obataya, I., Ohgushi, H., Nagamune, T., et al., 2008. High-efficiency DNA injection into a single human mesenchymal stem cell using a nanoneedle and atomic force microscopy. Nanomedicine 4 (3), 215–225.

Hiraishi, Y., Nandakumar, S., Choi, S.O., Lee, J.W., Kim, Y.C., Posey, J.E., et al., 2011. Bacillus Calmette-Guerin vaccination using a microneedle patch. Vaccine 29 (14), 2626–2636.

Hirschberg, H.J., van de Wijdeven, G.G., Kelder, A.B., van den Dobbelsteen, G.P., Kersten, G.F., 2008. Bioneedles as vaccine carriers. Vaccine 26 (19), 2389–2397.

Hu, Y., Xu, B., Ji, Q., Shou, D., Sun, X., Xu, J., et al., 2014. A mannosylated cell-penetrating peptide-graft-polyethylenimine as a gene delivery vector. Biomaterials 35 (13), 4236–4246.

Ibey, B.L., Pakhomov, A.G., Gregory, B.W., Khorokhorina, V.A., Roth, C.C., Rassokhin, M.A., et al., 2010. Selective cytotoxicity of intense nanosecond-duration electric pulses in mammalian cells. Biochim. Biophys. Acta 1800 (11), 1210–1219.

Jiang, J., Gill, H.S., Ghate, D., McCarey, B.E., Patel, S.R., Edelhauser, H.F., et al., 2007. Coated microneedles for drug delivery to the eye. IOVS 48, 9.

Jonathan, A.E., Michael, R.H., Difei, Y., inventors; Altea Technologies, Inc., Spectrx, Inc., assignee, 2000. Microporation of Human Skin for Drug Delivery and Monitoring Applications. US5885211.

Kalluri, H., Banga, A.K., 2011. Formation and closure of microchannels in skin following microporation. Pharm. Res. 28 (1), 82–94.

Kim, J., Jang, J.H., Lee, J.H., Choi, J.K., Park, W.R., Bae, I.H., et al., 2012. Enhanced topical delivery of small hydrophilic or lipophilic active agents and epidermal growth factor by fractional radiofrequency microporation. Pharm. Res. 29 (7), 2017–2029.

Kim, N.W., Lee, M.S., Kim, K.R., Lee, J.E., Lee, K., Park, J.S., et al., 2014. Polyplex-releasing microneedles for enhanced cutaneous delivery of DNA vaccine. J. Control. Release 179, 11–17.

Kim, Y., Yoo, D., Compans, R., Kang, S., Prausnitz, M.R., 2013. Cross-protection by co-immunization with influenza hemagglutinin DNA and inactivated virus vaccine using coated microneedles. J. Control. Release 172 (2), 579–588.

Kim, Y.T., Ahn, S.I., Park, H.J., Jung, S.H., Hong, H.K., Lee, H.-K., et al., 2009. Microneedle-mediated transdermal delivery of chondroitin sulfate. Tissue Eng. Regen. Med. 6 (4–11), 756–761.

Koutsonanos, D.G., Martin, M.P., Zarnitsyn, V.G., Sullivan, S.P., Compans, R.W., Prausnitz, M.R., et al., 2009. Transdermal influenza immunization with vaccine-coated microneedle arrays. PLoS One 4 (3), e4773.

Lee, M., Chen, H., Ho, M., Chen, C., Chuang, S., Huang, S., et al., 2013. PPARγ silencing enhances osteogenic differentiation of human adipose-derived mesenchymal stem cells. J. Cell. Mol. Med. 17 (9), 1188–1193.

Lee, S.G., Jeong, J.H., Lee, K.M., Jeong, K.H., Yang, H., Kim, M., et al., 2014. Nanostructured lipid carrier-loaded hyaluronic acid microneedles for controlled dermal delivery of a lipophilic molecule. Int. J. Nanomedicine 9, 289–299.

Ma, Y., Tao, W., Krebs, S., Sutton, W., Haigwood, N., Gill, H.S., 2014. Vaccine delivery to the oral cavity using coated microneedles induces systemic and mucosal immunity. Pharm. Res. 31 (9), 2393–2403.

Madeira, C., Ribeiro, S.C., Pinheiro, I.S.M., Martins, S.A.M., Andrade, P.Z., da Silva, C.L., et al., 2011. Gene delivery to human bone marrow mesenchymal stem cells by microporation. J. Biotechnol. 151 (1), 130–136.

Madeira, C., Rodrigues, C.A.V., Reis, M.S.C., Ferreira, F.F.C.G., Correia, R.E.S.M., Diogo, M.M., et al., 2013. Nonviral gene delivery to neural stem cells with minicircles by microporation. Biomacromolecules 14 (5), 1379–1387.

Martanto, W., Davis, S.P., Holiday, N.R., Wang, J., Gill, H.S., Prausnitz, M.R., 2004. Transdermal delivery of insulin using microneedles in vivo. Pharm. Res. 21 (6), 947–952.

Michael Barry, R.M., Smith, L., inventors; Baylor College of Medicine, assignee, 2000. Needle-Free Injection of Formulated Nucleic Acid Molecules. WO1999031262A3.

Mikszta, J.A., Sullivan, V.J., Dean, C., Waterston, A.M., Alarcon, J.B., Dekker, J.P., et al., 2005. Protective immunization against inhalational anthrax: a comparison of minimally invasive delivery platforms. J. Infect. Dis. 191 (2), 278–288.

Mir, R.J., KoWarz, H.W., Sarbadhikari, K.K., Ashe, P.R., inventors; Infotonics Technology Center, Inc., assignee, 2012. MEMS Interstitial Prothrombin Test Time. US8328720 B2.

Modepalli, N., Jo, S., Repka, M.A., Murthy, S.N., 2013. Microporation and 'iron'tophoresis for treating iron deficiency anemia. Pharm. Res. 30 (3), 889–898.

Neumann, E., Kakorin, S., Toensing, K., 1999. Fundamentals of electroporative delivery of drugs and genes. Bioelectrochem. Bioenerg. 48 (1), 3–16.

Norman, J.J., Choi, S.O., Tong, N.T., Aiyar, A.R., Patel, S.R., Prausnitz, M.R., et al., 2013. Hollow microneedles for intradermal injection fabricated by sacrificial micromolding and selective electrode-position. Biomed. Microdevices 15 (2), 203–210.

Pakhomov, A.G., Bowman, A.M., Ibey, B.L., Andre, F.M., Pakhomova, O.N., Schoenbach, K.H., 2009. Lipid nanopores can form a stable, ion channel-like conduction pathway in cell membrane. Biochem. Biophys. Res. Commun. 385 (2), 181–186.

Patel, H., Joshi, A., Joshi, A., Stagni, G., 2015. Effect of microporation on passive and iontophoretic delivery of diclofenac sodium. Drug Dev. Ind. Pharm. 41 (12), 1962–1967.

Paudel, K.S., Milewski, M., Swadley, C.L., Brogden, N.K., Ghosh, P., Stinchcomb, A.L., 2010. Challenges and opportunities in dermal/transdermal delivery. Ther. Deliv. 1 (1), 109–131.

Peters, E.E., Ameri, M., Wang, X., Maa, Y.F., Daddona, P.E., 2012. Erythropoietin-coated ZP-microneedle transdermal system: preclinical formulation, stability, and delivery. Pharm. Res. 29 (6), 1618–1626.

Prausnitz, M.R., 1999. A practical assessment of transdermal drug delivery by skin electroporation. Adv. Drug Deliv. Rev. 35 (1), 61–76.

Prausnitz, M.R., 2003. Microneedles for transdermal drug delivery. Adv. Drug Deliv. Rev. 56 (5), 581–587.

Prausnitz, M.R., Allen, M.G., Gujral, I.J. inventors; Georgia Tech Research Corporation, assignee, 2003. Microneedle Drug Delivery Device. Google Patents. US7226439.

Prow, T.W., Grice, J.E., Lin, L.L., Faye, R., Butler, M., Becker, W., et al., 2011. Nanoparticles and micropar-ticles for skin drug delivery. Adv. Drug Deliv. Rev. 63 (6), 470–491.

Raphael, A.P., Primiero, C.A., Ansaldo, A.B., Keates, H.L., Soyera, H.P., Prowa, T.W., 2013. Elongate mic-roparticles for enhanced drug delivery to ex vivo and in vivo pig skin. J. Control. Release 172 (1), 96–104.

Rebersek, M., Faurie, C., Kanduser, M., Corović, S., Teissié, J., Rols, M.P., et al., 2007. Electroporator with automatic change of electric field direction improves gene electrotransfer in-vitro. Biomed. Eng. Online 6, 25.

Ross, R.F., inventor; Kimberly-Clark Worldwide, assignee, 2014. Transdermal Patch Containing Microneedles. Google Patents. US8696637 B2.

Roth, C.C., Tolstykh, G.P., Payne, J.A., Kuipers, M.A., Thompson, G.L., DeSilva, M.N., et al., 2013. Nanosecond pulsed electric field thresholds for nanopore formation in neural cells. J. Biomed. Opt. 18 (3), 035005.

Sachdeva, V., Zhou, Y., 2013. In vivo transdermal delivery of leuprolide using microneedles and iontopho-resis. Curr. Pharm. Biotechnol. 14 (2), 180–193.

Scarlett, U.K., Rutkowski, M.R., Rauwerdink, A.M., Fields, J., Escovar-Fadul, X., Baird, J., et al., 2012. Ovarian cancer progression is controlled by phenotypic changes in dendritic cells. J. Exp. Med. 209 (3), 495–506.

Scheiblhofer, S., Thalhamer, J., Weiss, R., 2013. Laser microporation of the skin: Prospects for painless appli-cation of protective and therapeutic vaccines. Expert Opin. Drug Deliv. 10 (6), 761–773.

Singh, T.R., Garland, M.J., Cassidy, C.M., Migalska, K., Demir, Y.K., Abdelghany, S., et al., 2010. Microporation techniques for enhanced delivery of therapeutic agents. Recent Pat. Drug Deliv. Formul. 4 (1), 1–17.

Sintov, A.C., Krymberk, I., Daniel, D., Hannan, T., Sohn, Z., Levin, G., 2003. Radiofrequency-driven skin microchanneling as a new way for electrically assisted transdermal delivery of hydrophilic drugs. J. Control. Release 89 (2), 311–320.

Stinchcomb, A.L., Banks, S.L., Pinninti, R.R., inventors; University of Kentucky Research Foundation, assignee, 2008. Transdermal Delivery of Naltrexone Hydrochloride, Naltrexone Hydrochloride, and bis(Hydroxy-Methyl)Propionyl-3-0 Ester Naltrexone Using Microneedles. US20080008745 A1.

Stinchcomb, A.L., Banks, S.L., Golinski, M.J., Howard, J.L., Hammell, D.C., inventors; Alltranz Inc., assignee, 2011. Use of Cannabidiol Prodrugs in Topical and Transdermal Administration with Microneedles. US20110052694 A1.

Thomas, B.J., Finnin, B.C., 2004. The transdermal revolution. Drug Discov. Today 9 (16), 697–703.

Vandervoort, J., Ludwig, A., 2008. Microneedles for transdermal drug delivery: a minireview. Front. Biosci. 13, 1711–1715.

Verrips, G.H., Hirasing, R.A., Fekkes, M., Vogels, T., Verloove-Vanhorick, S.P., Delemarre-Van de Waal, H.A., 1998. Psychological responses to the needle-free Medi-Jector or the multidose disetronic injection pen in human growth hormone therapy. Acta Paediatr. 87 (2), 154–158.

Wang, Y., Ho, M., Chang, J., Chu, J., Lai, S., Wang, G., 2009. Microporation is a valuable transfection method for gene expression in human adipose tissue–derived stem cells. Mol. Ther. 17 (2), 302–308.

Weiss, R., Hessenberger, M., Kitzmuller, S., Bach, D., Weinberger, E.E., Krautgartner, W.D., et al., 2012. Transcutaneous vaccination via laser microporation. J. Control. Release 162 (2), 391–399.

Xiang, Z., Wang, H., Pant, A., Pastorin, G., Lee, C., 2013. Development of vertical SU-8 microtubes integrated with dissolvable tips for transdermal drug delivery. Biomicrofluidics 7 (2), 26502.

Xiaoyun, H.W.L., Fei, W., Zaozhan, W.U., Lizhu, C., Zhenguo, L., et al., 2013. Dissolving and biodegradable microneedle technologies for transdermal sustained delivery of drug and vaccine. Drug Des. Dev. Ther. 7, 945–952.

Yu, J., Bachhav, Y.G., Summer, S., Heinrich, A., Bragagna, T., Böhler, C., et al., 2010. Using controlled laser-microporation to increase transdermal delivery of prednisone. J. Control. Release 148 (1), e71–e73.

Yum, K., Wang, N., Yu, M.F., 2010. Nanoneedle: a multifunctional tool for biological studies in living cells. Nanoscale 2 (3), 363–372.

Zahn, R., Moll, J., Paiva, M., Garrido, G., Krueger, F., Huey, E.D., et al., 2009. The neural basis of human social values: evidence from functional MRI. Cereb. Cortex 19 (2), 276–283.

Zhang, Y., Brown, K., Siebenaler, K., Determan, A., Dohmeier, D., Hansen, K., 2011. Development of lidocaine-coated microneedle product for rapid, safe, and prolonged local analgesic action. Pharm. Res. 29 (1), 170–177.

INDEX

Note: Page numbers followed by "*f*" and "*t*" refer to figures and tables, respectively.